Convex Analysis
— in —
General Vector Spaces

Convex Analysis in General Vector Spaces

C Zălinescu

Faculty of Mathematics
University "Al. I. Cuza" Iasi, Romania

World Scientific
New Jersey • London • Singapore • Hong Kong

Published by

World Scientific Publishing Co. Pte. Ltd.
5 Toh Tuck Link, Singapore 596224
USA office: 27 Warren Street, Suite 401-402, Hackensack, NJ 07601
UK office: 57 Shelton Street, Covent Garden, London WC2H 9HE

Library of Congress Cataloging-in-Publication Data
Zălinescu, C., 1952–
Convex analysis in general vector spaces / C. Zălinescu
p. cm
Includes bibliographical references and index.
ISBN-13 978-981-238-067-8 (alk. paper)
ISBN-10 981-238-067-1 (alk. paper)
1. Convex functions. 2. Convex sets. 3. Functional analysis. 4. Vector spaces.
I. Title.
QA331.5.Z34 2002 2002069000
515'.8--dc21

British Library Cataloguing-in-Publication Data
A catalogue record for this book is available from the British Library.

To the memory of my parents

Casandra and **Vasile Zălinescu**

Preface

The text of this book has its origin in a course we delivered to students for Master Degree at the Faculty of Mathematics of the University "Al. I. Cuza" Iaşi, Romania.

One can ask if another book on Convex Analysis is needed when there are many excellent books dedicated to this discipline like those written by R.T. Rockafellar (1970), J. Stoer and C. Witzgall (1970), J.-B. Hiriart-Urruty and C. Lemaréchal (1993), J.M. Borwein and A. Lewis (2000) for finite dimensional spaces and by P.-J. Laurent (1972), I. Ekeland and R. Temam (1974), R.T. Rockafellar (1974), A.D. Ioffe and V.M. Tikhomirov (1974), V. Barbu and Th. Precupanu (1978, 1986), J.R. Giles (1982), R.R. Phelps (1989, 1993), D. Azé (1997) for infinite dimensional spaces.

We think that such a book is necessary for taking into consideration new results concerning the validity of the formulas for conjugates and subdifferentials of convex functions constructed from other convex functions by operations which preserve convexity, results obtained in the last 10–15 years. Also, there are classes of convex functions like uniformly convex, uniformly smooth, well behaving, well conditioned functions that are not studied in other books. Characterizations of convex functions using other types of derivatives or subdifferentials than usual directional derivatives or Fenchel subdifferential are quite recent and deserve being included in a book. All these themes are treated in this book.

We have chosen for studying convex functions the framework of locally convex spaces and the most general conditions met in the literature; even when restricted to normed vector spaces many results are stated in more general conditions than the corresponding ones in other books. To make this possible, in the first chapter we introduce several interiority and closedness conditions and state two strong open mapping theorems.

In the second chapter, besides the usual characterizations and properties of convex functions we study new classes of such functions: cs-convex, cs-closed, cs-complete, lcs-closed, ideally convex, bcs-complete and li-convex functions, respectively; note that the classes of li-convex and lcs-closed functions have very good stability properties. This will give the possibility to have a rich calculus for the conjugate and the subdifferential of convex functions under mild conditions. In obtaining these results we use the method of perturbation functions introduced by R.T. Rockafellar. The main tool is the fundamental duality formula which is stated under very general conditions by using open mapping theorems.

The framework of the third chapter is that of infinite dimensional normed vector spaces. Besides some classical results in convex analysis we give characterizations of convex functions using abstract subdifferentials and study differentiability of convex functions. Also, we introduce and study well-conditioned convex functions, uniformly convex and uniformly smooth convex functions and their applications to the study of the geometry of Banach spaces. In connection with well-conditioned functions we study the sets of weak sharp minima, well-behaved convex functions and global error bounds for convex inequality systems. The chapter ends with the study of monotone operators by using convex functions.

Every chapter ends with exercises and bibliographical notes; there are more than 80 exercises. The statements of the exercises are generally extracted from auxiliary results in recent articles, but some of them are known results that deserve being included in a textbook, but which do not fit very well our aims. The complete solutions of all exercises are given. The book ends with an index of terms and a list of symbols and notations.

Even if all the results with the exception of those in the first section are given with their complete proofs, for a successful reading of the book a good knowledge of topology and topological vector spaces is recommended.

Finally I would like to thank Prof. J.-P. Penot and Prof. A. Göpfert for reading the manuscript, for their remarks and encouragements.

C. Zălinescu
March 1, 2002
Iaşi, Romania

Contents

Introduction

The primary aim of this book is to present the conjugate and subdifferential calculus using the method of perturbation functions in order to obtain the most general results in this field. The secondary aim is to give important applications of this calculus and of the properties of convex functions. Such applications are: the study of well-conditioned convex functions, uniformly convex and uniformly smooth convex functions, best approximation problems, characterizations of convexity, the study of the sets of weak sharp minima, well-behaved functions and the existence of global error bounds for convex inequalities, as well as the study of monotone multifunctions by using convex functions.

The method of perturbation functions is based on the "fundamental duality theorem" which says that under certain conditions one has

$$\inf_{x \in X} \Phi(x,0) = \max_{y^* \in Y^*} \big(-\Phi^*(0,y^*) \big). \tag{FDF}$$

For many problems in convex optimization one can associate a useful perturbation function. We give here four examples; see [Rockafellar (1974)] for other interesting ones.

Example 1 (Convex programming; see Section 2.9) Let $f, g_1, \ldots, g_n : X \to \overline{\mathbb{R}}$ be proper convex functions with $\operatorname{dom} f \cap \bigcap_{i=1}^n \operatorname{dom} g_i \neq \emptyset$. The problem of minimizing $f(x)$ over the set of those $x \in X$ satisfying $g_i(x) \leq 0$ for all $i = 1, \ldots, n$ is equivalent to the minimization of $\Phi(x,0)$ for $x \in X$, where

$$\Phi : X \times Y \to \overline{\mathbb{R}}, \quad \Phi(x,y) := \begin{cases} f(x) & \text{if} \quad g_i(x) \leq y_i \ \forall\, 1 \leq i \leq n, \\ +\infty & \qquad \text{otherwise,} \end{cases}$$

and $Y := \mathbb{R}^n$; the element y^* obtained from the right-hand side of (FDF) will furnish the Lagrange multipliers.

Example 2 (Control problems) Let $F : X \times Y \to \overline{\mathbb{R}}$ be a proper convex function and $A : X \to Y$ a linear operator. A control problem (in its abstract form) is to minimize $F(x, y)$ for $x \in X$ and $y = Ax + y_0$. The perturbation function to be considered is $\Phi : X \times Y \to \overline{\mathbb{R}}$ defined by $\Phi(x, y) := F(x, Ax + y_0 + y)$.

Example 3 (Semi-infinite programming) We are as in Example 1 but $\{1, \ldots, n\}$ is replaced by a general nonempty set I. In this case $Y = \mathbb{R}^I$ and $\Phi(x, y) := f(x)$ if $g_i(x) \leq y_i$ for all $i \in I$, $\Phi(x, y) := \infty$ otherwise.

Formula (FDF), or more precisely the Fenchel–Rockafellar duality formula, can also be used for deriving results similar to that in the next example.

Example 4 ([Simons (1998b)]; see Exercise 2.37) Let X be a linear space, $(Y, \|\cdot\|)$ be a normed linear space, $A : X \to Y$ be a linear operator, $y_0 \in Y$ and $f : X \to \overline{\mathbb{R}}$ be a proper convex function. Then $f(x) + \|Ax + y_0\|^2 \geq 0$ for all $x \in X$ if and only if there exists $y^* \in Y^*$ such that $f(x) - 2\langle Tx + y_0, y^*\rangle - \|y^*\|^2 \geq 0$ for all $x \in X$.

It is worth mentioning that adequate perturbation functions can be used for deriving formulas for the conjugate and ε-subdifferential for many types of convex functions; this method is used by Rockafellar (1974) for $f \circ A$ and $f_1 + \cdots + f_n$, but we use it for almost all the operations which preserve convexity (see Section 2.8).

The formula (FDF) is automatically valid when $\inf_{x \in X} \Phi(x, 0) = -\infty$ and is equivalent to the subdifferentiability at $0 \in Y$ of the marginal function $h : Y \to \overline{\mathbb{R}}$, $h(y) := \inf_{x \in X} \Phi(x, y)$, when $\inf_{x \in X} \Phi(x, 0) \in \mathbb{R}$. A sufficient condition for this is the continuity of the restriction of h to the affine hull of its domain at 0; note that 0 is in the relative algebraic interior of the domain of h in this case (without this condition one can give simple examples in which the subdifferential of h at 0 is empty).

Considering the multifunction $\mathcal{R} : X \times \mathbb{R} \rightrightarrows Y$ whose graph is the set $\operatorname{gr} \mathcal{R} = \{(x, t, y) \mid (x, y, t) \in \operatorname{epi} \Phi\}$, the continuity of $h|_{\operatorname{aff}(\operatorname{dom} h)}$ at 0 is ensured if $\mathcal{R}$ is relatively open at some (x_0, t_0) with $(x_0, t_0, 0) \in \operatorname{gr} \mathcal{R}$. This fact was observed for the first time by Robinson (1976). This remark shows the importance of open mapping theorems for convex multifunctions in convex analysis. In Banach spaces such a result is the well-known Robinson–

Ursescu theorem. The preceding examples show that the consideration of more general spaces is natural: In Example 3 Y is a locally convex space while in Example 4 X can be endowed with the topology $\sigma(X, X')$. The original result of Ursescu (1975) is stated in very general topological vector spaces. The inconvenient of Ursescu's theorem is that one asks the multifunction to be closed, condition which is quite strong in certain situations. For example, when calculating the conjugate or subdifferential of $\max(f, g)$ with f, g proper lower semicontinuous convex functions one has to evaluate conjugate or the subdifferential of $0 \cdot f + 1 \cdot g$ which is not lower semicontinuous convex. Fortunately we dispose of another open mapping theorem in which the closedness condition is replaced by a weaker one, but one must pay for this by asking (slightly) more on the spaces involved.

As said above, the second aim of the book is to give some interesting applications of conjugate and subdifferential calculus, less treated in other books.

In many algorithms for the minimization problem (P) $\min f(x)$, s.t. $x \in X$, one obtains a sequence (x_n) which is minimizing (*i.e.* $(f(x_n)) \to \inf f$) or stationary (*i.e.* $\left(d_{\partial f(x_n)}(0)\right) \to 0$). It is important to know if such a sequence converges to a solution of (P). Assuming that $S := \operatorname{argmin} f := \{x \mid f(x) = \inf f\} \neq \emptyset$, one says that f is well-conditioned if $(d_S(x_n)) \to 0$ whenever (x_n) is a minimizing sequence, and f is well-behaved (asymptotically) if (x_n) is minimizing whenever (x_n) is a stationary sequence; when S is a singleton well-conditioning reduces to the well-known notion of well-posedness in the sense of Tikhonov. If f is well-conditioned with linear rate the set $\operatorname{argmin} f$ is a set of weak sharp minima. When f is convex, we establish several characterizations of well-conditioning using the conjugate or the subdifferential of f. When S is a singleton one of the characterizations is close to uniform convexity of f at a point.

One says that the proper function $f : (X, \|\cdot\|) \to \overline{\mathbb{R}}$ is strongly convex if

$$f(\lambda x + (1-\lambda)y) \leq \lambda f(x) + (1-\lambda)f(y) - \tfrac{c}{2}\lambda(1-\lambda)\|x-y\|^2$$

for some $c > 0$ and for all $x, y \in \operatorname{dom} f$, $\lambda \in [0,1]$. This notion is not very adequate for non-Hilbert spaces; for general normed spaces, one says that f is uniformly convex if there exists $\rho : \mathbb{R}_+ \to \overline{\mathbb{R}}_+$ with $\rho(0) = 0$ such that

$$f(\lambda x + (1-\lambda)y) \leq \lambda f(x) + (1-\lambda)f(y) - \lambda(1-\lambda)\rho\left(\|x-y\|\right)$$

for all $x, y \in \operatorname{dom} f$ and all $\lambda \in [0,1]$. From a numerical point of view the class of uniformly convex functions is important because on a Banach space every uniformly convex and lower semicontinuous proper function has a unique minimum point and the corresponding minimization problem is well-conditioned. It turns out that uniformly convex functions have very nice characterizations using their conjugates and subdifferentials. The dual notion for uniformly convex function is that of uniformly smooth convex function. An important fact is that f is uniformly convex (uniformly smooth) if and only if f^* is uniformly smooth (uniformly convex).

Another interesting application of convex analysis is in the study of monotone operators. This became possible by using a convex function associated to a multifunction introduced by M. Coodey and S. Simons. So, one obtains quite easily characterizations of maximal monotone operators, local boundedness of monotone operators and maximal monotonicity of the sum of two maximal monotone operators using continuity properties of convex functions, the formula for the subdifferential of a sum of convex functions and a minimax theorem (whose proof is also included).

A more detailed presentation of the book follows.

The book is divided into three chapters, every chapter ending with exercises and bibliographical notes; there are more than 80 exercises. It also includes the complete solutions of the exercises, the bibliography, the list of notations and the index of terms.

No prior knowledge of convex analysis is assumed, but basic knowledge of topology, linear spaces, topological (locally convex) linear spaces and normed spaces is needed.

In Chapter 1, as a preliminary, we introduce the notions and results of functional analysis we need in the rest of the book. For easy reference, in Section 1.1 we recall several notions, notations and results (without proofs) which can be found in almost all books on functional analysis; let us mention four separation theorems for convex sets, the Dieudonné and Alaoglu–Bourbaki theorems, as well as the bipolar theorem.

In Section 1.2 we introduce cs-closed, cs-complete, lcs-closed (*i.e.* lower cs-closed) and ideally convex, bcs-complete, li-convex (*i.e.* lower ideally convex) sets and prove several results concerning them. We point out the good stability properties of li-convex and lcs-closed sets. We also introduce two conditions denoted (Hx) and (Hwx) which refer to sets in product spaces that are stronger than the cs-closedness and ideal convexity, but weaker than the cs-completeness and bcs-completeness of the sets, respectively. Then, besides the classical algebraic interior A^i and relative algebraic interior ${}^{i}A$ of a subset A of a linear space X, we introduce, when X is a topological vector space, the sets ${}^{ic}A$ and ${}^{ib}A$, which reduce to ${}^{i}A$ when the affine hull aff A of A is closed or barreled, respectively, and are the empty set otherwise. The quasi interior of a set and united sets are also studied.

In Section 1.3 we state and prove the famous Ursescu's theorem as well as a slight amelioration of Simons' open mapping theorem. As application of these results one reobtain the Banach–Steinhaus theorem and the closed graph theorem as well as two results of O. Cârjă which are useful in controllability problems. Because the notions (with the exception of cs-closed and ideally convex sets) and results from Sections 1.2 and 1.3 are not treated in many books (to our knowledge only [Kusraev and Kutateladze (1995)] contains some similar material), we give complete proofs of the results.

The chapter ends with Section 1.4 in which we state and prove the Ekeland's variational principle, the smooth variational principle of Borwein and Preiss, as well as two (dual) results of Ursescu which generalize Baire's theorem.

Chapter 2 is dedicated, mainly, to conjugate and ε-subdifferential calculus. Because no prior knowledge of convex analysis is assumed, we introduce in Section 2.1 convex functions, give several characterizations using the epigraph, or the gradients in case of differentiability, point out the operations which preserve convexity and study the important class of convex functions of one variable; the existence of the (ε-)directional derivative and some of its properties are also studied. We close this section with a characterization of convex functions using the upper Dini directional derivative.

Section 2.2 is dedicated to the study of continuity properties of convex functions. To the classes of sets introduced in Section 1.2 correspond cs-closed, cs-complete, lcs-closed, ideally convex, bcs-complete and li-convex functions. We mention the fact that almost all operations which preserve convexity also preserve the lcs-closedness and the li-convexity of functions as seen in Proposition 2.2.19. The most part of the results of this section are not present in other books; among them we mention the result on the convexity of a quasiconvex positively homogeneous function and the results on cs-closed, cs-complete, cs-convex, lcs-closed, ideally convex, bcs-complete and li-convex functions.

Section 2.3 concerns conjugate functions; all the results are classical.

Section 2.4 is dedicated to the introduction and study of direct properties of the subdifferential. Using such properties one obtains easily the formulas for the subdifferentials of Af and $f_1 \square \cdots \square f_n$ which are valid without additional hypothesis. The classical theorem which states that the ε-subdifferential of a proper convex function is nonempty and w^*-compact at a continuity point of its domain, as well as the formula for the ε-directional derivative as the support function of the ε-subdifferential is also established. The less classical result which states that the same formula holds for $\varepsilon > 0$ when the function is not necessarily continuous (but is lower semicontinuous) is established, too. We mention also Theorem 2.4.14 related to the subdifferential of sublinear functions; some of its statements are not very spread. Other interesting results are introduced for completeness or further use.

In Section 2.5 we introduce the general problem of convex programming and establish sufficient conditions for the existence and uniqueness of solutions, respectively. We mention especially Theorems 2.5.2 and 2.5.5; Theorem 2.5.2 ameliorates a result of Polyak (1966), which shows that the reflexivity of the space, needed in proving the existence of solutions, is almost necessary, while Theorem 2.5.5 shows that the coercivity condition is essential for the existence of solutions.

Section 2.6 is dedicated to perturbed functions. One introduces primal and dual problems, the marginal function, and give some direct properties of them. Then one obtains the formula for the ε-subdifferential of the marginal function using the $(\varepsilon + \eta)$-subdifferentials (with $\eta > 0$) of the perturbed function. Applying this result one obtains formulas for the ε-subdifferential of several types of convex functions.

In the main result of Section 2.7 we provide nine (non-independent) sufficient conditions which ensure the validity of the fundamental duality formula (FDF). The most known of them is that $(x_0, 0) \in \operatorname{dom} \Phi$ and $\Phi(x_0, \cdot)$ is continuous at 0 for some $x_0 \in X$. For the proof of the sufficiency of some conditions one uses the open mapping theorems established in Section 1.3. A related result involves also a convex multifunction; this will be useful for obtaining the formulas for the conjugate and the ε-subdifferential of a function of the forms $g \circ A$ with A a densely defined and closed linear operator and of $g \circ H$ with g being increasing and H convex.

Section 2.8 is dedicated entirely to conjugate and ε-subdifferential calculus for convex functions. The considered functions φ have the form: $\varphi(x) = F(x, A(x))$ and $\varphi = f + g \circ A$ with A a continuous linear operator, $\varphi = f + g$, $\varphi(x) = \inf\{g(y) \mid y \in \mathcal{C}(x)\}$ with $\mathcal{C}$ a convex process, $\varphi = g \circ H$ with H a convex operator and g an increasing convex function, $\varphi = \max\{f_1, \ldots, f_n\}$ and $\varphi = f_1 \Diamond f_2$. Besides classical conditions one points out very recent ones. For the proof one constructs an adequate perturbation function and uses the fundamental duality theorem.

In Section 2.9 we apply the fundamental duality theorem for obtaining necessary and sufficient optimality conditions in convex optimization problems with constraints. These conditions involve the subdifferentials of the functions considered or the corresponding Lagrangian. The results are well-known. However we mention the formula for the normal cone to a level set stated in Corollary 2.9.5 for not necessarily finite-valued functions which is quite new.

The minimax theorem presented in Section 2.10 will be used in the section dedicated to monotone multifunctions.

Throughout Chapter 3 the involved spaces are normed spaces. In Section 3.1 besides the classical theorems of Borwein, Brøndsted–Rockafellar, Bishop–Phelps and Rockafellar (on the maximal monotonicity of the subdifferential of a convex function) we present a recent theorem of Simons and use it for a very short proof of Rockafellar's theorem (mentioned before). As a consequence of the Brøndsted–Rockafellar theorem we obtain other three conditions for the validity of the formulas for the conjugate and subdifferential of the function $F(\cdot, A(\cdot))$ (and therefore for the functions $f + g \circ A$ and $f + g$).

The aim of Section 3.2 is to characterize the convex functions using other types of subdifferentials. In fact we use an abstract subdifferential. An example of such subdifferential is Clarke's one for which we establish several properties. The main tool for such characterizations is the well-known Zagrodny's approximate mean value theorem; the version we present subsumes several results met in the literature. We present also an integration theorem of Thibault and Zagrodny which yields the fact that two lower semicontinuous convex functions on a Banach space which has the same Fenchel subdifferential coincide up to an additive constant.

In Section 3.3 we introduce the class $\mathcal{A}$ of functions $\varphi : \mathbb{R}_+ \to \overline{\mathbb{R}}_+$ with $\varphi(0) = 0$ and several useful subclasses. To any $\varphi \in \mathcal{A}$ we associate $\varphi^\# \in \mathcal{A}$ defined by $\varphi^\#(t) = \sup\{ts - \varphi(s) \mid s \geq 0\}$. These classes of functions turn out to be useful in studying well-conditioned convex functions, uniformly convex and uniformly smooth convex functions, as well as in the study of the geometry of normed spaces. As an illustration of the use of these classes of functions we study the differentiability of convex functions with respect to arbitrary bornologies. Using one of the characterizations and the Brøndsted–Rockafellar theorem one obtains the following interesting result of Asplund and Rockafellar: Let X be a Banach space and $f : X \to \overline{\mathbb{R}}$ a proper lower semicontinuous convex function; if f^* is Fréchet differentiable at $\overline{x}^* \in \operatorname{int}(\operatorname{dom} f^*)$ then $\nabla f^*(\overline{x}^*) \in X$.

In Section 3.4 we introduce the well-conditioned convex functions and give several characterizations of this notion using the conjugate and the subdifferential of the function. An important special case is that of well-conditioning with linear rate. This situation is studied in Section 3.10.

In Section 3.5 we study uniformly convex and uniformly smooth convex functions, respectively. To any convex function f one associates the gages ρ_f and σ_f of uniform convexity and uniform smoothness, respectively. The gage ρ_f has an important property: the mapping $0 < t \mapsto t^{-2}\rho_f(t)$ is nondecreasing. Because $\sigma_{f^*} = (\rho_f)^\#$ and $\sigma_f = (\rho_{f^*})^\#$ for any proper lower semicontinuous convex function f, the mapping $0 < t \mapsto t^{-2}\sigma_f(t)$ is non-increasing for such a function; moreover, one obtains that for such an f, f is uniformly convex if and only if f^* is uniformly smooth and f is uniformly smooth if and only if f^* is uniformly convex. Then one establishes many characterizations of uniformly convex functions and of uniformly smooth convex functions. In these characterizations appear functions (gages or moduli) belonging to different subclasses of $\mathcal{A}$ introduced in Section 3.3. These gages and moduli are sharp enough in order to obtain that f is c-strongly convex if and only if f^* is Fréchet differentiable on X^* and ∇f^* is c^{-1}-Lipschitz. Even if the results are established in general Banach spaces the natural framework for uniformly convex and uniformly smooth convex function is that of reflexive Banach spaces. This is due to the fact that when there exists a proper lower semicontinuous and uniformly convex function on a Banach space whose domain has nonempty interior, the space is necessarily reflexive.

Section 3.6 is dedicated to the study of those convex functions which are uniformly convex on bounded sets and uniformly smooth on bounded sets, respectively. Under strong coercivity of the function one shows that these notions are dual.

In Section 3.7 we study the function $f_\varphi : X \to \mathbb{R}$, $f_\varphi(x) = \int_0^{\|x\|} \varphi(t)\, dt$, where φ is a weight function, in connection with the geometric properties of the norm. So, one establishes characterizations of the strict convexity, the smoothness and the reflexivity of X by the strict convexity, the Gâteaux differentiability of f_φ and the surjectivity of ∂f_φ, respectively.

One obtains also characterizations of (local) uniform convexity and (local) uniform smoothness of X with the help of the properties of f_φ. For example one obtains: X is uniformly convex $\Leftrightarrow X^*$ is uniformly smooth $\Leftrightarrow f_\varphi$ is uniformly convex on bounded sets $\Leftrightarrow (f_\varphi)^*$ is uniformly smooth on bounded sets $\Leftrightarrow (f_\varphi)^*$ is Fréchet differentiable and $\nabla(f_\varphi)^*$ is uniformly continuous on bounded sets. Note that a part of the results of this section can be found in the book [Cioranescu (1990)], but the proofs are different; note also that some notions are introduced differently in Cioranescu's book.

Another application of convex analysis is emphasized in Section 3.8; here we apply the results on the existence, the uniqueness and the characterizations of optimal solutions of convex programs to the problem of the best approximation with elements of a convex subset of a normed space.

In Section 3.9 it is shown that there exists a strong relationship between the well-posedness of the minimization problem $\min f(x)$ s.t. $x \in X$, and the differentiability at 0 of the conjugate f^* of f; when f is convex these properties are equivalent. Using this result we establish a very interesting characterization of Chebyshev sets in Hilbert spaces and show that the class of weakly closed Chebyshev sets coincides with the class of closed convex sets in Hilbert spaces.

Section 3.10 deals with sets of weak sharp minima, well-behaved convex functions and the study of the existence of global error bounds for convex inequalities. These notions were studied separately for a time, but they are intimately related. As noted above, $\operatorname{argmin} f$ is a set of weak sharp minima for f exactly when f is well-conditioned with linear rate. But the inequality $f(x) \leq 0$ has a global error bound exactly when $\operatorname{argmin}[f]_+$ is a set of weak sharp minima for $[f]_+ := \max(f, 0)$. We give several characterizations of the fact that $\operatorname{argmin} f$ is a set of weak sharp minima for f, one of them being the fact that up to a constant, the conjugate f^* is sublinear on a neighborhood of the origin. Several numbers associated to a convex function are introduced which are related to the conditioning number from numerical analysis. Although the most part of the results from this section are stated in the literature in finite dimensional spaces, we present them in infinite dimensions.

The last section of this book, Section 3.11, is dedicated to the study of monotone multifunctions on Banach spaces. We use in the presentation two recent articles of Simons. The proofs are quite technical and use the lower semicontinuous convex function χ_M associated to the multifunction M :

$X \rightrightarrows X^*$, the minimax theorem and a few results of convex analysis. One obtains: two characterizations of maximal monotone multifunctions; the fact that the condition $0 \in \operatorname{int}(\operatorname{dom} T_1 - \operatorname{dom} T_2)$ is equivalent to other three conditions involving $\operatorname{dom} T_i$ and $\operatorname{dom} \chi_{T_i}$, and is sufficient for the maximal monotonicity of $T_1 + T_2$; $\overline{\operatorname{dom} T}$ and $\overline{\operatorname{Im} T}$ are convex if X is reflexive and T is maximal monotone; $\overline{\operatorname{dom} T}$ is convex if $\operatorname{int}(\operatorname{dom} T) \neq \emptyset$ and T is maximal monotone; T is locally bounded at $x_0 \in (\operatorname{co}(\operatorname{dom} T))^i$ if T is a monotone multifunction; Rockafellar's theorem on the local boundedness of maximal monotone multifunctions. The result stating that for a maximal monotone multifunction T on the Banach space X for which $\overline{\operatorname{dom} T}$ is convex the local boundedness of T at $\overline{x} \in \overline{\operatorname{dom} T}$ implies that $\overline{x} \in \operatorname{int}\left(\overline{\operatorname{dom} T}\right)$ seems to be new. When applied to the subdifferential of a proper lower semicontinuous convex function f on the Banach space X, this result gives (for example): f is continuous $\Leftrightarrow$ $\operatorname{dom} f$ is open $\Leftrightarrow$ ∂f is locally bounded at any $x \in \operatorname{dom} f$.

The exercises are intended to exemplify the topics treated in the book. Many exercises are auxiliary results spread in recent articles, although some of them are extracted from other books. Some exercises are important results which could be parts of textbooks, but which do not fit very well with the aim of the present book. Among them we mention Exercise 3.11 on the Moreau regularization.

Chapter 1

Preliminary Results on Functional Analysis

1.1 Preliminary Notions and Results

In this section we introduce several notions and results on separation of sets as well as some properties of topological vector spaces and locally convex spaces which are frequently used throughout the book, for easy reference.

Let X be a real linear (vector) space. Throughout this work we shall use the following notation (x, y being elements of X): $[x, y] := \{(1-\lambda)x + \lambda y \mid \lambda \in [0,1]\}$, $[x, y[:= \{(1-\lambda)x + \lambda y \mid \lambda \in [0,1[\}$, $]x, y[:= \{(1-\lambda)x + \lambda y \mid \lambda \in]0,1[\}$, called *closed*, *semi-closed* and *open segment*, respectively. Note that $[x, x] =]x, x[= \{x\}$!

If $\emptyset \neq A, B \subset X$, the *Minkowski sum* of A and B is $A + B := \{a + b \mid a \in A, b \in B\}$. Moreover, if $x \in X$, $\lambda \in \mathbb{R}$ and $\emptyset \neq \Gamma \subset \mathbb{R}$, then $x + A := A + x := A + \{x\}$, $\Lambda \cdot A = \{\gamma a \mid \lambda \in \Lambda, a \in A\}$ and $\lambda A := \{\lambda\} \cdot A$. We shall consider that $A + \emptyset = \emptyset$ and $\lambda \cdot \emptyset = \emptyset \cdot A = \emptyset$.

A nonempty set $A \subset X$ is **star-shaped** at a ($\in A$) if $[a, x] \subset A$ for all $x \in A$; A is **convex** if $[x, y] \subset A$ for all $x, y \in A$; A is a **cone** if $\mathbb{R}_+ \cdot A \subset A$ (in particular $0 \in A$ when A is a cone), $\mathbb{R}_+$ is the set of nonnegative reals; A is **affine** if $\lambda x + (1-\lambda)y \in A$ for all $x, y \in A$ and $\lambda \in \mathbb{R}$; A is **balanced** if $\lambda x \in A$ for all $x \in A$ and $\lambda \in [-1, 1]$; A is **symmetric** if $A = -A$. Hence A is balanced if and only if A is symmetric and star-shaped at 0. We consider that the empty set is convex and affine. It is easy to prove that

$$\begin{aligned} A \text{ is affine} &\Leftrightarrow \exists\, a \in X,\ \exists\, X_0 \text{ linear subspace of } X \ :\ A = a + X_0 \\ &\Leftrightarrow \forall\, a \in A\ (\exists\, a \in A)\ :\ A - a \text{ is a linear subspace.} \end{aligned}$$

When A is affine and $a \in A$, the linear space $X_0 := A - a$ is called the *linear space parallel to* A; we consider that the dimension of A is $\dim X_0$.

Note that if $(A_i)_{i\in I}$ is a family of affine, convex, balanced subsets or cones of X then $\bigcap_{i\in I} A_i$ has the same property (Exercise!); we use the usual convention that $\bigcap_{i\in\emptyset} A_i = X$. Taking into account this remark, we can introduce the notions of affine, convex and conic hull of a set. So, the **affine, convex** and **conic hull** of the subset A of X are:

$$\operatorname{aff} A := \bigcap\{V \subset X \mid A \subset V,\ V \text{ affine}\},$$
$$\operatorname{co} A := \bigcap\{C \subset X \mid A \subset C,\ C \text{ convex}\},$$
$$\operatorname{cone} A := \bigcap\{C \subset X \mid A \subset C,\ C \text{ cone}\},$$

respectively. Of course, the **linear hull** of the subset A of X is the linear subspace spanned by A:

$$\operatorname{lin} A := \bigcap\{X_0 \subset X \mid A \subset X_0,\ X_0 \text{ linear subspace of } X\}.$$

It is easy to verify (Exercise!) that

$$\operatorname{aff} A = \Big\{\sum\nolimits_{i=1}^n \lambda_i x_i \,\Big|\, n \in \mathbb{N},\ (\lambda_i)_{1\le i\le n} \subset \mathbb{R},\ \sum\nolimits_{i=1}^n \lambda_i = 1,\ (x_i)_{1\le i\le n} \subset A\Big\},$$
$$\operatorname{co} A = \Big\{\sum\nolimits_{i=1}^n \lambda_i x_i \,\Big|\, n \in \mathbb{N},\ (\lambda_i) \subset \mathbb{R}_+,\ (x_i) \subset A,\ \sum\nolimits_{i=1}^n \lambda_i = 1\Big\},$$
$$\operatorname{cone} A = \{\lambda x \mid \lambda \ge 0,\ x \in A\} = \mathbb{R}_+ \cdot A,$$

where $\mathbb{N}$ is the set of positive integers.

Let us mention some properties of the affine and convex hulls. Consider Y another linear space, $T : X \to Y$ a linear operator, $A, B \subset X$, $C \subset Y$ nonempty sets. Then: (i) $\operatorname{aff}(A\times C) = \operatorname{aff} A \times \operatorname{aff} C$; (ii) $\operatorname{aff} T(A) = T(\operatorname{aff} A)$; (iii) $\operatorname{aff}(A + B) = \operatorname{aff} A + \operatorname{aff} B$; (iv) $\forall a \in A\ :\ \operatorname{aff} A = a + \operatorname{aff}(A - A)$; (v) $\operatorname{aff}(A - A) = \cup_{\lambda>0}\lambda(A - A)$ if A is convex; (vi) $\operatorname{aff} A = \operatorname{lin} A$ if $0 \in A$; (vii) $\operatorname{co}(A \times C) = \operatorname{co} A \times \operatorname{co} C$; (viii) $\operatorname{co} T(A) = T(\operatorname{co} A)$; (ix) $\operatorname{co}(A + B) = \operatorname{co} A + \operatorname{co} B$; (x) $\operatorname{co}(\operatorname{cone} A) = \operatorname{cone}(\operatorname{co} A)$ (Exercises!).

Let $M \subset X$ be a linear subspace, and let $A \subset X$ be nonempty; the **algebraic interior** of A **with respect to** M is

$$\operatorname{aint}_M A := \{a \in X \mid \forall x \in M,\ \exists \delta > 0,\ \forall \lambda \in [0, \delta]\ :\ a + \lambda x \in A\}.$$

It is clear that $\operatorname{aint}_M A \subset A$ and that $M \subset \operatorname{aff}(A - A)$ when $\operatorname{aint}_M A \neq \emptyset$.

We distinguish two important cases: (i) $M = X$; in this case we write A^i instead of $\operatorname{aint}_M A$; A^i is called the **algebraic interior** of A, (ii) $M = \operatorname{aff}(A - A)$; in this case $\operatorname{aint}_M A$ is denoted by ${}^iA$ and is called the **relative algebraic interior** of A. Therefore $a \in A^i$ if and only if $\operatorname{aff} A = X$ and $a \in {}^iA$ (Exercise!).

When the set A is convex we have (Exercise!) that:

$$\forall a \in A \ : \ \operatorname{lin}(A - a) = \operatorname{cone}(A - A),$$

whence $\operatorname{aff} A = a + \operatorname{cone}(A - A)$ for every $a \in A$ (hence $\operatorname{cone}(A - A)$ is the linear subspace parallel to $\operatorname{aff} A$),

$$a \in A^i \Leftrightarrow \forall x \in X, \ \exists \lambda > 0 \ : \ a + \lambda x \in A \Leftrightarrow \operatorname{cone}(A - a) = X,$$

and

$$\begin{aligned} a \in {}^iA &\Leftrightarrow \forall x \in A, \ \exists \lambda > 0 \ : \ (1 + \lambda)a - \lambda x \in A \\ &\Leftrightarrow \operatorname{cone}(A - a) = \operatorname{cone}(A - A) \\ &\Leftrightarrow \operatorname{cone}(A - a) \text{ is a linear subspace} \\ &\Leftrightarrow \textstyle\bigcup_{n \in \mathbb{N}} n(C - a) \text{ is a linear subspace;} \end{aligned} \tag{1.1}$$

the dimension of the convex set A is $\dim A := \dim(\operatorname{aff} A) = \dim\big(\operatorname{cone}(A - A)\big)$.

Some properties of the algebraic interior are listed below. Let $\emptyset \neq A, B \subset X$, $x \in X$ and $\lambda \in \mathbb{R} \setminus \{0\}$; then: (i) ${}^i(x + A) = x + {}^iA$; (ii) ${}^i(\lambda A) = \lambda \cdot {}^iA$; (iii) $A + B^i \subset (A + B)^i$; (iv) $A + B^i = (A + B)^i$ if $B^i = B$; (v) ${}^iA + {}^iB \subset {}^i(A + B)$; (vi) ${}^i(A + B) = {}^iA + {}^iB$ if A, B are convex, ${}^iA \neq \emptyset$ and ${}^iB \neq \emptyset$; (vii) ${}^iA \neq \emptyset$ if A is convex and $\dim A < \infty$; (viii) if A is convex then $[a, x[\subset {}^iA$ for all $a \in {}^iA$ and $x \in A$.

In the sequel the results will be established for real topological vector spaces (tvs for short) or real locally convex spaces (lcs for short). When X is a tvs it is well-known that the class $\mathcal{N}_X$ of closed and balanced neighborhoods of $0 \in X$ is a base of neighborhoods of 0; when X is a lcs then the class $\mathcal{N}_X^c$ of the closed, convex and balanced neighborhoods of $0 \in X$ is also a base of neighborhoods of 0.

If X, Y are real linear spaces, we denote by $L(X, Y)$ the real linear space of linear operators from X into Y. The space $L(X, \mathbb{R})$ is denoted by X' and is called the **algebraic dual** of X; an element of X' is called a *linear functional.* When X, Y are topological vector spaces, we denote by $\mathcal{L}(X, Y)$

the linear space of continuous linear operators from X into Y; the space $\mathcal{L}(X,\mathbb{R})$ is denoted by X^* and is called the **topological dual** of X.

Let now A be an *absorbing* subset of the linear space X, *i.e.* $0 \in A^i$; the **Minkowski gauge** of A is defined by

$$p_A : X \to \mathbb{R}, \qquad p_A(x) := \inf\{\lambda \geq 0 \mid x \in \lambda A\}.$$

It is obvious that $p_A = p_{[0,1]A}$. Moreover, if $A, B \subset X$ and $C \subset Y$ are absorbing and star-shaped sets, where Y is another linear space, one has (Exercise!):

$$\{x \in X \mid p_A(x) < 1\} \subset A \subset \{x \in X \mid p_A(x) \leq 1\},$$
$$\forall\, x \in X \;:\; p_{A\cap B}(x) = \max\{p_A(x), p_B(x)\},$$
$$\forall\, x \in X,\; y \in Y \;:\; p_{A\times C}(x,y) = \max\{p_A(x), p_C(y)\}.$$

Other useful properties of the Minkowski gauge are mentioned in the next result. Recall that $p : X \to \overline{\mathbb{R}}$ is **sublinear** if $p(0) = 0$, $p(x+y) \leq p(x) + p(y)$ [with the convention $(+\infty) + (-\infty) = +\infty$] and $p(\lambda x) = \lambda p(x)$ for all $x, y \in X$, $\lambda \in \mathbb{P} :=]0,\infty[$; p is a **semi-norm** if p is a finite, sublinear and even [*i.e.* $p(-x) = p(x)$ for every $x \in X$] function.

Proposition 1.1.1 *Let A be a convex and absorbing subset of the linear space X.*

(i) *Then p_A is finite, sublinear and $A^i = \{x \in X \mid p_A(x) < 1\}$; furthermore, if A is symmetric then p_A is a semi-norm, too.*

(ii) *Assume, moreover, that X is a topological vector space and V is a neighborhood of $0 \in X$. Then p_V is continuous and*

$$\operatorname{int} V = \{x \in X \mid p_V(x) < 1\}, \quad \operatorname{cl} V = \{x \in X \mid p_V(x) \leq 1\}.$$

The following result will be useful, too.

Theorem 1.1.2 *Let C be a convex subset of the topological vector space X. Then*

(i) $\operatorname{cl} C$ *is convex;*

(ii) *if $a \in \operatorname{int} C$ and $x \in \operatorname{cl} C$, then $[a, x[\subset \operatorname{int} C$;*

(iii) $\operatorname{int} C$ *is convex;*

(iv) *if $\operatorname{int} C \neq \emptyset$ then $\operatorname{cl}(\operatorname{int} C) = \operatorname{cl} C$ and $\operatorname{int}(\operatorname{cl} C) = \operatorname{int} C$;*

(v) *if $\operatorname{int} C \neq \emptyset$ then $C^i = \operatorname{int} C$.*

Using the Minkowski gauge one obtains the geometrical versions of the Hahn–Banach theorem, *i.e.* separation theorems. In the sequel we give several separation theorems for convex subsets of topological vector spaces or locally convex space.

Theorem 1.1.3 (Eidelheit) *Let A and B be two nonempty convex subsets of the topological vector space X. If $\operatorname{int} A \neq \emptyset$ and $B \cap \operatorname{int} A = \emptyset$ then there exist $x^* \in X^* \setminus \{0\}$ and $\alpha \in \mathbb{R}$ such that*

$$\forall x \in A,\ \forall y \in B \ :\ \langle x, x^* \rangle \leq \alpha \leq \langle y, x^* \rangle, \tag{1.2}$$

or equivalently, $\sup x^*(A) \leq \inf x^*(B)$.

The separation condition (1.2) can be given in a different manner. Let $x^* \in X^* \setminus \{0\}$ and $\alpha \in \mathbb{R}$. Consider the sets

$$H^{\leq}_{x^*,\alpha} := \{x \in X \mid \langle x, x^* \rangle \leq \alpha\},$$
$$H^{<}_{x^*,\alpha} := \{x \in X \mid \langle x, x^* \rangle < \alpha\},$$
$$H_{x^*,\alpha} := \{x \in X \mid \langle x, x^* \rangle = \alpha\};$$

similarly one defines $H^{\geq}_{x^*,\alpha}$ and $H^{>}_{x^*,\alpha}$. All these sets are convex and nonempty. The set $H_{x^*,\alpha}$ is called a **closed hyperplane**, $H^{<}_{x^*,\alpha}$ and $H^{>}_{x^*,\alpha}$ are called **open half-spaces**, while $H^{\leq}_{x^*,\alpha}$ and $H^{\geq}_{x^*,\alpha}$ are called **closed half-spaces.** $H_{x^*,\alpha}$, $H^{\leq}_{x^*,\alpha}$ and $H^{\geq}_{x^*,\alpha}$ are closed sets, while $H^{<}_{x^*,\alpha}$ and $H^{>}_{x^*,\alpha}$ are open sets; moreover, $\operatorname{cl} H^{<}_{x^*,\alpha} = H^{\leq}_{x^*,\alpha}$ and $(H^{\leq}_{x^*,\alpha})^i = \operatorname{int} H^{\leq}_{x^*,\alpha} = H^{<}_{x^*,\alpha}$ (Exercises!).

Theorem 1.1.3 states the existence of $x^* \in X^* \setminus \{0\}$ and $\alpha \in \mathbb{R}$ such that $A \subset H^{\leq}_{x^*,\alpha}$ and $B \subset H^{\geq}_{x^*,\alpha}$; in this situation we say that $H_{x^*,\alpha}$ **separates** A and B; the separation is **proper** when $A \cup B \not\subset H_{x^*,\alpha}$ and the separation is **strict** when $A \cap H_{x^*,\alpha} = \emptyset$ or $B \cap H_{x^*,\alpha} = \emptyset$.

When $x_0 \in A$ and $H_{x^*,\alpha}$ separates A and $\{x_0\}$ we say that $H_{x^*,\alpha}$ is a **supporting hyperplane** of A at x_0; x_0 is called a **support point** and x^* is called a **support functional**. Therefore $x^* \in X^* \setminus \{0\}$ is a support functional for A if and only if x^* attains its supremum on A. Generally, $H_{x^*,\alpha}$, with $x^* \neq 0$, is a supporting hyperplane for A if $A \subset H^{\leq}_{x^*,\alpha}$ (or $A \subset H^{\geq}_{x^*,\alpha}$) and $A \cap H_{x^*,\alpha} \neq \emptyset$.

Corollary 1.1.4 *Let A be a convex subset of the topological vector space X having nonempty interior and $x \in A \setminus \operatorname{int} A$. Then x is a support point of A.*

In the case of locally convex spaces one has the following result for the separation of two sets.

Theorem 1.1.5 *Let X be a locally convex space and $A, B \subset X$ be two nonempty convex sets. If A is closed, B is compact and $A \cap B = \emptyset$, then there exist $x^* \in X^* \setminus \{0\}$ and $\alpha_1, \alpha_2 \in \mathbb{R}$ such that*

$$\forall\, x \in A,\ \forall\, y \in B\ :\ \langle x, x^* \rangle \leq \alpha_1 < \alpha_2 \leq \langle y, x^* \rangle,$$

or equivalently, $\sup x^*(A) < \inf x^*(B)$.

The two preceding results can be stated in a more general setting.

Theorem 1.1.6 *Let A and B be two nonempty convex subsets of the topological vector space X such that* $\operatorname{int}(A - B) \neq \emptyset$. *Then*

$$0 \notin \operatorname{int}(A - B) \Leftrightarrow \exists\, x^* \in X^* \setminus \{0\}\ :\ \sup x^*(A) \leq \inf x^*(B).$$

Theorem 1.1.7 *Let A and B be two nonempty convex subsets of the locally convex space X. Then*

$$0 \notin \operatorname{cl}(A - B) \Leftrightarrow \exists\, x^* \in X^*\ :\ \sup x^*(A) < \inf x^*(B).$$

The preceding theorem shows the usefulness of having criteria for the closedness of the difference (or sum) of two convex sets. In order to give such a criterion, let A be a nonempty convex subset of the topological vector space X. The *recession cone* of A is defined by

$$\operatorname{rec} A := \{u \in X \mid \forall a \in A\ :\ a + u \in A\}.$$

It is easy to show that $\operatorname{rec} A$ is a convex cone and $A + \operatorname{rec} A = A$. When A is a closed convex set we have that

$$\operatorname{rec} A = \bigcap\nolimits_{t>0} t(A - a) \tag{1.3}$$

for every $a \in A$. In this case it is obvious that $\operatorname{rec} A$ is a closed convex cone which is also denoted by A_∞. It is easy to see that when X is a finite dimensional separated topological vector space and A is a closed convex nonempty subset of X, $A_\infty = \{0\}$ if, and only if, A is bounded. This is no longer true when $\dim X = \infty$.

Example 1.1.1 Let $X := \ell^p$ with $p \in [1, \infty]$ and $A := \{(x_n)_{n \geq 1} \in \ell^p \mid |x_n| \leq n\ \forall\, n \in \mathbb{N}\}$. It is obvious that A is a closed convex set which is not bounded because $ne_n \in A$ for every $n \in \mathbb{N}$, but $A_\infty = \{0\}$. Indeed, if

$u = (u_n) \in A_\infty$ then $tu \in A$ for every $t \geq 0$; so $|tu_n| \leq n$ for every $t \geq 0$, whence $u_n = 0$. Hence $u = 0$.

The following famous theorem was obtained in [Dieudonné (1966)].

Theorem 1.1.8 (Dieudonné) *Let A, B be nonempty closed convex subsets of the locally convex space X. If A or B is locally compact and $A_\infty \cap B_\infty$ is a linear subspace, then $A - B$ is closed.*

In convex analysis (as well as in functional analysis) one often uses the following sets associated to a nonempty subset A of the locally convex space X:

$$A^\circ := \{x^* \in X^* \mid \forall x \in A \;:\; \langle x, x^* \rangle \geq -1\},$$
$$A^+ := \{x^* \in X^* \mid \forall x \in A \;:\; \langle x, x^* \rangle \geq 0\},$$
$$A^\perp := \{x^* \in X^* \mid \forall x \in A \;:\; \langle x, x^* \rangle = 0\},$$

called the **polar**, the **dual cone** and the **orthogonal space** of A, respectively. One verifies easily that A° is a w^*-closed convex set which contains 0, that A^+ is a w^*-closed convex cone, and, finally, that $A^\perp$ is a w^*-closed linear subspace of X^*, where $w^* = \sigma(X^*, X)$ is the weak* topology on X^*.

Similarly, for $\emptyset \neq B \subset X^*$ we define the polar, the dual cone and the orthogonal space; for example, the polar of B is

$$B^\circ := \{x \in X \mid \forall x^* \in B \;:\; \langle x, x^* \rangle \geq -1\}.$$

It is obvious that B° is a closed convex set containing 0, B^+ is a closed convex cone, and, finally, $B^\perp$ is a closed linear subspace.

One verifies easily that when $A, B \subset X$ and $\lambda \in \mathbb{P}$ we have: (i) A° is convex and $0 \in A^\circ$; (ii) $A \cup \{0\} \subset (A^\circ)^\circ =: A^{\circ\circ}$; (iii) $A \subset B \Rightarrow A^\circ \supset B^\circ$; (iv) $(A \cup B)^\circ = A^\circ \cap B^\circ$; (v) if $0 \in A \cap B$ then $(A+B)^+ = (A \cup B)^+ = A^+ \cap B^+$; (vi) $(\lambda A)^\circ = \frac{1}{\lambda} A^\circ$; (vii) $A^\circ = A^+$ if A is a cone, and $A^\circ = A^+ = A^\perp$ if A is a linear subspace; (viii) $(T(A))^\circ = (T^*)^{-1}(A^\circ)$, if $T \in \mathcal{L}(X, Y)$, where Y is another locally convex space.

A very useful result is the bipolar's theorem. Let X be a topological vector space and $A \subset X$; the set $\overline{\text{co}}A := \text{cl}(\text{co}\, A)$ is called the **closed convex hull** of the set A; it is the smallest closed convex set containing A. Similarly, $\overline{\text{cone}}A := \text{cl}(\text{cone}\, A)$ is called the **closed conic hull** of A.

Theorem 1.1.9 (bipolar) *Let A be a nonempty subset of the locally con-*

vex space X. Then

$$A^{\circ\circ} = \overline{\text{co}}(A \cup \{0\}), \quad A^{++} = \overline{\text{cone}}(\text{co}\, A), \quad A^{\perp\perp} = \text{cl}(\text{lin}\, A).$$

It follows that for the nonempty subset A of the lcs X one has: (a) $A^{\circ\circ} = A \Leftrightarrow A$ is closed, convex and $0 \in A$; (b) $A^{++} = A \Leftrightarrow A$ is a closed convex cone; (c) $A^{\perp\perp} = A \Leftrightarrow A$ is a closed linear subspace.

Another famous result is the following.

Theorem 1.1.10 (Alaoglu–Bourbaki) *Let X be a locally convex space and $U \subset X$ be a neighborhood of the origin. Then U° is w^*-compact.*

We finish this preliminary section with some notions and results concerning completeness and metrizability of topological vector spaces.

The subset A of the topological vector space X is **complete (quasi-complete)** if every (bounded) Cauchy net $(x_i)_{i\in I} \subset A$ is convergent to an element $x \in A$. Of course, any complete set is closed and any closed subset of a complete set is complete (Exercise!). Recall that the topological space (X, τ) is **first countable** if every element of X has a (at most) countable base of neighborhoods. Note that a subset A of a first countable tvs X is complete if and only if every Cauchy sequence of A is convergent to an element of A; in particular, a first countable tvs is complete if and only if it is quasi-complete (Exercise!).

We shall use several times the hypothesis that a certain topological vector space is first countable. The next result refers to the first countability of locally convex spaces.

Proposition 1.1.11 *Let (X, τ) be a locally convex space. Then*

(i) *(X, τ) is first countable $\Leftrightarrow \exists\, \mathcal{P}$ a (at most) countable family of semi-norms on X such that $\tau = \tau_{\mathcal{P}} \Leftrightarrow \tau$ is semi-metrizable, i.e. there exists a semi-metric d on X such that $\tau = \tau_d$; the semi-metric d may be chosen to be invariant to translations (i.e. $d(x+z, y+z) = d(x, y)$ for all $x, y, z \in X$).*

(ii) *(X, τ) is separated and first countable if and only if τ is metrizable, i.e. there exists a metric d on X such that $\tau = \tau_d$.*

Note that when the topology τ of a locally convex space X coincides with the topology τ_d determined by a semi-metric d invariant to translations, the completeness with respect to τ and d are equivalent. One says that the locally convex space X is a **Fréchet space** if X is complete and metrizable. It is obvious that every closed linear subspace of a Fréchet space is Fréchet, too. One says that the topological vector space X is **barreled** if every absorbing, convex and closed subset of X is a neighborhood of $0 \in X$. As application of the Baire theorem one obtains that every Fréchet space is barreled.

It is well-known that in a finite dimensional separated topological vector space any convex and absorbing set is a neighborhood of the origin.

1.2 Closedness and Interiority Notions

Consider X a real topological vector space. We say that the series $\sum_{n\geq 1} x_n$ is **convergent** (resp. **Cauchy**) if the sequence $(S_n)_{n\in\mathbb{N}}$ is convergent (resp. Cauchy), where $S_n := \sum_{k=1}^{n} x_k$ for every $n \in \mathbb{N}$; of course, any convergent series is Cauchy.

Let $A \subset X$; by a **convex series** with elements of A we mean a series of the form $\sum_{m\geq 1} \lambda_m x_m$ with $(\lambda_m) \subset \mathbb{R}_+$, $(x_m) \subset A$ and $\sum_{m\geq 1} \lambda_m = 1$; if, furthermore, the sequence (x_m) is bounded we speak about a **b-convex series**. We say that A is **cs-closed** if any convergent convex series with elements of A has its sum in A;* A is **cs-complete** if any Cauchy convex series with elements of A is convergent and its sum is in A. Similarly, the set A is called **ideally convex** if any convergent b-convex series with elements of A has its sum in A and A is **bcs-complete** if any Cauchy b-convex series with elements of A is convergent and its sum is in A. It is obvious that any cs-closed set is ideally convex, every ideally convex set is convex, every cs-complete set is cs-closed and every complete convex set is cs-complete; if X is complete, then $A \subset X$ is cs-complete (bcs-complete) if and only if A is cs-closed (ideally convex). If X_0 is a linear subspace of X and A is a cs-closed (ideally convex) subset of X, then $X_0 \cap A$ is a cs-closed (ideally convex) subset of X_0 (endowed with the induced topology). Moreover, if $A \subset X$ and $B \subset Y$ are nonempty, then $A \times B$ is cs-closed (cs-complete, ideally convex, bcs-complete) if and only if A and B are cs-closed (cs-complete, ideally convex, bcs-complete). If X is first countable, a linear subspace X_0 of X is cs-closed (cs-complete) if and only if it is closed (complete); moreover, if X is a locally convex space, X_0 is closed if and only if X_0 is ideally convex. Note also that $T(A)$ is cs-closed (cs-complete, ideally convex, bcs-complete) if $A \subset X$ is cs-closed (cs-complete, ideally convex,

*Because X may not be separated, in fact we ask that every limit of (S_n) is in A.

bcs-complete) and $T : X \to Y$ is an isomorphism of topological vector spaces (Exercise!), Y being another tvs. We consider that the empty set is convex, ideally convex, bcs-complete, cs-complete and cs-closed.

It is worth to point out that when X is a locally convex space, every b-convex series with elements of X is Cauchy (Exercise!).

The class of cs-closed sets (and consequently that of ideally convex sets) is larger than the class of closed convex sets, as the next result shows.

Proposition 1.2.1 *Let $A \subset X$ be a nonempty convex set.*

(i) *If A is closed or open then A is cs-closed.*

(ii) *If X is separated and* $\dim A < \infty$ *then A is cs-closed.*

Proof. (i) Let $\sum_{n\geq 1} \lambda_n x_n$ be a convergent convex series with elements of A; denote by x its sum.

Suppose that A is closed and fix $a \in A$. Then, for every $n \in \mathbb{N}$ we have that $\sum_{k=1}^{n} \lambda_k x_k + \left(1 - \sum_{k=n+1}^{\infty} \lambda_k\right) a \in A$. Taking the limit for $n \to \infty$, we obtain that $x \in \operatorname{cl} A = A$.

Suppose now that A is open. Assume that $x \notin A$. By Theorem 1.1.3, there exists $x^* \in X^*$ such that $\langle a - x, x^* \rangle > 0$ for every $a \in A$. In particular $\langle x_n - x, x^* \rangle > 0$ for every $n \in \mathbb{N}$. Multiplying by $\lambda_n \geq 0$ and adding for $n \in \mathbb{N}$ we get (since $\lambda_n > 0$ for some n) the contradiction

$$0 < \sum\nolimits_{n\geq 1} \lambda_n \langle x_n - x, x^* \rangle = \left\langle \sum\nolimits_{n\geq 1} \lambda_n x_n, x^* \right\rangle - \left(\sum\nolimits_{n\geq 1} \lambda_n\right) \langle x, x^* \rangle = 0.$$

Therefore $x \in A$. So in both cases A is cs-closed.

(ii) We prove the statement by mathematical induction on $n := \dim A$. If $n = 0$ A reduces to a point; it is obvious that A is cs-closed in this case. Suppose that the statement is true if $\dim A \leq n \in \mathbb{N} \cup \{0\}$ and show it for $\dim A = n + 1$. Without any loss of generality we suppose that $0 \in A$; then $X_0 := \operatorname{aff} A$ is a linear subspace with $\dim X_0 = n + 1$. Because on a finite dimensional linear space there exists a unique separated linear topology and in such spaces the interior and the algebraic interior coincide for convex sets, we have that ${}^{i}A = \operatorname{int}_{X_0} A \neq \emptyset$. Let $\sum_{n\geq 1} \lambda_n x_n$ be a convergent convex series with elements of A and sum $\overline{x}$. Assume that $\overline{x} \notin A$. Because A is convex, the set $P := \{n \in \mathbb{N} \mid \lambda_n > 0\}$ is infinite, and so we may assume that $P = \mathbb{N}$. Applying now Theorem 1.1.3 in X_0, there exists $x_0^* \in X_0^* \setminus \{0\}$ such that $\langle x - \overline{x}, x_0^* \rangle \geq 0$ for every $x \in A$. But $\sum_{n\geq 1} \lambda_n \langle x_n - \overline{x}, x_0^* \rangle = 0$. Since $\langle x_n - \overline{x}, x_0^* \rangle \geq 0$ and $\lambda_n > 0$ for every n, we obtain that $(x_n) \subset A_0 := A \cap H_{x_0^*,\lambda}$, where $\lambda := \langle \overline{x}, x_0^* \rangle$. Since

$\dim A_0 \leq \dim H_{x_0^*,\lambda} = n$, from the induction hypothesis we obtain the contradiction $\overline{x} \in A_0 \subset A$. Therefore $\overline{x} \in A$. The proof is complete. □

Other properties of cs-closed and ideally convex sets are given in the following result.

Proposition 1.2.2 (i) *If $A_i \subset X$ is cs-closed (resp. ideally convex) for every $i \in I$ then $\bigcap_{i\in I} A_i$ is cs-closed (resp. ideally convex).*

(ii) *If X_i is a topological vector space and $A_i \subset X_i$ is cs-closed (resp. ideally convex) for every $i \in I$, then $\prod_{i\in I} A_i$ is cs-closed (resp. ideally convex) in $\prod_{i\in I} X_i$ (which is endowed with the product topology).*

Proof. The proof of (i) is immediate, while for (ii) one must take into account that a sequence $(x_n)_{n\in\mathbb{N}} \subset X := \prod_{i\in I} X_i$ converges to $x \in X$ (resp. is bounded) if and only if (x_n^i) converges to x^i in X_i (resp. is bounded) for every $i \in I$. □

We say that the subset C of Y is **lower cs-closed** (lcs-closed for short) if there exist a Fréchet space X and a cs-closed subset B of $X \times Y$ such that $C = \Pr_Y(B)$. Similarly, the subset C of Y is **lower ideally convex** (li-convex for short) if there exist a Fréchet space X and an ideally convex subset B of $X \times Y$ such that $C = \Pr_Y(B)$. It is obvious that any cs-closed (resp. ideally convex) set is lcs-closed (resp. li-convex), any lcs-closed set is li-convex and any li-convex set is convex, but the converse implications are not true, generally. Note also that $T(A)$ is lcs-closed (resp. li-convex) if $A \subset X$ is lcs-closed (resp. li-convex) and $T : X \to Y$ is an isomorphism of topological vector spaces (Exercise!). The classes of lcs-closed and li-convex sets have very good stability properties as the following results show. We give only the proofs for the "li-convex" case, that for the "lcs-closed" case being similar.

Proposition 1.2.3 *Suppose that Y is a Fréchet space and $C \subset Y \times Z$ is a li-convex (lcs-closed) set. Then $\Pr_Z(C)$ is a li-convex (lcs-closed) set.*

Proof. By hypothesis, there exists a Fréchet space X and an ideally convex subset $B \subset X \times (Y \times Z)$ such that $C = \Pr_{Y\times Z}(B)$. Since $X \times Y$ is a Fréchet space and $\Pr_Z(C) = \Pr_Z(B)$, we have that $\Pr_Z(C)$ is a li-convex subset of Z. □

Proposition 1.2.4 *Let I be an at most countable nonempty set.*

(i) *If $C_i \subset Y$ is li-convex (lcs-closed) for every $i \in I$ then $\bigcap_{i\in I} C_i$ is li-convex (lcs-closed).*

(ii) *If Y_i is a topological vector space and $C_i \subset Y_i$ is li-convex (lcs-closed) for every $i \in I$, then $\prod_{i\in I} C_i$ is li-convex (lcs-closed) in $\prod_{i\in I} Y_i$.*

Proof. (i) For each $i \in I$ there exist X_i a Fréchet space and an ideally convex set $B_i \subset X_i \times Y$ such that $C_i = \Pr_Y(B_i)$. The space $X := \prod_{i\in I} X_i$ is a Fréchet space as the product of an at most countable family of Fréchet spaces. Let

$$\widehat{B}_i := \left\{ \left((x_j)_{j\in I}, y\right) \in X \times Y \mid (x_i, y) \in B_i \right\}.$$

Then $\widehat{B}_i$ is an ideally convex set by Proposition 1.2.2(ii). It follows that $B := \bigcap_{i\in I} \widehat{B}_i$ is ideally convex by Proposition 1.2.2(i). Since $\Pr_Y(B) = \bigcap_{i\in I} C_i$, it follows that $\bigcap_{i\in I} C_i$ is li-convex.

(ii) For each $i \in I$ there exist X_i a Fréchet space and an ideally convex set $B_i \subset X_i \times Y_i$ such that $C_i = \Pr_{Y_i}(B_i)$. By Proposition 1.2.2(ii), $\prod_{i\in I} B_i$ is an ideally convex subset of $\prod_{i\in I}(X_i \times Y_i)$. The space $X := \prod_{i\in I} X_i$ is a Fréchet space; let $Y := \prod_{i\in I} Y_i$. Consider the set

$$B := \left\{ \left((x_i)_{i\in I}, (y_i)_{i\in I}\right) \in X \times Y \mid (x_i, y_i) \in B_i \ \forall i \in I \right\}.$$

Since $T : \prod_{i\in I}(X_i \times Y_i) \to X \times Y$, $T\left((x_i, y_i)_{i\in I}\right) := \left((x_i)_{i\in I}, (y_i)_{i\in I}\right)$ is an isomorphism of topological vector spaces, $B = T\left(\prod_{i\in I} B_i\right)$ is ideally convex. As $C := \prod_{i\in I} C_i = \Pr_Y(B)$, C is li-convex. □

Before stating other properties of li-convex and lcs-closed sets, let us define some notions and notations related to multifunctions.

Let E, F be two nonempty sets; a mapping $\mathcal{R} : E \to 2^F$ is called a **multifunction**, and it will be denoted by $\mathcal{R} : E \rightrightarrows F$. The set $\operatorname{dom} \mathcal{R} := \{x \in E \mid \mathcal{R}(x) \neq \emptyset\}$ is called the **domain** of the multifunction $\mathcal{R}$; the **image** of $\mathcal{R}$ is $\operatorname{Im} \mathcal{R} := \bigcup_{x\in E} \mathcal{R}(x)$; the **graph** of $\mathcal{R}$ is the set $\operatorname{gr} \mathcal{R} := \{(x, y) \mid y \in \mathcal{R}(x)\} \subset E \times F$; the **inverse** of the multifunction $\mathcal{R}$ is the multifunction $\mathcal{R}^{-1} : F \rightrightarrows E$ defined by $\mathcal{R}^{-1}(y) := \{x \in E \mid y \in \mathcal{R}(x)\}$. Therefore $\operatorname{dom} \mathcal{R}^{-1} = \operatorname{Im} \mathcal{R}$, $\operatorname{Im} \mathcal{R}^{-1} = \operatorname{dom} \mathcal{R}$ and $\operatorname{gr} \mathcal{R}^{-1} = \{(y, x) \mid (x, y) \in \operatorname{gr} \mathcal{R}\}$. Frequently we shall identify a multifunction with its graph. For $A \subset E$ and $B \subset F$ one defines $\mathcal{R}(A) := \bigcup_{x\in A} \mathcal{R}(x)$ and $\mathcal{R}^{-1}(B) := \bigcup_{y\in B} \mathcal{R}^{-1}(y)$; in particular $\operatorname{Im} \mathcal{R} = \mathcal{R}(E)$ and $\operatorname{dom} \mathcal{R} = \mathcal{R}^{-1}(F)$. If $\mathcal{S} : F \rightrightarrows G$ is another multifunction, then the composition of the multifunctions $\mathcal{S}$ and $\mathcal{R}$ is the multifunction $\mathcal{S}\circ\mathcal{R} : E \rightrightarrows G$, $(\mathcal{S}\circ\mathcal{R})(x) := \bigcup_{y\in\mathcal{R}(x)} \mathcal{S}(y)$. If $\mathcal{R}, \mathcal{S} : E \rightrightarrows F$ and F is a linear space, the sum of $\mathcal{R}$ and $\mathcal{S}$ is the multifunction $\mathcal{R} + \mathcal{S} : E \rightrightarrows F$, $(\mathcal{R} + \mathcal{S})(x) := \mathcal{R}(x) + \mathcal{S}(x)$.

Let now $\mathcal{R} : X \rightrightarrows Y$; we say that $\mathcal{R}$ is **convex (closed, ideally convex, bcs-complete, cs-convex, cs-complete, li-convex, lcs-closed)** if its graph is a convex (closed, ideally convex, bcs-complete, cs-convex, cs-complete, li-convex, lcs-closed) subset of $X \times Y$. Note that $\mathcal{R}$ is convex if and only if

$$\forall x, x' \in X,\ \forall \lambda \in [0,1] \ :\ \lambda\mathcal{R}(x) + (1-\lambda)\mathcal{R}(x') \subset \mathcal{R}(\lambda x + (1-\lambda)x').$$

Proposition 1.2.5 *Let $A, B \subset X$, $\mathcal{R}, \mathcal{S} : X \rightrightarrows Y$ and $\mathcal{T} : Y \rightrightarrows Z$.*

(i) *If X is a Fréchet space and A, $\mathcal{R}$ are li-convex (resp. lcs-closed), then $\mathcal{R}(A)$ is li-convex (resp. lcs-closed).*

(ii) *If X is a Fréchet space and A, B are li-convex (resp. lcs-closed), then $A + B$ is li-convex (resp. lcs-closed).*

(iii) *If Y is a Fréchet space and $\mathcal{R}, \mathcal{T}$ are li-convex (resp. lcs-closed), then $\mathcal{T} \circ \mathcal{R}$ is li-convex (resp. lcs-closed).*

(iv) *If Y is a Fréchet space and $\mathcal{R}, \mathcal{S}$ are li-convex (resp. lcs-closed), then $\mathcal{R} + \mathcal{S}$ is li-convex (resp. lcs-closed).*

Proof. (i) We have that

$$\mathcal{R}(A) = \Pr_Y\left((A \times Y) \cap \operatorname{gr}\mathcal{R}\right).$$

Using successively Propositions 1.2.4(ii), 1.2.4(i) and 1.2.3, it follows that $\mathcal{R}(A)$ is li-convex.

(ii) Let $T : X \times X \to X$, $T(x, y) := x + y$. Since T is a continuous linear operator, $\operatorname{gr} T$ is a closed linear subspace; in particular T is a li-convex multifunction. Since $A \times B$ is li-convex, by (i) $A + B = T(A \times B)$ is a li-convex set.

(iii) We have that

$$\operatorname{gr}(\mathcal{T} \circ \mathcal{R}) = \Pr_{X \times Z}\big((\operatorname{gr}\mathcal{R} \times Z) \cap (X \times \operatorname{gr}\mathcal{T})\big).$$

The conclusion follows from Propositions 1.2.4(ii), 1.2.4(i) and 1.2.3.

(iv) The sets

$$T := \{(x, z, y, y') \mid x \in X,\ y, y' \in Y,\ z = y + y'\} \subset (X \times Y) \times (Y \times Y),$$

$A := \{(x, z, y, y') \mid (x, y) \in \operatorname{gr}\mathcal{R},\ y', z \in Y\}$ and $B := \{(x, z, y, y') \mid (x, y') \in \operatorname{gr}\mathcal{S},\ y, z \in Y\}$ are li-convex sets (the first being a closed linear subspace). Therefore $\operatorname{gr}(\mathcal{R} + \mathcal{S}) = \Pr_{X \times Y}(T \cap A \cap B)$ is li-convex. □

Let Y be another topological vector space and $A \subset X \times Y$; we introduce the conditions (Hx) and (Hwx) below, where x refers to the component $x \in X$:

(Hx) If the sequences $\left((x_n, y_n)\right)_{n\geq 1} \subset A$ and $(\lambda_n)_{n\geq 1} \subset \mathbb{R}_+$ are such that $\sum_{n\geq 1} \lambda_n = 1$, $\sum_{n\geq 1} \lambda_n y_n$ has sum y and $\sum_{n\geq 1} \lambda_n x_n$ is Cauchy, then the series $\sum_{n\geq 1} \lambda_n x_n$ is convergent and its sum $x \in X$ verifies $(x, y) \in A$.

(Hwx) If the sequences $\left((x_n, y_n)\right)_{n\geq 1} \subset A$ and $(\lambda_n)_{n\geq 1} \subset \mathbb{R}_+$ are such that $\left((x_n, y_n)\right)$ is bounded, $\sum_{n\geq 1} \lambda_n = 1$, $\sum_{n\geq 1} \lambda_n y_n$ has sum y and $\sum_{n\geq 1} \lambda_n x_n$ is Cauchy, then the series $\sum_{n\geq 1} \lambda_n x_n$ is convergent and its sum $x \in X$ verifies $(x, y) \in A$.

Of course, when X is a locally convex space, deleting "$\sum_{n\geq 1} \lambda_n x_n$ is Cauchy" in condition (Hwx) one obtains an equivalent statement.

In the next result we mention the relationships among conditions (Hx), (Hwx), ideal convexity, cs-closedness, cs-completeness and convexity. The proof being very easy we omit it.

Proposition 1.2.6 *Let $A \subset X \times Y$ and $B \subset X \times Y \times Z$ be nonempty sets.*

(i) *Assume that Y is complete. Then B satisfies* (H(x, y)) *if and only if B satisfies* (Hx)*; B satisfies* (Hwx) *if and only if B satisfies* (Hw(x, y)).

(ii) *Assume that X is complete. Then A satisfies* (Hx) *if and only if A is cs-closed; A satisfies* (Hwx) *if and only if A is ideally convex.*

(iii) *Assume that Y is complete. Then A satisfies* (Hx) *if and only if A is cs-complete; A satisfies* (Hwx) *if and only if A is bcs-complete.*

(iv) *If A satisfies* (Hx) *then A satisfies* (Hwx)*; if A satisfies* (Hwx) *then A is convex.*

(v) *Assume that X is a locally convex space and* $\operatorname{Pr}_X(A)$ *is bounded. If A satisfies* (Hx) *then* $\operatorname{Pr}_Y(A)$ *is cs-closed; if A satisfies* (Hwx) *then* $\operatorname{Pr}_Y(A)$ *is ideally convex.*

We define now several interiority notions. Let $\emptyset \neq A \subset X$. We denote by $\operatorname{rint} A$ the interior of A with respect to $\operatorname{aff} A$, *i.e.* $\operatorname{rint} A := \operatorname{int}_{\operatorname{aff} A} A$. Of course, $\operatorname{rint} A \subset {}^iA$. Consider also the sets

$$^{ic}A := \begin{cases} {}^iA & \text{if aff } A \text{ is a closed set,} \\ \emptyset & \text{otherwise,} \end{cases}$$

$$\operatorname{ri} A := \begin{cases} \operatorname{rint} A & \text{if aff } A \text{ is a closed set,} \\ \emptyset & \text{otherwise,} \end{cases}$$

$$^{ib}A := \begin{cases} {}^iA & \text{if } X_0 \text{ is a barreled linear subspace,} \\ \emptyset & \text{otherwise,} \end{cases}$$

where $X_0 = \operatorname{lin}(A - a)$ for some (every) $a \in A$; X_0 is the linear subspace, parallel to $\operatorname{aff} A$.

In the sequel, in this section, $A \subset X$ is a nonempty convex set. Taking into account the characterization (1.1) of ${}^iA$, we obtain that

$$\begin{aligned} x \in {}^{ic}A &\Leftrightarrow \operatorname{cone}(A - x) \text{ is a closed linear subspace of } X \\ &\Leftrightarrow \bigcup\nolimits_{\lambda>0} \lambda(A - x) \text{ is a closed linear subspace of } X, \end{aligned}$$

and

$$\begin{aligned} x \in {}^{ib}A &\Leftrightarrow \operatorname{cone}(A - x) \text{ is a barreled linear subspace of } X \\ &\Leftrightarrow \bigcup\nolimits_{n\in\mathbb{N}} n(A - x) \text{ is a barreled linear subspace of } X. \end{aligned}$$

If X is a Fréchet space and $\operatorname{aff} A$ is closed then ${}^{ic}A = {}^{ib}A$, but it is possible to have ${}^{ib}A \neq \emptyset$ and ${}^{ic}A = \emptyset$ (if $\operatorname{aff} A$ is not closed).

The **quasi relative interior** of A is the set

$$\operatorname{qri} A := \{x \in A \mid \overline{\operatorname{cone}}(A - x) \text{ is a linear subspace of } X\}.$$

Taking into account that in a finite dimensional separated topological vector space the closure of a convex cone C is a linear subspace if and only if C is a linear subspace (Exercise!), it follows that in this case $\operatorname{qri} A = {}^iA = {}^{ic}A = {}^{ib}A$.

Below we collect several properties of the quasi relative interior.

Proposition 1.2.7 *Let $A \subset X$ be a nonempty convex set and $T \in \mathcal{L}(X, Y)$. Then:*

$$\begin{aligned} a \in \operatorname{qri} A &\Leftrightarrow a \in A \text{ and } \overline{\operatorname{cone}}(A - a) = \overline{\operatorname{cone}}(A - A) \\ &\Leftrightarrow a \in A \text{ and } a - A \subset \overline{\operatorname{cone}}(A - a), \end{aligned} \tag{1.4}$$

$${}^iA \subset \operatorname{qri} A = A \cap \operatorname{qri} \overline{A}, \qquad \forall a \in \operatorname{qri} A,\ \forall x \in A\ :\ [a, x[\, \subset \operatorname{qri} A, \tag{1.5}$$

and $T(\operatorname{qri} A) \subset \operatorname{qri} T(A)$. *In particular* $\operatorname{qri} A$ *is a convex set.*

Assume that $\operatorname{qri} A \neq \emptyset$; *then* $\overline{\operatorname{qri} A} = \overline{A}$ *and*

$$^{i}(T(A)) \subset T(\operatorname{qri} A) \subset \operatorname{qri} T(A) \subset T(A) \subset \overline{T(\operatorname{qri} A)}. \tag{1.6}$$

Moreover, if Y *is separated and finite dimensional then*

$$T(\operatorname{qri} A) = {}^{i}(T(A)). \tag{1.7}$$

Proof. The first equivalence in Eq. (1.4) is immediate from the definition of $\operatorname{qri} A$. Of course, $\overline{\operatorname{cone}}(A - a) = \overline{\operatorname{cone}}(A - A)$ implies that $a - A \subset \overline{\operatorname{cone}}(A - a)$. Conversely, if $a - A \subset \overline{\operatorname{cone}}(A - a)$ then $-\overline{\operatorname{cone}}(A - a) = \overline{\operatorname{cone}}(a - A) \subset \overline{\operatorname{cone}}(A - a)$, which shows that $\overline{\operatorname{cone}}(A - a)$ is a linear subspace. Therefore Eq. (1.4) holds.

If $a \in {}^{i}A$, from Eq. (1.1) we have that $a \in A$ and $\operatorname{cone}(A - a)$ is a linear subspace, and so $\overline{\operatorname{cone}}(A - a)$ is a linear subspace, too. Hence $a \in \operatorname{qri} A$. The equality $\operatorname{qri} A = A \cap \operatorname{qri} \overline{A}$ follows immediately from the relation $\overline{\operatorname{cone}}\, C = \overline{\operatorname{cone}}\, \overline{C}$, valid for every nonempty subset C of X.

Let $a \in \operatorname{qri} A$, $x \in A$, $\lambda \in [0, 1[$ and $a_\lambda := (1 - \lambda)a + \lambda x$. Then

$$A - A \supset A - a_\lambda = (1 - \lambda)(A - a) + \lambda(A - x) \supset (1 - \lambda)(A - a),$$

and so, taking into account Eq. (1.4), we have

$$\begin{aligned} \overline{\operatorname{cone}}(A - A) &\supset \overline{\operatorname{cone}}(A - a_\lambda) \supset \overline{\operatorname{cone}}\left((1 - \lambda)(A - a)\right) \\ &= \overline{\operatorname{cone}}(A - a) = \overline{\operatorname{cone}}(A - A). \end{aligned}$$

Therefore $\overline{\operatorname{cone}}(A - A) = \overline{\operatorname{cone}}(A - a_\lambda)$. Since $a_\lambda \in A$, from Eq. (1.4) we obtain that $a_\lambda \in \operatorname{qri} A$. The proof of Eq. (1.5) is complete.

Let $a \in \operatorname{qri} A$; then, by Eq. (1.4), $a - A \subset \overline{\operatorname{cone}}(A - a)$, and so

$$\begin{aligned} Ta - T(A) = T(a - A) &\subset T\left(\overline{\operatorname{cone}}(A - a)\right) \subset \overline{T\left(\operatorname{cone}(A - a)\right)} \\ &= \overline{\operatorname{cone}}\left(T(A) - Ta\right), \end{aligned}$$

which shows that $Ta \in \operatorname{qri} T(A)$.

Assume now that $\operatorname{qri} A \neq \emptyset$ and fix $a \in \operatorname{qri} A$. It is sufficient to show Eq. (1.6); then the equality $\overline{\operatorname{qri} A} = \overline{A}$ follows immediately from the last inclusion in Eq. (1.6) for $T = \operatorname{Id}_X$ and from Eq. (1.5).

Let $y \in {}^{i}(T(A))$; using Eq. (1.1), there exists $\lambda > 0$ such that $(1 + \lambda)y - \lambda Ta \in T(A)$. So, $(1 + \lambda)y - \lambda Ta = Tx$ for some $x \in A$. It follows that $y = Tx_\lambda$, where $x_\lambda := (1 + \lambda)^{-1}(\lambda a + x)$. But, using Eq. (1.4), $x_\lambda \in \operatorname{qri} A$, and so $y \in T(\operatorname{qri} A)$.

The second inclusion in Eq. (1.6) was already proved, while the third is obvious. So, let $y \in T(A)$; there exists $x \in A$ such that $y = Tx$. By Eq. (1.5), $(1-\lambda)a + \lambda x \in \operatorname{qri} A$ for $\lambda \in]0,1[$, and so $(1-\lambda)Ta + \lambda y \in T\big(\operatorname{qri} A\big)$ for $\lambda \in]0,1[$. Taking the limit for $\lambda \to 1$, we obtain that $y \in \overline{T(\operatorname{qri} A)}$.

When Y is separated and finite dimensional we have (as already observed) that ${}^{i}(T(A)) = \operatorname{qri} T(A)$; then Eq. (1.7) follows immediately from Eq. (1.6). □

The notion of quasi relative interior is related to that of united sets. Let X be a locally convex space and $A, B \subset X$ be nonempty convex sets; we say that A and B are **united** if they cannot be properly separated, *i.e.* if every closed hyperplane which separates A and B contains both of them.

The next result is related to the above notions.

Proposition 1.2.8 *Let X be a locally convex space, $A, B \subset X$ be nonempty convex sets and $\overline{x} \in X$.*

(i) *A and B are united $\Leftrightarrow$ $\overline{\operatorname{cone}}(A-B)$ is a linear subspace $\Leftrightarrow$ $(A-B)^-$ is a linear subspace.*

(ii) *Assume that $\overline{\operatorname{cone}}(A-\overline{x})$ is a linear subspace. Then $\overline{x} \in \operatorname{cl} A$. Moreover, if $\operatorname{aff} A$ is closed and $\operatorname{rint} A \neq \emptyset$ then $\overline{x} \in \operatorname{rint} A$.*

Proof. (i) Assume that A and B are united but $C := \overline{\operatorname{cone}}(A-B)$ is not a linear subspace. Then there exists $x_0 \in (-C) \setminus C$. By Theorem 1.1.5 there exists $x^* \in X^*$ such that $\langle x_0, x^* \rangle < 0 \leq \langle z, x^* \rangle$ for every $z \in C$. Then $0 \leq \langle x - y, x^* \rangle$ for all $x \in A$, $y \in B$, and so $\langle x, x^* \rangle \leq \lambda \leq \langle y, x^* \rangle$ for all $x \in A$, $y \in B$, for some $\lambda \in \mathbb{R}$. Therefore $H_{x^*,\lambda}$ separates A and B. It follows that $\langle x, x^* \rangle = \lambda = \langle y, x^* \rangle$ for all $x \in A$, $y \in B$, and so $0 \leq \langle z, x^* \rangle$ for every $z \in C$. Thus we have the contradiction $\langle x_0, x^* \rangle = 0$. Therefore $\overline{\operatorname{cone}}(A-B)$ is a linear subspace.

Let $C := \overline{\operatorname{cone}}(A-B)$ be a linear subspace and $\langle x, x^* \rangle \leq \lambda \leq \langle y, x^* \rangle$ for all $x \in A$, $y \in B$, for some $x^* \in X^*$ and $\lambda \in \mathbb{R}$. Then $0 \leq \langle z, x^* \rangle$ for every $z \in A - B$, and so $0 \leq \langle z, x^* \rangle$ for $z \in C$. Then, since C is a linear subspace, $0 = \langle z, x^* \rangle$ for every $z \in C$ which implies immediately that $H_{x^*,\lambda}$ contains A and B. Therefore A and B are united.

The other equivalence is an immediate consequence of the bipolar theorem (Theorem 1.1.9).

(ii) By (i) we have that $\{\overline{x}\}$ and A are united. Assuming that $\overline{x} \notin \operatorname{cl} A$, using again Theorem 1.1.5, we obtain that $\{\overline{x}\}$ and A can be properly separated. This contradiction proves that $\overline{x} \in \operatorname{cl} A$.

Suppose now that $\operatorname{aff} A$ is closed and $\operatorname{rint} A \neq \emptyset$. Without loss of generality we suppose that $\overline{x} = 0$. By what was shown above we have that $\overline{x} = 0 \in \operatorname{cl} A \subset \operatorname{cl}(\operatorname{aff} A) = \operatorname{aff} A$. Thus $X_0 := \operatorname{aff} A$ is a linear space. Assuming now that $0 \notin \operatorname{int}_{X_0} A \neq \emptyset$, we obtain that $\{\overline{x}\}$ and A can be properly separated (in X_0, and therefore in X) using Theorem 1.1.3. This contradiction proves that $\overline{x} \in \operatorname{rint} A$. □

From the preceding proposition we obtain that when X is a locally convex space and $A \subset X$ is a nonempty convex set, the quasi relative interior of A is given by the formula

$$\operatorname{qri} A = A \cap \{x \in X \mid \{x\} \text{ and } A \text{ are united}\}.$$

The next result shows that the class of convex sets with nonempty quasi relative interior is large enough.

Proposition 1.2.9 *Let X be a first countable separable locally convex space and $A \subset X$ be a nonempty cs-complete set. Then* $\operatorname{qri} A \neq \emptyset$.

Proof. Since X is first countable, by Proposition 1.1.11, the topology of X is determined by a countable family $\mathcal{P} = \{p_n \mid n \in \mathbb{N}\}$ of semi-norms. Without loss of generality we suppose that $p_n \leq p_{n+1}$ for every $n \in \mathbb{N}$. Since $\tau_{\mathcal{P}}$ is semi-metrizable, the set A is separable, too. Let $A_0 = \{x_n \mid n \in \mathbb{N}\} \subset A$ be such that $A \subset \operatorname{cl} A_0$. Consider $\beta_n \in]0, 2^{-n}]$ such that $\beta_n p_n(x_n) \leq 2^{-n}$. The series $\sum_{n\geq 1} \beta_n x_n$ is Cauchy (since for $m \geq n$ and $p \in \mathbb{N}$ we have that $p_n(\sum_{k=m}^{m+p} \beta_k x_k) \leq \sum_{k=m}^{m+p} \beta_k p_n(x_k) \leq \sum_{k=m}^{m+p} \beta_k p_k(x_k) < 2^{-m+1}$). Taking $\lambda_n := (\sum_{n\geq 1} \beta_n)^{-1}\beta_n$, $\sum_{n\geq 1} \lambda_n x_n$ is a Cauchy convex series with elements of A. Because A is cs-complete, the series $\sum_{n\geq 1} \lambda_n x_n$ is convergent and its sum $\overline{x} \in A$. Suppose that $\overline{x} \notin \operatorname{qri} A$. Then there exists $x_0 \in (-C) \setminus C$, where $C := \overline{\operatorname{cone}}(A - \overline{x})$. Using Theorem 1.1.5, there exists $x^* \in X^*$ such that $\langle x_0, x^*\rangle < 0 \leq \langle z, x^*\rangle$ for all $z \in C$. In particular $\langle x - \overline{x}, x^*\rangle \geq 0$ for every $x \in A$. But $\sum_{n\geq 1} \lambda_n \langle x_n - \overline{x}, x_0^*\rangle = 0$. Since $\langle x_n - \overline{x}, x_0^*\rangle \geq 0$ and $\lambda_n > 0$ for every n, we obtain that $\langle x - \overline{x}, x^*\rangle = 0$ for every $x \in A_0$. Since A_0 is dense in A and x^* is continuous, we have that $\langle x - \overline{x}, x^*\rangle = 0$ for every $x \in A$, and so $\langle z, x^*\rangle = 0$ for every $z \in C$. Thus we get the contradiction $\langle x_0, x^*\rangle = 0$. Therefore $\operatorname{qri} A \neq \emptyset$. □

1.3 Open Mapping Theorems

Throughout this section the spaces X, Y are topological vector spaces if not stated otherwise. We begin with some auxiliary results.

Lemma 1.3.1 *Let X, Y be first countable topological vector spaces, $\mathcal{R} : X \rightrightarrows Y$ be a multifunction and $x_0 \in X$. Suppose that* $\operatorname{gr} \mathcal{R}$ *satisfies condition* (Hwx). *Then*

$$\bigcap\nolimits_{U\in\mathcal{N}_X(x_0)} \operatorname{int}\left(\operatorname{cl}\mathcal{R}(U)\right) \subset \bigcap\nolimits_{U\in\mathcal{N}_X(x_0)} \operatorname{int}\mathcal{R}(U), \tag{1.8}$$

where $\mathcal{N}_X(x_0)$ is the class of all neighborhoods of x_0.

Proof. Let $y_0 \in \bigcap_{U\in\mathcal{N}_X(x_0)} \operatorname{int}\left(\operatorname{cl}\mathcal{R}(U)\right)$. Replacing $\operatorname{gr}\mathcal{R}$ by $\operatorname{gr}\mathcal{R} - (x_0, y_0)$ if necessary, we may assume that $(x_0, y_0) = (0,0)$. Let $U \in \mathcal{N}_X$. Since X is first countable, there exists a base $(U_n)_{n\geq 1} \subset \mathcal{N}_X$ of neighborhoods of 0 such that $U_n + U_n \subset U_{n-1}$ for every $n \geq 1$, where $U_0 := U$. Because $0 \in \bigcap_{U\in\mathcal{N}_X} \operatorname{int}(\operatorname{cl}\mathcal{R}(U))$, there exists $(V_n)_{n\geq 1} \subset \mathcal{N}_Y$ such that $V_n \subset \operatorname{int}(\operatorname{cl}\mathcal{R}(U_n))$ for every $n \geq 1$. Since Y is first countable, we may suppose that $(V_n)_{n\geq 1}$ is a base of neighborhoods of $0 \in Y$ and, moreover, $V_{n+1} + V_{n+1} \subset V_n$ for every $n \geq 1$.

Consider $y' \in \operatorname{int}\left(\operatorname{cl}\mathcal{R}(U_1)\right)$; there exists $\mu \in]0,1[$ such that $y := (1-\mu)^{-1}y' \in \operatorname{cl}\mathcal{R}(U_1)$. There exists $(x_1, y_1) \in \operatorname{gr}\mathcal{R}$ such that $x_1 \in U_1$ and $y - y_1 \in \mu V_2$. It follows that $\mu^{-1}(y - y_1) \in V_2 \subset \operatorname{cl}\mathcal{R}(U_2)$. There exists $(x_2, y_2) \in \operatorname{gr}\mathcal{R}$ such that $x_2 \in U_2$ and $\mu^{-1}(y - y_1) - y_2 \in \mu V_3$. It follows that $\mu^{-2}y - \mu^{-2}y_1 - \mu^{-1}y_2 \in V_3 \subset \operatorname{cl}\mathcal{R}(U_3)$. Continuing in this way we find $\left((x_m, y_m)\right)_{m\geq 1} \subset \operatorname{gr}\mathcal{R}$ such that $x_m \in U_m$ and $\mu^{-m+1}y - \mu^{-m+1}y_1 - \mu^{-m+2}y_2 - \cdots - y_m \in \mu V_{m+1}$ for every $m \geq 1$. Therefore $v_m := y - y_1 - \mu y_2 - \cdots - \mu^{m-1}y_m \in \mu^m V_{m+1} \subset V_{m+1}$ for every $m \geq 1$. Moreover, $\mu^{m-1}y_m = v_{m-1} - v_m \in \mu^{m-1}V_m - \mu^m V_{m+1}$, and so $y_m \in V_m - \mu V_{m+1} \subset V_m + V_m \subset V_{m-1}$ for $m \geq 2$. It follows that $(x_m)_{m\geq 1} \to 0$, $(y_m)_{m\geq 1} \to 0$ and $(v_m)_{m\geq 1} \to 0$, whence $\sum_{m\geq 1} \mu^{m-1}y_m$ has sum y. Taking $\lambda_m := (1-\mu)\mu^{m-1}$, we have that $(\lambda_m)_{m\geq 1} \subset \mathbb{R}_+$, $\sum_{m\geq 1}\lambda_m = 1$, $\sum_{m\geq 1}\lambda_m y_m$ has sum $(1-\mu)y = y'$, the sequence $\left((x_m, y_m)\right)$ is bounded (being convergent) and, because $\sum_{m=n+1}^{n+p}\lambda_m x_m \in U_{n+1} + U_{n+2} + \cdots + U_{n+p} \subset U_{n+1} + U_{n+1} \subset U_n$ for every $n \geq 1$, the series $\sum_{m\geq 1}\lambda_m x_m$ is Cauchy. By hypothesis there exists $x' \in X$, sum of the series $\sum_{m\geq 1}\lambda_m x_m$, such that $(x', y') \in \operatorname{gr}\mathcal{R}$. Let $x := (1-\mu)^{-1}x'$. We have that $\sum_{m=1}^{n}\mu^{m-1}x_m \in U_1 + U_2 + \cdots + U_n \subset U_1 + U_1 \subset U$. Since U is closed we have that $x \in U$, and so $x' \in (1-\mu)U \subset$

U. Thus $y' \in \mathcal{R}(U)$. Therefore $\operatorname{int}\left(\operatorname{cl}\mathcal{R}(U_1)\right) \subset \mathcal{R}(U)$. In particular $0 \in \operatorname{int}\mathcal{R}(U)$. The proof is complete. □

Note that we didn't use the fact that X or Y is separated. Observe also that it is no need that $x_0 \in \operatorname{dom}\mathcal{R}$ when $y_0 \in \bigcap_{U \in \mathcal{N}_X(x_0)} \operatorname{int}(\operatorname{cl}\mathcal{R}(U))$, but, necessarily, $x_0 \in \operatorname{cl}(\operatorname{dom}\mathcal{R})$ and $y_0 \in \operatorname{int}(\operatorname{Im}\mathcal{R})$ in our conditions. Note also that condition (Hwx) may be weakened by asking that the sequence $\left((x_m, y_m)\right)_{m\geq 1} \subset A$ be convergent instead of being bounded.

In the case of normed spaces one has the following variant of the result in Lemma 1.3.1. Having the normed vector space (nvs) $(X, \|\cdot\|)$, we denote by U_X the *closed unit ball* $\{x \in X \mid \|x\| \leq 1\}$, by B_X the *open unit ball* $\{x \in X \mid \|x\| < 1\}$ and by S_X the *unit sphere* $\{x \in X \mid \|x\| = 1\}$.

Lemma 1.3.2 *Let $(X, \|\cdot\|)$, $(Y, \|\cdot\|)$ be two normed linear spaces and let $\mathcal{R} : X \rightrightarrows Y$ be a multifunction. Suppose that condition* (Hwx) *holds and $(x_0, y_0) \in X \times Y$. If*

$$y_0 + \eta U_Y \subset \operatorname{cl}\left(\mathcal{R}(x_0 + \rho U_X)\right),$$

where $\eta, \rho > 0$, then

$$y_0 + \eta B_Y \subset \mathcal{R}(x_0 + \rho U_X).$$

Proof. We may take $(x_0, y_0) = (0, 0)$. One follows the same argument as in the proof of the preceding lemma, but with $U_n := \rho U_X$ and $V_n := \eta B_Y$ for $n \geq 1$. Consider $y' \in \eta B_Y$ and take $\mu \in]0, 1[$ such that $y := (1 - \mu)^{-1} y' \in \operatorname{cl}\mathcal{R}(\rho U_X)$. We find the sequence $\left((x_n, y_n)\right)_{n\geq 1} \subset \operatorname{gr}\mathcal{R}$ such that $(x_n) \subset \rho U_X$ and $v_n := y - y_1 - \mu y_2 - \cdots - \mu^{n-1} y_n \in \mu^n \eta B_Y$ for $n \geq 1$. Hence $(v_n) \to 0$ and $\mu^{n-1} y_n = v_{n-1} - v_n$, whence $\|y_n\| \leq \eta(1 + \mu)$ for $n \geq 1$. Taking $\lambda_n := (1 - \mu)\mu^{n-1} > 0$ for $n \geq 1$, $\sum_{n\geq 1} \lambda_n = 1$, the series $\sum_{n\geq 1} \lambda_n x_n$ is Cauchy and the series $\sum_{n\geq 1} \lambda_n y_n$ is convergent with sum y'. Since $\mathcal{R}$ satisfies the condition (Hwx), we obtain that the series $\sum_{n\geq 1} \lambda_n x_n$ is convergent with sum x' and $(x', y') \in \operatorname{gr}\mathcal{R}$. Of course, $x' \in \rho U_X$. Hence $\eta B_Y \subset \mathcal{R}(\rho U_X)$. □

Lemma 1.3.3 *Let X, Y be topological vector spaces, $\mathcal{R} : X \rightrightarrows Y$ be a closed convex multifunction and $x_0 \in X$. Suppose that X is complete and first countable. Then condition* (1.8) *holds.*

Proof. Let $y_0 \in \bigcap_{U\in\mathcal{N}_X(x_0)} \operatorname{int}(\operatorname{cl}\mathcal{R}(U))$. Replacing $\operatorname{gr}\mathcal{R}$ by $\operatorname{gr}\mathcal{R}-(x_0,y_0)$ if necessary, we may assume that $(x_0,y_0)=(0,0)$. Let us first show that

$$\forall t \in]0,1[,\ \forall U,U' \in \mathcal{N}_X \ :\ t\operatorname{cl}\mathcal{R}(U) \subset \mathcal{R}(t(U+U')). \qquad (1.9)$$

Fix $t \in]0,1[$ and take $t_n := t^{2^{-n}}$ for $n \geq 1$. Of course, $\lim t_n \cdots t_1 = t$.

Consider $U' \in \mathcal{N}_X$. There exists a base of neighborhoods $(U_n)_{n\in\mathbb{N}} \subset \mathcal{N}_X$ such that $U_1 + U_1 \subset U'$ and $U_{n+1} + U_{n+1} \subset U_n$ for every $n \in \mathbb{N}$. Let $U_n' := (1-t_n)^{-1}t_n \cdots t_1 U_n$. For every $n \in \mathbb{N}$ there exists $V_n' \in \mathcal{N}_Y$ such that $V_n' \subset \operatorname{cl}\mathcal{R}(U'_n)$. Consider $V_n := t_n^{-1}(1-t_n)V_n'$.

Let $U \in \mathcal{N}_X$ and $y \in \operatorname{cl}\mathcal{R}(U)$: we intend to show that $ty \in \mathcal{R}(t(U+U'))$.

We construct a sequence $x = x_0 \in U$, $x_1 \in U_1, \ldots, x_n \in U_n, \ldots$ with the property:

$$\forall n \in \mathbb{N} \ :\ t_n \ldots t_1 y \in \operatorname{cl}\mathcal{R}(t_n \ldots t_1(x + \cdots + x_{n-1} + U_n)). \qquad (1.10)$$

Since $(y - V_1) \cap \mathcal{R}(U) \neq \emptyset$, there exist $y_1 \in V_1$ and $x \in U$ such that $y - y_1 \in \mathcal{R}(x)$; hence

$$\begin{aligned} t_1 y &= t_1(y-y_1) + (1-t_1)\left((1-t_1)^{-1}t_1 y_1\right) \in t_1\mathcal{R}(x) + (1-t_1)V_1' \\ &\subset t_1\mathcal{R}(x) + (1-t_1)\operatorname{cl}\mathcal{R}(U_1') \subset \operatorname{cl}\mathcal{R}(t_1(x+U_1)). \end{aligned}$$

Suppose that we already have $x \in U$, $x_1 \in U_1, \ldots, x_{n-1} \in U_{n-1}$ with the desired property. Since

$$(t_n \ldots t_1 y - V_{n+1}) \cap \mathcal{R}(t_n \ldots t_1(x + \cdots + x_{n-1} + U_n)) \neq \emptyset,$$

there exist $y_{n+1} \in V_{n+1}$ and $x_n \in U_n$ such that

$$t_n \ldots t_1 y - y_{n+1} \in \mathcal{R}(t_n \ldots t_1(x + \cdots + x_{n-1} + x_n)).$$

Therefore

$$\begin{aligned} t_{n+1} \ldots t_1 y &= t_{n+1}(t_n \ldots t_1 y - y_{n+1}) + (1-t_{n+1})\left((1-t_{n+1})^{-1}t_{n+1}y_{n+1}\right) \\ &\in t_{n+1}\mathcal{R}(t_n \ldots t_1(x + \cdots + x_{n-1} + x_n)) + (1-t_{n+1})V_{n+1}' \\ &\subset t_{n+1}\mathcal{R}(t_n \ldots t_1(x + \cdots + x_{n-1} + x_n)) + (1-t_{n+1})\operatorname{cl}\mathcal{R}(U_{n+1}') \\ &\subset \operatorname{cl}\mathcal{R}\big(t_{n+1} \ldots t_1(x + \cdots + x_{n-1} + x_n + U_{n+1})\big). \end{aligned}$$

So the desired sequence is obtained. Since $x_{n+1} + \cdots + x_{n+p} \in U_{n+1} + \cdots + U_{n+p} \subset U_n$ for all $n,p \in \mathbb{N}$, the series $\sum_{n\geq 1} x_n$ is convergent. Denote by x' its sum. Since $x_1 + \cdots + x_n \in x_1 + U_1$ and U_1 is closed we have that $x' \in x_1 + U_1 \subset U'$.

Let now $U'' \in \mathcal{N}_X$ and $V'' \in \mathcal{N}_Y$ be arbitrary. There exists $n \in \mathbb{N}$ such that $t_n \dots t_1(x+\dots+x_{n-1}+U_n) \subset t(x+x')+U''$ and $t_n \dots t_1 y \in ty+V''$. By Eq. (1.10) we obtain that $t_n \dots t_1 y \in \operatorname{cl} \mathcal{R}\,(t(x+x')+U'')$. It follows that $(ty+V'') \cap \mathcal{R}\,(t(x+x')+U'') \neq \emptyset$, and so there exist $x'' \in U''$ and $y'' \in V''$ such that $ty+y'' \in \mathcal{R}(t(x+x')+x'')$, *i.e.* $(t(x+x'),ty)+(x'',y'') \in \operatorname{gr} \mathcal{R}$. It follows that $(t(x+x'),ty) \in \operatorname{cl}(\operatorname{gr} \mathcal{R}) = \operatorname{gr} \mathcal{R}$. Therefore $ty \in \mathcal{R}\,(t(U+U'))$.

To complete the proof, let $U \in \mathcal{N}_X(0)$. Then there exists $U' \in \mathcal{N}_X$ such that $U'+U' \subset 2U$. From Eq. (1.9) we obtain that $\frac{1}{2} \operatorname{cl} \mathcal{R}(U') \subset \mathcal{R}\left(\frac{1}{2}(U'+U')\right) \subset \mathcal{R}(U)$. Since $0 \in \bigcap_{U \in \mathcal{N}_X(x_0)} \operatorname{int}(\operatorname{cl} \mathcal{R}(U))$, we have that $0 \in \operatorname{int} \mathcal{R}(U)$. □

Note that Lemma 1.3.3 is a particular case of Lemma 1.3.1 when Y is first countable; otherwise these results are independent.

Corollary 1.3.4 *Let Y be a first countable topological vector space and $C \subset Y$ be an ideally convex set. Then* $\operatorname{int} C = \operatorname{int}(\operatorname{cl} C)$.

Proof. Consider X an arbitrary Fréchet space (for example $X = \mathbb{R}$) and $\mathcal{R} : X \rightrightarrows Y$ defined by $\mathcal{R}(0) = C$, $\mathcal{R}(x) = \emptyset$ for $x \neq 0$. Of course condition (Hwx) is satisfied. Taking $y_0 \in \operatorname{int}(\operatorname{cl} C)$ and $x_0 = 0$ we have that $y_0 \in \bigcap_{U \in \mathcal{N}_X(x_0)} \operatorname{int}(\operatorname{cl} \mathcal{R}(U))$, and so $y_0 \in \operatorname{int} C$. □

Theorem 1.3.5 (Simons) *Let X and Y be first countable. Assume that X is a locally convex space, $\mathcal{R} : X \rightrightarrows Y$ satisfies condition* (Hwx), *$y_0 \in {}^{ib}(\operatorname{Im} \mathcal{R})$ and $x_0 \in \mathcal{R}^{-1}(y_0)$. Then $y_0 \in \operatorname{int}_{\operatorname{aff}(\operatorname{Im} \mathcal{R})} \mathcal{R}(U)$ for every $U \in \mathcal{N}_X(x_0)$. In particular ${}^{ib}(\operatorname{Im} \mathcal{R}) = \operatorname{rint}(\operatorname{Im} \mathcal{R})$ if ${}^{ib}(\operatorname{Im} \mathcal{R}) \neq \emptyset$.*

Proof. Once again we may consider that $(x_0,y_0) = (0,0)$; so $Y_0 := \operatorname{aff}(\operatorname{Im} \mathcal{R}) = \operatorname{lin}(\operatorname{Im} \mathcal{R})$. Replacing, if necessary, Y by Y_0, we may suppose that Y is barreled and $0 \in (\operatorname{Im} \mathcal{R})^i$. Let $U \in \mathcal{N}_X^c$; since $\operatorname{gr} \mathcal{R}$ is a convex set and $\mathcal{R}(U) = \operatorname{Pr}_Y(\operatorname{gr} \mathcal{R} \cap U \times Y)$, $\mathcal{R}(U)$ is convex, too. $\mathcal{R}(U)$ is also absorbing. Indeed, let $y \in Y$. Because $\operatorname{Im} \mathcal{R}$ is absorbing, there exists $\lambda > 0$ such that $\lambda y \in \operatorname{Im} \mathcal{R}$. Therefore there exists $x \in X$ such that $(x, \lambda y) \in \operatorname{gr} \mathcal{R}$. Since U is absorbing, there exists $\mu \in]0,1[$ such that $\mu x \in U$. As $\operatorname{gr} \mathcal{R}$ is convex we have that $(\mu x, \mu\lambda y) = \mu(x,\lambda y) + (1-\mu)(0,0) \in \operatorname{gr} \mathcal{R}$, whence $\mu\lambda y \in \mathcal{R}(U)$. Therefore $\mathcal{R}(U)$ is also absorbing. It follows that $\operatorname{cl}(\mathcal{R}(U))$ is an absorbing, closed and convex subset of the barreled space Y. Therefore $0 \in \operatorname{int}(\operatorname{cl} \mathcal{R}(U))$. Using Lemma 1.3.1 we obtain that $0 \in \operatorname{int} \mathcal{R}(U)$ for every neighborhood U of $0 \in X$. The last part is an immediate consequence of what was obtained above. □

Corollary 1.3.6 *Let X be a Fréchet space, Y be first countable and $\mathcal{R} : X \rightrightarrows Y$ be li-convex. Assume that $y_0 \in {}^{ib}(\operatorname{Im}\mathcal{R})$ and $x_0 \in \mathcal{R}^{-1}(y_0)$. Then $y_0 \in \operatorname{int}_{\operatorname{aff}(\operatorname{Im}\mathcal{R})} \mathcal{R}(U)$ for all $U \in \mathcal{N}_X(x_0)$. In particular ${}^{ib}(\operatorname{Im}\mathcal{R}) = \operatorname{rint}(\operatorname{Im}\mathcal{R})$ provided ${}^{ib}(\operatorname{Im}\mathcal{R}) \neq \emptyset$.*

Proof. There exist a Fréchet space Z and an ideally convex multifunction $\mathcal{S} : Z \times X \rightrightarrows Y$ such that $\operatorname{gr}\mathcal{R} = \operatorname{Pr}_{X\times Y}(\operatorname{gr}\mathcal{S})$. Then $\mathcal{S}$ verifies condition $(\mathrm{Hw}(z,x))$ by Proposition 1.2.6 (ii). Of course, there exists $z_0 \in Z$ such that $y_0 \in \mathcal{S}(z_0, x_0)$. Since $\operatorname{Im}\mathcal{S} = \operatorname{Im}\mathcal{R}$, by the preceding theorem, $y_0 \in \operatorname{int}_{\operatorname{aff}(\operatorname{Im}\mathcal{R})} \mathcal{R}(U) = \operatorname{int}_{\operatorname{aff}(\operatorname{Im}\mathcal{R})} \mathcal{S}(Z \times U)$ for every $U \in \mathcal{N}_X(x_0)$. □

Theorem 1.3.7 (Ursescu) *Let X be a complete semi-metrizable locally convex space and $\mathcal{R} : X \rightrightarrows Y$ be a closed convex multifunction. Assume that $y_0 \in {}^{ib}(\operatorname{Im}\mathcal{R})$ and $x_0 \in \mathcal{R}^{-1}(y_0)$. Then $y_0 \in \operatorname{int}_{\operatorname{aff}(\operatorname{Im}\mathcal{R})} \mathcal{R}(U)$ for every $U \in \mathcal{N}_X(x_0)$. In particular ${}^{ib}(\operatorname{Im}\mathcal{R}) = \operatorname{rint}(\operatorname{Im}\mathcal{R})$ if ${}^{ib}(\operatorname{Im}\mathcal{R}) \neq \emptyset$.*

Proof. If Y is first countable it is obvious that the conclusion follows from Simons' theorem. Otherwise the proof is exactly the same as that of Simons' theorem but using Lemma 1.3.3 instead of Lemma 1.3.1. □

An immediate consequence of Theorem 1.3.5 is the following corollary.

Corollary 1.3.8 *Let Y be a first countable barreled space and $C \subset Y$ be a lower ideally convex set. Then $C^i = \operatorname{int} C$.*

Proof. Let $y_0 \in C^i$. There exist a Fréchet space X and an ideally convex multifunction $\mathcal{R} : X \rightrightarrows Y$ such that $C = \operatorname{Im}\mathcal{R}$. The conclusion follows from Theorem 1.3.5 taking again $U = X$. □

In fact the conclusion of the above corollary holds if C is the projection on Y of a subset A of $X \times Y$ with (a′) X is a first countable locally convex space and A satisfies condition $(\mathrm{Hw}x)$ or (b′) X is a semi-metrizable complete locally convex space and A is closed and convex.

Putting together Corollaries 1.3.4 and 1.3.8 we get the next result.

Corollary 1.3.9 *Let Y be a first countable barreled space and $C \subset Y$ be an ideally convex set. Then $C^i = \operatorname{int} C = \operatorname{int}(\operatorname{cl} C) = (\operatorname{cl} C)^i$.* □

In normed spaces the following inversion mapping theorem holds.

Theorem 1.3.10 (Robinson) *Let $(X, \|\cdot\|)$ and $(Y, \|\cdot\|)$ be normed vector spaces, $\mathcal{R} : X \rightrightarrows Y$ be a convex multifunction and $(x_0, y_0) \in \operatorname{gr}\mathcal{R}$. Assume* *that* $y_0 + \eta U_Y \subset \mathcal{R}(x_0 + \rho U_X)$ *for some* $\eta, \rho > 0$. *Then*

$$d\left(x, \mathcal{R}^{-1}(y)\right) \leq \frac{\rho + \|x - x_0\|}{\eta - \|y - y_0\|} \cdot d\big(y, \mathcal{R}(x)\big) \quad \forall\, x \in X,\ \forall\, y \in y_0 + \eta B_Y.$$

Proof. Replacing $\operatorname{gr}\mathcal{R}$ by $\operatorname{gr}\mathcal{R} - (x_0, y_0)$, we may assume that $(x_0, y_0) = (0,0)$. Let $x \in X$ and $y \in \eta B_Y$. The conclusion is obvious if $x \notin \operatorname{dom}\mathcal{R}$ or $y \in \mathcal{R}(x)$, so suppose that neither is true. Choose $\theta > 0$ and find $y_\theta \in \mathcal{R}(x)$ such that $0 < \|y_\theta - y\| < d\big(y, \mathcal{R}(x)\big) + \theta$; define $\alpha := \eta - \|y\| > 0$ and take $\varepsilon \in]0, \alpha[$. Consider

$$y_\varepsilon := y + (\alpha - \varepsilon)\, \|y - y_\theta\|^{-1} (y - y_\theta);$$

thus $\|y_\varepsilon\| \leq \|y\| + (\alpha - \varepsilon) = \eta - \varepsilon$, and so $y_\varepsilon \in \eta B_Y$. Therefore there exists $x_\varepsilon \in \rho U_X$ with $y_\varepsilon \in \mathcal{R}(x_\varepsilon)$. Define $\lambda := \|y - y_\theta\| \left(\alpha - \varepsilon + \|y - y_\theta\|\right)^{-1} \in]0, 1[$. Then

$$y = (1 - \lambda) y_\theta + \lambda y_\varepsilon \in (1 - \lambda)\mathcal{R}(x) + \lambda \mathcal{R}(x_\varepsilon) \subset \mathcal{R}((1 - \lambda)x + \lambda x_\varepsilon).$$

Thus $(1 - \lambda)x + \lambda x_\varepsilon \in \mathcal{R}^{-1}(y)$, whence $d\left(x, \mathcal{R}^{-1}(y)\right) \leq \lambda \|x - x_\varepsilon\|$. As $\|x - x_\varepsilon\| \leq \|x\| + \|x_\varepsilon\| \leq \rho + \|x\|$ and $\lambda < (\alpha - \varepsilon)^{-1} \|y - y_\theta\|$, we obtain that

$$d\left(x, \mathcal{R}^{-1}(y)\right) \leq (\rho + \|x\|)\,(\alpha - \varepsilon)^{-1} \left(d(y, \mathcal{R}(x)) + \theta\right).$$

Letting $\theta, \varepsilon \to 0$, we obtain the conclusion. □

Combining the preceding result and Lemma 1.3.2 we obtain the following important result in normed spaces. The implication (i) ⇒ (ii) is met in the literature as the Robinson–Ursescu theorem.

Theorem 1.3.11 *Let* $(X, \|\cdot\|)$, $(Y, \|\cdot\|)$ *be two normed vector spaces and* $\mathcal{R} : X \rightrightarrows Y$ *be a multifunction. Suppose that* Y *is a barreled space,* $\operatorname{gr}\mathcal{R}$ *verifies condition (Hwx) and* $(x_0, y_0) \in \operatorname{gr}\mathcal{R}$. *Then the following conditions are equivalent:*

(i) $y_0 \in (\operatorname{Im}\mathcal{R})^i$;

(ii) $y_0 \in \operatorname{int}\mathcal{R}(x_0 + U_X)$;

(iii) $\exists\, \eta > 0,\ \forall\, \lambda \in [0, 1] \ :\ y_0 + \lambda\eta U_Y \subset \mathcal{R}(x_0 + \lambda U_X)$;

(iv) $\exists\, \gamma, \eta > 0,\ \forall\, x \in x_0 + \eta U_X,\ \forall\, y \in y_0 + \eta U_Y \ :\ d\left(x, \mathcal{R}^{-1}(y)\right) \leq \gamma \cdot d\big(y, \mathcal{R}(x)\big)$;

(v) $\exists\, \eta > 0,\ \forall\, y \in y_0 + \eta U_X,\ \forall\, x \in X \ :\ d\left(x, \mathcal{R}^{-1}(y)\right) \leq \frac{1 + \|x - x_0\|}{\eta - \|y - y_0\|} \cdot d\big(y, \mathcal{R}(x)\big)$.

Proof. The implications (iii) $\Rightarrow$ (ii) $\Rightarrow$ (i) are obvious. The implication (i) $\Rightarrow$ (ii) follows immediately from Simons' theorem, while the implication (ii) $\Rightarrow$ (v) is given by the preceding theorem.

(iv) $\Rightarrow$ (iii) Let $\gamma' > \gamma$; we may assume that $\eta\gamma' \le 1$, because, in the contrary case, we replace η by $\eta' := 1/\gamma' < \eta$. Let $y \in y_0 + \lambda\eta U_Y$, $y \ne y_0$, with $\lambda \in]0,1]$. Then $d\left(x_0, \mathcal{R}^{-1}(y)\right) \le \gamma \cdot d(y, \mathcal{R}(x_0)) < \gamma' \|y - y_0\|$. Hence there exists $x \in \mathcal{R}^{-1}(y)$ such that $\|x - x_0\| \le \gamma' \|y - y_0\| \le \gamma'\lambda\eta \le \eta$. Therefore $y_0 + \lambda\eta U_Y \subset \mathcal{R}(x_0 + \lambda U_X)$.

(v) $\Rightarrow$ (iv) Taking $x \in x_0 + \frac{\eta}{2}U_X$, $y \in y_0 + \frac{\eta}{2}U_Y$, we obtain that $d\left(x, \mathcal{R}^{-1}(y)\right) \le \gamma \cdot d(y, \mathcal{R}(x))$ with $\gamma := 1 + \eta/2$. □

Important consequences of the Ursescu theorem are: the closed graph theorem, the open mapping theorem and the uniform boundedness principle; we state the first two of them in Fréchet spaces.

Theorem 1.3.12 (closed graph) *Let X, Y be Fréchet spaces and $T : X \to Y$ be a linear operator. Then T is continuous if and only if $\operatorname{gr} T$ is closed in $X \times Y$.*

Proof. It is obvious that $\operatorname{gr} T$ is closed if T is continuous (even without being linear). Suppose that $\operatorname{gr} T$ is closed and consider the multifunction $\mathcal{R} := T^{-1} : Y \rightrightarrows X$. It is obvious that $\operatorname{gr} \mathcal{R}$ is closed and convex (even linear subspace). Moreover $\operatorname{Im} \mathcal{R} = X$. So we can apply Theorem 1.3.5 for $(Tx_0, x_0) \in Y \times X$. Therefore

$$\forall V \in \mathcal{N}_Y(Tx_0) \ : \ \mathcal{R}(V) = T^{-1}(V) \in \mathcal{N}_X(x_0),$$

which means that T is continuous at x_0. □

Corollary 1.3.13 (Banach–Steinhaus) *Let X, Y be Fréchet spaces and $T : X \to Y$ be a bijective linear operator. Then T is continuous if and only if T^{-1} is continuous; in particular, if T is continuous then T is an isomorphism of Fréchet spaces.*

Proof. Apply the closed graph theorem for T and T^{-1}, respectively. □

Theorem 1.3.14 (open mapping theorem) *Let X, Y be Fréchet spaces and $T \in \mathcal{L}(X, Y)$ be onto. Then T is open.*

Proof. Of course, T is a closed convex relation and $\operatorname{Im} T = Y$. Let $D \subset X$ be open and take $y_0 \in T(D)$; there exists $x_0 \in D$ such that $y_0 = Tx_0$. Applying the Ursescu theorem for this point, since D is a neighborhood of x_0, we have that $T(D)$ is a neighborhood of y_0. Therefore $T(D)$ is open.□

An interesting and useful result is the following.

Corollary 1.3.15 *Let X, Y be Fréchet spaces, $A \subset X$ and $T : X \to Y$ be a continuous linear operator. Suppose that* $\operatorname{Im} T$ *is closed. Then $T(A)$ is closed if and only if $A + \ker T$ is closed.*

Proof. Replacing, if necessary, Y by $\operatorname{Im} T$ and T by $T' : X \to \operatorname{Im} T$, $T'(x) := T(x)$ for $x \in X$, we may suppose that T is onto. Consider $\widehat{T} : X/\ker T \to Y$, $\widehat{T}(\widehat{x}) := T(x)$, where $\widehat{x}$ is the class of x. It is easy to verify that $\widehat{T}$ is well defined, linear and bijective. Let $q : X \to \mathbb{R}$ be a continuous semi-norm; since T is continuous, $p := q \circ T$ is a continuous semi-norm, too. For all $x \in X$ and $u \in \ker T$ we have that $(q \circ \widehat{T})(\widehat{x}) = q\left(T(x+u)\right) = p(x+u)$, whence $q \circ \widehat{T} \leq \widehat{p}$, where $\widehat{p}(\widehat{x}) := \inf\{p(x+u) \mid u \in \ker T\}$. Therefore $\widehat{T}$ is continuous. Since X is a Fréchet space, $X/\ker T$ is a Fréchet space, too. Using now the preceding corollary we have that $\widehat{T}$ is an isomorphism of Fréchet spaces. It is obvious that $T(A) = \widehat{T}(\widehat{A})$, where $\widehat{A} := \{\widehat{x} \mid x \in A\}$. Taking $\pi : X \to X/\ker T$, $\pi(x) := \widehat{x}$, we obtain that

$$T(A) \text{ is closed} \Leftrightarrow \pi(A) = \widehat{A} \text{ is closed} \Leftrightarrow \pi^{-1}\big(\widehat{A}\big) = A + \ker T \text{ is closed},$$

and so the conclusion holds. □

As an application of the Simons and Ursescu theorems we give the following two interesting results, useful in studying optimal control problems. We recall that a **process** is a multifunction $\mathcal{C} : X \rightrightarrows Y$ whose graph is a cone; when the graph of the process $\mathcal{C}$ is convex or closed one says that $\mathcal{C}$ is a **convex process** or **closed process**, respectively. The **adjoint** of the process $\mathcal{C}$ is the w^*-closed convex process $\mathcal{C}^* : Y^* \rightrightarrows X^*$ whose graph is the set $\{(y^*, x^*) \in Y^* \times X^* \mid (-x^*, y^*) \in (\operatorname{gr} \mathcal{C})^+\}$.

Theorem 1.3.16 *Let X, Y, Z be Banach spaces, $\mathcal{C} : X \rightrightarrows Y$ be an ideally convex process and $T \in \mathcal{L}(Z, Y)$. Consider the following statements:*

(i) $\operatorname{Im} T \subset \operatorname{Im} \mathcal{C}$;

(ii) $\exists \rho_1 > 0,\ \forall (y^*, x^*) \in \operatorname{gr} \mathcal{C}^* \ :\ \|T^* y^*\| \leq \rho_1 \|x^*\|$;

(iii) $\exists \rho_2 > 0 \ :\ T(U_Z) \subset \rho_2 \mathcal{C}(U_X)$;

(iv) $\exists \rho_3 > 0 \ :\ T(U_Z) \subset \rho_3 \operatorname{cl}(\mathcal{C}(U_X))$.

Then (i) $\Leftrightarrow$ (iii), (ii) $\Leftrightarrow$ (iv) *with $\rho_1 = \rho_3$ and* (iii) $\Rightarrow$ (iv) *with $\rho_3 = \rho_2$. Moreover, if T is onto then* (iv) $\Rightarrow$ (iii) *with (any) $\rho_2 > \rho_3$.*

Proof. The implications (iii) $\Rightarrow$ (i) and (iii) $\Rightarrow$ (iv) (with $\rho_3 = \rho_2$) are obvious.

(i) $\Rightarrow$ (iii) Let $\mathcal{R} := T^{-1} \circ \mathcal{C}$; one obtains immediately that $\mathcal{R}$ is an ideally convex process with $\operatorname{Im} \mathcal{R} = Z$. Applying the Simons theorem (Theorem 1.3.5) for $(x_0, y_0) = (0, 0)$, there exists $\rho > 0$ such that $\rho U_Z \subset \mathcal{R}(U_X)$, which means that $\rho T(U_Z) \subset \mathcal{C}(U_X)$. Taking $\rho_2 := \rho^{-1}$, (iii) holds.

(iv) $\Rightarrow$ (ii) Let $(y^*, x^*) \in \operatorname{gr} \mathcal{C}^*$, *i.e.* $(-x^*, y^*) \in (\operatorname{gr} \mathcal{C})^+$. Then

$$\begin{aligned}\|T^*y^*\| &= \sup_{\|z\| \le 1} \langle z, -T^*y^* \rangle = \sup_{z \in U_Z} \langle Tz, -y^* \rangle \le \sup_{y \in \rho_3 \operatorname{cl}(\mathcal{C}(U_X))} \langle y, -y^* \rangle \\ &= \rho_3 \sup_{y \in \mathcal{C}(U_X)} \langle y, -y^* \rangle \le \rho_3 \sup\{\langle x, -x^* \rangle \mid (x, y) \in \operatorname{gr} \mathcal{C},\ \|x\| \le 1\} \\ &\le \rho_2 \|x^*\|\end{aligned}$$

because $\langle y, -y^* \rangle \le \langle x, -x^* \rangle$ for $(x, y) \in \operatorname{gr} \mathcal{C}$. So (ii) holds with $\rho_1 = \rho_3$.

(ii) $\Rightarrow$ (iv) Suppose that (iv) does not hold and take $\overline{z} \in U_Z$ such that $T\overline{z} \notin \rho_3 \operatorname{cl}(\mathcal{C}(U_X))$. Since $\mathcal{C}(U_X)$ is convex, using Theorem 1.1.5, there exist $\overline{y}^* \in Y^*$ and $\lambda \in \mathbb{R}$ such that

$$\langle \overline{z}, T^*\overline{y}^* \rangle = \langle T\overline{z}, \overline{y}^* \rangle < \lambda < \langle \rho_3 y, \overline{y}^* \rangle \quad \forall y \in \mathcal{C}(U_X).$$

It follows that $\lambda < 0$, and so we can take $\lambda = -\rho_3$. Hence $-\rho_3 > \langle \overline{z}, T^*\overline{y}^* \rangle \ge -\|T^*\overline{y}^*\|$ and $-1 < \langle y, \overline{y}^* \rangle = \langle x, 0 \rangle + \langle y, \overline{y}^* \rangle$ for $(x, y) \in \operatorname{gr} \mathcal{C}$ with $x \in U_X$, whence $\|T^*\overline{y}^*\| > \rho_3$ and

$$\begin{aligned}(0, \overline{y}^*) \in (\operatorname{gr} \mathcal{C} \cap U_X \times Y)^\circ &= w^* - \operatorname{cl}\left((\operatorname{gr} \mathcal{C})^+ + U_{X^*} \times \{0\}\right) \\ &= (\operatorname{gr} \mathcal{C})^+ + U_{X^*} \times \{0\}\end{aligned}$$

because $U_{X^*} \times \{0\}$ is w^*-compact (we have used the Alaoglu–Bourbaki theorem). Therefore there exists $\overline{x}^* \in U_{X^*}$ such that $(\overline{y}^*, \overline{x}^*) \in \operatorname{gr} \mathcal{C}^*$ ($\Leftrightarrow (-\overline{x}^*, \overline{y}^*) \in (\operatorname{gr} \mathcal{C})^*$). Since $\|T^*\overline{y}^*\| > \rho_3 \ge \rho_3 \|\overline{x}^*\|$, (ii) does not hold for $\rho_1 = \rho_3$.

(iv) $\Rightarrow$ (iii) when T is onto. Since T is onto and Z, Y are Banach spaces, T is open (see Theorem 1.3.14). It follows that $\operatorname{int}(T(U_Z)) = T(B_Z)$. But $\mathcal{C}(U_X)$ is cs-closed. Indeed, by Proposition 1.2.2, $A := \operatorname{gr} \mathcal{C} \cap U_X \times Y$ is cs-closed, and so, X being complete, A satisfies condition (Hx) (see Proposition 1.2.6(ii)); by Proposition 1.2.6(v) we have that $\operatorname{Pr}_Y(A) = \mathcal{C}(U_X)$ is cs-closed. Using Corollary 1.3.4 we obtain that $\operatorname{int} \mathcal{C}(U_X) = \operatorname{int}(\operatorname{cl} \mathcal{C}(U_X))$. So, from (iv) we obtain that $T(B_Z) \subset \rho_3 \mathcal{C}(U_X)$, and so $T(U_Z) \subset \rho_2 \mathcal{C}(U_X)$ for every $\rho_2 > \rho_3$. □

A similar result holds in dual spaces.

Theorem 1.3.17 *Let X, Y, Z be normed spaces, $T \in \mathcal{L}(X, Y)$ and $\mathcal{C} : X \rightrightarrows Z$ be a convex process. Consider the following statements:*

(i) $\operatorname{Im} T^* \subset \operatorname{Im} \mathcal{C}^*$;

(ii) $\exists\, \rho > 0,\ \forall\, (x, z) \in \operatorname{gr} \mathcal{C} \ :\ \|Tx\| \le \rho\, \|z\|$;

(iii) $\exists\, \rho > 0 \ :\ T^*(U_{Y^*}) \subset \rho \mathcal{C}^*(U_{Z^*})$.

Then (i) $\Leftrightarrow$ (iii) *and* (ii) $\Leftrightarrow$ (iii) *with the same* ρ.

Proof. The equivalence of (i) and (iii) follows from the equivalence of (i) and (iii) of the preceding theorem; just apply it for the Banach spaces X^*, Y^*, Z^*, the continuous linear operator T^* and the closed convex process $\mathcal{C}^*$.

(iii) $\Rightarrow$ (ii) The proof follows the same lines as the proof of (iv) $\Rightarrow$ (ii) in the preceding theorem, so we omit it.

(ii) $\Rightarrow$ (iii) Even if the proof is similar to (ii) $\Rightarrow$ (iv) in the preceding theorem, we give the details. So, suppose that (iii) does not hold and take $\overline{y}^* \in U_{Y^*}$ such that $T^*\overline{y}^* \notin \rho \mathcal{C}^*(U_{Z^*})$. We may assume that $\overline{y}^* \in S_{X^*}$ (otherwise replace $\overline{y}^*$ by $\|\overline{y^*}\|^{-1}\overline{y}^*$). Using the fact that U_{Z^*} is w^*-compact and $\operatorname{gr} \mathcal{C}^*$ is w^*-closed it follows easily that $\mathcal{C}^*(U_{Z^*})$ is w^*-closed; being also convex, we can apply Theorem 1.1.5 in (X^*, w^*). Therefore there exist $\overline{x} \in X$ and $\lambda \in \mathbb{R}$ such that

$$\langle T\overline{x}, \overline{y}^* \rangle = \langle \overline{x}, T^*\overline{y}^* \rangle > \lambda > \langle \overline{x}, \rho x^* \rangle \quad \forall\, x^* \in \mathcal{C}^*(U_{Z^*}).$$

Hence $\lambda > 0$, and we can take $\lambda = \rho$. Hence $\|T\overline{x}\| > \rho$ and $1 > \langle \overline{x}, x^* \rangle$ for every $x^* \in \mathcal{C}^*(U_{Z^*})$, *i.e.* $-1 < \langle \overline{x}, x^* \rangle + \langle 0, z^* \rangle$ for $(x^*, z^*) \in (\operatorname{gr} \mathcal{C})^+$ with $z^* \in U_{Z^*}$. It follows that

$$\begin{aligned}(\overline{x}, 0) \in \big((\operatorname{gr} \mathcal{C})^+ \cap X \times U_{Z^*}\big)^\circ &= \operatorname{cl}\big((\operatorname{gr} \mathcal{C})^{++} + \{0\} \times U_Z\big) \\ &= \operatorname{cl}(\operatorname{gr} \mathcal{C} + \{0\} \times U_Z).\end{aligned}$$

Hence there exist $((x_n, z_n)) \subset \operatorname{gr} \mathcal{C}$ and $(z_n') \subset U_Z$ such that $(x_n) \to \overline{x}$ and $(w_n) := (z_n + z_n') \to 0$. It follows that $(\|Tx_n\|) \to \|T\overline{x}\|$ and $\|z_n\| = \|w_n - z_n'\| \le 1 + \|w_n\|$ for every n, whence $\rho \limsup \|z_n\| \le \rho < \|T\overline{x}\| \le \liminf \|Tx_n\|$. Therefore there exists n_0 such that $\|Tx_n\| > \rho\, \|z_n\|$ for $n \ge n_0$, which proves that (ii) does not hold for the same ρ. □

1.4 Variational Principles

A very important result in nonlinear analysis is the "Ekeland variational principle."

Theorem 1.4.1 (Ekeland) *Let (X, d) be a complete metric space and $f : X \to \overline{\mathbb{R}}$ be a proper lower semicontinuous and lower bounded function. Then for every $x_0 \in \operatorname{dom} f$ and $\varepsilon > 0$ there exists $x_\varepsilon \in X$ such that*

$$f(x_\varepsilon) \leq f(x_0) - \varepsilon d(x_0, x_\varepsilon)$$

and

$$f(x_\varepsilon) < f(x) + \varepsilon d(x_\varepsilon, x) \quad \forall x \in X \setminus \{x_\varepsilon\}.$$

Proof. Let $x_0 \in \operatorname{dom} f$ and $\varepsilon > 0$ be fixed. For every $x \in X$ consider the set $F(x) := \{y \in X \mid f(y) + \varepsilon d(x, y) \leq f(x)\}$. Note that the conclusion is equivalent to the existence of an $x_\varepsilon \in F(x_0)$ such that $F(x_\varepsilon) = \{x_\varepsilon\}$.

Since the function $X \ni y \mapsto f(y) + \varepsilon d(x, y) \in \overline{\mathbb{R}}$ is lower semicontinuous (lsc for short), $F(x)$ is closed. Note that $x \in F(x) \subset \operatorname{dom} f$ for every $x \in \operatorname{dom} f$ and $F(x) = X$ for $x \in X \setminus \operatorname{dom} f$. Also note that $F(y) \subset F(x)$ for every $y \in F(x)$. The inclusion is obvious for $x \notin \operatorname{dom} f$. So, let $x \in \operatorname{dom} f$, $y \in F(x)$ and $z \in F(y)$. Then

$$f(z) + \varepsilon d(y, z) \leq f(y), \quad f(y) + \varepsilon d(x, y) \leq f(x), \quad d(x, z) \leq d(x, y) + d(y, z).$$

Multiplying the last relation by ε, then adding all three relations we obtain that $f(z) + \varepsilon d(x, z) \leq f(x)$, *i.e.* $z \in F(x)$.

Since f is bounded from below, $s_0 := \inf\{f(x) \mid x \in F(x_0)\} \in \mathbb{R}$; take $x_1 \in F(x_0)$ such that $f(x_1) < s_0 + 2^{-1}$. Then consider $s_1 := \inf\{f(x) \mid x \in F(x_1)\} \in \mathbb{R}$ and take $x_2 \in F(x_1)$ such that $f(x_2) < s_1 + 2^{-2}$. Continuing in this way we obtain the sequences $(s_n)_{n \geq 0} \subset \mathbb{R}$ and $(x_n)_{n \geq 0} \subset X$ such that

$$s_n = \inf\{f(x) \mid x \in F(x_n)\}, \; x_{n+1} \in F(x_n), \; f(x_{n+1}) < s_n + 2^{-n-1}$$

for all $n \geq 0$. Because $F(x_{n+1}) \subset F(x_n)$, $s_{n+1} \geq s_n$ for $n \geq 0$. Moreover, as $x_{n+1} \in F(x_n)$,

$$\varepsilon d(x_{n+1}, x_n) \leq f(x_n) - f(x_{n+1}) \leq f(x_n) - s_n \leq f(x_n) - s_{n-1} < 2^{-n}$$

for $n \geq 1$, whence $d(x_{n+p}, x_n) < \varepsilon^{-1} 2^{1-n}$ for $n, p \geq 1$. This shows that (x_n) is a Cauchy sequence, and so (x_n) converges to some $x_\varepsilon \in X$. Since

$x_m \in F(x_n)$ for $m \geq n$, we have that $x_\varepsilon \in \operatorname{cl} F(x_n) = F(x_n)$ for every $n \geq 0$. In particular $x_\varepsilon \in F(x_0)$. Let $x \in F(x_\varepsilon)$; then $x \in F(x_n)$ for every $n \geq 0$, and so, as above, $\varepsilon d(x, x_n) \leq f(x_n) - f(x) \leq f(x_n) - s_n < 2^{-n}$, which shows that $(x_n) \to x$. As the limit is unique, we get $x = x_\varepsilon$, which shows that $F(x_\varepsilon) = \{x_\varepsilon\}$. The proof is complete. □

In applications one often uses the following variant of the Ekeland variational principle. In the sequel by $\inf_X f$, or simply $\inf f$, we mean $\inf\{f(x) \mid x \in X\}$, where $f : X \to \overline{\mathbb{R}}$.

Corollary 1.4.2 *Let (X, d) be a complete metric space and $f : X \to \overline{\mathbb{R}}$ be a bounded below lower semicontinuous proper function. Let also $\varepsilon > 0$ and $x_0 \in \operatorname{dom} f$ be such that $f(x_0) \leq \inf_X f + \varepsilon$. Then for every $\lambda > 0$ there exists $x_\lambda \in X$ such that*

$$f(x_\lambda) \leq f(x_0), \quad d(x_\lambda, x_0) \leq \lambda$$

and

$$f(x_\lambda) < f(x) + \varepsilon\lambda^{-1} d(x, x_\lambda) \quad \forall x \in X \setminus \{x_\lambda\}.$$

Proof. Applying the preceding theorem for x_0 and $\varepsilon\lambda^{-1}$, we get $x_\lambda \in X$ satisfying the second relation of the conclusion and $f(x_\lambda) + \varepsilon\lambda^{-1} d(x_0, x_\lambda) \leq f(x_0)$. Hence $f(x_\lambda) \leq f(x_0)$. Moreover, because $f(x_0) \leq \inf_X f + \varepsilon \leq f(x_\lambda) + \varepsilon$, we get also that $d(x_\lambda, x_0) \leq \lambda$. □

A good compromise is obtained by taking $\lambda = \sqrt{\varepsilon}$ in the preceding result. An interesting application of the Ekeland theorem is the following result.

Corollary 1.4.3 *Let $(X, \|\cdot\|)$ be a Banach space and $f : X \to \mathbb{R}$ be a lower semicontinuous function. Assume that f is Gâteaux differentiable and bounded from below. If $(x_n)_{n\in\mathbb{N}}$ is a minimizing sequence for f, i.e. $(f(x_n)) \to \inf_X f$, then there exists a minimizing sequence $(\overline{x}_n)$ for f such that $(\|x_n - \overline{x}_n\|) \to 0$ and $(\|\nabla f(\overline{x}_n)\|) \to 0$.*

Proof. Consider $\varepsilon_n := f(x_n) - \inf_X f \in \mathbb{R}_+$. If $f(x_n) = \inf_X f$ we take $\overline{x}_n := x_n$; otherwise, applying the preceding corollary for x_n, ε_n and $\lambda = \sqrt{\varepsilon_n}$ we get $\overline{x}_n \in X$ such that $f(\overline{x}_n) \leq f(x_n)$, $\|x - \overline{x}_n\| \leq \sqrt{\varepsilon_n}$ and

$$f(\overline{x}_n) \leq f(x) + \sqrt{\varepsilon_n}\,\|x - \overline{x}_n\| \quad \forall x \in X.$$

Note that these relations hold also in the case $\varepsilon_n = 0$ (with our choice). Taking $x := \overline{x}_n + tu$ with $t > 0$ and $u \in X$, we get $-\sqrt{\varepsilon_n}\,\|u\| \leq t^{-1}(f(\overline{x}_n +$

$tu) - f(\overline{x}_n))$. Taking the limit for $t \to 0$ we get $-\sqrt{\varepsilon_n}\,\|u\| \le \langle u, \nabla f(\overline{x}_n)\rangle$ for all $u \in X$. It follows that $\|\nabla f(\overline{x}_n)\| \le \sqrt{\varepsilon_n}$. The conclusion follows. □

The next result is met in the literature as the *smooth variational principle*. This name is due to the fact that, at least in Hilbert spaces, the perturbation function is smooth (for $p > 1$). The result will be stated in a Banach space $(X, \|\cdot\|)$ even if it holds in any complete metric space (with the same proof). Before stating it let us observe that, when $(u_n)_{n\ge 0} \subset X$ is bounded, $(\mu_n)_{n\ge 0} \subset \mathbb{R}_+$ is such that $\sum_{n\ge 0} \mu_n = 1$ and $p \in [1, \infty[$, the function

$$\Theta_p : X \to \mathbb{R}, \ \Theta_p(x) := \sum\nolimits_{n\ge 0} \mu_n \|x - u_n\|^p, \tag{1.11}$$

is well defined and Lipschitz on bounded sets; in particular Θ_p is continuous.

Theorem 1.4.4 (Borwein–Preiss) *Let $(X, \|\cdot\|)$ be a Banach space, $f : X \to \overline{\mathbb{R}}$ be a proper lower semicontinuous and bounded from below function and $p \in [1, \infty[$. If $x_0 \in X$ and $\varepsilon > 0$ are such that $f(x_0) < \inf_X f + \varepsilon$ then for every $\lambda > 0$ there exists a sequence $(u_n)_{n\ge 0} \subset B(x_0, \lambda)$ converging to some $u \in X$ such that*

$$f(u) < \inf\nolimits_X f + \varepsilon, \quad \|x_0 - u\| < \lambda$$

and

$$f(u) + \varepsilon\lambda^{-p}\Theta_p(u) \le f(x) + \varepsilon\lambda^{-p}\Theta_p(x) \quad \forall\, x \in X, \tag{1.12}$$

where Θ_p is defined by Eq. (1.11).

Proof. Fix $\lambda > 0$ and consider γ, δ, η, μ, $\theta > 0$ such that

$$f(x_0) - \inf\nolimits_X f < \eta < \gamma < \varepsilon, \quad \mu < 1 - \gamma\varepsilon^{-1}, \quad \theta < \mu\left(1 - (\eta\gamma^{-1})^{1/p}\right)^p, \tag{1.13}$$

and $\delta := (1 - \mu)\varepsilon\lambda^{-p}$. Let $u_0 := x_0$ and $f_0 := f$. Taking $f_1(x) := f_0(x) + \delta\|x - u_0\|^p$ for $x \in X$, we have that $f_1(u_0) = f_0(u_0) \ge \inf_X f_1$. Hence there exists $u_1 \in X$ such that $f_1(u_1) \le \theta f_0(u_0) + (1 - \theta)\inf_X f_1$. Taking $f_2(x) := f_1(x) + \delta\mu\|x - u_1\|^p$ for $x \in X$, we have that $f_2(u_1) = f_1(u_1) \ge \inf_X f_2$. Hence there exists $u_2 \in X$ such that $f_2(u_2) \le \theta f_1(u_1) + (1 - \theta)\inf_X f_2$. Continuing in this way we obtain a sequence $(u_n)_{n\ge 0} \subset X$ and a sequence of functions $(f_n)_{n\ge 0}$ such that

$$f_{n+1}(x) = f_n(x) + \delta\mu^n\|x - u_n\|^p \quad \forall\, x \in X, \ \forall\, n \ge 0, \tag{1.14}$$

and

$$f_{n+1}(u_{n+1}) \le \theta f_n(u_n) + (1-\theta)\inf_X f_{n+1} \quad \forall n \ge 0. \tag{1.15}$$

It is clear that $f_n \le f_{n+1}$; taking $s_n := \inf_X f_n$, $(s_n)_{n\ge 0} \subset \mathbb{R}$ is a non-decreasing sequence. Let $a_n := f_n(u_n)$; because $f_{n+1}(u_n) = f_n(u_n) \ge \inf_X f_{n+1}$, from Eq. (1.15) we obtain that $a_{n+1} \le a_n$ for every $n \ge 0$. Using again Eq. (1.15), we get

$$s_n \le s_{n+1} \le a_{n+1} \le \theta a_n + (1-\theta)s_{n+1} \le a_n,$$

and so

$$a_{n+1} - s_{n+1} \le \theta a_n + (1-\theta)s_{n+1} - s_{n+1} = \theta(a_n - s_{n+1}) \le \theta(a_n - s_n),$$

whence

$$a_n - s_n \le \theta^n(a_0 - s_0) \quad \forall n \ge 0. \tag{1.16}$$

It follows that $\lim s_n = \lim a_n \in \mathbb{R}$. Taking $x := u_{n+1}$ in Eq. (1.15), we get

$$a_n \ge a_{n+1} = f_n(u_{n+1}) + \delta\mu^n \|u_{n+1} - u_n\|^p \ge s_n + \delta\mu^n \|u_{n+1} - u_n\|^p,$$

which, together with Eqs. (1.13) and (1.16), yields

$$\delta\mu^n \|u_{n+1} - u_n\|^p \le a_n - s_n \le \theta^n(a_0 - s_0) \le \theta^n \eta \quad \forall n \ge 0.$$

It follows that $\|u_{n+1} - u_n\| \le (\eta/\delta)^{1/p}(\theta/\mu)^{n/p}$, whence

$$\|u_{n+m} - u_n\| \le (\eta/\delta)^{1/p}(\theta/\mu)^{n/p}\left(1 - (\theta/\mu)^{1/p}\right)^{-1} \quad \forall n, m \ge 0.$$

Since $0 < \theta/\mu < 1$, the sequence $(u_n)_{n\ge 0}$ is Cauchy, and so it converges to some $u \in X$. Moreover, the inequality above and (1.13) imply that

$$\|u_m - u_n\| \le (\eta/\delta)^{1/p}(\eta/\gamma)^{-1/p} = (\gamma/\delta)^{1/p} < (\varepsilon/\delta)^{1/p}(1-\mu)^{1/p} = \lambda \tag{1.17}$$

for all $n, m \ge 0$. In particular $(u_n) \subset B(x_0, \lambda)$. Letting $m \to \infty$ in the above inequality we get $\|u - u_n\| < \lambda$ for $n \ge 0$, and so $\|u - x_0\| < \lambda$. Consider $\mu_n := \mu^n(1-\mu) > 0$; hence $\sum_{n\ge 0} \mu_n = 1$. Let Θ_p be defined by Eq. (1.11). From Eq. (1.14) we obtain that

$$f_{n+1}(x) = f(x) + \varepsilon\lambda^{-p} \sum_{k=0}^{n} \mu_k \|x - u_k\|^p \quad \forall n \ge 0,\ \forall x \in X,$$

and so $f(x) + \varepsilon\lambda^{-p}\Theta_p(x) = \lim f_n(x) \geq \lim s_n$. Using Eq. (1.17) we get

$$f(u_n) + \varepsilon\lambda^{-p}\Theta_p(u_n) = f_n(u_n) + \varepsilon\lambda^{-p}\sum\nolimits_{k=n+1}^{\infty} \mu_k \|u_n - u_k\|^p$$
$$\leq a_n + \varepsilon\lambda^{-p}\sum\nolimits_{k=n+1}^{\infty} \mu_k\lambda^p = a_n + \varepsilon\sum\nolimits_{k=n+1}^{\infty} \mu_k$$

for every $n \geq 1$. Taking into account the continuity of Θ_p and the lower semicontinuity of f, the preceding inequality yields $f(u) + \varepsilon\lambda^{-p}\Theta_p(u) \leq \lim a_n = \lim s_n$. Therefore (1.12) holds.

From Eq. (1.17) we obtain that

$$\Theta_p(x_0) \leq \sum\nolimits_{n\geq 1} \mu_n \|u_0 - u_n\|^p \leq \sum\nolimits_{n\geq 1} \mu_k\gamma\delta^{-1} = \mu\gamma\delta^{-1} < \mu\lambda^p,$$

and so, by Eq. (1.12),

$$f(u) \leq f(x_0) + \varepsilon\lambda^{-p}\Theta_p(x_0) < f(x_0) + \varepsilon\mu < \inf\nolimits_X f + \gamma + \varepsilon\mu < \inf\nolimits_X f + \varepsilon.$$

The proof is complete. □

Note that $|\Theta_1(x) - \Theta_1(y)| \leq \|x - y\|$, and so Eq. (1.12) becomes $f(u) \leq f(x) + \varepsilon\lambda^{-1}\|x - u\|$ for every $x \in X$ when $p = 1$. So, under the slight stronger condition $f(x_0) < \inf_X f + \varepsilon$ one recovers the conclusion of Corollary 1.4.2, but the condition $f(x_\lambda) \leq f(x_0)$.

Although not directly related to what follows, we give the next two dual interesting results which have not been published by their author, C. Ursescu.

Theorem 1.4.5 *Let (X, d) be a complete metric space and $(F_n)_{n\in\mathbb{N}}$ be a sequence of closed subsets of X. Then*

$$\operatorname{cl}\left(\bigcup\nolimits_{n\in\mathbb{N}} \operatorname{int} F_n\right) = \operatorname{cl}\operatorname{int}\left(\bigcup\nolimits_{n\in\mathbb{N}} F_n\right). \tag{1.18}$$

Theorem 1.4.6 *Let X be a complete metric space and $(D_n)_{n\in\mathbb{N}}$ be a sequence of open subsets of X. Then*

$$\operatorname{int}\left(\bigcap\nolimits_{n\in\mathbb{N}} \operatorname{cl} D_n\right) = \operatorname{int}\operatorname{cl}\left(\bigcap\nolimits_{n\in\mathbb{N}} D_n\right). \tag{1.19}$$

Note that Theorem 1.4.5 can be obtained immediately from Theorem 1.4.6 by taking $D_n = X \setminus F_n$ (and Theorem 1.4.6 is obtained from Theorem 1.4.5 by taking $F_n = X \setminus D_n$). We give only the proof of Theorem 1.4.6.

Proof. Since the inclusion "$\supset$" in Eq. (1.19) is obvious, we show the converse one. So let $x \in \operatorname{int}\left(\bigcap_{n\in\mathbb{N}} \operatorname{cl} D_n\right)$ and $r > 0$; take $x_1 := x$.

Then there exists $r_1 \in]0, r[$ such that $D(x_1, r_1) \subset \bigcap_{n\in\mathbb{N}} \operatorname{cl} D_n$; therefore $D(x_1, r_1) \subset \operatorname{cl} D_n$ for every $n \in \mathbb{N}$. It follows that $B(x_1, r_1) \cap D_1$ is an open nonempty set. There exist $x_2 \in X$ and $r_2 \in]0, r_1/2]$ such that $D(x_2, r_2) \subset B(x_1, r_1) \cap D_1$. It follows that $D(x_2, r_2) \subset D(x_1, r_1) \subset \operatorname{cl} D_2$, and so $B(x_2, r_2) \cap D_2$ is an open nonempty set. Continuing in this way we find the sequences $(x_n)_{n\in\mathbb{N}} \subset X$ and $(r_n)_{n\in\mathbb{N}} \subset \mathbb{P}$ such that $(r_n)_{n\in\mathbb{N}} \to 0$ and $D(x_{n+1}, r_{n+1}) \subset B(x_n, r_n) \cap D_n$ for every n. In particular $D(x_{n+1}, r_{n+1}) \subset D(x_n, r_n)$ for every n. Since X is a complete metric space, using Cantor's theorem, $\bigcap_{n\in\mathbb{N}} D(x_n, r_n) = \{x'\}$ for some $x' \in X$. It follows that $x' \in D_n$ for every n. Since $x' \in D(x_1, r_1) \subset B(x, r)$, we have that $B(x,r) \cap \bigcap_{n\in\mathbb{N}} D_n$. As $r > 0$ is arbitrary, $x \in \operatorname{cl}\left(\bigcap_{n\in\mathbb{N}} D_n\right)$. Therefore $\operatorname{int}\left(\bigcap_{n\in\mathbb{N}} \operatorname{cl} D_n\right) \subset \operatorname{cl}\left(\bigcap_{n\in\mathbb{N}} D_n\right)$, whence the inclusion "$\subset$" in Eq. (1.19) holds, too. □

From Theorem 1.4.5 we obtain immediately the famous Baire's theorems:

Theorem 1.4.7 (Baire) *Let (X, d) be a complete metric space. Then any countable intersection of dense open subsets of X is dense.*

Proof. Let $A = \bigcap_{n\in\mathbb{N}} D_n$ with D_n open and dense for every $n \in \mathbb{N}$. From Eq. (1.19) it follows that $X = \operatorname{int}(\operatorname{cl} A)$, and so $\operatorname{cl} A = X$. □

Remind that an intersection of a countable family of open subsets of the topological space (X, τ) is called a G_δ set.

Theorem 1.4.8 (Baire) *If $(F_n)_{n\in\mathbb{N}}$ is a sequence of closed subsets of the complete metric space (X, d) such that $X = \bigcup_{n\in\mathbb{N}} F_n$, then at least one of F_n's has nonempty interior.*

Proof. The conclusion is immediate from Eq. (1.18). □

1.5 Exercises

Exercise 1.1 (Carathéodory) Let X be a linear space of dimension $n \in \mathbb{N}$ and let $A \subset X$ be nonempty. Prove that

$$\operatorname{co} A = \left\{ \sum\nolimits_{i=1}^{n+1} \lambda_i x_i \;\middle|\; (\lambda_i)_{i\in\overline{1,n+1}} \subset \mathbb{R}_+, (x_i)_{i\in\overline{1,n+1}} \subset A, \sum\nolimits_{i=1}^{n+1} \lambda_i = 1 \right\}.$$

Exercise 1.2 Let X be a finite dimensional normed space and $A \subset X$ be a nonempty convex set. Prove that ${}^{i}A \neq \emptyset$, ${}^{i}(\operatorname{cl} A) = {}^{i}A$, $\operatorname{cl}({}^{i}A) = \operatorname{cl} A$ and for every $a \in {}^{i}A$ and $z \in \operatorname{cl} A$ we have $]a, z[\subset {}^{i}A$.

Exercise 1.3 Let X be a separated locally convex space, $H \subset X$ be a closed hyperplane and $A \subset X$ be a nonempty convex set. If $\operatorname{int}_H(A \cap H) \neq \emptyset$ and $A \not\subset H$ prove that $\operatorname{int} A \neq \emptyset$. Moreover, if $M \subset X$ is a closed affine set with finite codimension and $\operatorname{int}_M(A \cap M) \neq \emptyset$, prove that $\operatorname{rint} A \neq \emptyset$.

Exercise 1.4 Let X be a separated locally convex space and $A, C \subset X$ be nonempty convex sets with $\operatorname{int} C \neq \emptyset$. Prove that $\operatorname{int}(A + C) = A + \operatorname{int} C$. Moreover, if $A \cap \operatorname{int} C \neq \emptyset$, then $\operatorname{cl}(A \cap C) = \operatorname{cl} A \cap \operatorname{cl} C$. In particular, if C is a convex cone with nonempty interior, then $\operatorname{cl} C + \operatorname{int} C = \operatorname{int} C$.

Exercise 1.5 Let X be a separated locally convex space, $X_0 \subset X$ be a linear subspace, $a_1, \ldots, a_p \in X$, $\varphi_1, \ldots, \varphi_k \in X^*$ and $\alpha_1, \ldots, \alpha_k \in \mathbb{R}$, where $p, k \in \mathbb{N}$. Prove that:

(a) if X_0 is closed and $C = \{\sum_{i=1}^{p} \lambda_i a_i \mid \forall i \in \overline{1,p} \; : \; \lambda_i \geq 0\}$, then $X_0 + C$ is a closed convex cone. In particular C is a closed convex cone;

(b) $X_0 + \{x \in X \mid \forall i \in \overline{1,k} \; : \; \langle x, \varphi_i \rangle \leq \alpha_i\}$ is a convex closed set.

Exercise 1.6 Let $(X, (\cdot \mid \cdot))$ be a Hilbert space, $x_0 \in X \setminus \{0\}$ and $\alpha \in [0, \pi/2]$. Prove that

$$P(\alpha) := \{x \in X \mid \angle(x, x_0) \leq \alpha\}$$

is a closed convex cone, where $\angle(x, y) := \operatorname{Arccos} \dfrac{(x \mid y)}{\|x\| \cdot \|y\|}$. Moreover, $P(\alpha)^\circ = P(\pi/2 - \alpha)$.

Exercise 1.7 Let $n \in \mathbb{N} \setminus \{1\}$, $\rho > 0$ and $\alpha_1, \ldots, \alpha_n \in]0, 1[$ be such that $\alpha_1 + \ldots + \alpha_n = 1$. Consider

$$K_\rho := \left\{(x_1, \ldots, x_n, x_{n+1}) \in \mathbb{R}^{n+1} \mid x_1, \ldots, x_n \geq 0, \; |x_{n+1}| \leq \rho x_1^{\alpha_1} \cdots x_n^{\alpha_n}\right\}.$$

Prove that $(K_\rho)^+ = K_{\rho'}$, where $\rho' := (\rho \alpha_1^{\alpha_1} \cdots \alpha_n^{\alpha_n})^{-1}$. In particular, K_ρ is a closed convex cone.

Exercise 1.8 Prove that

$$P := \left\{(x, y, z) \in \mathbb{R}^3 \;\middle|\; x, z \geq 0, \; 2xz \geq y^2\right\}$$

is a closed convex cone and $P^+ = P$. (P is called the "ice cream" cone.)

Exercise 1.9 Let X be a normed space, $\varphi \in X^* \setminus \{0\}$ and $0 < \alpha < \|\varphi\|$. Prove that the set $C := \{x \in X \mid \varphi(x) \geq a\|x\|\}$ is a pointed ($C \cap -C = \{0\}$) closed convex cone with nonempty interior and $C^+ = \mathbb{R}_+ \cdot (\varphi + \alpha U^*) = \mathbb{R}_+ \cdot D(\varphi, \alpha)$.

Exercise 1.10 Let X, Y be topological vector spaces and $\mathcal{R} : X \rightrightarrows Y$ be a convex multifunction. Assume that there exists $\overline{x} \in X$ such that $\operatorname{int} \mathcal{R}(\overline{x}) \neq \emptyset$. Prove that for every $(x_0, y_0) \in \operatorname{gr} \mathcal{R}$ with $y_0 \in (\operatorname{Im} \mathcal{R})^i$ there exists $u \in X$ such that $y_0 \in \operatorname{int} \mathcal{R}(x)$ for every $x \in]x_0, u]$.

Exercise 1.11 Let X, Y be two separated locally convex spaces and $\mathcal{C} : X \rightrightarrows Y$ be a multifunction whose graph is a closed linear subspace of $X \times Y$. Prove that $\operatorname{dom} \mathcal{C}$ is dense if and only if $\mathcal{C}^*(y^*)$ is a singleton for every $y^* \in \operatorname{dom} \mathcal{C}^*$. In particular, $\operatorname{dom} \mathcal{C}$ is dense and $\mathcal{C}(x)$ is a singleton for every $x \in \operatorname{dom} \mathcal{C}$ if and only if $\operatorname{dom} \mathcal{C}^*$ is w^*-dense and $\mathcal{C}^*(y^*)$ is a singleton for every $y^* \in \operatorname{dom} \mathcal{C}^*$.

Exercise 1.12 Let (X, d) be a metric space. Prove that (X, d) is complete if for any Lipschitz function $f : X \to \mathbb{R}_+$ there exists $\overline{x} \in X$ such that $f(\overline{x}) \leq f(x) + d(x, \overline{x})$ for every $x \in X$. This shows that the completeness assumption in Ekeland's variational principle is essential.

Exercise 1.13 Let (X, d) be a complete metric space and $f : X \to \overline{\mathbb{R}}$ be a proper lsc and lower bounded function. Suppose that for every $x \in X$ such that $f(x) > \inf f$, there exists $\overline{x} \in X \setminus \{x\}$ such that $f(\overline{x}) + d(x, \overline{x}) \leq f(x)$. Prove that $\operatorname{argmin} f \neq \emptyset$ and $d(x, \operatorname{argmin} f) \leq f(x) - \inf f$ for every $x \in X$.

Exercise 1.14 Let (X, d) be a complete metric space, $f : X \to \overline{\mathbb{R}}$ be a proper lsc lower bounded function and $\mathcal{R} : X \rightrightarrows X$. Assume that for every $x \in X$ there exists $y \in \mathcal{R}(x)$ such that $d(x, y) + f(y) \leq f(x)$. Prove that $\mathcal{R}$ has fixed points, *i.e.* there exists $x_0 \in X$ such that $x_0 \in \mathcal{R}(x_0)$.

Exercise 1.15 Let $(X, \|\cdot\|)$ be a normed space and $f : X \to \overline{\mathbb{R}}$. Prove that

(i) $\lim_{\|x\| \to \infty} f(x) = \infty$ if and only if $[f \leq \lambda]$ is bounded for every $\lambda \in \mathbb{R}$.

(ii) Assume that $X = \mathbb{R}^k$, f is lsc and $\lim_{\|x\| \to \infty} f(x) = \infty$ (*i.e.* f is coercive). Prove that there exists $\overline{x} \in \mathbb{R}^k$ such that $f(\overline{x}) \leq f(x)$ for every $x \in \mathbb{R}^k$.

1.6 Bibliographical Notes

Section 1.1: For the notions and results on topology and topological vector spaces not recalled in this section one can consult many classical books (see [Kelley (1955); Willard (1971); Bourbaki (1964); Holmes (1975)]). The proofs of the results mentioned in this section can be found in [Holmes (1975)] or [Bourbaki (1964)].

Section 1.2: Ideally convex sets were introduced by Lifšic (1970), cs-closed sets by Jameson (1972), cs-complete sets by Simons (1990), lower cs-closed sets by Amara and Ciligot-Travain (1999), while condition (Hx) was used by Zălinescu (1992b). The properties stated in Proposition 1.2.1 are mentioned in [Kusraev and Kutateladze (1995)]. All the results concerning lcs-closed sets, as well as Proposition 1.2.2 (for cs-closed sets), are from [Amara and Ciligot-Travain (1999)]. The set ${}^{ic}A$, introduced by Zălinescu (1987), is also introduced by Jeyakumar and Wolkowicz (1992) under the name of strong quasi relative interior of A. The notation ${}^{ib}A$ was introduced in [Zălinescu (1992b)] but the condition $0 \in {}^{ib}A$ was intensively used by Simons (1990) and Zălinescu (1992b). The quasi relative interior was introduced by Borwein and Lewis (1992); they stated the properties mentioned in Proposition 1.2.7 in locally convex spaces. Proposition 1.2.9 is stated in [Borwein and Lewis (1992)] for A a cs-closed set and X a Fréchet space. Joly and Laurent (1971) introduced the notion of united sets, but Proposition 1.2.8 is mainly established by Moussaoui and Volle (1997).

Section 1.3: Lemma 1.3.1 is proved by Amara and Ciligot-Travain (1999) for X a metrizable lcs, Y a Fréchet space and $\mathcal{R}$ cs-closed, while the statement and proof of Lemma 1.3.3 are those of Ursescu (1975); Corollary 1.3.4 (for Banach spaces) is due to Lifšic (1970). The statement of Theorem 1.3.5 is very close to that of [Kusraev and Kutateladze (1995), Th. 3.1.18]; when X, Y are Fréchet spaces our statement is slightly more general. For X, Y metrizable locally convex spaces and $\mathcal{R}$ satisfying (Hx) Theorem 1.3.5 is equivalent to the open mapping theorem of Simons (1990). Theorem 1.3.7 is obtained by Ursescu (1975) for $Y_0 := \operatorname{aff}(\operatorname{Im}\mathcal{R}) = Y$. It is established in [Robinson (1976)] for X, Y Banach spaces and $Y_0 = Y$; in this case the above result is met in the literature under the name of Robinson–Ursescu theorem. Corollary 1.3.8 was obtained by Amara and Ciligot-Travain (1999) for lcs-closed sets. Theorem 1.3.10 is due to Robinson (1976), while Theorem 1.3.11 can be found in [Li and Singer (1998)]. Theorems 1.3.16 and 1.3.17 are due to Cârjă (1989) (with different proofs).

Section 1.4: The Ekeland variational principle [Ekeland (1974)] is met in the literature, generally, in the form given in Corollary 1.4.2; the statement and proof of Theorem 1.4.1 are from [Aubin and Frankowska (1990)]. The statement and proof of the Borwein–Preiss variational principle are from [Borwein and Preiss (1987)]. Theorems 1.4.5 and 1.4.6 can be found in the recent book [Cârjă (1998)].

Exercises: The exercises are generally known results or auxiliary results from various papers. So Exercise 1.1 is the famous Carathéodory theorem, Exercise 1.3 is from [Joly and Laurent (1971)], Exercise 1.5 is from [Laurent (1972)], Exercise 1.7 (for $n = 2$ and $\rho = \rho'$) is from [Bauschke *et al.* (2000)], Exercise 1.10 is stated in [Li and Singer (1998)] in Banach spaces in a slightly weaker form, Exercise 1.12 is the main result of [Sullivan (1981)], Exercise 1.13 is from [Hamel (1994)], Exercise 1.14 is the known Caristi's fixed point theorem (see [Caristi (1976)]).

Chapter 2

Convex Analysis in Locally Convex Spaces

2.1 Convex Functions

We begin this section by introducing some notions and notations. For a function $f : X \to \overline{\mathbb{R}}$ and $\lambda \in \mathbb{R}$ consider:

$$\begin{aligned} \operatorname{dom} f &:= \{x \in X \mid f(x) < -\infty\}, \\ \operatorname{epi} f &:= \{(x,t) \in X \times \mathbb{R} \mid f(x) \le t\}, \\ \operatorname{epi}_s f &:= \{(x,t) \mid f(x) < t\}, \\ [f \le \lambda] &:= \{x \in X \mid f(x) \le \lambda\}, \\ [f < \lambda] &:= \{x \in X \mid f(x) < \lambda\}; \end{aligned}$$

the sets $\operatorname{dom} f$, $\operatorname{epi} f$ and $\operatorname{epi}_s f$ are called the **domain**, the **epigraph** and the **strict epigraph** of the function f, respectively, while the sets $[f \le \lambda]$ and $[f < \lambda]$ are the **level set** and **strict level set** of f at height λ. One says that the function f is **proper** if $\operatorname{dom} f \ne \emptyset$ and $f(x) > -\infty$ for every $x \in X$. It is evident that $\operatorname{dom} f = \Pr_X(\operatorname{epi} f)$.

Let X be a real linear space and $f : X \to \overline{\mathbb{R}}$; we say that f is **convex** if

$$\forall\, x,y \in X,\ \forall\, \lambda \in [0,1] \ :\ f(\lambda x + (1-\lambda)y) \le \lambda f(x) + (1-\lambda) f(y), \tag{2.1}$$

with the conventions: $(+\infty)+(-\infty) = +\infty$, $0\cdot(+\infty) = +\infty$, $0\cdot(-\infty) = 0$. Note that when $x = y$, or $\lambda \in \{0,1\}$, or $\{x,y\} \not\subset \operatorname{dom} f$, the inequality (2.1) is automatically satisfied. Therefore the function f is convex if

$$\forall\, x,y \in \operatorname{dom} f,\ x \ne y,\ \forall\, \lambda \in\,]0,1[\, :\ f(\lambda x+(1-\lambda)y) \le \lambda f(x)+(1-\lambda) f(y). \tag{2.2}$$

If relation (2.2) holds with "$<$" instead of "$\leq$" we say that f is **strictly convex**. Similarly, we say that f is *(strictly) concave* if $-f$ is (strictly) convex. Since every property of convex functions can be transposed easily to concave functions, in the sequel we consider, practically, only convex functions.

To avoid multiplication with 0, taking into account the above remark, we shall limit ourselves to $\lambda \in]0,1[$ in the sequel.

In the following theorem we establish some characterizations of convex functions.

Theorem 2.1.1 *Let $f : X \to \overline{\mathbb{R}}$. The following statements are equivalent:*

(i) *f is (strictly) convex;*

(ii) *the functions $\varphi_{x,y} : \mathbb{R} \to \overline{\mathbb{R}}$, $\varphi_{x,y}(t) := f((1-t)x+ty)$, are (strictly) convex for all $x, y \in X$ ($x \neq y$);*

(iii) $\operatorname{dom} f$ *is a convex set and*

$$\forall\, x, y \in \operatorname{dom} f,\ \forall\, \lambda \in]0,1[\ :\ f(\lambda x + (1-\lambda)y) \leq \lambda f(x) + (1-\lambda) f(y)$$

(the inequality being strict for $x \neq y$);

(iv) $\forall n \in \mathbb{N},\ \forall x_1, \ldots, x_n \in X,\ \forall \lambda_1, \ldots, \lambda_n \in]0,1[,\ \lambda_1 + \cdots + \lambda_n = 1\ :$

$$f(\lambda_1 x_1 + \cdots + \lambda_n x_n) \leq \lambda_1 f(x_1) + \cdots + \lambda_n f(x_n) \tag{2.3}$$

(the inequality being strict when $x_1, \ldots, x_n$ are not all equal);

(v) $\operatorname{epi} f$ *is a convex subset of* $X \times \mathbb{R}$;

(vi) $\operatorname{epi}_s f$ *is a convex subset of* $X \times \mathbb{R}$.

Proof. The implications (i) $\Rightarrow$ (ii), (ii)$\Rightarrow$(i), (i) $\Rightarrow$ (iii), (iii) $\Rightarrow$ (i) and (iv) $\Rightarrow$ (i) are obvious.

(i) $\Rightarrow$ (v) Let $(x_1, t_1), (x_2, t_2) \in \operatorname{epi} f$ and $\lambda \in]0,1[$. Then $x_1, x_2 \in \operatorname{dom} f$, $f(x_1) \leq t_1$ and $f(x_2) \leq t_2$. Since f is convex, from Eq. (2.1) we obtain that

$$f(\lambda x_1 + (1-\lambda)x_2) \leq \lambda f(x_1) + (1-\lambda) f(x_2) \leq \lambda t_1 + (1-\lambda) t_2,$$

whence $\lambda(x_1, t_1) + (1-\lambda)(x_2, t_2) \in \operatorname{epi} f$. Therefore $\operatorname{epi} f$ is convex.

(v) $\Rightarrow$ (iv) Let $k \in \mathbb{N}$, $\lambda_1, \ldots, \lambda_k \in]0,1[$, $\lambda_1 + \cdots + \lambda_k = 1$, $x_1, \ldots, x_k \in X$. If there exists i such that $f(x_i) = \infty$, then Eq. (2.3) is obviously true. If $f(x_i) \in \mathbb{R}$ for every i, $1 \leq i \leq k$, then $(x_i, f(x_i)) \in \operatorname{epi} f$ for every i; since $\operatorname{epi} f$ is convex, $\sum_{i=1}^{k} \lambda_i (x_i, f(x_i)) \in \operatorname{epi} f$, which proves that the inequality

(2.3) is verified. Finally, suppose that $f(x_i) < \infty$ for every i and that there exists i_0 with $f(x_{i_0}) = -\infty$; consider $t_i \in \mathbb{R}$ such that $(x_i, t_i) \in \operatorname{epi} f$ for every $i \neq i_0$; since $(x_{i_0}, -n) \in \operatorname{epi} f$ for every $n \in \mathbb{N}$, we have

$$f(\lambda_1 x_1 + \cdots + \lambda_k x_k) \leq \lambda_1 t_1 + \cdots + \lambda_{i_0-1} t_{i_0-1} + \lambda_{i_0}(-n) + \lambda_{i_0+1} t_{i_0+1} + \ldots + \lambda_k t_k$$

for $n \in \mathbb{N}$. Letting $n \to \infty$ in the above inequality we get $f(\sum_{i=1}^{k} \lambda_i x_i) = -\infty$; hence Eq. (2.3) is verified.

The proofs for (i) $\Rightarrow$ (vi) and (vi) $\Rightarrow$ (iv) are similar to those of (i) $\Rightarrow$ (v) and (v) $\Rightarrow$ (iv), respectively.

The proof for the "strictly convex" case is similar. □

Note that in (v) and (vi) there are no counterparts corresponding to f strictly convex.

Proposition 2.1.2 *If $f : X \to \overline{\mathbb{R}}$ is sublinear then f is convex. Moreover, f is sublinear if and only if* $\operatorname{epi} f$ *is a convex cone with* $(0, -1) \notin \operatorname{epi} f$.

Proof. Let f be sublinear. It is obvious that f is convex, and so $\operatorname{epi} f$ is a convex set which contains $(0, 0)$. If $(x, t) \in \operatorname{epi} f$ and $\lambda > 0$ then $f(\lambda x) = \lambda f(x) \leq \lambda t$, and so $\lambda(x, t) \in \operatorname{epi} f$; hence $\operatorname{epi} f$ is a convex cone. Since $f(0) = 0 > -1$, we have that $(0, -1) \notin \operatorname{epi} f$.

Assume now that $\operatorname{epi} f$ is a convex cone with $(0, -1) \notin \operatorname{epi} f$. It is immediate that f is convex (epi f being convex) and $f(\lambda x) = \lambda f(x)$ for $\lambda > 0$ and $x \in X$. So, for $x, y \in X$ we have that $f(x + y) = 2f(\frac{1}{2}x + \frac{1}{2}y) \leq f(x) + f(y)$. If $f(0) < 0$ then $(0, -t) \in \operatorname{epi} f$ for some $t > 0$, whence the contradiction $(0, -1) \in \operatorname{epi} f$. Therefore $f(0) = 0$, and so f is sublinear. □

The **indicator function** of the subset A of X is

$$\iota_A : X \to \overline{\mathbb{R}}, \qquad \iota_A(x) := \begin{cases} 0 & \text{if } x \in A, \\ +\infty & \text{if } x \in X \setminus A. \end{cases}$$

Note that $\operatorname{dom} \iota_A = A$ and $\operatorname{epi} \iota_A = A \times \mathbb{R}_+$. From the preceding theorem we obtain that ι_A is convex if and only if A is convex.

If f is convex the sets $[f \leq \lambda]$ and $[f < \lambda]$ are convex for every $\lambda \in \mathbb{R}$. The converse is generally false. A function f with the property that "$[f \leq \lambda]$ is convex for every $\lambda \in \mathbb{R}$" is said to be **quasi-convex**. So, $f : X \to \overline{\mathbb{R}}$ is quasi-convex if and only if

$$\forall\, x, y \in X,\ \forall\, \lambda \in [0, 1] \ :\ f((1 - \lambda)x + \lambda y) \leq \max\{f(x), f(y)\}.$$

The notion of convex function is extended naturally to mappings with values in ordered linear spaces. Let Y be a real linear space and $Q \subset Y$ be a convex cone; Q generates an order relation on Y: $y_1 \leq y_2$ (or $y_1 \leq_Q y_2$ if there is any risk of confusion) if $y_2 - y_1 \in Q$. By analogy with $\overline{\mathbb{R}}$, consider $Y^\bullet := Y \cup \{\infty\}$, where $\infty \notin Y$; we put $y \leq \infty$ for all $y \in Y$ and $\lambda \cdot \infty = \infty$ for every $\lambda \in \mathbb{P}$ (note that $y < \infty$ if $y \neq \infty$). The element $-\infty$ is introduced similarly. To point out that Y is ordered by Q we write (Y, Q) or $(Y, \leq)$.

Let (Y, Q) be an ordered linear space and $H : X \to Y^\bullet$; the operator H is **Q-convex** if

$$\forall\, x, y \in X,\ \forall\, \lambda \in]0,1[\, :\ H(\lambda x + (1-\lambda)y) \leq \lambda H(x) + (1-\lambda)H(y).$$

The operator $H : X \to Y \cup \{-\infty\}$ is *Q-concave* if $-H : X \to Y^\bullet$ is Q-convex.

If $A \in L(X, Y)$ then A is Q-convex for every convex cone $Q \subset Y$.

As in the case $Y = \mathbb{R}$, the *domain* and the *epigraph* are defined by: $\operatorname{dom} H := \{x \in X \mid H(x) < \infty\}$ and $\operatorname{epi} H := \{(x, y) \in X \times Y \mid H(x) \leq y\}$. The characterizations of convex functions given in Theorem 2.1.1 are valid for a Q-convex operator.

A function $f : (Y, Q) \to \overline{\mathbb{R}}$ is **Q-increasing** if $y_1 \leq_Q y_2 \Rightarrow f(y_1) \leq f(y_2)$. For such a function we consider that $f(\infty) = +\infty$. It is obvious that every function is $\{0\}$-increasing, and that a linear functional $\varphi : Y \to \mathbb{R}$ is Q-increasing if and only if $\varphi(y) \geq 0$ for every $y \in Q$. One defines similarly the *Q-decreasing* functions.

In the next result we mention some methods for deriving new convex functions from known ones. For $\alpha, \beta \in \overline{\mathbb{R}}$ we set $\alpha \vee \beta := \max\{\alpha, \beta\}$ and $\alpha \wedge \beta := \min\{\alpha, \beta\}$.

Theorem 2.1.3 *Let X, Y be linear spaces and $Q \subset Y$ be a convex cone.*

(i) *If $f_i : X \to \overline{\mathbb{R}}$ is convex for every $i \in I$ ($I \neq \emptyset$) then $\sup_{i\in I} f_i$ is convex. Moreover, $\operatorname{epi}(\sup_{i\in I} f_i) = \bigcap_{i\in I} \operatorname{epi} f_i$.*

(ii) *If $f_1, f_2 : X \to \overline{\mathbb{R}}$ are convex and $\lambda \in \mathbb{R}_+$, then $f_1 + f_2$ and λf_1 are convex, where $0 \cdot f_1 := \iota_{\operatorname{dom} f_1}$. Moreover,*

$$\operatorname{dom}(f_1 + f_2) = \operatorname{dom} f_1 \cap \operatorname{dom} f_2, \quad \operatorname{dom}(\lambda f_1) = \operatorname{dom} f_1.$$

(iii) *If $f_n : X \to \overline{\mathbb{R}}$ is convex for every $n \in \mathbb{N}$ and $f : X \to \overline{\mathbb{R}}$ is such that $f(x) = \limsup_{n\to\infty} f_n(x)$ for every $x \in X$, then f is convex.*

(iv) *If $A \subset X \times \mathbb{R}$ is a convex set then the function φ_A is convex, where*

$$\varphi_A : X \to \overline{\mathbb{R}}, \qquad \varphi_A(x) := \inf\{t \mid (x,t) \in A\}. \tag{2.4}$$

(v) *If $F : X \times Y \to \overline{\mathbb{R}}$ is convex, then the* ***marginal function*** *h associated to F is convex, where*

$$h : Y \to \overline{\mathbb{R}}, \quad h(y) := \inf_{x \in X} F(x,y). \tag{2.5}$$

(vi) *Let $g : Y \to \overline{\mathbb{R}}$ be a convex function. If $H : X \to Y^\bullet$ is Q-convex and g is Q-increasing, then $g \circ H$ is convex; this conclusion holds also if $H : X \to Y \cup \{-\infty\}$ is Q-concave and g is Q-decreasing. In particular, if $A \in L(X,Y)$ then $g \circ A$ is convex.*

(vii) *Let $f : X \to \overline{\mathbb{R}}$, $g : Y \to \overline{\mathbb{R}}$ be proper convex functions, and let*

$$\Phi, \Psi : X \times Y \to \overline{\mathbb{R}}, \quad \Phi(x,y) := f(x) + g(y), \quad \Psi(x,y) := f(x) \vee g(y).$$

Then Φ and Ψ are convex and proper. Moreover, $\operatorname{dom} \Phi = \operatorname{dom} \Psi = \operatorname{dom} f \times \operatorname{dom} g$, $\inf \Phi = \inf f + \inf g$ *and* $\inf \Psi = \inf f \vee \inf g$.

(viii) *If $f : X \to \overline{\mathbb{R}}$ is convex and $A \in L(X,Y)$ then the function*

$$Af : Y \to \overline{\mathbb{R}}, \qquad (Af)(y) := \inf\{f(x) \mid Ax = y\},$$

is convex. Moreover $\operatorname{dom}(Af) = A(\operatorname{dom} f)$ *and* $\inf Af = \inf f$.

(ix) *If $f_1, f_2 : X \to \overline{\mathbb{R}}$ are convex and proper, their* ***convolution*** *and* ***max-convolution****, defined by*

$$f_1 \Box f_2 : X \to \overline{\mathbb{R}}, \quad (f_1 \Box f_2)(x) := \inf\{f_1(x_1) + f_2(x_2) \mid x_1 + x_2 = x\},$$
$$f_1 \Diamond f_2 : X \to \overline{\mathbb{R}}, \quad (f_1 \Diamond f_2)(x) := \inf\{f_1(x_1) \vee f_2(x_2) \mid x_1 + x_2 = x\},$$

are convex. Moreover $\operatorname{dom}(f_1 \Box f_2) = \operatorname{dom}(f_1 \Diamond f_2) = \operatorname{dom} f_1 + \operatorname{dom} f_2$, $\inf f_1 \Box f_2 = \inf f_1 + \inf f_2$, $\inf f_1 \Diamond f_2 = \inf f_1 \vee \inf f_2$,

$$\operatorname{epi}_s(f_1 \Box f_2) = \operatorname{epi}_s f_1 + \operatorname{epi}_s f_2 \tag{2.6}$$

and

$$\forall \lambda \in \mathbb{R} \ : \ [f_1 \Diamond f_2 < \lambda] = [f_1 < \lambda] + [f_2 < \lambda]. \tag{2.7}$$

Proof. (i) It is clear that $\operatorname{epi}(\sup_{i \in I} f_i) = \bigcap_{i \in I} \operatorname{epi} f_i$. Since f_i is convex for every i, using Theorem 2.1.1, we have that $\operatorname{epi} f_i$ is convex, and so $\operatorname{epi}(\sup_{i \in I} f_i)$ is convex. The conclusion follows using again Theorem 2.1.1.

(ii) is immediate.

(iii) Let $\lambda \in]0,1[$ and $x, y \in \operatorname{dom} f$. Since $\limsup f_n(x)$, $\limsup f_n(y) < \infty$, there exists $n_0 \in \mathbb{N}$ such that $f_n(x)$, $f_n(y) < \infty$ for every $n \geq n_0$. Since f_n is convex, we have

$$f_n(\lambda x + (1-\lambda)y) \leq \lambda f_n(x) + (1-\lambda) f_n(y).$$

Taking the limit superior we obtain

$$f(\lambda x + (1-\lambda)y) \leq \lambda f(x) + (1-\lambda) f(y).$$

Therefore f is convex.

(iv) Let $(x_1, t_1), (x_2, t_2) \in \operatorname{epi}_s \varphi_A$ and $\lambda \in]0,1[$. Then there exist $s_1, s_2 \in \mathbb{R}$ such that $(x_1, s_1), (x_2, s_2) \in A$ and $s_1 < t_1$, $s_2 < t_2$. Since A is convex, $\lambda(x_1, s_1) + (1-\lambda)(x_2, s_2) = (\lambda x_1 + (1-\lambda)x_2, \lambda s_1 + (1-\lambda)s_2) \in A$. Therefore, $\varphi_A(\lambda x_1 + (1-\lambda)x_2) \leq \lambda s_1 + (1-\lambda)s_2 < \lambda t_1 + (1-\lambda)t_2$, and so $\lambda(x_1, t_1) + (1-\lambda)(x_2, t_2) \in \operatorname{epi}_s \varphi_A$. Hence φ_A is convex.

(v) Note that

$$\operatorname{epi}_s h = \operatorname{Pr}_{Y \times \mathbb{R}}(\operatorname{epi}_s F), \quad \operatorname{dom} h = \operatorname{Pr}_Y(\operatorname{dom} F). \tag{2.8}$$

The conclusion follows from Theorem 2.1.1(vi).

(vi) We observe that $(f \circ H)(x) = \inf_{y \in Y} F(x, y)$, where $F(x, y) := g(y) + \iota_{\operatorname{epi} H}(x, y)$, because f is Q-increasing. Since obviously F is convex, by (v) we obtain that $(f \circ H)$ is convex, too.

The other case is proved similarly. The second part is immediate taking into account that A is $\{0\}$-convex and g is $\{0\}$-increasing.

(vii) One obtains immediately that $\operatorname{dom} \Phi = \operatorname{dom} \Psi = \operatorname{dom} f \times \operatorname{dom} g$ and that Φ and Ψ are convex. The formulas for $\inf \Phi$ and $\inf \Psi$ are well-known.

(viii) We have that $(Af)(y) = \inf_{x \in X} F(x, y)$, where $F(x, y) := f(x) + \iota_{\operatorname{gr} A}(x, y)$. It is obvious that F is convex. From (v) we obtain that Af is convex. As $\operatorname{dom} F = \{(x, Ax) \mid x \in \operatorname{dom} f\}$, we have that $\operatorname{dom}(Af) = A(\operatorname{dom} f)$. The formula for $\inf Af$ is obvious.

(ix) Let us consider the functions $\Phi, \Psi : X \times X \to \overline{\mathbb{R}}$ defined by

$$\Phi(x_1, x_2) := f_1(x_1) + f_2(x_2), \qquad \Psi(x_1, x_2) := f_1(x_1) \vee f_2(x_2),$$

and $A : X \times X \to X$, $A(x_1, x_2) := x_1 + x_2$. Then $(f_1 \Box f_2)(x) = (A\Phi)(x)$ and $(f_1 \Diamond f_2)(x) = (A\Psi)(x)$. By (v) we have that Φ and Ψ are convex; using (viii) we obtain that $f_1 \Box f_2$ and $f_1 \Diamond f_2$ are convex, too. The relations on the domains, epigraph and level sets follow by easy verification. The formulas for $\inf f_1 \Box f_2$ and $\inf f_1 \Diamond f_2$ follow immediately from (vii) and (viii). $\square$

The formulas (2.6) and (2.7) motivate the use of "epi-sum convolution" and "level-sum convolution" for the convolution and max-convolution, respectively. The convolution and max-convolution are extended in an obvious way to a finite number of functions; from Eqs. (2.6) and (2.7) one obtains that these operations are associative.

The convex functions which take the value $-\infty$ are rather special.

Proposition 2.1.4 *Let $f : X \to \overline{\mathbb{R}}$ be a convex function. If there exists $x_0 \in X$ such that $f(x_0) = -\infty$ then $f(x) = -\infty$ for every $x \in {}^i(\operatorname{dom} f)$. In particular, if f is sublinear and $0 \in {}^i(\operatorname{dom} f)$, then f is proper.*

Proof. Let $x \in {}^i(\operatorname{dom} f)$; since $x_0 \in \operatorname{dom} f$, there exists $\mu > 0$ such that $y := (1+\mu)x - \mu x_0 \in \operatorname{dom} f$. Taking $\lambda := (\mu+1)^{-1} \in]0,1[$, $x = (1-\lambda)x_0 + \lambda y$, and so

$$f(x) \leq (1-\lambda)f(x_0) + \lambda f(y) = -\infty.$$

The case of f sublinear is immediate from the first part. $\square$

In the sequel we denote by $\Lambda(X)$ the class of proper convex functions defined on X.

Let now $\emptyset \neq C \subset X$ and $f : C \to \mathbb{R}$; we say that f is convex if C is convex and

$$\forall\, x, y \in C,\ \forall\, \lambda \in]0,1[\, :\ f(\lambda x + (1-\lambda)y) \leq \lambda f(x) + (1-\lambda) f(y).$$

It is easy to see (Exercise!) that the function f above is convex if and only if

$$\widehat{f} : X \to \overline{\mathbb{R}}, \quad \widehat{f}(x) := \begin{cases} f(x) & \text{if } x \in C, \\ \infty & \text{if } x \in X \setminus C, \end{cases}$$

is convex in the sense of the definition given at the beginning of this section. Of course, we can proceed in the opposite direction; so a proper function $f : X \to \overline{\mathbb{R}}$ is convex if and only if $f|_{\operatorname{dom} f}$ is convex in the above sense. The consideration of (convex) functions with values in $\overline{\mathbb{R}}$ has certain advantages, as we shall often see in the sequel.

Taking into account the equivalence (i) $\Leftrightarrow$ (ii) of Theorem 2.1.1, it is useful to know properties of convex functions of (only) one variable. The following result collects some properties of such convex functions.

Theorem 2.1.5 *Let $f \in \Lambda(\mathbb{R})$ be such that* $\operatorname{int}(\operatorname{dom} f) \neq \emptyset$.

(i) *Let $t_1, t_2 \in \operatorname{dom} f$, $t_1 < t_2$. Suppose that there exists $\lambda_0 \in]0,1[$ such that $f((1-\lambda_0)t_1 + \lambda_0 t_2) = (1-\lambda_0)f(t_1) + \lambda_0 f(t_2)$. Then*

$$\forall \lambda \in [0,1] \ : \ f((1-\lambda)t_1 + \lambda t_2) = (1-\lambda)f(t_1) + \lambda f(t_2), \tag{2.9}$$

$$\forall t \in [t_1, t_2] \ : \ f(t) = \frac{t_2 - t}{t_2 - t_1} f(t_1) + \frac{t - t_1}{t_2 - t_1} f(t_2). \tag{2.10}$$

(ii) *Let $t_0 \in \operatorname{dom} f$. The function*

$$\varphi_{t_0} : \operatorname{dom} f \setminus \{t_0\} \to \mathbb{R}, \quad \varphi_{t_0}(t) := \frac{f(t) - f(t_0)}{t - t_0},$$

is nondecreasing; if f is strictly convex then φ_{t_0} is increasing.

(iii) *Let $t_0 \in \operatorname{dom} f$. The following limits exist:*

$$f'_+(t_0) := \lim_{t \downarrow t_0} \frac{f(t) - f(t_0)}{t - t_0} = \inf_{t > t_0} \frac{f(t) - f(t_0)}{t - t_0} \in \overline{\mathbb{R}}, \tag{2.11}$$

$$f'_-(t_0) := \lim_{t \uparrow t_0} \frac{f(t) - f(t_0)}{t - t_0} = \sup_{t < t_0} \frac{f(t) - f(t_0)}{t - t_0} \in \overline{\mathbb{R}}, \tag{2.12}$$

and

$$f'_-(t_0) \leq f'_+(t_0); \tag{2.13}$$

moreover $f'_-(t_0), f'_+(t_0) \in \mathbb{R}$ whenever $t_0 \in \operatorname{int}(\operatorname{dom} f)$. Therefore f is left and right derivable at any point of $\operatorname{int}(\operatorname{dom} f)$. *Moreover*

$$\tau \in [f'_-(t_0), f'_+(t_0)] \cap \mathbb{R} \Leftrightarrow \forall t \in \mathbb{R} \ : \ \tau(t - t_0) \leq f(t) - f(t_0). \tag{2.14}$$

If f is strictly convex, in relation (2.14) the inequality is strict for $t \neq t_0$.

(iv) *Let $t_1, t_2 \in \operatorname{dom} f$, $t_1 < t_2$. Then $f'_+(t_1) \leq f'_-(t_2)$; if f is strictly convex then $f'_+(t_1) < f'_-(t_2)$. Therefore the functions f'_- and f'_+ are nondecreasing on* $\operatorname{dom} f$. *Furthermore, f is strictly convex if and only if f'_- [resp. f'_+] is increasing on* $\operatorname{int}(\operatorname{dom} f)$.

(v) *The function f is Lipschitz on every compact interval included in* $\operatorname{int}(\operatorname{dom} f)$, *and so f is continuous on* $\operatorname{int}(\operatorname{dom} f)$; *moreover, for every $t_0 \in$* $\operatorname{int}(\operatorname{dom} f)$ *we have:*

$$\lim_{t \uparrow t_0} f'_-(t) = \lim_{t \uparrow t_0} f'_+(t) = f'_-(t_0), \quad \lim_{t \downarrow t_0} f'_+(t) = \lim_{t \downarrow t_0} f'_-(t) = f'_+(t_0).$$

(vi) *The function* f *is monotone on* $\operatorname{int}(\operatorname{dom} f)$, *or there exists* $t_0 \in \operatorname{int}(\operatorname{dom} f)$ *such that* f *is nonincreasing on* $]-\infty, t_0] \cap \operatorname{dom} f$ *and nondecreasing on* $[t_0, \infty[\, \cap \operatorname{dom} f$.

(vii) *Let* $t_0 \in \operatorname{dom} f$ *and* $\lambda \in]0,1[$. *Then the mapping* $\psi : \operatorname{dom} f \to \mathbb{R}$, $\psi(t) := \lambda f(t) - f(t_0 + \lambda(t - t_0))$, *is nonincreasing on* $I_l :=]-\infty, t_0] \cap \operatorname{dom} f$ *and nondecreasing on* $I_r := [t_0, \infty[\, \cap \operatorname{dom} f$. *If* f *is strictly convex then* ψ *is decreasing on* I_l *and increasing on* I_r.

Proof. To begin with, let $t_1, t_2, t_3 \in \operatorname{dom} f$ be such that $t_1 < t_2 < t_3$. Then $t_2 = (1-\lambda)t_1 + \lambda t_3$ with $\lambda = (t_3 - t_2)/(t_3 - t_1) \in]0,1[$; therefore

$$f(t_2) \leq \frac{t_3 - t_2}{t_3 - t_1} f(t_1) + \frac{t_2 - t_1}{t_3 - t_1} f(t_3), \tag{2.15}$$

the inequality being strict if f is strictly convex. Subtracting successively $f(t_1)$, $f(t_2)$, $f(t_3)$ from both members of Eq. (2.15), multiplying then by $\frac{1}{t_2 - t_1}$, $\frac{t_3 - t_1}{(t_2 - t_1)(t_3 - t_2)}$ and $\frac{1}{t_3 - t_2}$, respectively, we obtain

$$\frac{f(t_2) - f(t_1)}{t_2 - t_1} \leq \frac{f(t_3) - f(t_1)}{t_3 - t_1}, \quad \frac{f(t_1) - f(t2)}{t_1 - t_2} \leq \frac{f(t_3) - f(t_2)}{t_3 - t_2},$$
$$\frac{f(t_1) - f(t_3)}{t_1 - t_3} \leq \frac{f(t_2) - f(t_3)}{t_2 - t_3}, \tag{2.16}$$

these inequalities being strict if f is strictly convex.

(i) Let $t_1, t_2 \in \operatorname{dom} f$ and $\lambda_0 \in]0,1[$ be such that $f((1-\lambda_0)t_1 + \lambda_0 t_2) = (1-\lambda_0) f(t_1) + \lambda_0 f(t_2)$. Suppose that there exists $\lambda \in]0,1[$ such that $f((1-\lambda)t_1 + \lambda t_2) < (1-\lambda) f(t_1) + \lambda t_2$; assume that $\lambda < \lambda_0$ (one proceeds similarly for $\lambda > \lambda_0$). Taking $\theta := (1-\lambda_0)/(1-\lambda) \in]0,1[$, we have that $\lambda_0 = \theta\lambda + (1-\theta) \cdot 1$ and $(1-\lambda_0)t_1 + \lambda_0 t_2 = \theta((1-\lambda)t_1 + \lambda t_2) + (1-\theta)t_2$; so we get the contradiction

$$\begin{aligned} f((1-\lambda_0)t_1 + \lambda t_2) &\leq \theta f((1-\lambda)t_1 + \lambda t_2) + (1-\theta) f(t_2) \\ &< \theta[(1-\lambda) f(t_1) + \lambda f(t_2)] + (1-\theta) f(t_2) \\ &= (1-\lambda_0) f(t_1) + \lambda_0 f(t_2). \end{aligned}$$

Therefore Eq. (2.9) holds. Taking $t \in [t_1, t_2]$ and $\lambda = (t - t_1)/(t_2 - t_1) \in [0,1]$ in Eq. (2.9), we obtain Eq. (2.10).

(ii) Let $t_0 \in \operatorname{dom} f$ and $t_1, t_2 \in \operatorname{dom} f \setminus \{t_0\}$, $t_1 < t_2$. Considering successively the cases $t_1 < t_2 < t_0$, $t_1 < t_0 < t_2$, $t_0 < t_1 < t_2$, from Eq. (2.16) we obtain that φ_{t_0} is nondecreasing; if f is strictly convex, φ_{t_0} is even increasing.

(iii) To begin with, let $t_0 \in \operatorname{int}(\operatorname{dom} f)$. Taking into account that φ_{t_0} is nondecreasing on everyone of the nonempty sets $\operatorname{dom} f \cap]t_0, \infty[$ and $\operatorname{dom} f \cap]-\infty, t_0[$, the following limits exist:

$$\lim_{t \downarrow t_0} \varphi_{t_0}(t) := \lim_{t \downarrow t_0} \frac{f(t) - f(t_0)}{t - t_0} = \inf_{t > t_0} \frac{f(t) - f(t_0)}{t - t_0} < \infty,$$
$$\lim_{t \uparrow t_0} \varphi_{t_0}(t) := \lim_{t \uparrow t_0} \frac{f(t) - f(t_0)}{t - t_0} = \sup_{t < t_0} \frac{f(t) - f(t_0)}{t - t_0} > -\infty,$$

and so $f'_+(t_0)$ and $f'_-(t_0)$ exist. Since φ_{t_0} is nondecreasing on $\operatorname{dom} f \setminus \{t_0\}$, the inequality (2.13) holds. What was proved above shows that $f'_-(t_0)$, $f'_+(t_0) \in \mathbb{R}$.

If $t_0 = \max(\operatorname{dom} f)$ then $f(t) = \infty$ for $t > t_0$, whence $f'_+(t_0) = \infty$; of course $f'_-(t_0) \leq \infty$ (the existence of $f'_-(t_0)$ follows from the monotonicity of φ_{t_0}). In the case $t_0 = \min(\operatorname{dom} f)$ we have $f'_-(t_0) = -\infty \leq f'_+(t_0)$.

Let now $\tau \in [f'_-(t_0), f'_+(t_0)] \cap \mathbb{R}$ and $t < t_0 < t'$; from Eqs. (2.11) and (2.12) we have that

$$\frac{f(t) - f(t_0)}{t - t_0} \leq f'_-(t_0) \leq \tau \leq f'_+(t_0) \leq \frac{f(t') - f(t_0)}{t' - t_0}.$$

Note that the first inequality, in the above relation, is strict if the first quantity is finite and the function f is strictly convex; similarly for the last inequality and the last quantity. From these inequalities we have immediately Eq. (2.14), with strict inequality if f is strictly convex. Conversely, assume that $\tau \in \mathbb{R}$ and $\tau(t - t_0) \leq f(t) - f(t_0)$ for every $t \in \mathbb{R}$; dividing by $t - t_0$ in each of the cases $t > t_0$ and $t < t_0$ and taking the limit for $t \to t_0$, we obtain that $\tau \in [f'_-(t_0), f'_+(t_0)]$.

(iv) Let $t_1, t_2 \in \operatorname{dom} f$, $t_1 < t_2$ and consider $t \in]t_1, t_2[$; using Eqs. (2.11), (2.12) and (2.16) we obtain that

$$f'_+(t_1) \leq \frac{f(t) - f(t_1)}{t - t_1} \leq \frac{f(t_2) - f(t_1)}{t_2 - t_1} = \frac{f(t_1) - f(t_2)}{t_1 - t_2} \leq f'_-(t_2),$$

the inequalities being strict if f is strictly convex. Using Eq. (2.13) we get that f'_- and f'_+ are nondecreasing, even increasing if f is strictly convex.

Suppose that f is not strictly convex; then there exists $t_1, t_2 \in \operatorname{dom} f$, $t_1 < t_2$, and $\lambda_0 \in]0, 1[$ such that $f((1 - \lambda_0)t_1 + \lambda_0 t_2) = (1 - \lambda_0)f(t_1) + \lambda_0 f(t_2)$. Therefore Eq. (2.10) holds, whence

$$\forall\, t \in]t_1, t_2[\,:\; f'_-(t) = f'_+(t) = \frac{f(t_2) - f(t_1)}{t_2 - t_1}.$$

Because $]t_1, t_2[\subset \operatorname{int}(\operatorname{dom} f)$, we obtain that f'_- and f'_+ are not increasing on $\operatorname{int}(\operatorname{dom} f)$.

(v) Let $t_1, t_2 \in \operatorname{int}(\operatorname{dom} f)$, $t_1 < t_2$ and $t, t' \in]t_1, t_2[$, $t < t'$. From Eqs. (2.11), (2.12) and (2.16) we get

$$f'_+(t_1) \le \frac{f(t_1) - f(t)}{t_1 - t} \le \frac{f(t') - f(t)}{t' - t} \le \frac{f(t_2) - f(t)}{t_2 - t} \le f'_-(t_2),$$

whence $|f(t') - f(t)| \le M|t' - t|$, where $M := \max\{|f'_+(t_1)|, |f'_-(t_2)|\} \in \mathbb{R}$. Therefore f is Lipschitz on $]t_1, t_2[$. Since every compact sub-interval of $\operatorname{int}(\operatorname{dom} f)$ is contained in an interval $]t_1, t_2[$ with $[t_1, t_2] \subset \operatorname{int}(\operatorname{dom} f)$, we get the desired conclusion.

Let now $t_0 \in \operatorname{dom} f$ be such that f is left-continuous at t_0 (therefore $t_0 > \inf(\operatorname{dom} f)$); for example $t_0 \in \operatorname{int}(\operatorname{dom} f)$. We already know that

$$\forall t \in \operatorname{dom} f \cap]-\infty, t_0[: \ f'_-(t) \le f'_+(t) \le f'_-(t_0) \in]-\infty, \infty].$$

Let $\lambda \in \mathbb{R}$, $\lambda < f'_-(t_0)$; from the definition of the least upper bound and relation (2.12), there exists $t_1 \in \operatorname{dom} f$, $t_1 < t_0$, such that $\lambda < (f(t_1) - f(t_0))/(t_1 - t_0)$. By the left-continuity of f at t_0, there exists $t_2 \in]t_1, t_0[$ such that $\lambda < (f(t_1) - f(t_2))/(t_1 - t_2)$. Let $t \in]t_2, t_1[$. Using again Eq. (2.16) we get

$$\lambda < \frac{f(t) - f(t_2)}{t - t_2} = \frac{f(t_2) - f(t)}{t_2 - t} \le f'_-(t).$$

Therefore $\lim_{t \uparrow t_0} f'_-(t) = \lim_{t \uparrow t_0} f'_+(t) = f'_-(t_0)$. The other formula is obtained similarly.

(vi) To begin with, note that f is nondecreasing (resp. increasing) on $[t_0, \infty[\cap \operatorname{dom} f$ if $f'_+(t_0) \ge (>) 0$. Indeed, if $t_1, t_2 \in \operatorname{dom} f$, $t_0 \le t_1 < t_2$, $0 (<) \le f'_+(t_0) \le f'_+(t_1) \le (f(t_2) - f(t_1))/(t_2 - t_1)$. Similarly, if $f'_-(t_0) \le (<) 0$ then f is (decreasing) nonincreasing on $]-\infty, t_0] \cap \operatorname{dom} f$.

So, from the above arguments, if $f'_+(t) \ge 0$ for every $t \in \operatorname{int}(\operatorname{dom} f)$ then f is nondecreasing on $\operatorname{int}(\operatorname{dom} f)$; if $f'_+(t) \le 0$ for every $t \in \operatorname{int}(\operatorname{dom} f)$, by Eq. (2.13), $f'_-(t) \le 0$ for every $t \in \operatorname{int}(\operatorname{dom} f)$, whence f is nonincreasing on $\operatorname{int}(\operatorname{dom} f)$. If none of these conditions is satisfied, there exist $t_1, t_2 \in \operatorname{int}(\operatorname{dom} f)$, $t_1 < t_2$, such that $f'_+(t_1) < 0 < f'_+(t_2)$. Let $t_0 := \inf\{t \in \operatorname{dom} f \mid f'_+(t) \ge 0\} \in]t_1, t_2]$. It follows that $f'_-(t) < 0$ for every $t \in \operatorname{dom} f$, $t < t_0$, whence f is decreasing on $]-\infty, t_0] \cap \operatorname{dom} f$. By (v) we have that $f'_+(t_0) \ge 0$, whence f is nondecreasing on $[t_0, \infty[\cap \operatorname{dom} f$.

(vii) Let $t_1, t_2 \in I_r$ be such that $t_1 < t_2$. Then

$$t_0 + \lambda(t_1 - t_0) \le t_1 < t_2 \quad \text{and} \quad t_0 + \lambda(t_1 - t_0) < t_0 + \lambda(t_2 - t_0) < t_2.$$

So, from the convexity of f, we have

$$f(t_1) \le \tfrac{t_2 - t_1}{t_2 - (1-\lambda)t_0 - \lambda t_1} f(t_0 + \lambda(t_1 - t_0)) + \tfrac{(1-\lambda)(t_1 - t_0)}{t_2 - (1-\lambda)t_0 - \lambda t_1} f(t_2),$$
$$f(t_0 + \lambda(t_2 - t_0)) \le \tfrac{(1-\lambda)(t_2 - t_0)}{t_2 - (1-\lambda)t_0 - \lambda t_1} f(t_0 + \lambda(t_1 - t_0)) + \tfrac{\lambda(t_2 - t_1)}{t_2 - (1-\lambda)t_0 - \lambda t_1} f(t_2),$$

the second inequality being strict if f is strictly convex. Multiplying the first inequality by λ, then adding them, we get

$$\lambda f(t_1) + f(t_0 + \lambda(t_2 - t_0)) \le f(t_0 + \lambda(t_1 - t_0)) + \lambda f(t_2),$$

i.e. $\psi(t_1) \le \psi(t_2)$, the inequality being strict if f is strictly convex. The proof is similar on I_l. □

The preceding theorem shows that $f|_{\operatorname{dom} f}$ is continuous on $\operatorname{int}(\operatorname{dom} f)$ when $f : \mathbb{R} \to \overline{\mathbb{R}}$ is a proper lower semicontinuous convex function. We have even more.

Proposition 2.1.6 *Let $f \in \Lambda(\mathbb{R})$. Then $f|_{\operatorname{cl}(\operatorname{dom} f)}$ is upper semicontinuous (on $\operatorname{cl}(\operatorname{dom} f)$). Moreover, if f is lower semicontinuous then $f|_{\operatorname{cl}(\operatorname{dom} f)}$ is continuous.*

Proof. As mentioned above, f is continuous at any $t_0 \in \operatorname{int}(\operatorname{dom} f)$. Let $t_0 \in \operatorname{dom} f$. If $\operatorname{dom} f = \{t_0\}$ it is nothing to prove. In the contrary case we can take $t_0 = \inf(\operatorname{dom} f)$ and some $\bar{t} \in \operatorname{dom} f$ with $\bar{t} > t_0$. Let $(t_n) \subset \operatorname{dom} f \setminus \{t_0\}$ converge to t_0. We may assume that $t_n < \bar{t}$ for every $n \in \mathbb{N}$. Suppose first that $t_0 \in \operatorname{dom} f$; then, by Eq. (2.15), $f(t_n) \le \frac{\bar{t} - t_n}{\bar{t} - t_0} f(t_0) + \frac{t_n - t_0}{\bar{t} - t_0} f(\bar{t})$ for every n. Passing to limit superior we get $\limsup f(t_n) \le f(t_0)$. If $t_0 \notin \operatorname{dom} f$ the preceding inequality is obvious. Hence f is usc at t_0. If f is lsc then $f|_{\operatorname{cl}(\operatorname{dom} f)}$ is lsc, too. Hence $f|_{\operatorname{cl}(\operatorname{dom} f)}$ is continuous. □

The next theorem furnishes a useful representation of lsc convex functions on $\mathbb{R}$. This result will motivate the following conventions for a function $f \in \Lambda(\mathbb{R})$:

$$f'_-(t) := f'_+(t) := -\infty \quad \forall t \in \mathbb{R} \setminus \operatorname{dom} f,\ t \le \inf(\operatorname{dom} f),$$
$$f'_-(t) := f'_+(t) := +\infty \quad \forall t \in \mathbb{R} \setminus \operatorname{dom} f,\ t \ge \sup(\operatorname{dom} f).$$

Theorem 2.1.7 *Let $\varphi : \mathbb{R} \to \overline{\mathbb{R}}$ be a non-decreasing function and $a \in \mathbb{R}$ be such that $\varphi(a) \in \mathbb{R}$. Then*

$$f : \mathbb{R} \to \overline{\mathbb{R}}, \quad f(t) := \int_a^t \varphi(s)\, ds,$$

is a proper lower semicontinuous convex function with $I := \{s \in \mathbb{R} \mid \varphi(s) \in \mathbb{R}\} \subset \operatorname{dom} f \subset \operatorname{cl} I$ and

$$f'_- = \varphi_- \le \varphi \le \varphi_+ = f'_+, \tag{2.17}$$

where the integral is taken in Lebesgue sense and

$$\varphi_-(t) := \sup\{\varphi(t') \mid t' < t\}, \quad \varphi_+(t) := \inf\{\varphi(t') \mid t' > t\},$$

for $t \in \mathbb{R}$. Moreover, if $g : \mathbb{R} \to \overline{\mathbb{R}}$ is a proper lower semicontinuous convex function such that $g'_- \le \varphi \le g'_+$ then $g = f + \alpha$ for some $\alpha \in \mathbb{R}$.

Proof. As usual, $\int_a^b \varphi(s)\, ds := -\int_b^a \varphi(s)\, ds$ if $a > b$. The statement is obvious if I is a singleton. So we assume that $\operatorname{int} I \neq \emptyset$. If $t \in \mathbb{R} \setminus \operatorname{cl} I$ it is obvious that $f(t) = \infty$. Hence $I \subset \operatorname{dom} f \subset \operatorname{cl} I$. If $\varphi(b) \in \mathbb{R}$, the integral may be taken in Riemann sense, while for $\bar{t} = \sup I \in \mathbb{R}$ we have $f(\bar{t}) = \lim_{t \uparrow \bar{t}} f(t)$, and similarly for $\underline{t} = \inf I$. Therefore $f|_{\operatorname{cl} I}$ is continuous, whence f is lower semicontinuous. Let $t_0, t_1 \in I$, $t_0 < t_1$ and $\lambda \in]0,1[$. Taking $t_\lambda := (1-\lambda)t_0 + \lambda t_1$, we have that

$$f(t_\lambda) - f(t_0) = \int_{t_0}^{t_\lambda} \varphi(s)\, ds \le (t_\lambda - t_0)\varphi(t_\lambda) = \lambda(t_1 - t_0)\varphi(t_\lambda),$$
$$f(t_1) - f(t_\lambda) = \int_{t_\lambda}^{t_1} \varphi(s)\, ds \ge (t_1 - t_\lambda)\varphi(t_\lambda) = (1-\lambda)(t_1 - t_0)\varphi(t_\lambda).$$

Multiplying the first relation by $(1-\lambda)$ and the second one by $-\lambda$, then adding them, we obtain that $f(t_\lambda) \le (1-\lambda)f(t_0) + \lambda f(t_1)$. Since $f|_{\operatorname{cl} I}$ is continuous we obtain this inequality also for $t_0, t_1 \in \operatorname{dom} f$ and $\lambda \in]0,1[$; hence f is convex. Let $t_0 \in \operatorname{dom} f \cap \operatorname{cl} I$ and $t \in \mathbb{R}$, $t > t_0$; because $\varphi(s) \ge \varphi_+(t_0)$ for $s > t_0$, we have that

$$\varphi(t) \ge \frac{f(t) - f(t_0)}{t - t_0} = \frac{1}{t - t_0} \int_{t_0}^t \varphi(s)\, ds \ge \varphi_+(t_0),$$

and so $f'(t_0) = \varphi_+(t_0)$. Similarly, $f'_-(t_0) = \varphi_-(t_0)$. Since these relations are obvious for $t_0 \notin \operatorname{dom} f \cap \operatorname{cl} I$, Eq. (2.17) holds.

Let now $g : \mathbb{R} \to \overline{\mathbb{R}}$ be a proper lsc convex function such that $g'_- \le \varphi \le g'_+$. It follows that $\operatorname{int}(\operatorname{dom} g) = \operatorname{int} I$. Taking into account Theorem 2.1.5(v) we obtain that $g'_-(t) = \varphi_-(t)$ and $g'_+(t) = \varphi_+(t)$ for $t \in \mathbb{R} \setminus \{\underline{t}, \bar{t}\}$, and so $g'_-(t) = f'_-(t)$ and $g'_+(t) = f'_+(t)$ for every $t \in \operatorname{int} I$. Taking $h : \operatorname{int} I \to \mathbb{R}$, $h(t) := f(t) - g(t)$, we have that h is derivable and $h' = 0$. It follows that h is a constant function. Therefore there exists $\alpha \in \mathbb{R}$ such that $g(t) = f(t) + \alpha$ for all $t \in \operatorname{int} I$. Using the preceding proposition it follows that $g = f + \alpha$. □

The following consequence will be used several times.

Corollary 2.1.8 *Let $f : \mathbb{R} \to \overline{\mathbb{R}}$ be a proper convex function. Then*

$$\forall t, t' \in \operatorname{int}(\operatorname{dom} f) \ : \ f(t') - f(t) = \int_t^{t'} f'_+(s)\, ds = \int_t^{t'} f'_-(s)\, ds.$$

Moreover, if f is lower semicontinuous, then the preceding equalities hold for all $t, t' \in \operatorname{dom} f$.

Proof. Assume that $\operatorname{int}(\operatorname{dom} f) \neq \emptyset$. When $\overline{t} := \sup(\operatorname{dom} f) \in \mathbb{R}$ or $\underline{t} := \inf(\operatorname{dom} f) \in \mathbb{R}$, we replace, if necessary, $f(\overline{t})$ by $\lim_{t \uparrow \overline{t}} f(t)$ and $f(\underline{t})$ by $\lim_{t \downarrow \underline{t}} f(t)$; then f is lower semicontinuous. Applying the preceding theorem for $\varphi = f'_+$ and $\varphi = f'_-$ the conclusion follows. □

The following result is used frequently to show that a function of one variable is convex.

Theorem 2.1.9 *Let $I \subset \mathbb{R}$ be a nonempty open interval and $f : I \to \mathbb{R}$ be a derivable function. The following statements are equivalent:*

(i) *f is convex;*

(ii) $\forall t, s \in I \ : \ f'(s) \cdot (t - s) \leq f(t) - f(s)$;

(iii) $\forall t, s \in I \ : \ \big(f'(t) - f'(s)\big) \cdot (t - s) \geq 0$, *i.e. f' is nondecreasing;*

(iv) *(if f is twice derivable on I)* $\forall t \in I \ : \ f''(t) \geq 0$.

Proof. (i) ⇒ (ii) Let f be convex. Since f is derivable we have that $f'(s) = f'_-(s) = f'_+(s)$ for every $s \in I$. The conclusion follows from Eq. (2.14) taking $\tau = f'(s)$.

(ii) ⇒ (iii) Let $s, t \in I$. By hypothesis we have that

$$f'(s) \cdot (t - s) \leq f(t) - f(s), \quad f'(t) \cdot (s - t) \leq f(s) - f(t).$$

Adding these two inequalities side by side we obtain the conclusion.

(iii) $\Rightarrow$ (i) Let $t_1, t_2 \in I$, $t_1 < t_2$, $\lambda \in]0,1[$ and $t_\lambda := (1-\lambda)t_1 + \lambda t_2 \in]t_1, t_2[$. Then

$$\begin{aligned}
&(1-\lambda)f(t_1) + \lambda f(t_2) - f(t_\lambda) \\
&\quad = (1-\lambda)[f(t_1) - f(t_\lambda)] + \lambda(f(t_2) - f(t_\lambda)) \\
&\quad = (1-\lambda)f'(\tau_1)(t_1 - t_\lambda) + \lambda f'(\tau_2)(t_2 - t_\lambda) \\
&\quad = \lambda(1-\lambda)f'(\tau_1)(t_1 - t_2) + \lambda(1-\lambda)f'(\tau_2)(t_2 - t_1) \\
&\quad = \lambda(1-\lambda)(t_1 - t_2)(f'(\tau_1) - f'(\tau_2)) \geq 0,
\end{aligned}$$

with $\tau_1 \in]t_1, t_\lambda[$, $\tau_2 \in]t_\lambda, t_2[$ (obtained by using the Lagrange theorem); of course, we have used the fact that f' is nondecreasing.

Suppose now that f is derivable of order 2 on I. In this case (iii) $\Leftrightarrow$ (iv) by a known consequence of the Lagrange theorem. □

Theorem 2.1.10 *Let $I \subset \mathbb{R}$ be an open interval and $f : I \to \mathbb{R}$ be a derivable function. The following statements are equivalent:*

(i) *f is strictly convex;*

(ii) $\forall t, s \in I,\ t \neq s\ :\ f'(s) \cdot (t-s) < f(t) - f(s)$;

(iii) $\forall t, s \in I,\ t \neq s\ :\ \big(f'(t) - f'(s)\big) \cdot (t-s) > 0$, *i.e. f' is increasing;*

(iv) *(if f is twice derivable on I) $\forall t \in I\ :\ f''(t) \geq 0$ and $\{t \in I \mid f''(t) = 0\}$ does not contain any nontrivial interval.*

Proof. The proof is completely similar to that of the preceding theorem; just replace, where necessary, the inequalities by strict inequalities and, of course, use the properties of increasing functions. □

Similar characterizations to those of the above theorems can be immediately formulated for concave and strictly concave functions.

As immediate applications of the above two theorems we obtain that: (i) $f_1 : \mathbb{R} \to \mathbb{R}$, $f_1(t) := |t|^p$, where $p \in]1, \infty[$, is strictly convex; (ii) $f_2 : \mathbb{R}_+ \to \mathbb{R}$, $f_2(t) := t^p$, where $p \in]0,1[$, is strictly concave and increasing; (iii) $f_3 : \mathbb{P} \to \mathbb{R}$, $f_3(t) := t^p$, where $p \in \mathbb{R}^*_-$, is strictly convex and decreasing; (iv) $f_4 : \mathbb{R} \to \mathbb{R}$, $f_4(t) := t^p$ if $t \geq 0$, $f_4(t) := 0$ if $t < 0$, where $p \in]1, \infty[$, is convex and nondecreasing; (v) $f_5 : \mathbb{R} \to \mathbb{R}$, $f_5(t) := \exp(t)$, is strictly convex and increasing; (vi) $f_6 : \mathbb{P} \to \mathbb{R}$, $f_6(t) := \ln t$, is strictly concave and increasing; (vii) $f_7 : \mathbb{R}_+ \to \mathbb{R}$, $f_7(t) := t \ln t$ if $t > 0$, $f_7(0) := 0$, is strictly convex; (viii) $f_8 : \mathbb{R} \to \mathbb{R}$, $f_8(t) := \sqrt{1+t^2}$, is strictly convex.

In the following two theorems we characterize the convexity of Gâteaux differentiable functions defined on linear normed spaces.

Theorem 2.1.11 *Let $D \subset (X, \|.\|)$ be a nonempty, convex and open set, and $f : D \to \mathbb{R}$ be a Gâteaux differentiable function (on D). The following statements are equivalent:*

(i) *f is convex;*

(ii) $\forall\, x, y \in D \;:\; \langle y - x, \nabla f(x)\rangle \le f(y) - f(x)$;

(iii) $\forall\, x, y \in D \;:\; \langle y - x, \nabla f(y) - \nabla f(x)\rangle \ge 0$;

(iv) $\forall\, x \in D,\ \forall\, u \in X \;:\; \nabla^2 f(x)(u, u) \ge 0$ *(when f is twice Gâteaux differentiable on D).*

Proof. For $x, y \in D$ consider $I_{x,y} := \{t \in \mathbb{R} \mid (1 - t)x + ty \in D\}$; one proves without difficulty that $I_{x,y}$ is an open (since D is open) interval (since D is convex) and that $[0, 1] \subset I_{x,y}$. Let us consider the function

$$\varphi_{x,y} : I_{x,y} \to \mathbb{R}, \quad \varphi_{x,y}(t) := f((1 - t)x + ty). \tag{2.18}$$

It is well-known that

$$\varphi'_{x,y}(t) = \nabla f((1-t)x+ty)(y-x), \quad \varphi''_{x,y}(t) = \nabla^2 f((1-t)x+ty)(y-x, y-x) \tag{2.19}$$

for every $t \in I_{x,y}$; of course, the second formula is valid when f is twice Gâteaux differentiable on D.

(i) $\Rightarrow$ (ii) Let $x, y \in D$. By Theorem 2.1.1 we have that $\varphi_{x,y}$ is convex; using then Theorem 2.1.9 we obtain that $\varphi'_{x,y}(0)(1-0) \le \varphi_{x,y}(1) - \varphi_{x,y}(0)$, whence the conclusion.

(ii) $\Rightarrow$ (iii) Writing the hypothesis for the couples (x, y) and (y, x) we obtain immediately the conclusion.

(iii) $\Rightarrow$ (i) Let $x, y \in D$ and $t, s \in I_{x,y}$, $s < t$; then

$$\begin{aligned} 0 &\le (t - s)\big(\varphi'(t) - \varphi'(s)\big) \\ &= \langle (1 - t)x + ty - (1 - s)x - sy, \nabla f((1 - t)x + ty) - \nabla f((1 - s)x + sy)\rangle. \end{aligned}$$

Using Theorem 2.1.9 we obtain that $\varphi_{x,y}$ is convex, whence, by Theorem 2.1.1, f is convex.

Suppose now that f is twice Gâteaux differentiable on D.

(i) $\Rightarrow$ (iv) Let $x \in D$ and $u \in X$. Taking into account that D is open, $D - x$ is absorbing; hence there exists $\alpha > 0$ such that $y := x + \alpha u \in D$. From the hypothesis we have that $\varphi_{x,y}$ is convex, whence by Theorem 2.1.9, $\varphi''_{x,y}(t) \ge 0$ for every $t \in I_{x,y}$. In particular

$$\varphi''_{x,y}(0) = \nabla^2 f(x)(y - x, y - x) = \alpha^2\, \nabla^2 f(x)(u, u) \ge 0.$$

Therefore the conclusion holds.

(iv) $\Rightarrow$ (i) Let $x, y \in D$. For every $t \in I_{x,y}$, using Eq. (2.19), we have $\varphi''_{x,y}(t) \geq 0$. Applying again Theorem 2.1.9 we obtain that $\varphi_{x,y}$ is convex, and so f is convex. □

In the next result we give characterizations for strictly convex functions.

Theorem 2.1.12 *Let $D \subset (X, \|\cdot\|)$ be a nonempty, convex and open set, and $f : D \to \mathbb{R}$ be Gâteaux differentiable (on D). Then* (i) $\Leftrightarrow$ (ii) $\Leftrightarrow$ (iii) $\Leftarrow$ (iv), *where*

(i) *f is strictly convex;*

(ii) $\forall x, y \in D,\ x \neq y\ :\ \langle y - x, \nabla f(x) \rangle < f(y) - f(x)$;

(iii) $\forall x, y \in D, x \neq y\ :\ \langle y - x, \nabla f(y) - \nabla f(x) \rangle > 0$;

(iv) $\forall x \in D,\ \forall u \in X \setminus \{0\}\ :\ \nabla^2 f(x)(u, u) > 0$ *(when f is twice Gâteaux differentiable on D).*

Proof. Of course, using Theorem 2.1.1, we know that f is strictly convex if and only if the function $\varphi_{x,y}$ given by Eq. (2.18) is strictly convex for all $x, y \in D$, $x \neq y$. The proof is similar to that of the preceding theorem, using Theorem 2.1.10 instead of Theorem 2.1.9. □

A remarkable property of convex functions is given below.

Theorem 2.1.13 *Let $f \in \Lambda(X)$ and $x \in \operatorname{dom} f$. Then for every $u \in X$ there exists*

$$f'(x, u) := \lim_{t \downarrow 0} \frac{f(x + tu) - f(x)}{t} = \inf_{t > 0} \frac{f(x + tu) - f(x)}{t} \in \overline{\mathbb{R}}, \tag{2.20}$$

and

$$\forall u \in X\ :\ f'(x, u) \leq f(x + u) - f(x), \tag{2.21}$$

the inequality being strict if f is strictly convex, $u \neq 0$ and $f'(x, u) < \infty$. Moreover $f'(x, \cdot)$ is sublinear and $\operatorname{dom} f'(x, \cdot) = \operatorname{cone}(\operatorname{dom} f - x)$. *If $x \in {}^{i}(\operatorname{dom} f)$ then $f'(x, \cdot)$ is proper, while if $x \in (\operatorname{dom} f)^{i}$ then $f'(x, u) \in \mathbb{R}$ for every $u \in X$.*

Proof. Let $u \in X$ and $\psi : \mathbb{R} \to \overline{\mathbb{R}}$, $\psi(t) := f(x + tu)$. Taking into account that $\psi = \varphi_{x,x+u}$ ($\varphi_{x,y}$ being constructed in Theorem 2.1.1), ψ is convex (even strictly convex if f is strictly convex and $u \neq 0$). From Theorem

2.1.5 we obtain the existence of

$$\psi'_+(0) = \lim_{t \downarrow 0} \frac{\psi(t) - \psi(0)}{t - 0} = \inf_{t > 0} \frac{\psi(t) - \psi(0)}{t - 0},$$

i.e. Eq. (2.20) holds. Using again Theorem 2.1.5 we obtain Eq. (2.21), with the corresponding variant for f strictly convex and $u \neq 0$.

It is obvious that $f'(x, 0) = 0$ and that $f'(x, \lambda u) = \lambda f'(x, u)$ for every $u \in X$ and $\lambda > 0$. Let now $u, v \in X$. We have

$$f(x + t(u + v)) = f\left(\tfrac{1}{2}(x + 2tu) + \tfrac{1}{2}(x + 2tv)\right) \leq \tfrac{1}{2} f(x + 2tu) + \tfrac{1}{2} f(x + 2tv),$$

and so

$$\frac{f(x + t(u + v)) - f(x)}{t} \leq \frac{f(x + 2tu) - f(x)}{2t} + \frac{f(x + 2tv) - f(x)}{2t}.$$

Letting $t \downarrow 0$ we obtain

$$\forall u, v \in X \ : \ f'(x, u + v) \leq f'(x, u) + f'(x, v).$$

Therefore $f'(x, \cdot)$ is sublinear.

Note that $\operatorname{dom} f'(x, \cdot) = \mathbb{R}_+ \cdot (\operatorname{dom} f - x) = \operatorname{cone}(\operatorname{dom} f - x)$. If $x \in {}^i(\operatorname{dom} f)$ then $\operatorname{dom} f'(x, \cdot) = \operatorname{lin}(\operatorname{dom} f - x)$, whence $0 \in {}^i(\operatorname{dom} f'(x, \cdot))$. From Proposition 2.1.4 we have that $f'(x, \cdot)$ is proper. If $x \in (\operatorname{dom} f)^i$ then $\operatorname{dom} f'(x, \cdot) = X$ and $f'(x, \cdot)$ is proper, which proves that $f'(x, u) \in \mathbb{R}$ for every $u \in X$. □

The number $f'(x, u) \in \overline{\mathbb{R}}$ is called the **directional derivative** of f at x in the direction u. It is possible that $f'(x, \cdot)$ take the value $-\infty$. Consider the function $f : \mathbb{R} \to \overline{\mathbb{R}}$, $f(t) := -\sqrt{1 - t^2}$ if $|t| \leq 1$, $f(t) := \infty$ if $|t| > 1$; we have $f'(-1, t) = -\infty$ for every $t > 0$ (Exercise!).

The preceding result can be extended in a significant way.

Theorem 2.1.14 *Let $f \in \Lambda(X)$, $x \in \operatorname{dom} f$ and $\varepsilon \in \mathbb{R}_+$. Then the function*

$$f'_\varepsilon(x, \cdot) : X \to \overline{\mathbb{R}}, \quad f'_\varepsilon(x, u) := \inf_{t > 0} \frac{f(x + tu) - f(x) + \varepsilon}{t},$$

is sublinear,

$$\operatorname{dom} f'_\varepsilon(x, \cdot) = \operatorname{cone}(\operatorname{dom} f - x), \tag{2.22}$$

and

$$f'_\varepsilon(x,u) \le f(x+u) - f(x) + \varepsilon \quad \forall\, u \in X, \tag{2.23}$$
$$f'(x,u) = \lim_{\varepsilon \downarrow 0} f'_\varepsilon(x,u) = \inf_{\varepsilon>0} f'_\varepsilon(x,u) \quad \forall\, u \in X.$$

Furthermore, if $x \in {}^i(\operatorname{dom} f)$ *then* $f'_\varepsilon(x,\cdot)$ *is proper, while if* $x \in (\operatorname{dom} f)^i$ *then* $f'_\varepsilon(x,u) \in \mathbb{R}$ *for every* $u \in X$.

Proof. From the definition of $f'_\varepsilon(x,\cdot)$ it is clear that $f'_\varepsilon(x,u) < \infty$ if and only if there exists $t > 0$ such that $x + tu \in \operatorname{dom} f$, *i.e.* $u \in \operatorname{cone}(\operatorname{dom} f - x)$. Hence Eq. (2.22) holds.

It is obvious that $f'_\varepsilon(x,0) = 0$ and for $\lambda > 0$

$$f'_\varepsilon(x,\lambda u) = \inf_{t>0} \frac{f(x+t\lambda u) - f(x) + \varepsilon}{t\lambda} \cdot \lambda = \lambda f'_\varepsilon(x,u).$$

Let now $u, v \in X$ and $s, t > 0$. Then

$$\begin{aligned} f\left(x + \tfrac{st}{s+t}(u+v)\right) &= f\left(\tfrac{t}{s+t}(x+su) + \tfrac{s}{s+t}(x+tv)\right) \\ &\le \tfrac{t}{s+t} f(x+su) + \tfrac{s}{s+t} f(x+tv). \end{aligned}$$

It follows that

$$\begin{aligned} &\left(f\left(x + \tfrac{st}{s+t}(u+v)\right) - f(x) + \varepsilon\right) \Big/ \tfrac{st}{s+t} \\ &\qquad \le \frac{f(x+su) - f(x) + \varepsilon}{s} + \frac{f(x+tv) - f(x) + \varepsilon}{t}. \end{aligned}$$

Therefore, for all $s, t > 0$ we have

$$f'_\varepsilon(x,u+v) \le \frac{f(x+su) - f(x) + \varepsilon}{s} + \frac{f(x+tv) - f(x) + \varepsilon}{t}.$$

Taking the infimum in the right-hand side, successively with respect to s and t, we obtain that

$$f'_\varepsilon(x,u+v) \le f'_\varepsilon(x,u) + f'_\varepsilon(x,v).$$

Letting $t = 1$ in the definition of $f'_\varepsilon(x,u)$ we get Eq. (2.23). On the other hand, observing that for $0 \le \varepsilon_1 \le \varepsilon_2 < \infty$ the inequalities

$$\forall\, u \in X \;:\; f'_0(x,u) = f'(x,u) \le f'_{\varepsilon_1}(x,u) \le f'_{\varepsilon_2}(x,u)$$

hold, we obtain

$$\lim_{\varepsilon\downarrow 0} f'_\varepsilon(x,u) = \inf_{\varepsilon>0} f'_\varepsilon(x,u) = \inf_{\varepsilon>0}\inf_{t>0} \frac{f(x+tu)-f(x)+\varepsilon}{t}$$
$$= \inf_{t>0}\inf_{\varepsilon>0} \frac{f(x+tu)-f(x)+\varepsilon}{t} = \inf_{t>0} \frac{f(x+tu)-f(x)}{t}$$
$$= f'(x,u)$$

for every $u \in X$. The other conclusions are proved like in the preceding theorem. □

The number $f'_\varepsilon(x,u) \in \overline{\mathbb{R}}$ is called the **ε-directional derivative** of f at x in the direction u.

We end this section with a characterization of convex functions in arbitrary topological vector spaces using a generalized directional derivative. We shall give other characterizations in Section 3.2 using generalized subdifferentials. So, let X be a topological vector space and $f : X \to \overline{\mathbb{R}}$ a proper function. We define the **upper Dini directional derivative** of f at $x \in X$ in the direction $u \in X$ by

$$\overline{D}f(x,u) := \begin{cases} \limsup_{t\downarrow 0} t^{-1}\big(f(x+tu)-f(x)\big) & \text{if} \quad x \in \operatorname{dom} f, \\ -\infty & \text{otherwise.} \end{cases}$$

When f is convex and $x \in \operatorname{dom} f$ from Theorem 2.1.13 we have that $\overline{D}f(x,u) = f'(x,u)$ for all $x \in \operatorname{dom} f$ and $u \in X$.

For the proof of the announced characterization we need the following auxiliary result which is interesting in itself, too.

Lemma 2.1.15 *Let $a, b \in \mathbb{R}$, $a < b$, and $\varphi : [a,b] \to \overline{\mathbb{R}}$ be a lower semicontinuous proper function with $\varphi(a) < \varphi(b)$. Then there exists $c \in [a,b[$ such that $\overline{D}\varphi(c,1) > 0$.*

Proof. By contradiction, assume that $\overline{D}\varphi(t,1) \le 0$ for every $t \in [a,b[$. Fix $\alpha > 0$; we have that $\varphi(t) \le \varphi(a) + \alpha(t-a)$ for all $t \in [a,b]$. Indeed, because $\overline{D}\varphi(a,1) \le 0 < \alpha$, there exists $t' \in]a,b]$ such that $\varphi(t) \le \varphi(a) + \alpha(t-a)$ for all $t \in [a,t'[$. Let

$$\bar{t} := \sup\left\{\tilde{t} \in]a,b] \mid \varphi(t) \le \varphi(a) + \alpha(t-a)\ \forall\, t \in [a,\tilde{t}[\right\} \in]a,b].$$

Because φ is lsc, it follows that $\varphi(\bar{t}) \le \varphi(a) + \alpha(\bar{t}-a)$. Assume that $\bar{t} < b$. Then, as above, there exists $t'' \in]\bar{t},b]$ such that $\varphi(t) \le \varphi(\bar{t}) + \alpha(t-\bar{t})$ for all $t \in [\bar{t},t''[$. Using the preceding inequalities we obtain that $\varphi(t) \le$

$\varphi(a) + \alpha(\bar{t} - a) + \alpha(t - \bar{t}) = \varphi(a) + \alpha(t - a)$ for $t \in [\bar{t}, t''[$, contradicting the choice of $\bar{t}$. Hence $\bar{t} = b$, and so desired inequality is proved. As the inequality $\varphi(b) \le \varphi(a) + \alpha(b - a)$ holds for all $\alpha > 0$, we obtain the contradiction $\varphi(b) \le \varphi(a)$. The conclusion holds. □

Theorem 2.1.16 *Let X be a topological vector space and $f : X \to \overline{\mathbb{R}}$ be a proper lower semicontinuous function. Then f is convex if and only if*

$$\overline{D}f(x, y - x) + \overline{D}f(y, x - y) \le 0$$

for all $x, y \in X$ for which the sum makes sense.

Proof. Assume first that f is convex. If $x \notin \operatorname{dom} f$ then $\overline{D}f(x, y - x) = -\infty$, and so $\overline{D}f(x, y-x) + \overline{D}f(y, x-y)$ equals $-\infty$ or does not make sense. Assume that $x, y \in \operatorname{dom} f$. Then

$$\overline{D}f(x, y - x) = f'(x, y - x) \le f(y) - f(x),$$
$$\overline{D}f(y, x - y) = f'(y, x - y) \le f(x) - f(y);$$

summing the inequalities side by side we get the conclusion.

We prove the sufficiency by contradiction. Assume that f is not convex. Then there exists $x, y \in \operatorname{dom} f$, $x \ne y$, and $\lambda \in]0, 1[$ such that $f((1 - \lambda)x + \lambda y) > (1 - \lambda)f(x) + \lambda f(y)$. Let $\varphi : [0, \lambda] \to \overline{\mathbb{R}}$ and $\psi : [0, 1 - \lambda] \to \overline{\mathbb{R}}$ be defined by

$$\varphi(t) := f((1 - t)x + ty) - (1 - t)f(x) - tf(y),$$
$$\psi(t) := f(tx + (1 - t)y) - tf(x) - (1 - t)f(y).$$

Of course, φ and ψ are lsc, f being so. Then $\varphi(0) = 0 < \varphi(\lambda)$, $\psi(0) = 0 < \psi(1 - \lambda) = \varphi(\lambda)$, and

$$\overline{D}\varphi(t, 1) = \overline{D}f((1 - t)x + ty, y - x) + f(x) - f(y) \quad \forall t \in [0, \lambda[,$$
$$\overline{D}\psi(t, 1) = \overline{D}f(tx + (1 - t)y, x - y) + f(y) - f(x) \quad \forall t \in [0, 1 - \lambda[.$$

From the preceding lemma we get $t_1 \in [0, \lambda[$ and $t_2 \in [0, 1 - \lambda[$ such that $\overline{D}\varphi(t_1, 1) > 0$ and $\overline{D}\psi(t_2, 1) > 0$. Taking $u := (1 - t_1)x + t_1 y$ and $v := t_2 x + (1 - t_2)y$, we have that $u - v = s(x - y)$ with $s := 1 - t_1 - t_2 \in]0, 1]$. Moreover,

$$\begin{aligned}
&\overline{D}f(u, v - u) + \overline{D}f(v, u - v) \\
&\quad = s^{-1}\left(\overline{D}\varphi(t_1, 1) - f(x) + f(y)\right) + s^{-1}\left(\overline{D}\psi(t_2, 1) + f(x) - f(y)\right) \\
&\quad = s^{-1}\left(\overline{D}\varphi(t_1, 1) + \overline{D}\psi(t_2, 1)\right) > 0,
\end{aligned}$$

a contradiction. The proof is complete. □

2.2 Semi-Continuity of Convex Functions

In this section X is a separated locally convex space if not stated explicitly otherwise. We begin with some characterizations of lower semicontinuous convex functions.

Theorem 2.2.1 *Let $f : X \to \overline{\mathbb{R}}$. The following conditions are equivalent:*

(i) *f is convex and lower semicontinuous;*

(ii) *f is convex and w-lower semicontinuous;*

(iii) $\operatorname{epi} f$ *is convex and closed;*

(iv) $\operatorname{epi} f$ *is convex and w-closed.*

Proof. It is well-known that: f is lsc $\Leftrightarrow$ $\operatorname{epi} f$ is closed in $X \times \mathbb{R}$ $\Leftrightarrow$ $[f \leq \lambda]$ is closed $\forall \lambda \in \mathbb{R}$. The equivalence of conditions (i)–(iv) follows immediately applying Theorem 2.1.1. □

In the sequel we shall denote by $\Gamma(X)$ the class of proper lower semicontinuous convex functions on X.

The following criterion for convexity is sometimes useful.

Theorem 2.2.2 *Let $f : X \to \overline{\mathbb{R}}$ be a proper function satisfying the following conditions:* (i) $f(0) = 0$, (ii) $\forall x \in X$, $\forall \lambda \in \mathbb{P}$: $f(\lambda x) = \lambda f(x)$, (iii) *$f$ is quasi-convex. Suppose that either* (a) *or* (b) *holds, where* (a) $\forall x \in X$: $f(x) \geq 0$, (b) $\operatorname{dom} f \subset \operatorname{cl}\{x \in X \mid f(x) < 0\}$ *and f is lsc at every point $x \in \operatorname{dom} f$. Then f is sublinear.*

Proof. Note first that $f(x + y) \leq f(x) + f(y)$ if $x, y \in \operatorname{dom} f$ and either $f(x) \cdot f(y) \geq 0$ or $0 = f(x) \leq f(y)$. Indeed, if $f(x) \cdot f(y) > 0$ then

$$f\left(\tfrac{f(x)}{f(x)+f(y)} \cdot \tfrac{f(y)}{f(x)} \cdot x + \tfrac{f(y)}{f(x)+f(y)} \cdot y\right) \leq \max\left\{f\left(\tfrac{f(y)}{f(x)}x\right), f(y)\right\} = f(y),$$

whence $f(x+y) \leq f(x) + f(y)$. If $0 = f(x) \leq f(y)$, for $n \in \mathbb{N}$ we have that

$$\tfrac{n}{n+1} f(x+y) = f\left(\tfrac{1}{n+1} nx + \tfrac{n}{n+1} y\right) \leq \max\{f(nx), f(y)\} = f(x) + f(y).$$

Taking the limit for $n \to \infty$ we obtain again that $f(x+y) \leq f(x) + f(y)$.

From the hypothesis we have that $\operatorname{dom} f$ is a convex cone. So, it is sufficient to prove that $f(x+y) \leq f(x) + f(y)$ for all $x, y \in \operatorname{dom} f$. This is already done in case (a).

Suppose that (b) holds. Let $A := \{x \in X \mid f(x) < 0\} \subset \operatorname{dom} f$. If $x \in \operatorname{dom} f \setminus A$, by hypothesis, there exists a net $(x_i)_{i\in I} \subset A$, with $(x_i) \to x$. Moreover,

$$0 \leq f(x) \leq \liminf_{i\in I} f(x_i) \leq \limsup_{i\in I} f(x_i) \leq 0,$$

whence $(f(x_i)) \to f(x) = 0$. Therefore $f(x) \leq 0$ for every $x \in \operatorname{dom} f$.

Let $x, y \in \operatorname{dom} f$. Since the roles of x, y are symmetric, we have to consider the following three situations: α) $x, y \in A$, β) $x, y \in \operatorname{dom} f \setminus A$ and γ) $x \in \operatorname{dom} f \setminus A$ and $y \in A$. In the case α) we have that $f(x) \cdot f(y) > 0$, and so $f(x+y) \leq f(x) + f(y)$. In the case β) we have that $f(x) = f(y) = 0$ and $x + y \in \operatorname{dom} f$, whence $f(x + y) \leq 0 = f(x) + f(y)$. In the case γ) there exists a net $(x_i)_{i\in I} \subset A$ such that $(x_i) \to x$; then $(f(x_i)) \to 0$. Since $x + y \in \operatorname{dom} f$ and $f(x_i + y) \leq f(x_i) + f(y)$ for every $i \in I$ [by α)], taking the limit inferior we obtain that $f(x + y) \leq f(x) + f(y)$. □

Corollary 2.2.3 *Let $f : X \to \overline{\mathbb{R}}$ be a quasi-convex, proper, positively homogeneous and lsc function, $f_+ := f \vee 0$ and $g := f + \iota_{\operatorname{cl} A}$, where $A := \{x \in X \mid f(x) < 0\} \cup \{0\}$. Then f_+ and g are sublinear and $f = f_+ \wedge g$.*

Proof. It is obvious that f_+ and g are quasi-convex, lsc and positively homogeneous. The sublinearity of f_+ and g follows applying the preceding theorem. Because $f(x) \leq 0$ for $x \in \operatorname{cl} A$, we have also that $f = f_+ \wedge g$. □

A result of the same type is given by the following corollary.

Corollary 2.2.4 *Let $f : X \to \mathbb{R}$ be a quasi-convex and positively homogeneous function. If f is upper bounded on a neighborhood of the origin or* $\dim X < \infty$, *then f_+ is sublinear.*

Proof. Suppose that there exists $\lambda_0 > 0$ such that $[f \leq \lambda_0]$ is a neighborhood of 0. Since $[f \leq \lambda] = \frac{\lambda}{\lambda_0}[f \leq \lambda_0]$ for every $\lambda > 0$, we have $0 \in \operatorname{int}[f \leq \lambda]$ for every $\lambda > 0$. Let $0 < \lambda < \mu$ and $x \in \operatorname{cl}[f \leq \lambda]$. Since $0 \in \operatorname{int}[f \leq \lambda]$, from Theorem 1.1.2 we have that $\frac{\lambda}{\mu}x \in \operatorname{int}[f \leq \lambda]$. Hence

$$x \in \tfrac{\mu}{\lambda}\operatorname{int}[f \leq \lambda] = \operatorname{int}\left(\tfrac{\mu}{\lambda}[f \leq \lambda]\right) = \operatorname{int}[f \leq \mu] \subset [f \leq \mu].$$

Therefore

$$\forall \lambda > 0 \ : \ \operatorname{cl}[f \leq \lambda] \subset \bigcap\nolimits_{\mu>\lambda}[f \leq \mu] = [f \leq \lambda].$$

Hence $[f \leq \lambda]$ is a closed set for every $\lambda > 0$. Since $[f \leq 0] = \bigcap_{\mu>0}[f \leq \mu]$, the set $[f \leq 0]$ is closed, too. Since $[f_+ \leq \lambda] = [f \leq \lambda]$ for $\lambda \geq 0$ and

$[f_+ \le \lambda] = \emptyset$ for every $\lambda < 0$, it follows that f_+ is lsc. Taking into account that f_+ is quasi-convex, f_+ is sublinear (applying Theorem 2.2.2). Suppose now that $\dim X < \infty$. By hypothesis, $[f \le 1]$ is convex and absorbing. Since $\dim X < \infty$, as observed at the end of Section 1.1, we obtain that $0 \in \operatorname{int}[f \le 1]$. The conclusion follows as above. □

Note that, under the conditions of the preceding theorem, f_+ is continuous (see Theorems 2.2.9 and 2.2.21).

A result similar to that of Proposition 2.1.4 is the following.

Proposition 2.2.5 *Let $f : X \to \overline{\mathbb{R}}$ be a lsc convex function. If there exists $x_0 \in X$ such that $f(x_0) = -\infty$, then $f(x) = -\infty$ for every $x \in \operatorname{dom} f$. In particular, if f is sublinear then f is proper.*

Proof. Suppose that there exists $x \in \operatorname{dom} f$ such that $f(x) =: t \in \mathbb{R}$. Then $(x, t) \in \operatorname{epi} f$ and $(x_0, t - n) \in \operatorname{epi} f$ for every $n \in \mathbb{N}$. It follows that

$$\forall n \in \mathbb{N} \; : \; \tfrac{1}{n}(x_0, t - n) + (1 - \tfrac{1}{n})(x, t) = \left(\tfrac{1}{n}x_0 + \tfrac{n-1}{n}x, t - 1\right) \in \operatorname{epi} f;$$

therefore $(x, t - 1) \in \operatorname{cl}(\operatorname{epi} f) = \operatorname{epi} f$, whence the contradiction $f(x) \le f(x) - 1$. □

Propositions 2.1.4 and 2.2.5 motivate the consideration, in the sequel, of (almost only) proper convex functions.

It is useful to consider the **lower semicontinuous envelope** or **lower semicontinuous regularization** $\overline{f} := \varphi_{\operatorname{cl}(\operatorname{epi} f)}$ of the function $f : X \to \overline{\mathbb{R}}$ (see Eq. (2.4)). Because $\operatorname{cl}(\operatorname{epi} f)$ is closed we have that $\operatorname{epi} \overline{f} = \operatorname{cl}(\operatorname{epi} f)$.

Theorem 2.2.6 *Let $f : X \to \overline{\mathbb{R}}$ be a convex function. Then*

(i) *$\overline{f}$ is convex;*

(ii) *if $g : X \to \overline{\mathbb{R}}$ is convex, lsc and $g \le f$ then $g \le \overline{f}$;*

(iii) *the function $\overline{f}$ does not take the value $-\infty$ if and only if f is bounded from below by a continuous affine function;*

(iv) *if there exists $x_0 \in X$ such that $\overline{f}(x_0) = -\infty$ (in particular if $f(x_0) = -\infty$) then $\overline{f}(x) = -\infty$ for every $x \in \operatorname{dom} \overline{f} \supset \operatorname{dom} f$.*

Proof. (i) Since $\operatorname{epi} \overline{f} = \operatorname{cl}(\operatorname{epi} f)$ and $\operatorname{epi} f$ is convex, we have that $\operatorname{epi} \overline{f}$ is convex; hence $\overline{f}$ is convex.

(ii) Let g be lsc, convex and $g \le f$; then $\operatorname{epi} f \subset \operatorname{epi} g$. Taking the closure of the sets, the conclusion follows.

(iii) Suppose that $\overline{f}$ does not take the value $-\infty$. If $\overline{f}(x) = \infty$ for every $x \in X$ then $f(x) \geq \langle x, 0\rangle + 0$ for every $x \in X$. Consider $\operatorname{dom} \overline{f} \neq \emptyset$ and let $\overline{x} \in \operatorname{dom} \overline{f}$. Then $(\overline{x}, \overline{t}) \notin \operatorname{epi} \overline{f}$, where $\overline{t} := f(\overline{x}) - 1$. Since $\operatorname{epi} \overline{f}$ is a convex, closed and nonempty set, using Theorem 1.1.5 we obtain $(x^*, \alpha) \in X^* \times \mathbb{R}$ such that

$$\forall (x,t) \in \operatorname{epi} \overline{f} \ : \ \langle x, x^*\rangle + \alpha t < \langle \overline{x}, x^*\rangle + \alpha \overline{t}.$$

Taking $x = \overline{x}$ and $t = \overline{f}(\overline{x}) + n$, $n \in \mathbb{N}$, we obtain that $\alpha < 0$. Dividing, by $-\alpha > 0$, we can suppose that $\alpha = -1$. Thus

$$\forall x \in \operatorname{dom} \overline{f} \supset \operatorname{dom} f \ : \ f(x) \geq \overline{f}(x) \geq \langle x, x^*\rangle + \gamma,$$

where $\gamma := \overline{t} - \langle \overline{x}, x^*\rangle$. Therefore f is bounded from below by a continuous affine function.

Conversely, if $f(x) \geq \langle x, x^*\rangle + \gamma =: g(x)$, where $x^* \in X^*$ and $\gamma \in \mathbb{R}$, then g is convex and lsc; by (ii) we have that $g \leq \overline{f}$. Therefore $\overline{f}$ does not take the value $-\infty$.

(iv) If $\overline{f}(x_0) = -\infty$, using the preceding theorem, $\overline{f}(x) = -\infty$ for every $x \in \operatorname{dom} \overline{f}$. It is obvious that $\operatorname{dom} \overline{f} \supset \operatorname{dom} f$. □

Proposition 2.2.7 *Assume that $f : X \to \overline{\mathbb{R}}$ is sublinear. Then $\overline{f}$ is sublinear $\Leftrightarrow$ f is lsc at 0 $\Leftrightarrow$ $\overline{f}$ is proper.*

Proof. Because $\operatorname{epi} \overline{f} = \operatorname{cl}(\operatorname{epi} f)$ is a convex cone, the first equivalence follows by Proposition 2.1.2. If $\overline{f}$ is sublinear, by Proposition 2.2.5 we have that $\overline{f}$ is proper. Conversely, if $\overline{f}$ is proper then $0 \geq \overline{f}(0) > -\infty$ which implies that $\overline{f}(0) = 0$. □

To an arbitrary function $f : X \to \overline{\mathbb{R}}$ one associates, naturally, a lower semicontinuous and convex function; this function is denoted by $\overline{\text{co}} f$ and has the property that $\operatorname{epi}(\overline{\text{co}} f) = \operatorname{cl}(\operatorname{co}(\operatorname{epi} f))$ and is called the **lsc convex hull**. It is obvious that $\overline{\text{co}} f \leq \overline{f} \leq f$.

An application of the lsc convex hull of a function is the property in the next example used in the study of Hamilton-Jacobi equations.

Example 2.2.1 Let X be a locally convex space, $f \in \Gamma(X)$, $g : X \to \overline{\mathbb{R}}$ be a proper function and $\alpha, \beta > 0$. Then $\overline{\text{co}}\big(\overline{\text{co}}(f + \alpha g) + \beta g\big) = \overline{\text{co}}\big(f + (\alpha + \beta) g\big)$. (See the solution of Exercise 2.19 for details.)

Related to the upper semi-continuity of a convex function, we remark that when $f : X \to \overline{\mathbb{R}}$ is upper semicontinuous (usc for short) at $x_0 \in \operatorname{dom} f$,

f is upper bounded (by a real constant) on a neighborhood of x_0; therefore $x_0 \in \operatorname{int}(\operatorname{dom} f) \neq \emptyset$. In the sequel we prove that the converse is true for every convex function.

Let us first establish the following auxiliary result.

Lemma 2.2.8 *Let $f \in \Lambda(X)$ and $x_0 \in \operatorname{dom} f$. Suppose there exist $U \in \mathcal{N}_X^c$ and $m \in \mathbb{R}_+$ such that*

$$\forall x \in x_0 + U \; : \; f(x) \leq f(x_0) + m. \tag{2.24}$$

Then

$$\forall x \in x_0 + U \; : \; |f(x) - f(x_0)| \leq m p_U(x - x_0), \tag{2.25}$$

where p_U is the Minkowski gauge of U. In particular f is continuous at x_0.

Proof. Replacing f by $g : X \to \overline{\mathbb{R}}$, $g(x) = f(x_0 + x) - f(x_0)$, we may suppose that $x_0 = 0$ and $f(x_0) = 0$. Therefore $f(x) \leq m$ for every $x \in U$. From Proposition 1.1.1 we have that p_U is a continuous semi-norm and $U = \{x \in X \mid p_U(x) \leq 1\}$. Let $x \in U$. Suppose first that $t := p_U(x) > 0$. Then $y := t^{-1}x \in U$. Since f is convex and $t \leq 1$, we obtain that $f(x) \leq tf(y) + (1-t)f(0) \leq tm = mp_U(x)$. Suppose now that $p_U(x) = 0$. Then $nx \in U$ for every $n \in \mathbb{N}$. It follows that $f(x) \leq \frac{1}{n}f(nx) + \frac{n-1}{n}f(0) \leq \frac{1}{n}m$ for every $n \in \mathbb{N}$. Therefore, once again, $f(x) \leq mp_U(x)$. Since $0 = f\left(\frac{1}{2}x' + \frac{1}{2}(-x')\right) \leq \frac{1}{2}f(x') + \frac{1}{2}f(-x')$, we obtain that $-f(x') \leq f(-x')$ for every $x' \in X$. Using this inequality we obtain that $|f(x)| \leq mp_U(x)$ for every $x \in U$. Therefore Eq. (2.25) holds. □

Using the preceding lemma we get the following important result.

Theorem 2.2.9 *If the convex function $f : X \to \overline{\mathbb{R}}$ is bounded above on a neighborhood of a point of its domain then f is continuous on the interior of its domain. Moreover, if f is not proper then f is identically $-\infty$ on* $\operatorname{int}(\operatorname{dom} f)$.

Proof. Suppose that there exist $x_0 \in \operatorname{dom} f$, $V \in \mathcal{N}(x_0)$ and $m \in \mathbb{R}$ such that $f(x) \leq m$ for every $x \in V$. Then $V \subset \operatorname{dom} f$, and so $x_0 \in \operatorname{int}(\operatorname{dom} f) \neq \emptyset$. If f takes the value $-\infty$ then, by Proposition 2.1.4, $f(x) = -\infty$ for every $x \in \operatorname{int}(\operatorname{dom} f)$. Therefore the conclusion holds in this case. Let f be proper and $\overline{x} \in \operatorname{int}(\operatorname{dom} f)$. Then there exists $\mu > 0$ such that $x_1 := (1+\mu)\overline{x} - \mu x_0 \in \operatorname{dom} f$. Taking $\lambda := (1+\mu)^{-1} \in]0,1[$, we have that

$$f(\lambda x_1 + (1-\lambda)x) \leq \lambda f(x_1) + (1-\lambda)f(x) \leq \lambda f(x_1) + (1-\lambda)m =: m_1 \in \mathbb{R}$$

for every $x \in V$. But $V_1 := \lambda x_1 + (1-\lambda)V \in \mathcal{N}(\overline{x})$. Applying the preceding lemma we obtain that f is continuous at $\overline{x}$. □

Corollary 2.2.10 *Let $f : X \to \overline{\mathbb{R}}$ be a convex function. Then f is continuous on* $\operatorname{int}(\operatorname{dom} f)$ *if and only if* $\operatorname{int}(\operatorname{epi} f)$ *is nonempty in $X \times \mathbb{R}$.*

Proof. Suppose that f is continuous at some $x_0 \in \operatorname{int}(\operatorname{dom} f)$. Then, for $m \in]f(x_0), \infty[$, there exists $U \in \mathcal{N}_X$ such that $f(x) < m$ for every $x \in x_0 + U$. So $(x_0, m+1) \in \operatorname{int}(\operatorname{epi} f)$ because $(x_0 + U) \times [m, \infty[\subset \operatorname{epi} f$. Conversely, assume that $(x_0, t_0) \in \operatorname{int}(\operatorname{epi} f)$; then there exist $U \in \mathcal{N}_X$ and $\varepsilon > 0$ such that $(x_0 + U) \times [t_0 - \varepsilon, t_0 + \varepsilon[\subset \operatorname{epi} f$, and so $f(x) \leq m := t_0 + \varepsilon$ for every $x \in x_0 + U$. By Theorem 2.2.9 we have that f is continuous on $\operatorname{int}(\operatorname{dom} f)$ $(\neq \emptyset)$. □

Under the conditions of Lemma 2.2.8 we have a stronger conclusion.

Theorem 2.2.11 *Let $f \in \Lambda(X)$ and $x_0 \in \operatorname{dom} f$. Suppose that there exist $U \in \mathcal{N}_X^c$ and $m \in \mathbb{R}_+$ such that condition* (2.24) *is satisfied. Then for every $\rho \in]0,1[$,*

$$\forall\, x, y \in x_0 + \rho U \;:\; |f(x) - f(y)| \leq m \frac{1+\rho}{1-\rho} p_U(x-y).$$

Proof. As in the proof of Lemma 2.2.8, we may assume that $x_0 = 0$ and $f(x_0) = 0$. Let $\rho \in]0,1[$ and consider $x, y \in \rho U$. We consider two cases: a) $p_U(x-y) \neq 0$ and b) $p_U(x-y) = 0$. In the proof we shall use the fact that p_U is a continuous semi-norm, $U = \{x \mid p_U(x) \leq 1\}$ and $\operatorname{int} U = \{x \mid p_U(x) < 1\}$ (see Proposition 1.1.1).

a) Let $\varphi : [1, \infty[\to \mathbb{R}$, $\varphi(t) = p_U(tx + (1-t)y)$; since p_U is continuous, φ is continuous, too. Moreover, $\varphi(t) \geq t p_U(x-y) - p_U(-y)$, and so $\lim_{t\to\infty} \varphi(t) = \infty$. As $\varphi(1) = p_U(x) \leq \rho < 1$, there exists $\hat{t} > 1$ such that $\varphi(\hat{t}) = 1$. Then $z := \hat{t}x + (1-\hat{t})y \in U$. It follows that $x = (1-\hat{t}^{-1})y + \hat{t}^{-1}z$, and so $f(x) \leq (1-\hat{t}^{-1})f(y) + \hat{t}^{-1}f(z)$. From Lemma 2.2.8 we have that $|f(y)| \leq m p_U(y) \leq m\rho$. Therefore

$$f(x) - f(y) \leq \hat{t}^{-1}\left(f(z) - f(y)\right) \leq \hat{t}^{-1}(m + m\rho).$$

But $\hat{t}(x-y) = z-y$, whence $\hat{t} p_U(x-y) = p_U(z-y) \geq p_U(z) - p_U(y) \geq 1-\rho$. The preceding inequalities imply that

$$f(x) - f(y) \leq m \frac{1+\rho}{1-\rho} p_U(x-y). \tag{2.26}$$

b) In this case we have, using again Lemma 2.2.8, that $f(t(x-y)) = 0$ for every $t \in \mathbb{R}$. So,

$$\forall t \in]0,1[\; : \; f(tx) = f\big(ty + (1-t)\tfrac{t}{1-t}(x-y)\big) \le tf(y).$$

Since $x \in \operatorname{int} U \subset \operatorname{int}(\operatorname{dom} f)$, by Theorem 2.2.9, f is continuous at x. From the above inequality we obtain that $f(x) \le f(y)$ taking the limit for $t \to 1$. Hence (2.26) also holds in this case.

The conclusion follows from Eq. (2.26) changing x and y. □

Corollary 2.2.12 *Let $(X, \|\cdot\|)$ be a normed space and $f \in \Lambda(X)$. Suppose that $x_0 \in \operatorname{dom} f$ and for some $\rho > 0$ and $m \ge 0$,*

$$\forall x \in D(x_0, \rho) \; : \; f(x) \le f(x_0) + m.$$

Then

$$\forall \rho' \in]0, \rho[, \; \forall x, y \in D(x_0, \rho') \; : \; |f(x) - f(y)| \le \frac{m}{\rho} \cdot \frac{\rho + \rho'}{\rho - \rho'} \cdot \|x - y\|.$$

Proof. Consider $U := D(0;\rho) = \rho U_X$. Then $p_U(x) = \rho^{-1}\|x\|$ for any $x \in X$. The conclusion is immediate from the preceding theorem. □

The conclusion of Theorem 2.2.11 says, in fact, that f is Lipschitz on a neighborhood of x_0. We say that the function $f : X \to \overline{\mathbb{R}}$ is *Lipschitz* on a set $A \subset X$ if f is finite on A and there exists a continuous semi-norm p on X such that $|f(x) - f(y)| \le p(x-y)$ for all $x, y \in A$; we say that f is *locally Lipschitz* on A if for every $x \in A$ there exists a neighborhood V of x such that f is Lipschitz on V.

Corollary 2.2.13 *If $f \in \Lambda(X)$ is bounded above on a neighborhood of a point of its domain then f is locally Lipschitz on the interior of its domain.*

Proof. Let $x \in \operatorname{int}(\operatorname{dom} f)$. Applying Theorem 2.2.9 we obtain that f is continuous at x, and so f is bounded above on a neighborhood of x. Applying now Theorem 2.2.11 we obtain that f is Lipschitz on a neighborhood of x. □

Recall that in Theorem 2.1.5 we have already proved that a function $f \in \Lambda(\mathbb{R})$ is locally Lipschitz on $\operatorname{int}(\operatorname{dom} f)$, while in Proposition 2.1.6 we proved that $f|_{\operatorname{dom} f}$ is continuous if f is, moreover, lsc.

Another consequence of Theorem 2.2.9 is the following result.

Corollary 2.2.14 *Let $f_i \in \Lambda(X)$ for $1 \le i \le n$ and set $f := f_1 \Box \cdots \Box f_n$, $g := f_1 \Diamond \cdots \Diamond f_n$. If f_1 is continuous at a point of its domain then*

$$\operatorname{int}(\operatorname{dom} f) = \operatorname{int}(\operatorname{dom} g) = \operatorname{int}(\operatorname{dom} f_1) + \operatorname{dom} f_2 + \cdots + \operatorname{dom} f_n$$

and either f (resp. g) is identically $-\infty$ on $\operatorname{int}(\operatorname{dom} f)$ (resp. $\operatorname{int}(\operatorname{dom} g)$), or f (resp. g) is proper and continuous on $\operatorname{int}(\operatorname{dom} f)$ (resp. $\operatorname{int}(\operatorname{dom} g)$).□

Recall that $f_1 \Box f_2$ and $f_1 \Diamond f_2$ are defined in Theorem 2.1.3.

From Theorem 2.2.9 (or Corollary 2.2.10) we obtain that g is continuous on $\operatorname{int}(\operatorname{dom} g)$ if f, g are convex, $g \le f$ and f is continuous on int(dom) (supposed to be nonempty). A similar result is true for a larger class of convex functions.

The convex function $f : X \to \overline{\mathbb{R}}$ is said to be **quasi-continuous** if $\operatorname{aff}(\operatorname{dom} f)$ is closed and has finite codimension (*i.e.* its parallel linear subspace has finite codimension), $\operatorname{rint}(\operatorname{dom} f) \neq \emptyset$ and $f|_{\operatorname{aff}(\operatorname{dom} f)}$ is continuous on $\operatorname{rint}(\operatorname{dom} f)$. The set $A \subset X$ is *quasi-continuous* if ι_A is quasi-continuous; it follows that A is quasi-continuous exactly when $\operatorname{aff} A$ is a closed affine set of finite codimension and $\operatorname{rint} A \neq \emptyset$. The following result holds.

Proposition 2.2.15 *Let $f, g : X \to \overline{\mathbb{R}}$ be convex functions such that $g \le f$. If f is quasi-continuous, then g is quasi-continuous, too.*

Proof. Without loss of generality we assume that $0 \in \operatorname{dom} f$ and $f(0) \le 0$. Then $\operatorname{aff}(\operatorname{dom} f)$ is a linear subspace and $Y_0 := \operatorname{aff}(\operatorname{epi} f) = \operatorname{aff}(\operatorname{dom} f) \times \mathbb{R}$. Of course Y_0 has finite codimension and is closed. Moreover, by Corollary 2.2.10, we have that $\operatorname{int}_{Y_0}(\operatorname{epi} f) \neq \emptyset$. Since $\operatorname{epi} f \subset \operatorname{epi} g$, we have that $\operatorname{int}_{Y_0}(Y_0 \cap \operatorname{epi} g) \neq \emptyset$. It follows (see Exercise 1.3) that $\operatorname{rint}(\operatorname{epi} g) \neq \emptyset$. Since $\operatorname{aff}(\operatorname{epi} g) = \operatorname{aff}(\operatorname{dom} g) \times \mathbb{R}$, using again Corollary 2.2.10, we obtain that $g|_{\operatorname{aff}(\operatorname{dom} g)}$ is continuous, and so g is quasi-continuous. □

Applying the preceding result to indicator functions one obtains

Corollary 2.2.16 *Let $A \subset B \subset X$. If A is quasi-continuous then so is B.* □

There are several classes of convex functions, larger than the class of lower semicontinuous convex functions, which will reveal themselves to be useful in the sequel. We introduce them now.

Let $f : X \to \overline{\mathbb{R}}$. We say that f is **cs-convex** if

$$f(x) \le \liminf_{n\to\infty} \sum\nolimits_{k=1}^{n} \lambda_k f(x_k)$$

whenever $\sum_{n\geq 1} \lambda_n x_n$ is a convex series with elements of X and sum $x \in X$. Of course, if f is cs-convex then f is convex, while if f is lsc and convex, f is cs-convex (Exercise!). Also, we say that f is **ideally convex, bcs-complete, cs-closed, cs-complete, li-convex** or **lcs-closed** if $\operatorname{epi} f$ is ideally convex, bcs-complete, cs-closed, cs-complete, li-convex or lcs-closed, respectively. Of course, taking into account the relationships among these notions for sets and Proposition 2.2.17 (i) below, we have:

$$\begin{array}{ccccccc}
f \text{ lsc, convex} & \Rightarrow & f \text{ sc-convex} & & & & \\
 & & \Downarrow & & & & \\
f \text{ cs-complete} & \Rightarrow & f \text{ cs-closed} & \Rightarrow & f \text{ lcs-closed} & & \\
\Downarrow & & \Downarrow & & \Downarrow & & \\
f \text{ bcs-complete} & \Rightarrow & f \text{ ideally convex} & \Rightarrow & f \text{ li-convex} & \Rightarrow & f \text{ convex,}
\end{array}$$

the reversed implications being not true, in general.

Taking into account Proposition 1.2.3 and that for $f : X \to \overline{\mathbb{R}}$ we have

$$[f \leq \lambda] = \operatorname{Pr}_X (\operatorname{epi} f \cap (X \times]-\infty, \lambda])),$$
$$[f < \lambda] = \operatorname{Pr}_X (\operatorname{epi} f \cap (X \times]-\infty, \lambda[),$$

the sets $[f \leq \lambda]$ and $[f < \lambda]$ are li-convex (lcs-closed) for every $\lambda \in \mathbb{R}$ if f is li-convex (lcs-closed).

Let $A \subset X$; since $\operatorname{epi} \iota_A = A \times \mathbb{R}_+$ and $\mathbb{R}$ is a Fréchet space, we have that A is ideally convex (bcs-complete, cs-closed, cs-complete, li-convex, lcs-closed) if and only if ι_A is so.

Remark 2.2.1 When $\varphi : X \to \mathbb{R}$ is a continuous affine functional (that is $\varphi = x^* + \alpha$ for some $x^* \in X^*$ and $\alpha \in \mathbb{R}$), $f : X \to \overline{\mathbb{R}}$ is cs-convex (ideally convex, cs-closed) if and only if $f + \varphi$ is so (Exercise!).

Proposition 2.2.17 *Let $f, g : X \to \overline{\mathbb{R}}$ have nonempty domains.*

(i) *If f is cs-convex then f is cs-closed. Conversely, if f is cs-closed and is minorized by a continuous affine functional then f is cs-convex.*

(ii) *If f and g are ideally convex (resp. cs-closed) and are minorized by continuous affine functionals then $f + g$ is ideally convex (resp. cs-closed). Moreover, if g is bcs-complete (resp. cs-complete) then $f+g$ is bcs-complete (resp. cs-complete).*

Proof. Taking into account the Remark 2.2.1 above, when f or/and g is bounded from below by a continuous affine function we may assume that even $f \geq 0$ or/and $g \geq 0$.

(i) The proof of the first part is immediate. Suppose that f is cs-closed and $f \geq 0$. Let $\sum_{n\geq 1} \lambda_n x_n$ be a convergent convex series with elements of X and sum $x \in X$. Since f is convex ($\operatorname{epi} f$ being convex), we may suppose that $\lambda_n > 0$ for every $n \in \mathbb{N}$; moreover, we may assume that $x_n \in \operatorname{dom} f$ for every n. Because $f(x_n) \geq 0$, there exists $\tau := \lim_{n\to\infty} \sum_{k=1}^{n} \lambda_k f(x_k) \in \mathbb{R} \cup \{\infty\}$. If $\tau < \infty$, because $(x_n, f(x_n)) \in \operatorname{epi} f$ and $\sum_{n\geq 1} \lambda_n (x_n, f(x_n)) = (x, \tau)$, and f is cs-closed, we have that $(x, \tau) \in \operatorname{epi} f$, whence $f(x) \leq \tau$. If $\tau = \infty$, the preceding inequality is obvious. Hence f is cs-convex.

(ii) We prove only the second part of the "ideally convex" case, the rest of the proof being similar.

So, let f be ideally convex, g be bcs-complete and $f, g \geq 0$. Let $\sum_{n\geq 1} \lambda_n (x_n, r_n)$ be a Cauchy b-convex series with elements of $\operatorname{epi}(f + g)$. Then $r_n = s_n + t_n$ with $(x_n, s_n) \in \operatorname{epi} f_n$ and $(x_n, t_n) \in \operatorname{epi} g$ for every n. Because $0 \leq s_n, t_n \leq r_n$, we have that (s_n), (t_n) are bounded, $s := \sum_{n\geq 1} \lambda_n s_n \in \mathbb{R}_+$, $t := \sum_{n\geq 1} \lambda_n t_n \in \mathbb{R}_+$ and $r = s + t$. Because g is bcs-complete we obtain that the b-convex series $\sum_{n\geq 1} \lambda_n (x_n, t_n)$ with elements of $\operatorname{epi} g$ is convergent with sum $(x, t) \in \operatorname{epi} g$ for some $x \in X$. Hence the b-convex series $\sum_{n\geq 1} \lambda_n (x_n, s_n)$ with elements of $\operatorname{epi} f$ is convergent with sum $(x, s) \in \operatorname{epi} f$. Therefore $(x, r) \in \operatorname{epi}(f + g)$. Thus $f + g$ is bcs-complete. □

The classes of li-convex and lcs-closed functions have good stability properties. Let us begin with the following characterizations.

Proposition 2.2.18 *Let $f : X \to \overline{\mathbb{R}}$ have nonempty domain. Then the following statements are equivalent:*

(i) *f is li-convex (resp. lcs-closed);*

(ii) *$\operatorname{epi}_s f$ is li-convex (resp. lcs-closed);*

(iii) *there exist a Fréchet space Y and an ideally convex (resp. cs-closed) function $F : X \times Y \to \overline{\mathbb{R}}$ such that $f(x) = \inf_{y\in Y} F(x, y)$ for every $x \in X$ (i.e. f is the marginal function associated to an ideally convex (resp. cs-closed) function).*

Proof. We prove the "li-convex" case, the proof for the other case being similar.

(i) ⇒ (ii) Taking $\mathcal{R}, \mathcal{S} : X \rightrightarrows \mathbb{R}$ such that $\operatorname{gr} \mathcal{R} = \operatorname{epi} f$ and $\operatorname{gr} \mathcal{S} = X \times \mathbb{P}$, we have that $\mathcal{R}$ and $\mathcal{S}$ are li-convex multifunctions. Since $\mathbb{R}$ is a Fréchet space, using Proposition 1.2.5 (iv), $\operatorname{epi}_s f = \operatorname{epi} f + \{0\} \times \mathbb{P} = \operatorname{gr}(\mathcal{R} + \mathcal{S})$ is li-convex.

(ii) $\Rightarrow$ (i) Since $\operatorname{epi}_s f$ is li-convex, as above, the set $A_n := \operatorname{epi}_s f + \{0\}\times]-\frac{1}{n}, \infty[$ is li-convex for every $n \in \mathbb{N}$. Since $\operatorname{epi} f = \bigcap_{n\in\mathbb{N}} A_n$, from Proposition 1.2.4 (i) we obtain that epi f is li-convex.

(i) $\Rightarrow$ (iii) Since f is li-convex, there exist a Fréchet space Y and an ideally convex set $A \subset Y \times X \times \mathbb{R}$ such that $\operatorname{epi} f = \operatorname{Pr}_{X\times\mathbb{R}}(A)$. Consider the function $F : X \times Y \times \mathbb{R} \to \overline{\mathbb{R}}$ defined by $F(x, y, r) := r + \iota_A(y, x, r)$. Then

$$f(x) = \inf\nolimits_{(x,r)\in \operatorname{epi} f} r = \inf\nolimits_{(y,x,r)\in A} r = \inf\nolimits_{(y,r)\in Y\times\mathbb{R}} F(x, y, r)$$

for every $x \in X$. Since A is ideally convex, so is ι_A; using Proposition 2.2.17 (ii) and Remark 2.2.1 we obtain that F is ideally convex. The conclusion follows because $Y \times \mathbb{R}$ is a Fréchet space.

(iii) $\Rightarrow$ (ii) Let $f(x) = \inf_{y\in Y} F(x, y)$ for every $x \in X$, where Y is a Fréchet space and F is an ideally convex function. From (i)$\Rightarrow$(ii) it follows that $\operatorname{epi}_s F$ is li-convex. Then, by Eq. (2.8), $\operatorname{epi}_s f$ is li-convex. □

Other useful properties of li-convex and lcs-closed functions are collected in the following result.

Proposition 2.2.19 (i) *If $f_n : X \to \overline{\mathbb{R}}$ is li-convex (resp. lcs-closed) for every $n \in \mathbb{N}$, then* $\sup_{n\in\mathbb{N}} f_n$ *is li-convex (resp. lcs-closed).*

(ii) *If $f_1, f_2 : X \to \overline{\mathbb{R}}$ are li-convex (resp. lcs-closed) functions and $\lambda \in \mathbb{R}_+$, then $f_1 + f_2$ and λf_1 are li-convex (resp. lcs-closed).*

(iii) *If $F : X \times Y \to \overline{\mathbb{R}}$ is li-convex (resp. lcs-closed) and X is a Fréchet space, then $h : Y \to \overline{\mathbb{R}}$, $h(y) := \inf_{x\in X} F(x, y)$, is li-convex (resp. lcs-closed).*

(iv) *Let Y be a Fréchet space and $g : Y \to \overline{\mathbb{R}}$. If $Q \subset Y$ is a convex cone, $H : X \to (Y^\bullet, Q)$ has li-convex (resp. lcs-closed) epigraph and g is li-convex (resp. lcs-closed) and Q-increasing, then $g \circ H$ is li-convex (resp. lcs-closed). In particular, if $A : X \to Y$ is a linear operator with li-convex (resp. lcs-closed) graph and g is li-convex (resp. lcs-closed), then $g \circ A$ is li-convex (resp. lcs-closed).*

(v) *If X is a Fréchet space, $A : X \to Y$ is a linear operator with li-convex (resp. lcs-closed) graph and $f : X \to \overline{\mathbb{R}}$ is li-convex (resp. lcs-closed), then Af is li-convex (resp. lcs-closed).*

(vi) *If X is a Fréchet space and $f_1, f_2 : X \to \overline{\mathbb{R}}$ are li-convex (resp. lcs-closed) functions then $f_1 \Box f_2$ and $f_1 \Diamond f_2$ are li-convex (resp. lcs-closed).*

Proof. Again, we treat only the "li-convex" case.

(i) Because $\operatorname{epi}(\sup_{n\in\mathbb{N}} f_n) = \bigcap_{n\in\mathbb{N}} \operatorname{epi} f_n$, the conclusion follows using Proposition 1.2.4 (i).

(ii) Taking $\mathcal{R}_i : X \rightrightarrows \mathbb{R}$, $\operatorname{gr}\mathcal{R}_i := \operatorname{epi} f_i$ $(i = 1, 2)$, we have that $\operatorname{epi}(f_1 + f_2) = \operatorname{gr}(\mathcal{R}_1 + \mathcal{R}_2)$. The conclusion follows from Proposition 1.2.5 (iv). Also, $\operatorname{epi}(\lambda f_1) = T(\operatorname{epi} f_1)$ for $\lambda > 0$, where $T : X\times\mathbb{R} \to X\times\mathbb{R}$ is the isomorphism of topological vector spaces given by $T(x,t) = (x, \lambda t)$; hence λf_1 is li-convex in this case. If $\lambda = 0$, $\lambda f_1 = \iota_{\operatorname{dom} f_1}$. But $\operatorname{dom} f_1 = \operatorname{Pr}_X(\operatorname{epi} f_1)$, and so $\operatorname{dom} f_1$ is li-convex by Proposition 1.2.3, whence $0f_1$ is li-convex.

(iii) We have, by Eq. (2.8), that $\operatorname{epi}_s f = \operatorname{Pr}_{Y\times\mathbb{R}}(\operatorname{epi}_s F)$. Since $\operatorname{epi}_s F$ is li-convex and X is a Fréchet space, we get from Proposition 1.2.3 that $\operatorname{epi}_s f$ is li-convex.

(iv)–(vi) Using the same constructions as in the proofs of Theorem 2.1.3 (vi), (viii) and (ix), the conclusions follows from (iii). □

If X is a barreled space, the lower semi-continuity of a convex function (even weaker conditions) ensures its continuity on the interior of its domain.

Theorem 2.2.20 *Let X be a barreled space and $f : X \to \overline{\mathbb{R}}$ be convex. Suppose that either* (a) *X is first countable and f is li-convex or* (b) *f is lower semicontinuous. Then* $(\operatorname{dom} f)^i = \operatorname{int}(\operatorname{dom} f)$ *and f is continuous on* $\operatorname{int}(\operatorname{dom} f)$.

Proof. (a) By Proposition 2.2.18, there exist a Fréchet space Y and a li-convex function $F : X\times Y \to \overline{\mathbb{R}}$ such that $f(x) = \inf_{y\in Y} F(x,y)$ for every $x \in X$. Consider the multifunction $\mathcal{R} : Y\times\mathbb{R} \rightrightarrows X$ with $\operatorname{gr}\mathcal{R} := \{(y,t,x) \mid (x,y,t) \in \operatorname{epi} F\}$. Of course, $\mathcal{R}$ is ideally convex and $\operatorname{Im}\mathcal{R} = \operatorname{dom} f$. Let $x_0 \in (\operatorname{dom} f)^i$ (if this set is nonempty) and $(y_0, t_0) \in \mathcal{R}^{-1}(x_0)$. Applying Simons' theorem (Theorem 1.3.5), we have that $U := \mathcal{R}(Y\times\,]-\infty, t_0+1[)$ is a neighborhood of x_0. So, for every $x \in U$ there exist $y \in Y$ and $t < t_0 + 1 =: m$ such that $f(x) \le F(x,y) \le t < m$. So f is bounded above on the neighborhood U of x_0, whence $x_0 \in \operatorname{int}(\operatorname{dom} f)$ and f is continuous at x_0.

The case (b) follows similarly (taking $Y = \{0\}$ for example) and using Ursescu's theorem instead of Simons' theorem. □

In finite dimensional linear spaces the preceding result becomes:

Corollary 2.2.21 *Let X be a finite dimensional linear normed space and $f \in \Lambda(X)$ be such that* $(\operatorname{dom} f)^i \neq \emptyset$. *Then f is continuous on* $\operatorname{int}(\operatorname{dom} f) = (\operatorname{dom} f)^i$.

Proof. By Proposition 1.2.1 we have that $\operatorname{epi} f$ is cs-closed, and so f is lcs-closed. The conclusion follows from the assertion (a) of the preceding theorem. □

The application of Corollary 2.2.12 and Theorem 2.2.20 yields the following uniform boundedness principle for convex functions.

Theorem 2.2.22 *Let X be a Banach space and $C \subset X$ be an open convex set. Consider $(f_i)_{i\in I}$ a nonempty family of continuous convex functions from C into $\mathbb{R}$. If $\left(f_i(x)\right)_{i\in I}$ is bounded for every $x \in C$ then for every $x \in C$ there exist $r_x, L_x > 0$ such that $U_x := x + r_x U_X \subset C$ and $|f_i(y) - f_i(z)| \leq L_x\|y - z\|$ for all $y, z \in U_x$ and all $i \in I$ (i.e. (f_i) is* locally equi-Lipschitz on *C).*

Proof. By hypothesis, for every $x \in C$ there exists $M_x \in \mathbb{R}_+$ such that $|f_i(x)| \leq M_x$ for all $i \in I$. Let $r'_x > 0$ be such that $U'_x := x + r'_x U_X \subset C$ and consider $F_i : X \to \overline{\mathbb{R}}$ be defined by $F_i(y) := f_i(y)$ for $y \in U'_x$ and $F_i(y) := \infty$ otherwise. Then $F_i \in \Gamma(X)$. Consider $F := \sup_{i\in I} F_i \in \Gamma(X)$. The hypothesis shows that $\operatorname{dom} F = U_x$. By Theorem 2.2.20 we obtain that F is continuous on $\operatorname{int}(\operatorname{dom} F) = x + r_x B_X$. Therefore there exist $r''_x \in]0, r'_x[$ and $M > 0$ such that $F(y) \leq M$ for all $y \in x + r''_x U_X$. Then for $y \in x + r''_x U_X$ and $i \in I$ we obtain that $f_i(y) - f_i(x) \leq M - (-M_x) = M + M_x =: L'_x$. Using now Corollary 2.2.12 we have that f_i is Lipschitz with constant $L_x := 3L'_x/r''_x$ on $x + r_x U_X$, where $r_x := r''_x/2$. The conclusion follows. □

Taking X a Banach space, Y a normed space, and $\{T_i \mid i \in I\} \subset \mathcal{L}(X, Y)$ $(I \neq \emptyset)$ with $\{T_i x \mid i \in I\}$ bounded for every $x \in X$, the conditions of the preceding theorem are satisfied by the family of functions $(f_i)_{i\in I}$, where $f_i(x) := \|T_i(x)\|$, and $C := X$. Hence $\|T_i(x)\| = |f_i(x) - f_i(0)| \leq L\|x\|$ for $x \in rU_X$ and $i \in I$, for some $r, L > 0$; hence $\|T_i\| \leq L$ for every $i \in I$. So we obtained the classic uniform boundedness principle in Functional Analysis.

From the preceding result we obtain that pointwise convergence implies uniform convergence on compact sets.

Corollary 2.2.23 *Let X be a Banach space and $C \subset X$ be an open convex set. Consider (f_n) a sequence of continuous convex functions defined on C with values in $\mathbb{R}$ such that $\left(f_n(x)\right) \to f(x)$ for every $x \in C$ with $f : C \to \mathbb{R}$. Then (f_n) converges uniformly to f on the compact subsets of C*

(or, equivalently, $(f_n(x_n)) \to f(x)$ *for every sequence* $(x_n) \subset C$ *converging to* $x \in C$*).*

Proof. First of all observe that f is convex (by Theorem 2.1.3) and locally Lipschitz (by the preceding theorem).

Let $x, x_n \in C$ ($n \in \mathbb{N}$) be such that $(x_n) \to x$. By Theorem 2.2.22 there exist $r, L > 0$ such that $U_x := x + rU_X \subset C$ and $|f_n(y) - f_n(z)| \leq L\|y - z\|$ for all $n \in \mathbb{N}$ and $y, z \in U_x$. Since $(x_n) \to x$, $x_n \in U_x$ for $n \geq n_x$ for some $n_x \in \mathbb{N}$. Then

$$\begin{aligned} |f_n(x_n) - f(x)| &\leq |f_n(x_n) - f_n(x)| + |f_n(x) - f(x)| \\ &\leq M\|x_n - x\| + |f_n(x) - f(x)| \end{aligned}$$

for $n \geq n_x$, whence $(f_n(x_n))_n \to f(x)$.

Assume now that there exists some compact subset K of U such that (f_n) does not converge uniformly to f on K. Then there exist $\varepsilon > 0$, $P \subset \mathbb{N}$ an infinite set and a sequence $(x_n)_{n\in P} \subset K$ such that $|f_n(x_n) - f(x_n)| \geq \varepsilon$. Since K is compact, we may assume that $(x_n) \to x \in K \subset C$. Then, as shown above, $\left(f_n(x_n)\right)_{n\in P} \to f(x)$. Since f is continuous, we have also that $\left(f(x_n)\right)_{n\in P} \to f(x)$, which yields the contradiction $0 \geq \varepsilon$. □

The next result corresponds to a known theorem for continuous linear operators.

Proposition 2.2.24 *Let* X *be a Banach space and* $C \subset X$ *be an open convex set. Assume that* $f, f_n : C \to \mathbb{R}$ ($n \in \mathbb{N}$) *are continuous convex functions. Then* $\left(f_n(x)\right) \to f(x)$ *for every* $x \in C$ *if and only if* (a) $\left(f_n(x)\right)$ *is bounded for every* $x \in C$ *and* (b) $\left(f_n(x)\right) \to f(x)$ *for every* $x \in D$ *for some dense subset* D *of* C.

Proof. The necessity is obvious. Assume that conditions (a) and (b) above are satisfied but $(f_n(x))$ does not converge to $f(x)$ for some $x \in C$. Hence there exist $\varepsilon > 0$ and $P \subset \mathbb{N}$ an infinite subset such that $|f_n(x) - f(x)| \geq \varepsilon$ for every $n \in P$. By Theorem 2.2.22 we find $r, L > 0$ such that $U_x := x + rU_X \subset U$ and f, f_n ($n \in \mathbb{N}$) are L-Lipschitz on U_x. Since D is dense, there exists $(x_k) \subset D$ converging to x; we assume that $x_k \in U_x$ for every $k \in \mathbb{N}$. Then for $n \in P$ and $k \in \mathbb{N}$ we have

$$\begin{aligned} \varepsilon \leq |f_n(x) - f(x)| &\leq |f_n(x) - f_n(x_k)| + |f_n(x_k) - f(x_k)| + |f(x_k) - f(x)| \\ &\leq 2L\|x_k - x\| + |f_n(x_k) - f(x_k)|. \end{aligned}$$

Fixing $k \in \mathbb{N}$ such that $\|x_k - x\| < \varepsilon/(4L)$, we get $|f_n(x_k) - f(x_k)| > \varepsilon/2$ for $n \in P$, contradicting that $(f_n(x_k) - n \in \mathbb{N})$ converges to $f(x_k)$. □

Let $f \in \Gamma(X)$; we call the **recession function** of f the function $f_\infty : X \to \overline{\mathbb{R}}$ whose epigraph is $(\operatorname{epi} f)_\infty$. Let $x_0 \in \operatorname{dom} f$; taking into account formula (1.3), we have that

$$\begin{aligned}(u,\lambda) \in \operatorname{epi} f_\infty &\Leftrightarrow \forall\, t > 0 \ : \ (x_0, f(x_0)) + t(u,\lambda) \in \operatorname{epi} f \\ &\Leftrightarrow \frac{f(x_0+tu) - f(x_0)}{t} \le \lambda \\ &\Leftrightarrow \lim_{t\to\infty} \frac{f(x_0+tu) - f(x_0)}{t} = \sup_{t>0} \frac{f(x_0+tu) - f(x_0)}{t} \le \lambda.\end{aligned}$$

Therefore

$$\forall\, u \in X \ : \ f_\infty(u) = \lim_{t\to\infty} \frac{f(x_0+tu) - f(x_0)}{t}; \tag{2.27}$$

thus $f_\infty(u) > -\infty$ for every $u \in X$ and $f_\infty(0) = 0$. Because $\operatorname{epi} f_\infty$ is a closed convex cone we have that f_∞ is a lsc sublinear functional.

The preceding relations show that

$$f(x+u) \le f(x) + f_\infty(u) \quad \forall\, x \in \operatorname{dom} f,\ \forall\, u \in X, \tag{2.28}$$

$$[f \le \lambda]_\infty = \{u \in X \mid f_\infty(u) \le 0\} \quad \forall\, \lambda \in \mathbb{R} \text{ with } [f \le \lambda] \ne \emptyset, \tag{2.29}$$

when $f \in \Gamma(X)$. Taking into account the discussion on page 6, relation (2.29) shows that the function $f \in \Gamma(X)$ has bounded level sets when $\dim X < \infty$. This result is no longer true when $\dim X = \infty$; take f.i. $f := p_A$, the Minkowski functional associated to the set in Example 1.1.1 ($f_\infty = f$ in this case).

Remark 2.2.2 Note that, for $f \in \Gamma(X)$ and $x \in X$, the mapping $t \mapsto f(x+tu)$ is nonincreasing from $\mathbb{R}$ into $\overline{\mathbb{R}}$ when $f_\infty(u) \le 0$; in particular $\operatorname{dom} f + \mathbb{R}_+ u = \operatorname{dom} f$. Moreover, if $f_\infty(u) \le 0$ and $f_\infty(-u) \le 0$ then $f(x+tu) = f(x)$ for all $x \in X$ and $t \in \mathbb{R}$.

Indeed, let $t_1 < t_2$. If $f(x+t_1u) < \infty$ then

$$\begin{aligned}\frac{f(x+t_2u) - f(x+t_1u)}{t_2 - t_1} &= \frac{f(x+t_1u+(t_2-t_1)u) - f(x+t_1u)}{t_2 - t_1} \\ &\le f_\infty(u) \le 0,\end{aligned}$$

and so $f(x+t_2u) \le f(x+t_1u)$; the inequality is obvious if $f(x+t_1u) = \infty$. It follows that $\operatorname{dom} f + \mathbb{R}_+ u = \operatorname{dom} f$. When $f_\infty(u) \le 0$ and $f_\infty(-u) \le 0$ we can apply the previous result for u and $-u$ and obtain that the mapping $t \mapsto f(x+tu)$ is constant.

2.3 Conjugate Functions

In this section X and Y are separated locally convex spaces if it is not stated explicitly otherwise. Let $f : X \to \overline{\mathbb{R}}$; the function

$$f^* : X^* \to \overline{\mathbb{R}}, \quad f^*(x^*) := \sup\{\langle x, x^* \rangle - f(x) \mid x \in X\}, \tag{2.30}$$

is called the **conjugate** or *Fenchel conjugate* of f. Note that if there exists $x_0 \in X$ such that $f(x_0) = -\infty$ then $f^*(x^*) = \infty$ for every $x^* \in X^*$, while if $f(x) = \infty$ for every x then $f^*(x^*) = -\infty$ for every $x^* \in X^*$. When f is proper (but also for f not proper, using the convention $\inf \emptyset = \infty$), we have

$$f^*(x^*) = \sup\{\langle x, x^* \rangle - f(x) \mid x \in \operatorname{dom} f\}.$$

The conjugate of a function $h : X^* \to \overline{\mathbb{R}}$ is defined similarly:

$$h^* : X \to \overline{\mathbb{R}}, \quad h^*(x) := \sup\{\langle x, x^* \rangle - h(x^*) \mid x^* \in X^*\}.$$

In fact, considering the natural duality between X and X^*, it is the same definition; this distinction is useful in the case of normed spaces. The above remark concerning f^* is also valid for h^*.

In the following theorem we establish several simple properties of conjugate functions.

Theorem 2.3.1 *Let $f, g : X \to \overline{\mathbb{R}}$, $h : Y \to \overline{\mathbb{R}}$, $k : X^* \to \overline{\mathbb{R}}$ and $A \in \mathcal{L}(X,Y)$.*

(i) *f^* is convex and w^*-lsc, k^* is convex and lsc;*

(ii) *the **Young–Fenchel inequality** below holds:*

$$\forall x \in X,\ \forall x^* \in X^* \ : \ f(x) + f^*(x^*) \ge \langle x, x^* \rangle;$$

(iii) *$f \le g \Rightarrow g^* \le f^*$;*

(iv) *$f^* = \overline{f}^* = (\overline{\operatorname{co}} f)^*$ and $f^{**} := (f^*)^* \le \overline{\operatorname{co}} f \le \overline{f} \le f$;*

(v) *if* $\alpha > 0$ *then* $(\alpha f)^*(x^*) = \alpha f^*(\alpha^{-1}x^*)$ *for every* $x^* \in X^*$*; if* $\beta \neq 0$ *then* $\big(f(\beta\cdot)\big)^*(x^*) = f^*(\beta^{-1}x^*)$ *for every* $x^* \in X^*$*;*

(vi) *if* $g(x) = f(x + x_0)$ *for* $x \in X$*, then* $g^*(x^*) = f^*(x^*) - \langle x_0, x^*\rangle$ *for every* $x^* \in X^*$*;*

(vii) *if* $x_0^* \in X^*$ *then* $(f + x_0^*)^*(x^*) = f^*(x^* - x_0^*)$ *for every* $x^* \in X^*$*;*

(viii) *if* f, h *are proper and* $\Phi : X \times Y \to \overline{\mathbb{R}}$, $\Phi(x, y) := f(x) + h(y)$*, then* $\Phi^*(x^*, y^*) = f^*(x^*) + h^*(y^*)$ *for all* $(x^*, y^*) \in X^* \times Y^*$*;*

(ix) $(Af)^* = f^* \circ A^*$ *and* $(f\Box g)^* = f^* + g^*$.

Proof. (i) If f is not proper, we have already seen that f^* is constant, and so f^* is convex and w^*-continuous. If f is proper we have that $f^* = \sup_{x\in \operatorname{dom} f} \varphi_x$, where $\varphi_x : X^* \to \overline{\mathbb{R}}$, $\varphi_x(x^*) := \langle x, x^*\rangle - f(x)$. It is obvious that for every $x \in \operatorname{dom} f$, φ_x is affine (hence convex!) and w^*-continuous (hence w^*-lsc!). Therefore f^* is convex and w^*-lsc. For the statement about h we use the same arguments.

(ii) By Eq. (2.30) we have

$$\forall\, x \in X,\ \forall\, x^* \in X^* \ :\ f^*(x^*) \geq \langle x, x^*\rangle - f(x),$$

which gives immediately the Young–Fenchel inequality.

(iii) is an immediate consequence of the definition and of the relation $f \leq g$.

(iv) We already remarked that $\overline{\text{co}}f \leq \overline{f} \leq f$, whence, using (iii), we get $f^* \leq \overline{f}^* \leq (\overline{\text{co}}f)^*$. Let $x^* \in X^*$ and $\alpha \in \mathbb{R}$ be such that $f^*(x^*) \leq \alpha$. Then $\langle x, x^*\rangle - f(x) \leq \alpha$ for every $x \in X$, whence $\varphi(x) := \langle x, x^*\rangle - \alpha \leq f(x)$ for every x. Since $\varphi \in \Gamma(X)$, and $\operatorname{epi}\varphi \supset \operatorname{epi} f$, we have that $\operatorname{epi}\varphi \supset \operatorname{epi}(\overline{\text{co}}f) = \overline{\text{co}}(\operatorname{epi} f)$. Therefore $\varphi(x) \leq (\overline{\text{co}}f)(x)$ for every $x \in X$, and so $\langle x, x^*\rangle - \overline{\text{co}}f(x) \leq \alpha$ for every x; hence $(\overline{\text{co}}f)^*(x^*) \leq \alpha$. Thus $f^* = \overline{f}^* = (\overline{\text{co}}f)^*$.

(v), (vi), (vii) and (viii) are immediate.

(ix) We have

$$\begin{aligned}(Af)^*(y^*) &= \sup_{y\in Y}[\langle y, y^*\rangle - (Af)(y)] = \sup_{y\in Y}\left(\langle y, y^*\rangle - \inf_{\{x \mid Ax=y\}} f(x)\right)\\ &= \sup\{\langle y, y^*\rangle - f(x) \mid (x, y) \in X \times Y,\ Ax = y\}\\ &= \sup\{\langle x, A^*y^*\rangle - f(x) \mid x \in X\}\\ &= f^*(A^*y^*) = (f^* \circ A^*)(y^*).\end{aligned}$$

Similarly one proves the corresponding relation for $(f\Box g)^*$. □

I the sequel we denote by $\Gamma^*(X^*)$ the class of those functions in $\Lambda(X^*)$ which are w^*-lower semicontinuous. The discussion at the beginning of this section and the preceding theorem yields the next result.

Corollary 2.3.2 *Let $f : X \to \overline{\mathbb{R}}$ and $h : X^* \to \overline{\mathbb{R}}$. Then*

(i) $f^* \in \Gamma^*(X^*) \Leftrightarrow \operatorname{dom} f \neq \emptyset$ *and* $\exists\, x^* \in X^*,\ \alpha \in \mathbb{R},\ \forall x \in X :$ $f(x) \geq \langle x, x^* \rangle + \alpha$;

(ii) $h^* \in \Gamma(X) \Leftrightarrow \operatorname{dom} h \neq \emptyset$ *and* $\exists\, x \in X,\ \alpha \in \mathbb{R},\ \forall x^* \in X^* :$ $h(x^*) \geq \langle x, x^* \rangle + \alpha$. □

The following result is fundamental in duality theory.

Theorem 2.3.3 (of the biconjugate) *Let $f \in \Gamma(X)$. Then $f^* \in \Gamma^*(X^*)$ and $f^{**} := (f^*)^* = f$.*

Proof. Applying Theorem 2.2.6 we get $x_0^* \in X^*$ and $\alpha \in \mathbb{R}$ such that

$$\forall x \in X \ :\ f(x) \geq \langle x, x_0^* \rangle + \alpha. \tag{2.31}$$

Thus, by the preceding corollary, we have that $f^* \in \Gamma^*(X^*)$. Moreover, Theorem 2.3.1 shows that $f^{**} \leq f$.

Let $\overline{x} \in X$ be fixed and consider $\overline{t} \in \mathbb{R}$ such that $\overline{t} < f(\overline{x})$; therefore $(\overline{x}, \overline{t}) \notin \operatorname{epi} f$. Applying Theorem 1.1.5 for $\{(\overline{x}, \overline{t})\}$ and $\operatorname{epi} f$, there exist $(\overline{x}^*, \alpha) \in X^* \times \mathbb{R}$ and $\lambda \in \mathbb{R}$ such that

$$\forall (x, t) \in \operatorname{epi} f \ :\ \langle x, \overline{x}^* \rangle + t\alpha < \lambda < \langle \overline{x}, \overline{x}^* \rangle + \overline{t}\alpha. \tag{2.32}$$

Taking $(x, t) = (\tilde{x}, f(\tilde{x}) + n)$, $n \in \mathbb{N}$, with $\tilde{x} \in \operatorname{dom} f$, we obtain that

$$\forall n \in \mathbb{N} \ :\ \langle \tilde{x}, \overline{x}^* \rangle + \alpha f(\tilde{x}) + n\alpha < \lambda < \langle \overline{x}, \overline{x}^* \rangle + \overline{t}\alpha.$$

Letting $n \to \infty$ we obtain that $\alpha \leq 0$. Take first $\alpha < 0$. Dividing eventually by $-\alpha > 0$, in Eq. (2.32) we can suppose that $\alpha = -1$. Thus

$$\langle x, \overline{x}^* \rangle - f(x) < \lambda \quad \forall x \in \operatorname{dom} f.$$

Hence $f^*(\overline{x}^*) \leq \lambda < \langle \overline{x}, \overline{x}^* \rangle - \overline{t}$, and so

$$\overline{t} < \langle \overline{x}, \overline{x}^* \rangle - f^*(\overline{x}^*) \leq f^{**}(\overline{x}).$$

Take now $\alpha = 0$; using relation (2.32) we get $c > 0$ such that

$$\forall x \in \operatorname{dom} f \ :\ \langle x, \overline{x}^* \rangle + c \leq \langle \overline{x}, \overline{x}^* \rangle.$$

This together with Eq. (2.31) yields

$$f(x) \geq \langle x, x_0^* \rangle + \alpha \geq \langle x, x_0^* \rangle + \alpha + t\langle x, \overline{x}^* \rangle + tc - t\langle \overline{x}, \overline{x}^* \rangle$$

for all $x \in \operatorname{dom} f$ and all $t > 0$; this implies successively:

$$\forall x \in X,\ \forall t > 0 \ :\ -tc + t\langle \overline{x}, \overline{x}^* \rangle - \alpha \geq \langle x, x_0^* + t\overline{x}^* \rangle - f(x),$$
$$\forall t > 0 \ :\ -tc + t\langle \overline{x}, \overline{x}^* \rangle - \alpha \geq f^*(x_0^* + t\overline{x}^*),$$
$$\forall t > 0 \ :\ f^{**}(x) \geq \langle x, x_0^* + t\overline{x}^* \rangle - f^*(x_0^* + t\overline{x}^*) \geq \alpha + tc + \langle \overline{x}, x_0^* \rangle.$$

But there exists $t > 0$ such that $\alpha + tc + \langle \overline{x}, x_0^* \rangle > \overline{t}$; hence $f^{**}(\overline{x}) > \overline{t}$ in this case, too. Thus we obtained $f^{**}(\overline{x}) \geq f(\overline{x})$. Therefore $f^{**} = f$. □

The preceding theorem shows that for any function $f : X \to \overline{\mathbb{R}}$ we have $\left((f^*)^*\right)^* = f^*$ (for the duality (X, X^*)) which shows that there is no interest to consider conjugates of order greater than 2. It also shows that the conjugation is an isomorphism between $\Gamma(X)$ and $\Gamma^*(X^*)$. More precise information on the biconjugate of an arbitrary function is furnished by the following result.

Theorem 2.3.4 *Let $f : X \to \overline{\mathbb{R}}$ have nonempty domain.*

(i) *If $\overline{\text{co}} f$ is proper, then $f^{**} = \overline{\text{co}} f$; if $\overline{\text{co}} f$ is not proper, then $f^{**} = -\infty$.*

(ii) *Suppose that f is convex. If f is lsc at $\overline{x} \in \operatorname{dom} f$, then $f(\overline{x}) = f^{**}(\overline{x})$; moreover, if $f(\overline{x}) \in \mathbb{R}$, then $f^{**} = \overline{f}$ and $\overline{f}$ is proper.*

Proof. (i) The function $\overline{\text{co}} f$ is convex and lsc. If $\overline{\text{co}} f$ is proper, using the preceding theorem and Theorem 2.3.1(iv), we have that

$$\overline{\text{co}} f = (\overline{\text{co}} f)^{**} = (f^*)^* = f^{**}.$$

If $\overline{\text{co}} f$ is not proper, since $\operatorname{dom}(\overline{\text{co}} f) \supset \operatorname{dom} f \neq \emptyset$, $\overline{\text{co}} f$ takes the value $-\infty$; hence $f^* = (\overline{\text{co}} f)^* = \infty$, and so $f^{**} = -\infty$.

(ii) Taking into account that f is convex, $\overline{\text{co}} f = \overline{f}$. Since f is lsc at $\overline{x}$, we have that $\overline{f}(\overline{x}) = f(\overline{x})$. It is obvious that $f^{**}(\overline{x}) = f(\overline{x})$ if $f(\overline{x}) = -\infty$. Let $f(\overline{x}) \in \mathbb{R}$; then $\overline{f}(\overline{x}) \in \mathbb{R}$, and so $\overline{f}$ is proper. From the first part we have that $f^{**} = \overline{f}$, whence $f^{**}(\overline{x}) = \overline{f}(\overline{x}) = f(\overline{x})$. □

Corollary 2.3.5 *Let $f, g \in \Lambda(X)$. If $\overline{f \Box g}$ is proper, then $(f^* + g^*)^* = \overline{f \Box g} = \overline{\overline{f} \Box \overline{g}}$, while in the contrary case $(f^* + g^*)^* = -\infty$. Furthermore, if f is continuous at a point of its domain then*

$$\forall x \in \operatorname{int}(\operatorname{dom}(f \Box g)) = \operatorname{int}(\operatorname{dom} f) + \operatorname{dom} g \ :\ (f \Box g)(x) = (f^* + g^*)^*(x).$$

Proof. From Theorem 2.3.1 we have that

$$(f\Box g)^* = f^* + g^* = \overline{f}^* + \overline{g}^* = (\overline{f}\Box\overline{g})^*;$$

since $f\Box g$ is convex, the conclusion of the first part follows from Theorem 2.3.4.

If f is continuous at $x_0 \in \operatorname{dom} f$, then f is upper bounded on a neighborhood of x_0, whence $f\Box g$ is upper bounded on a neighborhood of x_0+y_0, where $y_0 \in \operatorname{dom} g$. Applying Corollary 2.2.14 we get that $f\Box g$ is continuous on $\operatorname{int}(\operatorname{dom}(f\Box g))$, and so it is lsc on this set. The conclusion follows now from Theorem 2.3.4(ii). □

Consider $\emptyset \neq A \subset X$; the **support function** of A is defined as being

$$s_A : X^* \to \overline{\mathbb{R}}, \quad s_A(x^*) := \sup\{\langle x, x^*\rangle \mid x \in A\}; \tag{2.33}$$

for $\emptyset \neq B \subset X^*$ the support function $s_B : X \to \overline{\mathbb{R}}$ is defined similarly. It is obvious that s_A is w^*-lsc and sublinear while s_B is lsc and sublinear. Furthermore, if $C \subset X$ is another nonempty set then $s_{A+C} = s_A + s_C$ and $s_{A\cup C} = s_A \vee s_C$. Note that

$$(\iota_A)^*(x^*) = \sup_{x\in A}\langle x, x^*\rangle = s_A(x^*) = \sup_{x\in\overline{\operatorname{co}}A}\langle x, x^*\rangle = (\iota_{\overline{\operatorname{co}}A})^*(x^*) = s_{\overline{\operatorname{co}}A}.$$

Moreover we have that $\iota_{\overline{\operatorname{co}}A} = \overline{\operatorname{co}}\iota_A$ and

$$\operatorname{dom} s_A = \operatorname{dom}(\iota_A)^* = \{x^* \in X^* \mid x^* \text{ is upper bounded on } A\};$$

therefore $\operatorname{dom} s_A = X^*$ if and only if A is w-bounded. In the next section we shall see that ι_A is useful in determining the normal cone of C.

2.4 The Subdifferential of a Convex Function

We have already seen in Section 2.1 that if the proper convex function $f : (X, \|\cdot\|) \to \overline{\mathbb{R}}$ is Gâteaux differentiable at $\overline{x} \in \operatorname{int}(\operatorname{dom} f)$ then

$$\forall\, x \in \operatorname{dom} f \ (\forall\, x \in X) \ : \ \langle x - \overline{x}, \nabla f(\overline{x})\rangle \le f(x) - f(\overline{x}).$$

Taking into account this relation, it is quite natural to consider the elements $x^* \in X^*$ which satisfy the inequality

$$\forall\, x \in X \ : \ \langle x - \overline{x}, x^*\rangle \le f(x) - f(\overline{x}), \tag{2.34}$$

even if f is not Gâteaux differentiable at $\overline{x}$.

In this section the spaces under consideration are separated locally convex spaces if not stated otherwise.

Let $f : X \to \overline{\mathbb{R}}$ and $\overline{x} \in X$ be such that $f(\overline{x}) \in \mathbb{R}$. An element $x^* \in X^*$ is called a **subgradient** of the function f at $\overline{x}$ if relation (2.34) is satisfied; the set of all the subgradients of the function f at $\overline{x}$ is denoted by $\partial f(\overline{x})$ and is called the **subdifferential** or *Fenchel subdifferential* of f at $\overline{x}$. We consider that $\partial f(\overline{x}) = \emptyset$ if $f(\overline{x}) \notin \mathbb{R}$; of course we can have $\partial f(\overline{x}) = \emptyset$ even if $f(\overline{x}) \in \mathbb{R}$. Thus we obtain a multifunction $\partial f : X \rightrightarrows X^*$. By the preceding considerations we have that $\operatorname{dom} \partial f \subset \operatorname{dom} f$. We say that f is **subdifferentiable** at $\overline{x} \in X$ if $\partial f(\overline{x}) \neq \emptyset$.

Note that if $x^* \in \partial f(\overline{x})$, the affine function $\varphi : X \to \mathbb{R}$, $\varphi(x) := \langle x, x^* \rangle - \langle \overline{x}, x^* \rangle + f(\overline{x})$ minorizes f and coincides with f at $\overline{x}$; this proves that

$$\forall\, (x,t) \in \operatorname{epi} f \; : \; \langle x, x^* \rangle - t \leq \alpha := \langle \overline{x}, x^* \rangle - f(\overline{x}),$$

which shows that the hyperplane $\{(x,t) \in X \times \mathbb{R} \mid \langle x, x^* \rangle - t \cdot 1 = \alpha\}$ is a non vertical (since the coefficient of t is $\neq 0$) supporting hyperplane (see p. 5 for the definition).

Recall that in Theorem 2.1.5 we have already determined the subdifferential of a function $f \in \Lambda(\mathbb{R})$ at $t_0 \in \operatorname{dom} f$:

$$\partial f(t_0) = [f'_-(t_0), f'_+(t_0)] \cap \mathbb{R}.$$

In the sequel we shall establish properties of the subdifferential and methods for computing it also for $X \neq \mathbb{R}$.

In the next result we collect several easy properties of the subdifferential.

Theorem 2.4.1 *Let $f : X \to \overline{\mathbb{R}}$ and $\overline{x} \in X$ be such that $f(\overline{x}) \in \mathbb{R}$. Then:*

(i) *$\partial f(\overline{x}) \subset X^*$ is a convex and w^*-closed (eventually empty) set.*

(ii) *If $\partial f(\overline{x}) \neq \emptyset$ then*

$$(\overline{\operatorname{co}} f)(\overline{x}) = \overline{f}(\overline{x}) = f(\overline{x}) \quad \text{and} \quad \partial(\overline{\operatorname{co}} f)(\overline{x}) = \partial \overline{f}(\overline{x}) = \partial f(\overline{x});$$

in particular $f^{**} = \overline{\operatorname{co}} f$ and f is proper and lsc at $\overline{x}$.

(iii) *If f is proper, $\operatorname{dom} f$ is a convex set and f is subdifferentiable at every $x \in \operatorname{dom} f$, then f is convex.*

Proof. (i) Let $x_1^*, x_2^* \in \partial f(\overline{x})$ and $\lambda \in]0,1[$. Then

$$\forall\, x \in X \; : \; \langle x - \overline{x}, x_1^* \rangle \leq f(x) - f(\overline{x}), \quad \langle x - \overline{x}, x_2^* \rangle \leq f(x) - f(\overline{x}).$$

Multiplying the first relation with $\lambda > 0$, the second with $1-\lambda > 0$, then adding them, we obtain

$$\forall\, x \in X \;:\; \langle x - \overline{x}, \lambda x_1^* + (1-\lambda)x_2^* \rangle \le f(x) - f(\overline{x}),$$

whence $\lambda x_1^* + (1-\lambda)x_2^* \in \partial f(\overline{x})$.

Let $\overline{x}^* \in X^* \setminus \partial f(\overline{x})$. Then there exists $x_0 \in X$ such that $\langle x_0 - \overline{x}, \overline{x}^* \rangle > f(x_0) - f(\overline{x})$. Let $\alpha \in \mathbb{R}$ be such that $\langle x_0 - \overline{x}, \overline{x}^* \rangle > \alpha > f(x_0) - f(\overline{x})$. Then $V := \{x^* \mid \langle x_0 - \overline{x}, x^* \rangle > \alpha\}$ is a neighborhood of $\overline{x}^*$ for the topology $w^* = \sigma(X^*, X)$. It is obvious that $V \cap \partial f(\overline{x}) = \emptyset$. So $\partial f(\overline{x})$ is w^*-closed.

(ii) We already know that $\overline{\text{co}} f \le \overline{f} \le f$. Let $\overline{x}^* \in \partial f(\overline{x})$ and

$$\varphi : \mathbb{R} \to \mathbb{R}, \quad \varphi(x) := \langle x - \overline{x}, \overline{x}^* \rangle + f(\overline{x});$$

φ is convex, continuous and $\varphi \le f$. Therefore $\varphi \le \overline{\text{co}} f \le \overline{f} \le f$. Since $\varphi(\overline{x}) = f(\overline{x})$, we have that $\overline{\text{co}} f(\overline{x}) = \overline{f}(\overline{x}) = f(\overline{x})$. This relation proves that the functions $f, \overline{f}, \overline{\text{co}} f$ are proper and f is lsc at $\overline{x}$. From the above inequality we obtain that $\partial(\overline{\text{co}} f)(\overline{x}) \supset \partial \overline{f}(\overline{x}) \supset \partial f(\overline{x})$. If $x^* \in \partial(\overline{\text{co}} f)(\overline{x})$, then

$$\forall\, x \in X \;:\; \langle x - \overline{x}, x^* \rangle \le (\overline{\text{co}} f)(x) - (\overline{\text{co}} f)(\overline{x}) \le f(x) - f(\overline{x}),$$

whence $\partial(\overline{\text{co}} f)(\overline{x}) \subset \partial f(\overline{x})$. Therefore $\partial(\overline{\text{co}} f)(\overline{x}) = \partial \overline{f}(\overline{x}) = \partial f(\overline{x})$.

(iii) By (ii) we have that $f(x) = \overline{\text{co}} f(x)$ for every $x \in \operatorname{dom} f$. Since $\overline{\text{co}} f$ is a convex function and $\operatorname{dom} f$ is convex, it is obvious that f is convex. □

The property (ii) of Theorem 2.4.1 justifies why we consider only proper convex functions when discussing subdifferentiability (in the sense of convex analysis!).

In a similar way we introduce the subdifferential of a function $h : X^* \to \overline{\mathbb{R}}$ at a point $\overline{x}^* \in X^*$ with $h(\overline{x}^*) \in \mathbb{R}$:

$$\partial h(\overline{x}^*) = \{x \in X \mid \forall\, x^* \in X^* \;:\; \langle x, x^* - \overline{x}^* \rangle \le h(x^*) - h(\overline{x}^*)\}.$$

(As for conjugation, this distinction is important only when working with normed spaces; in the case of locally convex spaces we always consider the natural duality between X and X^*, when the dual of X^* is X.)

Similar results to those of Theorem 2.4.1 hold; more precisely $\partial h(\overline{x}^*)$ is a closed convex (eventually empty) set and if $\partial h(\overline{x}^*) \ne \emptyset$ then h is proper and w^*-lsc at $\overline{x}^*$; the other statements are also valid if one takes the closures with respect to the weak* topology.

In practice (for example for solving numerically problems using computers) it is possible to determine the subgradients only approximately. In this sense the following notion of subgradient reveals itself to be useful. Let $f : X \to \overline{\mathbb{R}}$, $\overline{x} \in X$ with $f(\overline{x}) \in \mathbb{R}$ and $\varepsilon \in \mathbb{R}_+$; the element $x^* \in X^*$ is called an ε-**subgradient**, of the function f at $\overline{x}$ if

$$\forall x \in X \ : \ \langle x - \overline{x}, x^* \rangle \le f(x) - f(\overline{x}) + \varepsilon; \tag{2.35}$$

the set of ε-subgradients of f at $\overline{x}$ is denoted by $\partial_\varepsilon f(\overline{x})$ and is called the ε-**subdifferential** of f at $\overline{x}$. As for the subdifferential, if $f(\overline{x}) \notin \mathbb{R}$ we consider that $\partial_\varepsilon f(\overline{x}) = \emptyset$; we obtain a multifunction $\partial_\varepsilon f : X \rightrightarrows X^*$ with $\operatorname{dom}(\partial_\varepsilon f) \subset \operatorname{dom} f$. Note that f is proper if $\partial_\varepsilon f(\overline{x}) \neq \emptyset$ for some $\varepsilon \ge 0$; if $0 \le \varepsilon_1 \le \varepsilon_2 < \infty$ then

$$\partial f(\overline{x}) = \partial_0 f(\overline{x}) \subset \partial_{\varepsilon_1} f(\overline{x}) \subset \partial_{\varepsilon_2} f(\overline{x}).$$

Moreover

$$\forall \varepsilon \in \mathbb{R}_+ \ : \ \partial_\varepsilon f(\overline{x}) = \bigcap\nolimits_{\eta > \varepsilon} \partial_\eta f(\overline{x}).$$

The ε-subdifferential of a function $h : X^* \to \overline{\mathbb{R}}$ at $\overline{x}^* \in X^*$ with $h(\overline{x}^*) \in \mathbb{R}$ is introduced similarly.

In the following theorem we collect several properties of the subdifferential and of the ε-subdifferential. Before stating this theorem, let us introduce some notions: we say that the set $M \subset X \times X^*$ is **monotone** if

$$\forall\, (x, x^*),\, (y, y^*) \in M \ : \ \langle x - y, x^* - y^* \rangle \ge 0;$$

M is **strictly monotone** if

$$\forall\, (x, x^*),\, (y, y^*) \in M,\ x \neq y \ : \ \langle x - y, x^* - y^* \rangle > 0;$$

M is **maximal monotone** if a) M is monotone and b) $M' \subset X \times X^*$ monotone and $M \subset M'$ imply $M = M'$, *i.e.* if M is a maximal element of the class of monotone subsets ordered by inclusion.

We say that the multifunction $T : X \rightrightarrows X^*$ is **monotone, strictly monotone** or **maximal monotone** if $\operatorname{gr} T$ is monotone, strictly monotone or maximal monotone, respectively. Of course, when $X = \mathbb{R}$ and T is single-valued (*i.e.* $T(x)$ is a singleton for every $x \in \operatorname{dom} T$), T is (strictly) monotone if and only if the function $T|_{\operatorname{dom} T}$ is nondecreasing (increasing).

In the next result we do not assume that the functions are convex.

Theorem 2.4.2 *Let $f, g : X \to \overline{\mathbb{R}}$, $h : Y \to \overline{\mathbb{R}}$ be proper functions, $A \in \mathcal{L}(X,Y)$, $\overline{x} \in \operatorname{dom} f \cap \operatorname{dom} g$, $\overline{y} \in \operatorname{dom} h$ and $\varepsilon \in \mathbb{R}_+$. Then:*

(i) *$\partial_\varepsilon f(\overline{x})$ is a convex w^*-closed set.*

(ii) $x^* \in \partial_\varepsilon f(\overline{x}) \Leftrightarrow f(\overline{x}) + f^*(x^*) \le \langle \overline{x}, x^* \rangle + \varepsilon \Rightarrow \overline{x} \in \partial_\varepsilon f^*(\overline{x}^*)$.

(iii) $x^* \in \partial f(\overline{x}) \Leftrightarrow f(\overline{x}) + f^*(x^*) \le \langle \overline{x}, x^* \rangle \Leftrightarrow f(\overline{x}) + f^*(x^*) = \langle \overline{x}, x^* \rangle \Rightarrow \overline{x} \in \partial f^*(\overline{x}^*)$.

(iv) $\operatorname{dom}(\partial_\varepsilon f) \subset \operatorname{dom} f$, $\operatorname{Im}(\partial_\varepsilon f) \subset \operatorname{dom} f^*$ *and ∂f is monotone.*

(v) $\partial f(\overline{x}) \ne \emptyset \Leftrightarrow f(\overline{x}) = \max_{x^* \in X^*} \left(\langle \overline{x}, x^* \rangle - f^*(x^*)\right)$.

(vi) $\partial_\varepsilon (f + \overline{x}^*)(\overline{x}) = \overline{x}^* + \partial_\varepsilon f(\overline{x})$ *for* $\overline{x}^* \in X^*$; $\partial_\varepsilon(\lambda f)(\overline{x}) = \lambda \partial_{\varepsilon/\lambda} f(\overline{x})$ *and* $\partial(\lambda f)(\overline{x}) = \lambda \partial f(\overline{x})$ *for* $\lambda > 0$; *if* $g(x) = f(x + x_0)$ *for* $x \in X$ *then* $\partial_\varepsilon g(\overline{x}) = \partial_\varepsilon f(\overline{x} + x_0)$.

(vii) *Assume that $\overline{y} = A\overline{x}$; then $A^*\left(\partial_\varepsilon h(\overline{y})\right) \subset \partial_\varepsilon (h \circ A)(\overline{x})$.*

(viii) $\bigcup_{\eta \in [0,\varepsilon]} \left(\partial_\eta f(\overline{x}) + \partial_{\varepsilon - \eta} g(\overline{x})\right) \subset \partial_\varepsilon (f + g)(\overline{x})$.

(ix) *Suppose X is a normed space, f is lsc, $\varepsilon_i \ge 0$ and $(x_i, x_i^*) \in \operatorname{gr} \partial_{\varepsilon_i} f$ for every $i \in I$. If $(\varepsilon_i)_{i \in I} \to \varepsilon < \infty$ and either* (a) *$(x_i)_{i \in I} \xrightarrow{\|\cdot\|} x$, $(x_i^*)_{i \in I} \xrightarrow{w^*} x^*$ and $(x_i^*)_{i \in I}$ is norm-bounded or* (b) *$(x_i)_{i \in I} \xrightarrow{w} x$, $(x_i)_{i \in I}$ is norm-bounded and $(x_i^*) \xrightarrow{\|\cdot\|} x^*$, then $(x, x^*) \in \operatorname{gr} \partial_\varepsilon f$. In particular $\operatorname{gr} \partial_\varepsilon f$ is closed in $X \times X^*$ (for the norm topology).*

Proof. (i) The fact that $\partial_\varepsilon f(\overline{x})$ is convex and w^*-closed is shown similarly to the first part of Theorem 2.4.1.

(ii) We have that

$$\begin{aligned} x^* \in \partial_\varepsilon f(\overline{x}) &\Leftrightarrow \forall x \in X \ : \ \langle x - \overline{x}, x^* \rangle \le f(x) - f(\overline{x}) + \varepsilon \\ &\Leftrightarrow \forall x \in X \ : \ \langle x, x^* \rangle - f(x) \le \langle \overline{x}, x^* \rangle - f(\overline{x}) + \varepsilon \\ &\Leftrightarrow f^*(x^*) \le \langle \overline{x}, x^* \rangle - f(\overline{x}) + \varepsilon \\ &\Leftrightarrow f(\overline{x}) + f^*(x^*) \le \langle \overline{x}, x^* \rangle + \varepsilon. \end{aligned}$$

Assume now that $\overline{x}^* \in \partial_\varepsilon f(\overline{x})$. Then, by what precedes,

$$f^{**}(\overline{x}) + f^*(\overline{x}^*) \le f(\overline{x}) + f^*(\overline{x}^*) \le \langle \overline{x}, \overline{x}^* \rangle + \varepsilon,$$

and so $\overline{x} \in \partial_\varepsilon f^*(\overline{x}^*)$.

(iii) We already know that $\langle \overline{x}, x^* \rangle \le f(\overline{x}) + f^*(x^*)$ (the Young–Fenchel inequality!); the equivalences follow now from (ii) taking $\varepsilon = 0$.

(iv) It is obvious that $\operatorname{dom}(\partial_\varepsilon f) \subset \operatorname{dom} f$, while the inclusion $\operatorname{Im} \partial_\varepsilon f \subset \operatorname{dom} f^*$ follows from (ii).

Let (x, x^*), $(y, y^*) \in \operatorname{gr} \partial f$. Then $x, y \in \operatorname{dom} f$ and

$$\langle y - x, x^* \rangle \leq f(y) - f(x), \quad \langle x - y, y^* \rangle \leq f(x) - f(y).$$

Adding these relations we get $\langle y - x, x^* - y^* \rangle \leq 0$, and so ∂f is monotone.

(v) If $\overline{x}^* \in \partial f(\overline{x})$, taking into account (iii) and the Young–Fenchel inequality, we have

$$\forall\, x^* \in X^* \ :\ \langle \overline{x}, \overline{x}^* \rangle - f^*(\overline{x}^*) = f(\overline{x}) \geq \langle \overline{x}, x^* \rangle - f^*(x^*),$$

and so the implication "$\Rightarrow$" is true. The converse implication is an immediate consequence of the equivalences of (iii).

(vi)–(viii) follow easily from the definition of the ε-subdifferential.

(ix) Suppose that X is a normed space. Let $(\varepsilon_i)_{i\in I} \to \varepsilon < \infty$ and $\left((x_i, x_i^*)\right)_{i\in I}$ have the properties from (a) or (b). Taking into account that

$$\langle x_i, x_i^* \rangle - \langle x, x^* \rangle = \langle x_i - x, x_i^* \rangle + \langle x, x_i^* - x^* \rangle = \langle x_i, x_i^* - x^* \rangle + \langle x_i - x, x^* \rangle$$

for every i, we get $(\langle y - x_i, x_i^* \rangle) \to \langle y - x, x^* \rangle$ for every $y \in X$ in both cases. But

$$\forall\, i \in I,\ \forall\, y \in X \ :\ f(x_i) + \langle y - x_i, x_i^* \rangle \leq f(y) + \varepsilon_i;$$

taking the limit inferior, we obtain $x^* \in \partial_\varepsilon f(x)$. □

In the preceding theorem we obtained that ∂f is a monotone multifunction. In fact ∂f is cyclically monotone. One says that $T : X \rightrightarrows X^*$ is **cyclically monotone** if for all $n \in \mathbb{N}$ and $\left((x_i, x_i^*)\right)_{i=0}^n \subset \operatorname{gr} T$ one has

$$\sum\nolimits_{i=0}^{n} \langle x_{i+1} - x_i, x_i^* \rangle \leq 0, \tag{2.36}$$

where $x_{n+1} := x_0$. Taking $n = 1$ in Eq. (2.36) we obtain that $\langle x_1 - x_0, x_0^* \rangle + \langle x_0 - x_1, x_1^* \rangle \leq 0$, *i.e.* $\langle x_1 - x_0, x_1^* - x_0^* \rangle \geq 0$, for all (x_0, x_0^*), $(x_1, x_1^*) \in \operatorname{gr} T$. Hence every cyclically monotone multifunction is monotone. When $f : X \to \overline{\mathbb{R}}$ is a proper function, ∂f is a cyclically monotone multifunction. Indeed, let $n \in \mathbb{N}$ and $\left((x_i, x_i^*)\right)_{i=0}^n \subset \operatorname{gr} \partial f$; then $(x_i)_{i=0}^n \subset \operatorname{dom} f$ and

$$\langle x_{i+1} - x_i, x_i^* \rangle \leq f(x_{i+1}) - f(x_i) \quad \forall\, i,\ 0 \leq i \leq n,$$

where, as above, $x_{n+1} := x_0$. Adding these relations side by side for $i = 0, 1, \ldots, n$, we get Eq. (2.36). Note that to a nonempty cyclically monotone multifunction one can associate a function $f \in \Gamma(X)$.

Proposition 2.4.3 *Let $T : X \rightrightarrows X^*$ be a cyclically monotone multifunction and $(x_0, x_0^*) \in \operatorname{gr} T$. Consider $f_T : X \to \overline{\mathbb{R}}$ defined by*

$$f_T(x) := \sup \left(\langle x - x_n, x_n^* \rangle + \sum\nolimits_{i=0}^{n-1} \langle x_{i-1} - x_i, x_i^* \rangle \right), \tag{2.37}$$

where the supremum is taken for all families $\left((x_i, x_i^*)\right)_{i=0}^{n} \subset \operatorname{gr} T$ with $n \in \mathbb{N}$. Then $f_T \in \Gamma(X)$, $f_T(x_0) = 0$ and $T(x) \subset \partial f_T(x)$ for every $x \in X$.

Proof. Because f_T is a supremum of continuous affine functions, f_T is lsc, convex and nowhere $-\infty$. Note that $\operatorname{dom} T \subset \operatorname{dom} f_T$. Indeed, let $(\overline{x}, \overline{x}^*) \in \operatorname{gr} T$. Taking $k \in \mathbb{N}$, $\left((x_i, x_i^*)\right)_{i=0}^{k} \subset \operatorname{gr} T$, $n := k+1$ and $(x_n, x_n^*) := (\overline{x}, \overline{x}^*)$, from Eq. (2.36) we obtain that

$$\sum\nolimits_{i=0}^{k-1} \langle x_{i+1} - x_i, x_i^* \rangle + \langle \overline{x} - x_k, x_k^* \rangle + \langle x_0 - \overline{x}, \overline{x}^* \rangle \leq 0,$$

whence $f_T(\overline{x}) \leq \langle \overline{x} - x_0, \overline{x}^* \rangle$. Hence $\overline{x} \in \operatorname{dom} f_T$. In particular the preceding inequality shows that $f_T(x_0) \leq 0$. Taking $n = 1$ and $(x_1, x_1^*) = (x_0, x_0^*)$ in the definition of $f_T(x_0)$, we get $f_T(x_0) \geq 0$. Hence $f_T(x_0) = 0$. Therefore $f_T \in \Gamma(X)$.

Let now $(\overline{x}, \overline{x}^*) \in \operatorname{gr} T$ and $x \in X$. Consider an arbitrary $\mu < f_T(\overline{x}) + \langle x - \overline{x}, \overline{x}^* \rangle$. Then there exist $k \in \mathbb{N}$ and $\left((x_i, x_i^*)\right)_{i=0}^{k} \subset \operatorname{gr} T$ such that

$$\mu - \langle x - \overline{x}, \overline{x}^* \rangle < \langle \overline{x} - x_k, x_k^* \rangle + \sum\nolimits_{i=0}^{k-1} \langle x_{i+1} - x_i, x_i^* \rangle .$$

Taking $n = k + 1$ and $(x_n, x_n^*) = (\overline{x}, \overline{x}^*)$, the preceding inequality yields

$$\mu < \langle x - x_n, x_n^* \rangle + \sum\nolimits_{i=0}^{n-1} \langle x_{i+1} - x_i, x_i^* \rangle \leq f(x).$$

Letting $\mu \to f_T(\overline{x}) + \langle x - \overline{x}, \overline{x}^* \rangle$, we obtain that $f_T(\overline{x}) + \langle x - \overline{x}, \overline{x}^* \rangle \leq f_T(x)$, and so $\overline{x}^* \in \partial f_T(\overline{x})$. □

In the next theorem we use the convexity of the function f.

Theorem 2.4.4 *Let $f \in \Lambda(X)$, $\overline{x} \in \operatorname{dom} f$ and $\varepsilon \in \mathbb{R}_+$. Then:*

(i) *$\partial_\varepsilon f(\overline{x}) = \partial f'_\varepsilon(\overline{x}, \cdot)(0)$; moreover, if $\overline{x} \in (\operatorname{dom} f)^i$ and f is Gâteaux differentiable at $\overline{x}$ then $\partial f(\overline{x}) = \{\nabla f(\overline{x})\}$.*

(ii) *If f is strictly convex then ∂f is strictly monotone.*

(iii) *f is lsc at $\overline{x}$ if and only if $\partial_\varepsilon f(\overline{x}) \neq \emptyset$ for every $\varepsilon > 0$; $\partial_\varepsilon f^*(\overline{x}^*) \neq \emptyset$ if $f^*(\overline{x}^*) \in \mathbb{R}$ and $\varepsilon > 0$.*

(iv) *Suppose that f is lsc at $\overline{x}$. Then $x^* \in \partial_\varepsilon f(\overline{x}) \Leftrightarrow \overline{x} \in \partial_\varepsilon f^*(x^*)$.*

Proof. (i) If $x^* \in \partial_\varepsilon f(\overline{x})$, replacing x by $\overline{x}+tu$, with $t > 0$, in Eq. (2.35), then dividing by t we obtain

$$\forall u \in X \ : \ \langle u, x^* \rangle \leq \inf_{t>0} \frac{f(\overline{x}+tu) - f(\overline{x}) + \varepsilon}{t} = f'_\varepsilon(\overline{x}, u),$$

and so $x^* \in \partial f'_\varepsilon(\overline{x}, \cdot)(0)$. The converse inclusion follows from the inequality $f'_\varepsilon(\overline{x}, u) \leq f(\overline{x}+u) - f(\overline{x}) + \varepsilon$, valid for every $u \in X$.

If f is Gâteaux differentiable at $\overline{x}$ then $f'_+(\overline{x}, \cdot) = \nabla f(\overline{x})$; the conclusion is obvious.

(ii) Let $x, y \in \operatorname{dom} f$, $x \neq y$, and $x^* \in \partial f(x)$, $y^* \in \partial f(y)$. From (i) we have that

$$\langle y - x, x^* \rangle \leq f'_+(x, y - x) < f(y) - f(x),$$
$$\langle x - y, y^* \rangle \leq f'_+(y, x - y) < f(x) - f(y),$$

because f is strictly convex. Adding the above two relations we obtain that $\langle y - x, x^* - y^* \rangle < 0$. This shows that ∂f is strictly monotone.

(iii) Suppose that f is lsc at $\overline{x}$ and let us take $\varepsilon > 0$; using Theorem 2.3.4 we have

$$f(\overline{x}) = f^{**}(\overline{x}) = \sup\{\langle \overline{x}, x^* \rangle - f^*(x^*) \mid x^* \in X^*\} > f(\overline{x}) - \varepsilon.$$

Therefore there exists $x^* \in X^*$ such that $\langle \overline{x}, x^* \rangle - f^*(x^*) > f(\overline{x}) - \varepsilon$, whence $x^* \in \partial_\varepsilon f(\overline{x})$.

Conversely, suppose that $\partial_\varepsilon f(\overline{x}) \neq \emptyset$ for every $\varepsilon > 0$. Then

$$\forall \varepsilon > 0, \ \exists x^* \in X^* \ : \ f(\overline{x}) - \varepsilon \leq \langle \overline{x}, x^* \rangle - f^*(x^*) \leq f^{**}(\overline{x}),$$

whence $f(\overline{x}) \leq f^{**}(\overline{x})$. Since $f^{**} \leq \overline{f} \leq f$, it follows that $\overline{f}(\overline{x}) = f(\overline{x})$, *i.e.* f is lsc at $\overline{x}$.

Let $\overline{x}^* \in X^*$ be such that $f^*(\overline{x}^*) \in \mathbb{R}$ and $\varepsilon > 0$. By the definition of f^*, there exists $\overline{x} \in X$ such that $f^*(\overline{x}^*) < \langle \overline{x}, \overline{x}^* \rangle - f(\overline{x}) + \varepsilon$, whence

$$\langle \overline{x}, x^* - \overline{x}^* \rangle < \langle \overline{x}, x^* \rangle - f(\overline{x}) - f^*(\overline{x}^*) + \varepsilon \leq f^*(x^*) - f^*(\overline{x}^*) + \varepsilon,$$

i.e. $\overline{x} \in \partial_\varepsilon f^*(\overline{x}^*)$.

(iv) Since f is lsc at $\overline{x}$, we have that $f^{**}(\overline{x}) = f(\overline{x})$; the conclusion is immediate using Theorem 2.4.2 (ii). □

Remark 2.4.1 Note that for $f \in \Lambda(X)$, $\overline{x} \in \operatorname{dom} f$ and $U \in \mathcal{N}^c_X(\overline{x})$ we have that $\partial f(\overline{x}) = \partial(f + \iota_U)(\overline{x})$, but $\partial_\varepsilon f(\overline{x})$ and $\partial_\varepsilon(f + \iota_U)(\overline{x})$ may be different for $\varepsilon > 0$, *i.e.* ∂f is a local notion while $\partial_\varepsilon f$ (with $\varepsilon > 0$) is a global one for convex functions.

Indeed, since $f'(\overline{x}, u) = (f + \iota_U)'(\overline{x}, u)$ for every $u \in X$, the first remark follows from assertion (i) of the preceding theorem. For the second remark let us consider $f \in \Lambda(\mathbb{R})$ given by $f(x) = x^2$ and $U = [-1, 1]$. Then $\partial_\varepsilon(f + \iota_U)(0) = \partial_\varepsilon f(0) = [-2\sqrt{\varepsilon}, 2\sqrt{\varepsilon}]$ for $\varepsilon \in [0, 1]$ and $\partial_\varepsilon(f + \iota_U)(0) = [-1 - \varepsilon, 1 + \varepsilon] \neq [-2\sqrt{\varepsilon}, 2\sqrt{\varepsilon}] = \partial_\varepsilon f(0)$ for $\varepsilon > 1$.

Generally, the formula $\partial(\lambda f)(x) = \lambda \partial f(x)$ is not true for $\lambda = 0$; this formula, however, is true if $x \in (\operatorname{dom} f)^i$. In assertions (vii) and (viii) of Theorem 2.4.2 we have equality only under supplementary conditions called generally "constraint qualification conditions;" in Section 2.8 we shall give such conditions.

Let $A \subset X$ be a nonempty convex set and $a \in A$; then

$$\begin{aligned}
\partial \iota_A(a) &= \{x^* \in X^* \mid \forall\, x \in A \ : \ \langle x - a, x^* \rangle \leq 0\} \\
&= \{x^* \mid \forall\, x \in A \ : \ \langle x, x^* \rangle \leq \langle a, x^* \rangle\} \\
&= \{x^* \mid \forall\, x \in \operatorname{cone}(A - a) \ : \ \langle x, x^* \rangle \leq 0\} \\
&= -(\operatorname{cone}(A - a))^+ = -\overline{\operatorname{cone}(A - a)}^{\,+}.
\end{aligned}$$

The set $\partial \iota_A(a)$ is denoted by $N(A; a)$ and is called the **normal cone** of A at $a \in A$, while the set $\overline{\operatorname{cone}}(A - a)$ is denoted by $C(A; a)$ (even if A is not convex); evidently, the set $C(A; a)$ is a closed (convex if A is so) cone. Taking into account the relation established above and the bipolar theorem (Theorem 1.1.9), we have that

$$N(A; a) = -\big(C(A; a)\big)^- \quad \text{and} \quad C(A; a) = -\big(N(A; a)\big)^+.$$

It is obvious that $N(A; a) = \{0\}$ if $a \in A^i$. We observe that $x^* \in N(A; a) \setminus \{0\}$ if and only if $H_{x^*, \langle a, x^* \rangle}$ is a supporting hyperplane to A at a and $A \subset H^{\leq}_{x^*, \langle a, x^* \rangle}$.

The set $\partial f(\overline{x})$ may be empty even if f is lsc at $\overline{x}$. For example, the function $f : \mathbb{R} \to \overline{\mathbb{R}}$, $f(x) := -\sqrt{1 - x^2}$ for $|x| \leq 1$, $f(x) := \infty$ for $|x| > 1$ (already considered in Section 2.1) is lsc, finite at 1, but $\partial f(1) = \emptyset$. Moreover $\partial(0 \cdot f)(1) = \mathbb{R}_-$.

In the preceding section we have met some situations in which it was easy to compute the conjugate of a function. We shall point such situations for the ε-subdifferential, too.

Corollary 2.4.5 *Let $f_i : X_i \to \overline{\mathbb{R}}$ for $i \in \overline{1,n}$ be proper functions and $\overline{x} = (\overline{x}_1, \ldots, \overline{x}_n) \in \prod_{i=1}^n \operatorname{dom} f_i$. Let us consider the function*

$$\varphi : \prod\nolimits_{i=1}^n X_i \to \overline{\mathbb{R}}, \quad \varphi(x_1, \ldots, x_n) := \sum\nolimits_{i=1}^n f_i(x_i),$$

and $\varepsilon \in \mathbb{R}_+$. Then

$$\partial_\varepsilon \varphi(\overline{x}) = \bigcup \left\{ \prod\nolimits_{i=1}^n \partial_{\varepsilon_i} f_i(\overline{x}_i) \,\middle|\, \varepsilon_i \geq 0,\ \varepsilon_1 + \cdots + \varepsilon_n = \varepsilon \right\}.$$

In particular

$$\partial \varphi(\overline{x}_1, \ldots, \overline{x}_n) = \prod\nolimits_{i=1}^n \partial f_i(\overline{x}_i).$$

Proof. We have already seen in the preceding section (for $n = 2$, but the extension is immediate) that $\varphi^*(x_1^*, \ldots, x_n^*) = \sum_{i=1}^n f_i^*(x_i^*)$. Therefore, by Theorem 2.4.2 (ii),

$$\begin{aligned}
x^* \in \partial_\varepsilon \varphi(\overline{x}) &\Leftrightarrow \varphi(\overline{x}) + \varphi^*(x^*) \leq \langle \overline{x}, x^* \rangle + \varepsilon = \langle \overline{x}_1, x_1^* \rangle + \cdots + \langle \overline{x}_n, x_n^* \rangle + \varepsilon \\
&\Leftrightarrow \sum\nolimits_{i=1}^n \left(f_i(\overline{x}_i) + f_i^*(x_i^*) - \langle \overline{x}_i, x_i^* \rangle \right) \leq \varepsilon \\
&\Leftrightarrow \exists \varepsilon_1, \ldots, \varepsilon_n \geq 0,\ \varepsilon_1 + \ldots + \varepsilon_n = \varepsilon, \\
&\qquad\qquad \forall i \in \overline{1,n} \ :\ f_i(\overline{x}_i) + f_i^*(x_i^*) - \langle \overline{x}_i, x_i^* \rangle \leq \varepsilon_i, \\
&\Leftrightarrow \exists \varepsilon_1, \ldots, \varepsilon_n \geq 0,\ \varepsilon_1 + \ldots + \varepsilon_n = \varepsilon, \\
&\qquad\qquad \forall i \in \overline{1,n} \ :\ x_i^* \in \partial_{\varepsilon_i} f_i(\overline{x}_i),
\end{aligned}$$

whence the conclusion. □

Corollary 2.4.6 *Let $f : X \to \overline{\mathbb{R}}$ be a proper function and $A \in \mathcal{L}(X, Y)$. If $x \in \operatorname{dom} f$ and $y \in Y$ are such that $y = Ax$ and $(Af)(y) = f(x)$, then for every $\varepsilon \in \mathbb{R}_+$ we have*

$$\partial_\varepsilon (Af)(y) = {A^*}^{-1}(\partial_\varepsilon f(x)).$$

Proof. Using Theorem 2.4.2 (ii) we have that

$$\begin{aligned}
y^* \in \partial_\varepsilon (Af)(y) &\Leftrightarrow (Af)(y) + (Af)^*(y^*) \leq \langle y, y^* \rangle + \varepsilon \\
&\Leftrightarrow f(x) + f^*(A^* y^*) \leq \langle Ax, y^* \rangle + \varepsilon = \langle x, A^* y^* \rangle + \varepsilon \\
&\Leftrightarrow A^* y^* \in \partial_\varepsilon f(x) \Leftrightarrow y^* \in {A^*}^{-1} \left(\partial_\varepsilon f(x) \right).
\end{aligned}$$

We used the relation $(Af)^* = f^* \circ A^*$, proved in Theorem 2.3.1 (ix). □

Corollary 2.4.7 *Let $f_1, \dots, f_n : X \to \overline{\mathbb{R}}$ $(n \in \mathbb{N})$ be proper functions. Suppose that for every $i \in \overline{1,n}$ there exists $\overline{x}_i \in \operatorname{dom} f_i$ such that*

$$(f_1 \square \cdots \square f_n)(\overline{x}_1 + \ldots + \overline{x}_n) = f_1(\overline{x}_1) + \cdots + f_n(\overline{x}_n). \tag{2.38}$$

Then for every $\varepsilon \in \mathbb{R}_+$

$$\partial_\varepsilon (f_1 \square \ldots \square f_n)(\overline{x}_1 + \cdots + \overline{x}_n)$$
$$= \bigcup \{\partial_{\varepsilon_1} f_1(\overline{x}_1) \cap \cdots \cap \partial_{\varepsilon_n} f_n(\overline{x}_n) \mid \varepsilon_1, \ldots, \varepsilon_n \geq 0,\ \varepsilon_1 + \ldots + \varepsilon_n = \varepsilon\}.$$

In particular

$$\partial(f_1 \square \ldots \square f_n)(\overline{x}_1 + \cdots + \overline{x}_n) = \partial f_1(\overline{x}_1) \cap \cdots \cap \partial f_n(\overline{x}_n).$$

Conversely, if $\partial f_1(\overline{x}_1) \cap \cdots \cap \partial f_n(\overline{x}_n) \neq \emptyset$, then relation (2.38) is valid.

Proof. We apply Corollary 2.4.5 to the functions $f_1, \dots, f_n$ and Corollary 2.4.6 to $f : X^n \to \overline{\mathbb{R}}$ defined by $f(x_1, \dots, x_n) := f_1(x_1) + \cdots + f_n(x_n)$ and to the operator $A : X^n \to X$ defined by $A(x_1, \dots, x_n) := x_1 + \cdots + x_n$.

If $x^* \in \partial f_1(\overline{x}_1) \cap \cdots \cap \partial f_n(\overline{x}_n)$, then

$$\forall i \in \overline{1,n},\ \forall x_i \in X \ :\ \langle x_i - \overline{x}_i, x^* \rangle \leq f_i(x_i) - f_i(\overline{x}_i);$$

hence $f_1(\overline{x}_i) + \cdots + f_n(\overline{x}_n) \leq f_1(x_1) + \cdots + f_n(x_n)$ for all $x_1, \dots, x_n \in X$ such that $x_1 + \cdots + x_n = \overline{x}_1 + \cdots + \overline{x}_n$, and so Eq. (2.38) holds. □

The next result shows that the convolution has a regularizing effect.

Corollary 2.4.8 *Let $f_1, f_2 \in \Lambda(X)$, $\overline{x}_i \in \operatorname{dom} f_i$ for $i \in \{1,2\}$ and $\overline{x} = \overline{x}_1 + \overline{x}_2$. Assume that $(f_1 \square f_2)(\overline{x}) = f_1(\overline{x}_1) + f_2(\overline{x}_2)$ and $f_1 \square f_2$ is subdifferentiable at $\overline{x}$. If f_1 is Gâteaux differentiable at $\overline{x}_1$ then $f_1 \square f_2$ is Gâteaux differentiable at $\overline{x}$ and $\nabla(f_1 \square f_2)(\overline{x}) = \nabla f_1(\overline{x}_1)$. Moreover, if X is a normed vector space and f_1 is Fréchet differentiable at $\overline{x}_1$ then $f_1 \square f_2$ is Fréchet differentiable at $\overline{x}$.*

Proof. By the preceding corollary we have that $\partial(f_1 \square f_2)(\overline{x}) = \partial f_1(\overline{x}_1) \cap \partial f_2(\overline{x}_2)$. Let $\overline{x}^* \in \partial(f_1 \square f_2)(\overline{x})$. Then

$$\begin{aligned} \langle u, \overline{x}^* \rangle &\leq (f_1 \square f_2)(\overline{x} + u) - (f_1 \square f_2)(\overline{x}) \\ &\leq f_1(\overline{x}_1 + u) + f_2(\overline{x}_2) - f_1(\overline{x}_1) - f_2(\overline{x}_2) \\ &\leq f_1(\overline{x}_1 + u) - f_1(\overline{x}_1) \end{aligned} \tag{2.39}$$

for every $u \in X$. If f_1 is Gâteaux differentiable at $\overline{x}_1$ then, from the preceding corollary and Theorem 2.4.4 (i), we have that $\overline{x}^* = \nabla f_1(\overline{x}_1)$.

Using (2.39) we obtain that $(f_1\Box f_2)'(\overline{x},u) = \langle u, \nabla f_1(\overline{x}_1)\rangle$ for every $u \in X$, which shows that $f_1\Box f_2$ is Gâteaux differentiable at $\overline{x}$ and $\nabla(f_1\Box f_2)(\overline{x}) = \nabla f_1(\overline{x}_1)$.

Assume now that X is a normed space and f_1 is Fréchet differentiable at $\overline{x}_1$. From the preceding situation we have that $f_1\Box f_2$ is Gâteaux differentiable at $\overline{x}$ and $\nabla(f_1\Box f_2)(\overline{x}) = \nabla f_1(\overline{x}_1)$. Using again Eq. (2.39), we have that

$$\begin{aligned} 0 &\le (f_1\Box f_2)(\overline{x}+u) - (f_1\Box f_2)(\overline{x}) - \langle u, \nabla f_1(\overline{x}_1)\rangle \\ &\le f_1(\overline{x}_1+u) - f_1(\overline{x}_1) - \langle u, \nabla f_1(\overline{x}_1)\rangle \quad \forall u \in X; \end{aligned}$$

hence $f_1\Box f_2$ is Fréchet differentiable at $\overline{x}$ and $\nabla(f_1\Box f_2)(\overline{x}) = \nabla f_1(\overline{x}_1)$. □

In Section 2.6 we shall extend the results of Corollaries 2.4.6 and 2.4.7 to the case where the infimum is not attained in y and $\overline{x}$, respectively.

The following result establishes a sufficient condition for the subdifferentiability of a convex function; this result is of exceptional importance.

Theorem 2.4.9 *Let $f \in \Lambda(X)$. If f is continuous at $\overline{x} \in \operatorname{dom} f$, then $\partial_\varepsilon f(\overline{x})$ is nonempty and w^*-compact for each $\varepsilon \in \mathbb{R}_+$. Furthermore, for every $\varepsilon \ge 0$, $f'_\varepsilon(\overline{x},\cdot)$ is continuous and*

$$\forall u \in X \;:\; f'_\varepsilon(\overline{x},u) = \max\{\langle u,x^*\rangle \mid x^* \in \partial_\varepsilon f(\overline{x})\}. \tag{2.40}$$

Proof. Suppose that f is continuous at $\overline{x} \in \operatorname{dom} f$ (from Theorem 2.4.4 we already know that $\partial_\varepsilon f(\overline{x}) \ne \emptyset$ for $\varepsilon > 0$!).

Let $\eta > 0$. Since f is continuous at $\overline{x}$, there exists $V \in \mathcal{N}_X$ such that

$$\forall x \in \overline{x}+V \;:\; f(x) \le f(\overline{x}) + \eta. \tag{2.41}$$

Therefore $(\overline{x}+V) \times [f(\overline{x})+\eta,\infty[\subset \operatorname{epi} f$, whence $\operatorname{int}(\operatorname{epi} f) \ne \emptyset$. The set $\operatorname{epi} f$ being convex and $(\overline{x}, f(\overline{x})) \notin \operatorname{int}(\operatorname{epi} f)$ (because $(\overline{x}, f(\overline{x})-\delta) \notin \operatorname{epi} f$ for every $\delta > 0$), we can apply a separation theorem; so, there exists $(x^*,\alpha) \in X^* \times \mathbb{R} \setminus \{(0,0)\}$ such that

$$\forall (x,t) \in \operatorname{epi} f \;:\; \langle x,x^*\rangle + \alpha t \le \langle \overline{x},x^*\rangle + \alpha f(\overline{x}). \tag{2.42}$$

Taking $x = \overline{x}$ and $t = f(\overline{x}) + n$, $n \in \mathbb{N}$, we get $\alpha \le 0$. If $\alpha = 0$, using Eq. (2.42), we obtain that $\langle x-\overline{x},x^*\rangle \le 0$ for every $x \in \operatorname{dom} f$, and so, by Eq. (2.41), we have that $\langle x,x^*\rangle \le 0$ for every $x \in V$; since V is a neighborhood of the origin, this implies that $x^* = 0$. Thus we obtain the contradiction:

$(x^*, \alpha) = (0,0)$. Therefore $\alpha < 0$; so we can consider $\alpha = -1$ in Eq. (2.42). This relation becomes

$$\forall x \in \operatorname{dom} f \; : \; \langle x, x^* \rangle - f(x) \leq \langle \overline{x}, x^* \rangle - f(\overline{x}),$$

i.e. $x^* \in \partial f(\overline{x})$.

Let now $\varepsilon \geq 0$. For every $x^* \in \partial_\varepsilon f(\overline{x}) = \partial f'_\varepsilon(\overline{x}, \cdot)(0)$, using Eq. (2.41), we have

$$\forall u \in V \; : \; \langle u, x^* \rangle \leq f'_\varepsilon(\overline{x}, u) \leq f(\overline{x} + u) - f(\overline{x}) + \varepsilon \leq \eta + \varepsilon, \tag{2.43}$$

whence $\langle u, x^* \rangle \geq -1$ for every $u \in (\eta + \varepsilon)^{-1} V$ (recall that V is symmetric) which proves that

$$\partial_\varepsilon f(\overline{x}) \subset \left((\eta + \varepsilon)^{-1} V\right)^\circ = (\eta + \varepsilon) V^\circ. \tag{2.44}$$

By the Alaoglu–Bourbaki theorem (Theorem 1.1.10) we have that V° is w^*-compact; since $\partial_\varepsilon f(\overline{x})$ is w^*-closed, it follows that $\partial_\varepsilon f(\overline{x})$ is w^*-compact.

Taking into account that $\overline{x} \in \operatorname{int}(\operatorname{dom} f)$, from Theorem 2.1.14, we have that $\operatorname{dom} f'_\varepsilon(\overline{x}, \cdot) = X$; hence, using Theorem 2.2.13 and Eq. (2.43), the function $f'_\varepsilon(\overline{x}, \cdot)$ is continuous. Furthermore, using again Eq. (2.43), we have that

$$f'_\varepsilon(\overline{x}, u) \geq \sup\{\langle u, x^* \rangle \mid x^* \in \partial_\varepsilon f(\overline{x})\}.$$

We intend to prove that in the above inequality we have equality and the supremum is attained. For this end let $u \in X \setminus \{0\}$. Consider $X_0 := \mathbb{R}u$ and $\varphi : X_0 \to \mathbb{R}$, $\varphi(tu) := t f'_\varepsilon(\overline{x}, u)$; it is obvious that $\varphi(u') \leq f'_\varepsilon(\overline{x}, u')$ for every $u' \in X_0$, and so, applying the Hahn–Banach theorem, there exists $x^* \in X'$ such that $\langle u, x^* \rangle = \varphi(u) = f'_\varepsilon(\overline{x}, u)$ and $\langle u', x^* \rangle \leq f'_\varepsilon(\overline{x}; u')$ for every $u' \in X$. Since $f'_\varepsilon(\overline{x}, \cdot)$ is continuous, x^* is continuous, too; hence $x^* \in \partial f'_\varepsilon(\overline{x}, \cdot)(0) = \partial_\varepsilon f(\overline{x})$. The proof is complete. □

Using the preceding result we obtain the following criterion for the Gâteaux differentiability of a continuous convex function.

Corollary 2.4.10 *Let $f \in \Lambda(X)$ be continuous at $\overline{x} \in \operatorname{dom} f$. Then f is Gâteaux differentiable at $\overline{x}$ if and only if $\partial f(\overline{x})$ is a singleton.*

Proof. The necessity was already observed in Theorem 2.4.4(i). Assume that $\partial f(\overline{x}) = \{\overline{x}^*\}$. Using Theorem 2.4.9 we obtain that $f'(\overline{x}, u) = \langle u, \overline{x}^* \rangle$ for every $u \in X$. Because $\lim_{t \uparrow 0} t^{-1} (f(\overline{x} + tu) - f(\overline{x})) = -f'(\overline{x}, -u) =$

$-\langle -u, \overline{x}^*\rangle = \langle u, \overline{x}^*\rangle$, we have that f is Gâteaux differentiable and $\nabla f(\overline{x}) = \overline{x}^*$. □

Related to Fréchet differentiability of convex functions see Theorem 3.3.2 in the next chapter.

When f is not continuous at $\overline{x} \in \operatorname{dom} f$, relation (2.40) may be false; see Corollary 2.4.15 and Exercise 2.30. However the following result holds.

Theorem 2.4.11 *Let $f \in \Gamma(X)$, $\overline{x} \in \operatorname{dom} f$, and $\varepsilon \in]0, \infty[$. Then*

$$\forall u \in X \ : \ f'_\varepsilon(\overline{x}, u) = \sup\{\langle u, x^*\rangle \mid x^* \in \partial_\varepsilon f(\overline{x})\}. \tag{2.45}$$

Therefore $f'_\varepsilon(\overline{x}, \cdot)$ is a lsc sublinear function. Furthermore, for every $u \in X$,

$$\begin{aligned} f'(\overline{x}, u) &= \lim_{\varepsilon \downarrow 0} \big(\sup\{\langle u, x^*\rangle \mid x^* \in \partial_\varepsilon f(\overline{x})\}\big) \\ &= \inf_{\varepsilon > 0} \big(\sup\{\langle u, x^*\rangle \mid x^* \in \partial_\varepsilon f(\overline{x})\}\big). \end{aligned} \tag{2.46}$$

Proof. In Theorem 2.4.4(i) we have seen that $\partial_\varepsilon f(\overline{x}) = \partial f'_\varepsilon(\overline{x}, \cdot)(0)$, whence

$$\forall x^* \in \partial_\varepsilon f(\overline{x}), \ \forall u \in X \ : \ \langle u, x^*\rangle \le f'_\varepsilon(\overline{x}, u).$$

Therefore the inequality " $\ge$ " holds in Eq. (2.45). For proving the converse inequality, let $u \in X$ and $\lambda \in \mathbb{R}$ with $\lambda < f'_\varepsilon(\overline{x}, u)$. From the definition of $f'_\varepsilon(\overline{x}, u)$, we have that

$$\forall t > 0 \ : \ \lambda < \frac{f(\overline{x} + tu) - f(\overline{x}) + \varepsilon}{t} \tag{2.47}$$

and

$$\lambda < \lim_{t \to \infty} \frac{f(\overline{x} + tu) - f(\overline{x}) + \varepsilon}{t} = \lim_{t \to \infty} \frac{f(\overline{x} + tu) - f(\overline{x})}{t} = f_\infty(u). \tag{2.48}$$

Let us consider $A := \operatorname{epi} f$ and $B := \{(\overline{x} + su, f(\overline{x}) + \lambda s - \varepsilon) \mid s \ge 0\}$. It is obvious that A and B are nonempty closed convex sets and, from Eq. (2.47), $A \cap B = \emptyset$, *i.e.* $(0,0) \notin A - B$. Moreover B is locally compact (as subset of a finite dimensional separated locally convex space). Since $A_\infty = \operatorname{epi} f_\infty$ and $B_\infty = \mathbb{R}_+ \cdot (u, \lambda)$, from Eq. (2.48) we obtain that $A_\infty \cap B_\infty = \{(0,0)\}$. Then, by Corollary 1.1.8, $A - B$ is a closed set.

Applying Theorem 1.1.7, there exists $(\overline{x}^*, \alpha) \in X^* \times \mathbb{R}$ such that

$$\forall (x, t) \in \operatorname{epi} f, \ \forall s \ge 0 \ : \ 0 > \langle x - \overline{x} - su, \overline{x}^*\rangle + \alpha(t - f(\overline{x}) - s\lambda + \varepsilon).$$

Taking $x = \overline{x}$ and letting $t \to \infty$ we obtain that $\alpha \leq 0$; if $\alpha = 0$, for $s = 0$ we get the contradiction $0 > 0$. Therefore $\alpha < 0$ and we can suppose that $\alpha = -1$. The above relation becomes

$$\forall x \in \operatorname{dom} f,\ \forall s \geq 0 \ :\ 0 > \langle x - \overline{x}, \overline{x}^* \rangle - (f(x) - f(\overline{x}) + \varepsilon) + s(\lambda - \langle u, \overline{x}^* \rangle).$$

For $s = 0$ we obtain that $\overline{x}^* \in \partial_\varepsilon f(\overline{x})$, while for $s \to \infty$ we obtain $\langle u, \overline{x}^* \rangle \geq \lambda$. Therefore $\lambda \leq \sup\{\langle u, x^* \rangle \mid x^* \in \partial_\varepsilon f(\overline{x})\}$. Since $\lambda < f'_\varepsilon(\overline{x}, u)$ is arbitrary, the relation (2.45) is true. The equality (2.46) follows from relation (2.45) and Theorem 2.1.14. □

The subdifferentiability criterion of Theorem 2.4.9 can be extended.

Theorem 2.4.12 *Let $f \in \Lambda(X)$ and $X_0 := \operatorname{aff}(\operatorname{dom} f)$. If $f|_{X_0}$ is continuous at $\overline{x} \in \operatorname{dom} f$, then $\partial f(\overline{x}) \neq \emptyset$. In particular, if $\dim X < \infty$, then $\partial f(x) \neq \emptyset$ for every $x \in {}^i(\operatorname{dom} f)$.*

Proof. Without any loss of generality, we can suppose that $\overline{x} = 0$; then $X_0 = \operatorname{lin}(\operatorname{dom} f)$. The function $g := f|_{X_0}$ is convex, proper and continuous at 0. By Theorem 2.4.9 we have that $\partial g(0) \neq \emptyset$. Let $\varphi \in \partial g(0)$; hence $\varphi \in X_0^*$ and

$$\forall x \in \operatorname{dom} g \ :\ \varphi(x) - \varphi(0) \leq g(x) - g(0).$$

Using the Hahn–Banach theorem we get $x^* \in X^*$ such that $x^*|_{X_0} = \varphi$. The above inequality shows that

$$\forall x \in \operatorname{dom} f = \operatorname{dom} g \ :\ \langle x - 0, x^* \rangle = \varphi(x) \leq g(x) - g(0) = f(x) - f(0),$$

whence $x^* \in \partial f(0)$.

If $\dim X < \infty$, then $\dim X_0 < \infty$. By Theorem 2.2.21, g is continuous on $\operatorname{int}(\operatorname{dom} f) = {}^i(\operatorname{dom} f)$. The conclusion follows from the first part. □

Note that under the conditions of Theorem 2.4.12 $\partial f(x)$ is, generally, not bounded, and so it is not w^*-compact. But, under the conditions of Theorem 2.4.9, in the case of normed spaces, ∂f has supplementary properties (mentioned in the next theorem). The local boundedness of monotone operators was proved by R.T. Rockafellar (see Theorem 3.11.14); the converse is true in Banach spaces for lower semicontinuous functions as shown by Corollary 3.11.16.

Theorem 2.4.13 *Let X be a normed space, $f \in \Lambda(X)$ be continuous on $\operatorname{int}(\operatorname{dom} f)$ and $\varepsilon \geq 0$. Then $\partial_\varepsilon f$ is locally bounded on $\operatorname{int}(\operatorname{dom} \partial f) =$*

int(dom f). Moreover, if f is bounded on bounded sets then f is Lipschitz on bounded sets and $\partial_\varepsilon f$ is bounded on bounded sets.

Proof. By Theorem 2.4.9 we have that $\operatorname{int}(\operatorname{dom} f) \subset \operatorname{dom} \partial f \subset \operatorname{dom} f$; hence $\operatorname{int}(\operatorname{dom} \partial f) = \operatorname{int}(\operatorname{dom} f)$. Let $x_0 \in \operatorname{int}(\operatorname{dom} f)$; since f is continuous at x_0, from Corollary 2.2.12 we get the existence of $m, \rho > 0$ such that

$$\forall x, y \in D(x_0, 2\rho) \ : \ |f(x) - f(y)| \leq m\|x - y\|.$$

Let us fix $x \in D(x_0, \rho)$. Then for $y \in x + \rho U_X$ we have that $f(y) \leq f(x) + m\rho$, *i.e.* Eq. (2.41) holds with $V := \rho U_X$ and $\eta := m\rho$. Using Eq. (2.44), we obtain that $\partial_\varepsilon f(x) \subset (\eta + \varepsilon)V^\circ = (m + \varepsilon/\rho)U_{X^*}$ for every $x \in D(x_0, \rho)$.

Assume now that f is bounded on bounded sets. Then $\operatorname{dom} f = X$. Taking $\rho > 0$, f is bounded above on $2\rho U_X$, and so Eq. (2.41) holds with $V := 2\rho U_X$ and some $\eta > 0$. By Corollary 2.2.12 f is Lipschitz on ρU_X, and, by Eq. (2.44), $\partial_\varepsilon f(x) \subset (\eta + \varepsilon)\rho^{-1} U_{X^*}$ for $x \in D(x_0, \rho)$. The proof is complete. □

For other relationships between the continuity of f and the local boundedness of ∂f see Corollary 3.11.16.

In Theorem 2.4.4 we have seen that finding the ε-subdifferential of a convex function at a point reduces to compute the subdifferential of a sublinear function at the origin. Other subdifferentials (the subdifferentials of Clarke, of Michel-Penot, etc.), for nonconvex functions are introduced through certain sublinear functions. So, we consider it is worth giving some properties of sublinear functions.

Theorem 2.4.14 *Let $f, g : X \to \overline{\mathbb{R}}$, $h : Y \to \overline{\mathbb{R}}$ be sublinear functions, $T \in \mathcal{L}(X, Y)$ and $B, C \subset X^*$ be nonempty sets. Then:*

(i) $f^* = \iota_{\partial f(0)}$;

(ii) $\partial f(0) \neq \emptyset \Leftrightarrow f$ *is lsc at* 0;

(iii) *for every $\overline{x} \in \operatorname{dom} f$ and for every $\varepsilon \in \mathbb{R}_+$ we have:*

$$\partial f(\overline{x}) = \{x^* \in \partial f(0) \mid \langle \overline{x}, x^* \rangle = f(\overline{x})\},$$

$$\partial_\varepsilon f(\overline{x}) = \{x^* \in \partial f(0) \mid \langle \overline{x}, x^* \rangle \geq f(\overline{x}) - \varepsilon\}, \quad \partial_\varepsilon f(0) = \partial f(0);$$

(iv) *if f is lsc at* 0 *then*

$$\forall x \in X \ : \ \overline{f}(x) = \sup\{\langle x, x^* \rangle \mid x^* \in \partial f(0)\}.$$

(v) *Suppose that f and g are lsc. Then $f \le g \Leftrightarrow \partial f(0) \subset \partial g(0)$.*

(vi) *The support function s_B of B is sublinear, lsc, $s_B = (\iota_B)^*$ and $\partial s_B(0) = \overline{\text{co}}B$, the closure being taken with respect to the weak* topology w^* on X^*; moreover, $s_B \le s_C$ if and only if $B \subset \overline{\text{co}}C$.*

(vii) *If h is lsc then $\partial(h \circ A)(0) = w^*\text{-cl}\,(A^*(\partial h(0)))$;*

(viii) *if f and g are lsc, then $\partial(f+g)(0) = w^*\text{-cl}\,(\partial f(0) + \partial g(0))$.*

Proof. (i) For every $x^* \in X^*$ we have

$$f^*(x^*) = \sup\{\langle x, x^*\rangle - f(x) \mid x \in \operatorname{dom} f\} \ge \langle 0, x^*\rangle - f(0) = 0.$$

If $x^* \in \partial f(0)$ then $\langle x, x^*\rangle \le f(x)$ for every $x \in \operatorname{dom} f$, whence $f^*(x^*) = 0$. If $x^* \notin \partial f(0)$, there exists $\overline{x} \in X$ such that $\langle \overline{x}, x^*\rangle > f(\overline{x})$. In this situation we have

$$f^*(x^*) \ge \sup\{\langle t\overline{x}, x^*\rangle - f(t\overline{x}) \mid t > 0\} = \sup\{t\langle \overline{x}, x^*\rangle - f(\overline{x}) \mid t > 0\} = \infty.$$

(ii) If $\partial f(0) \ne \emptyset$ then, from Theorem 2.4.1(ii), we have that f is lsc at 0. Suppose that f is lsc at 0. Then $\overline{f}(0) = f(0) = 0$, and so, by Theorem 2.3.4, $f(0) = f^{**}(0) = 0$. Assuming that $\partial f(0) = \emptyset$ we obtain that $f^* = \iota_{\partial f(0)} = +\infty$, and so the contradiction $f^{**} = -\infty$. Therefore $\partial f(0) \ne \emptyset$.

(iii) Let $\overline{x} \in \operatorname{dom} f$ and $\varepsilon \ge 0$. If $x^* \in \partial f(0)$ and $\langle \overline{x}, x^*\rangle \ge f(\overline{x}) - \varepsilon$ then

$$\forall x \in X \ : \ \langle x - \overline{x}, x^*\rangle = \langle x, x^*\rangle - \langle \overline{x}, x^*\rangle \le f(x) - f(\overline{x}) + \varepsilon,$$

i.e. $x^* \in \partial_\varepsilon f(\overline{x})$. Conversely, let $x^* \in \partial_\varepsilon f(\overline{x})$; then the above inequality holds. Taking $x = 0$ we get $\langle \overline{x}, x^*\rangle \ge f(\overline{x}) - \varepsilon$; taking now $x := \overline{x} + ty$, $t > 0$, $y \in X$, we obtain that

$$t\langle y, x^*\rangle \le f(\overline{x} + ty) - f(\overline{x}) + \varepsilon \le f(\overline{x}) + tf(y) - f(\overline{x}) + \varepsilon = tf(y) + \varepsilon.$$

Dividing by $t > 0$ and letting then $t \to \infty$, we obtain that $\langle y, x^*\rangle \ge f(y)$ for every $y \in X$, *i.e.* $x^* \in \partial f(0)$.

For $\varepsilon = 0$ we have

$$\partial f(\overline{x}) = \{x^* \in \partial f(0) \mid \langle \overline{x}, x^*\rangle \ge f(\overline{x})\} = \{x^* \in \partial f(0) \mid \langle \overline{x}, x^*\rangle = f(\overline{x})\},$$

while for $\overline{x} = 0$ we obtain that

$$\partial_\varepsilon f(0) = \{x^* \in \partial f(0) \mid \langle 0, x^*\rangle \ge f(0) - \varepsilon\} = \partial f(0).$$

(iv) Suppose that f is lsc at 0. Then $\overline{f}(0) = f(0) = 0$. Using Theorem 2.3.4, we have that $f^{**} = \overline{f}$. Therefore

$$\overline{f}(x) = \sup\{\langle x, x^*\rangle - f^*(x^*) \mid x^* \in X^*\} = \sup\{\langle x, x^*\rangle \mid x^* \in \partial f(0)\}.$$

(v) It is obvious that $\partial f(0) \subset \partial g(0)$ if $f \leq g$. The converse implication follows immediately from (iv).

(vi) At the end of Section 2.3 we noted that s_B is lsc, sublinear and $s_B = (\iota_B)^*$. From the obvious inclusion $B \subset \partial s_B(0)$ we get $\overline{\text{co}}B \subset \partial s_B(0)$ (because $\partial s_B(0)$ is convex and w^*-closed). Let $\overline{x}^* \notin \overline{\text{co}}B$. Using Theorem 1.1.5 in the space (X^*, w^*) and taking into account that $(X^*, w^*)^* = X$, there exist $\overline{x} \in X$ and $\lambda \in \mathbb{R}$ such that

$$\forall\, x^* \in \overline{\text{co}}B \supset B \ : \ \langle \overline{x}, \overline{x}^*\rangle > \lambda > \langle \overline{x}, x^*\rangle,$$

whence $\langle \overline{x}, \overline{x}^*\rangle > s_B(\overline{x})$, *i.e.* $\overline{x}^* \notin \partial s_B(0)$. The conclusion follows.

(vii) By Theorem 2.4.2 we have that $A^*(\partial h(0) \subset \partial(h \circ A)(0)$. Let $\overline{x}^* \notin w^*\text{-cl}\left(A^*(\partial h(0))\right)$. Using Theorem 1.1.5, there exist $\overline{x} \in Y$ and $\lambda \in \mathbb{R}$ such that $\langle \overline{x}, \overline{x}^*\rangle > \lambda \geq \langle \overline{x}, Ay^*\rangle = \langle A\overline{x}, y^*\rangle$ for every $y^* \in \partial h(0)$. From (iv) we get $\lambda \geq h(A\overline{x})$, and so $\overline{x}^* \notin \partial(h \circ A)(0)$.

(viii) Let $H : X \times X \to \overline{\mathbb{R}}$, $H(x, y) := f(x) + g(y)$, and $T : X \to X \times X$, $Tx := (x, x)$. It is obvious that H is sublinear and lsc, and T is continuous and linear; moreover, $\partial H(0,0) = \partial f(0) \times \partial g(0)$ and $T^*(x^*, y^*) = x^* + y^*$. By (vii) we obtain that

$$\partial(f + g)(0) = \partial(H \circ T)(0) = w^*\text{- cl}\,(T^*(\partial H(0,0))) = w^*\text{- cl}(\partial f(0) + \partial g(0)).$$

The proof is complete. □

Using the preceding result we have

Corollary 2.4.15 *Let $f \in \Lambda(X)$ and $x \in \text{dom}\, f$. Then $\partial f(x) \neq \emptyset$ if and only if $f'(x, \cdot)$ is lsc at* 0. *In this case*

$$\overline{f'(x, \cdot)}(u) = \sup\{\langle u, x^*\rangle \mid x^* \in \partial f(x)\} \quad \forall\, u \in X.$$

Proof. The conclusion follows from the formula $\partial f(x) = \partial f'(x, \cdot)(0)$ (see Theorem 2.4.4) and assertions (i) and (iv) of the preceding theorem. □

Note that, even for $X = \mathbb{R}^2$ and $f \in \Gamma(X)$ subdifferentiable at x, $f'(x, \cdot)$ may not be lsc; take for example $f := \iota_U$ and $x = (0, 1)$, where $U := \{(x, y) \in \mathbb{R}^2 \mid x^2 + y^2 \leq 1\}$.

Corollary 2.4.16 *Let $(X, \|\cdot\|)$ be a normed space, $f : X \to \mathbb{R}$, $f(x) := \|x\|$, and $\overline{x} \in X$. Then, denoting U_{X^*} by U^*, we have:*

$$f^* = \iota_{U^*}, \quad \partial f(0) = U^*, \quad \partial f(\overline{x}) = \{x^* \in U^* \mid \langle \overline{x}, x^* \rangle = \|\overline{x}\|\}.$$

Proof. The above formulas follow immediately from Theorem 2.4.14. □

Another example of sublinear function is given in the following result.

Corollary 2.4.17 *Let $f : \mathbb{R}^n \to \mathbb{R}$, $f(x) := x_1 \vee \cdots \vee x_n$, and $\overline{x} \in \mathbb{R}^n$, where $n \in \mathbb{N}$. Then*

$$\partial f(0) = \Delta_n, \quad f^* = \iota_{\Delta_n}, \quad \partial_\varepsilon f(\overline{x}) = \{y \in \Delta_n \mid \overline{x}_1 y_1 + \ldots + \overline{x}_n y_n \geq f(\overline{x}) - \varepsilon\},$$

where

$$\Delta_n := \{(\lambda_1, \ldots, \lambda_n) \in \mathbb{R}^n \mid \lambda_i \geq 0,\ \lambda_1 + \ldots + \lambda_n = 1\}.$$

Proof. It is obvious that f is sublinear and continuous. Moreover

$$y \in \partial f(0) \Leftrightarrow \forall x \in \mathbb{R}^n \ : \ x_1 y_1 + \cdots + x_n y_n \leq x_1 \vee \cdots \vee x_n.$$

Taking $x_j = 0$ for $j \neq i$ and $x_i = -1$, we obtain that $-y_i \leq 0$, *i.e.* $y_i \geq 0$ for every i. Taking then $x_i = t \in \mathbb{R}$ for every i, we obtain that $t(y_1 + \cdots + y_n) \leq t$ for every t, whence $y_1 + \cdots + y_n = 1$. Therefore $\partial f(0) \subset \Delta_n$. The converse inclusion is immediate. The other relations follow directly from Theorem 2.4.14. □

The following theorem furnishes a formula for the subdifferential of a supremum of convex functions. Other formulas will be given in Section 2.8.

Theorem 2.4.18 (Ioffe–Tikhomirov) *Let (A, τ) be a separated compact topological space and $f_\alpha : X \to \overline{\mathbb{R}}$ be a convex function for every $\alpha \in A$. Consider the function $f := \sup_{\alpha \in A} f_\alpha$ and $F(x) := \{\alpha \in A \mid f_\alpha(x) = f(x)\}$. Assume that the mapping $A \ni \alpha \mapsto f_\alpha(x) \in \overline{\mathbb{R}}$ is upper semicontinuous and $x_0 \in \operatorname{dom} f$ is such that f_α is continuous at x_0 for every $\alpha \in A$. Then*

$$\partial f(x_0) = \overline{\operatorname{co}}\left(\bigcup\nolimits_{\alpha \in F(x_0)} \partial f_\alpha(x_0)\right). \tag{2.49}$$

Proof. Since A is compact and $\alpha \mapsto f_\alpha(x)$ is upper semicontinuous we have that $F(x)$ is a nonempty compact subset of A for every $x \in X$. The inclusion $\supset$ being true without any condition on A and f (and easy to prove), let us show the converse inclusion. If $f(x_0) = -\infty$, then $f_\alpha(x_0) =$

$-\infty$ for all $\alpha \in A = F(x_0)$, and so $\partial f(x_0) = \partial f_\alpha(x_0) = \emptyset$; the conclusion holds. We assume now that $f(x_0) \in \mathbb{R}$.

Let us note first that in our conditions $x_0 \in (\operatorname{dom} f)^i$. Indeed, let $x \in X$. Consider a fixed $\gamma_0 > f(x_0)$ ($\geq f_\alpha(x_0)$). Because f_α is continuous at x_0, for every $\alpha \in A$ there exists $t_\alpha > 0$ such that $f_\alpha(x_0 + t_\alpha x) < \gamma_0$. Since $\beta \mapsto f_\beta(x_0 + t_\alpha x)$ is upper semicontinuous, the set $A_\alpha := \{\beta \in A \mid f_\beta(x_0 + t_\alpha x) < \gamma_0\}$ is an open set containing α. As $A = \bigcup_{\alpha\in A} A_\alpha$, there exist $\alpha_1, \ldots, \alpha_n \in A$ such that $A = \bigcup_{i=1}^n A_{\alpha_i}$. Let $t := \min\{t_{\alpha_1}, \ldots, t_{\alpha_n}\} > 0$. It follows that $f_\alpha(x_0 + tx) < \gamma_0$ for every $\alpha \in A$, and so $f(x_0 + tx) < \gamma_0$. Hence $x_0 \in (\operatorname{dom} f)^i$.

Since f_α is continuous at x_0, $\partial f_\alpha(x_0) \neq \emptyset$ for every $\alpha \in F(x_0)$, and so $Q \neq \emptyset$, where Q is the set on the right-hand side of Eq. (2.49). Suppose that there exists $\overline{x}^* \in \partial f(x_0) \setminus Q$. Using Theorem 1.1.5 in the space (X^*, w^*), there exist $\overline{x} \in X$ and $\varepsilon > 0$ such that

$$\forall \alpha \in F(x_0),\ \forall x^* \in \partial f_\alpha(x_0) \ :\ \langle \overline{x}, \overline{x}^*\rangle - \varepsilon \geq \langle \overline{x}, x^*\rangle, \tag{2.50}$$

or equivalently, by Theorem 2.4.9,

$$\forall \alpha \in F(x_0) \ :\ \langle \overline{x}, \overline{x}^*\rangle - \varepsilon \geq (f_\alpha)'(x_0; \overline{x}). \tag{2.51}$$

Because $x_0 \in (\operatorname{dom} f)^i$, we may suppose that $x_0 + \overline{x} \in \operatorname{dom} f$. Let $t \in]0,1[$; then $x_0 + t\overline{x} \in \operatorname{dom} f$. There exists $\alpha_t \in A$ such that $f_{\alpha_t}(x_0 + t\overline{x}) = f(x_0 + t\overline{x})$. Because $(1-t)f_{\alpha_t}(x_0) + tf_{\alpha_t}(x_0 + \overline{x}) \geq f_{\alpha_t}(x_0 + t\overline{x})$ and $\overline{x}^* \in \partial f(x_0)$, we obtain that

$$\begin{aligned}(1-t)f_{\alpha_t}(x_0) &\geq f(x_0 + t\overline{x}) - tf_{\alpha_t}(x_0 + \overline{x}) \geq f(x_0 + t\overline{x}) - tf(x_0 + \overline{x}) \\ &\geq f(x_0) - t\langle \overline{x}, \overline{x}^*\rangle - tf(x_0 + \overline{x}).\end{aligned}$$

It follows that $\lim_{t\to 0} f_{\alpha_t}(x_0) = f(x_0)$. The space (A, τ) being compact, there exists a convergent subnet $(\alpha_{\psi(j)})_{j\in J}$ of $(\alpha_t)_{t\in]0,1[}$ converging to $\alpha_0 \in A$; hence $f_{\alpha_0}(x_0) = f(x_0)$, i.e. $\alpha_0 \in F(x_0)$. Let $s \in]0,1[$ be fixed (for the moment) and take $t \in]0,s]$. Using Eq. (2.50) we obtain that

$$\begin{aligned}\frac{f_{\alpha_t}(x_0 + s\overline{x}) - f_{\alpha_t}(x_0)}{s} &\geq \frac{f_{\alpha_t}(x_0 + t\overline{x}) - f_{\alpha_t}(x_0)}{t} \geq \frac{f(x_0 + t\overline{x}) - f(x_0)}{t} \\ &\geq \langle \overline{x}, \overline{x}^*\rangle.\end{aligned}$$

We have that $\psi(j) < s$ for $j \succeq j_s$ for some $j_s \in J$. Taking $t = \psi(j)$ with

$j \succeq j_s$ and using the upper semi-continuity hypothesis, we obtain that

$$\forall s \in]0,1[: \frac{f_{\alpha_0}(x_0 + s\overline{x}) - f_{\alpha_0}(x_0)}{s} \geq \langle \overline{x}, \overline{x}^* \rangle ,$$

and so $(f_{\alpha_0})'(x_0;\overline{x}) \geq \langle \overline{x}, \overline{x}^* \rangle$, contradicting Eq. (2.51). □

Of course, for conjugates and for the ε-subdifferentials it is desirable to dispose of numerous formulas. There exist effectively a large set of such formulas we shall establish in Section 2.8.

2.5 The General Problem of Convex Programming

By **problem of convex programming** we mean the problem of minimizing a convex function $f : X \to \overline{\mathbb{R}}$, called **objective function** (or *cost function*) on a convex set $C \subset X$ called the set of **admissible solutions**, or set of *constraints*. We shall denote such a problem by

(P) $\quad \min\ f(x),\ x \in C.$

Of course, to consider this problem, X must be a linear space, however most of the results will be obtained in the framework of separated locally convex spaces or even normed spaces.

To problem (P) we can associate a problem (apparently) without constraints:

$(\widetilde{P})$ $\quad \min\ \tilde{f}(x),\ x \in X,$

where $\tilde{f} := f + \iota_C$.

In order for the problem (P) to be nontrivial, it is natural to assume that $C \cap \operatorname{dom} f \neq \emptyset$ ($\Leftrightarrow \operatorname{dom} \tilde{f} \neq \emptyset$) and that f does not take the value $-\infty$ on C (*i.e.* $\tilde{f}$ does not take the value $-\infty$).

We call **value** of problem (P) the extended real

$$v(P) := v(f, C) := \inf\{f(x) \mid x \in C\} \in \overline{\mathbb{R}};$$

we call **(optimal) solution** of problem (P) an element $\overline{x} \in C$ with the property that $f(\overline{x}) = v(P)$; this means that $\overline{x}$ is a (global) minimum point for the function $\tilde{f}$. We denote by $S(P)$ or $S(f, C)$ the set of optimal solutions of problem (P). Therefore

$$\begin{aligned} S(P) &= \{\overline{x} \in C \mid \forall x \in C \ :\ f(\overline{x}) \leq f(x)\} \\ &= \{\overline{x} \in X \mid \forall x \in X \ :\ \tilde{f}(\overline{x}) \leq \tilde{f}(x)\} = S(\widetilde{P}) \end{aligned}$$

if $C \cap \operatorname{dom} f \neq \emptyset$. The set $S(f, X)$ is denoted also by $\operatorname{argmin} f$.

Of course, an important problem is that of the existence of solutions for (P), resp. $(\widetilde{P})$. The most important result which assures the existence of solutions for (P) is the famous Weierstrass' theorem. Because the underlying spaces are not compact we have to use some coercivity conditions. We say that $f : X \to \overline{\mathbb{R}}$ is **coercive** if $\lim_{\|x\| \to \infty} f(x) = \infty$. It is obvious that f is coercive if and only if all the level sets $[f \leq \lambda]$ are bounded (see also Exercise 1.15); when f is convex then f is coercive if and only if the level set $[f \leq \lambda]$ is bounded for some $\lambda > \inf f$ (see Exercise 2.41). We have the following result.

Theorem 2.5.1 *Let $f \in \Gamma(X)$.*

(i) *If there exists $\lambda > v(f, X)$ such that $[f \leq \lambda]$ is w-compact, then $S(f, X) \neq \emptyset$.*

(ii) *If X is a reflexive Banach space and f is coercive then $S(f, X) \neq \emptyset$.*

Proof. (i) Of course, $v(f, X) = v(f, [f \leq \lambda])$. Since f is lsc and convex, f is w-lsc. The conclusion follows using the Weierstrass theorem applied to the function $f|_{[f \leq \lambda]}$.

(ii) Because f is coercive (see Exercise 1.15), $[f \leq \lambda]$ is bounded for every $\lambda \in \mathbb{R}$. Since $[f \leq \lambda]$ is w-closed and X is reflexive, we have that $[f \leq \lambda]$ is w-compact for every $\lambda \in \mathbb{R}$. The conclusion follows from (i). □

Of course, in the preceding theorem, the condition that f is convex can be replaced by the fact that f is quasi-convex. The next result shows that the reflexivity of the space X is almost necessary in Theorem 2.5.1.

Theorem 2.5.2 *Let $(X, \|\cdot\|)$ be a Banach space. Assume that there exists a proper function $f : X \to \overline{\mathbb{R}}$ satisfying the following conditions:*

(i) *$[f \leq f(x)]$ is closed, convex and bounded for every $x \in \operatorname{dom} f$;*

(ii) *f attains its infimum on every nonempty closed convex subset of X;*

(iii) *there exist $x_0, x_1 \in \operatorname{dom} f$ and $r > 0$ such that f is Lipschitz on $[f \leq f(x_1)]$ and $[f \leq f(x_0)] + rU_X \subset [f \leq f(x_1)]$, where $U_X := \{x \in X \mid \|x\| \leq 1\}$.*

Then X is reflexive.

Proof. Of course, $f(x_0) \leq f(x_1)$; if $f(x_0) = f(x_1)$ then the relation $[f \leq f(x_0)] + rU_X \subset [f \leq f(x_1)]$ contradicts the boundedness of $[f \leq$

$f(x_1)]$. Hence $f(x_0) < f(x_1)$. Since f is Lipschitz on $[f \le f(x_1)]$, there exists $L > 0$ such that $|f(x) - f(x')| \le L \|x - x'\|$ for all $x, x' \in [f \le f(x_1)]$. Since $f|_{[f\le f(x_1)]}$ is continuous, there exists $x_2 \in [x_0, x_1]$ such that $f(x_2) = (f(x_0) + f(x_1))/2$. The set $S := [f \le f(x_2)]$ is convex, bounded with nonempty interior. Indeed, $D(x_0, \delta_0) \subset S$, where $\delta_0 := \min\{r, (f(x_1) - f(x_0))/(2L)\}$; moreover $S + \delta_0 U_X \subset [f \le f(x_1)]$.

Consider $A := S - S$; A is a bounded, convex and symmetric set with nonempty interior, and so $0 \in \operatorname{int} A$. Therefore there exist $\alpha, \beta > 0$ such that $\alpha U_X \subset A \subset \beta U_X$. It follows that $\alpha^{-1}\|\cdot\| = p_{\alpha U_X} \ge p_A \ge p_{\beta U_X} = \beta^{-1}\|\cdot\|$, where p_A is the Minkowski gauge associated to A. By Theorem 1.1.1 p_A is a semi-norm, and so it is a norm equivalent to $\|\cdot\|$; moreover, by Proposition 1.1.1, the unit closed ball with respect to p_A is $\operatorname{cl} A$. To obtain that X is reflexive, by the famous James' theorem (see [Diestel (1975), Th. 1.6]), it is sufficient to show that every $x^* \in X^*$ attains its infimum on A.

Let $x^* \in X^* \setminus \{0\}$ and $\alpha^* := \inf\{\langle x, x^*\rangle \mid x \in S\}$. Because S is bounded, $\alpha^* \in \mathbb{R}$. Let $H := \{x \in X \mid \langle x, x^*\rangle = \alpha^*\}$. It is obvious that $(S+\varepsilon U_X)\cap H \ne \emptyset$ for every $\varepsilon > 0$. Taking $\varepsilon \in]0, \delta_0]$ and $x_\varepsilon \in (S+\varepsilon U_X)\cap H$, we obtain that $f(x_\varepsilon) \le f(x_2) + \varepsilon L$, and so $\inf_{x\in H} f(x) \le f(x_2)$. By (ii) there exists $\overline{x}_1 \in H$ such that $f(\overline{x}_1) = \inf_{x\in H} f(x)$, and so $\overline{x}_1 \in S \cap H$. Therefore $\langle \overline{x}_1, x^*\rangle \le \langle x, x^*\rangle$ for every $x \in S$. Similarly, there exists $\overline{x}_2 \in S$ such that $\langle \overline{x}_2, x^*\rangle \ge \langle x, x^*\rangle$ for every $x \in S$, whence there exists $\overline{x} := \overline{x}_1 - \overline{x}_2 \in A$ such that $\langle \overline{x}, x^*\rangle \le \langle x, x^*\rangle$ for every $x \in A$. □

Also the coercivity condition in Theorem 2.5.1(ii) is essential as the next theorem will show. In order to establish it we introduce some preliminary notations and results. Let $f \in \Gamma(X)$ and denote by Π_f the set

$$\{g \in \Gamma(X) \mid \operatorname{dom} g = \operatorname{dom} f,\ \sup_{x\in\operatorname{dom} f} |f(x) - g(x)| < \infty,\ \inf f = \inf g\}.$$

Since for $g \in \Pi_f$ we have that $f - \gamma \le f \le f + \gamma$ for some $\gamma \in \mathbb{R}_+$, we have that $f_\infty = g_\infty$ and f, g are simultaneously coercive or not coercive. Consider

$$d : \Pi_f \times \Pi_f \to \mathbb{R}_+, \quad d(g_1, g_2) := \sup_{x\in\operatorname{dom} f} |f(x) - g(x)|.$$

Lemma 2.5.3 *The mapping d is a metric on Π_f and (Π_f, d) is a complete metric space.*

Proof. It is obvious that d is a metric. Let $(g_n)_{n\ge1} \subset \Pi_f$ be a Cauchy sequence. Then $(g_n(x))_{n\ge1} \subset \mathbb{R}$ is a Cauchy sequence, and so $(g_n(x)) \to$

$g(x) \in \mathbb{R}$ for every $x \in \operatorname{dom} f$. Moreover, $\left(\sup_{x\in \operatorname{dom} f} |g_n(x) - g(x)|\right)_{n\geq 1} \to 0$. We extend g to X setting $g(x) := \infty$ for $x \in X \setminus \operatorname{dom} f$; hence $\operatorname{dom} g = \operatorname{dom} f$. Because $(g_n(x)) \to g(x)$ for every $x \in X$, from Theorem 2.1.3(ii) we have that g is convex. Let $\varepsilon > 0$; then there exists n_ε such that $g_n(x) - \varepsilon \leq g(x) \leq g_n(x) + \varepsilon$ for all $n \geq n_\varepsilon$ and $x \in \operatorname{dom} f$. It follows that g is proper (being minorized by $g_{n_\varepsilon} - \varepsilon$ even on X) and $\inf f - \varepsilon = \inf g_n - \varepsilon \leq \inf g \leq \inf f + \varepsilon = \inf g_n + \varepsilon$ for every $\varepsilon > 0$ which shows that $\inf g = \inf f$ (even if $\inf f = -\infty$). We must only show that g is lsc. Take first $x \in \operatorname{dom} g = \operatorname{dom} f$ and $\lambda < g(x)$. There exists $\varepsilon > 0$ such that $\lambda < g(x) - 2\varepsilon$. As above, there exists $n = n_\varepsilon$ such that $g(y) - \varepsilon < g_n(y) < g(y) + \varepsilon$ for every $y \in \operatorname{dom} f$. In particular $g(x) - \varepsilon < g_n(x)$. Because g_n is lsc at x, there exists $U \in \mathcal{N}_X(x)$ such that $g(x) - \varepsilon < g_n(y)$ for every $y \in U$. It follows that $g(x) - \varepsilon < g(y) + \varepsilon$ for every $y \in U \cap \operatorname{dom} f$, and so $\lambda < g(x) - 2\varepsilon < g(y)$ for $y \in U \cap \operatorname{dom} f$. As $g(y) = \infty$ for $y \in U \setminus \operatorname{dom} f$, we have that $\lambda < g(y)$ for $y \in U$, and so g is lsc at x. Consider now $x \in X \setminus \operatorname{dom} g$ and take $\lambda \in \mathbb{R}$. Let $n \in \mathbb{N}$ be such that $g(y) - 1 < g_n(y) < g(y) + 1$ for every $y \in \operatorname{dom} f$. Because g_n is lsc at x and $\lambda + 1 < g_n(x) = \infty$, there exists $U \in \mathcal{N}_X(x)$ such that $\lambda + 1 < g_n(y)$ for every $y \in U$. It follows that $\lambda < g_n(y) - 1 < g(y)$ for all $y \in U \cap \operatorname{dom} f$. Since $\lambda < g(y) = \infty$ for $y \in U \setminus \operatorname{dom} g$, we have that $\lambda < g(y)$ for $y \in U$, and so g is lsc at x. Hence $g \in \Pi_f$. □

Lemma 2.5.4 *Let $f \in \Gamma(X)$ be bounded from below and $K := \{u \in X \mid f_\infty(u) \leq 0\}$. Consider $\varepsilon > 0$ and $y \in X$. Then the set $\overline{\operatorname{co}}(\operatorname{epi} f \cup \{(y, \inf f - \varepsilon)\})$ is the epigraph of a function $f_{y,\varepsilon} \in \Gamma(X)$ with $\inf f_{y,\varepsilon} = \inf f - \varepsilon$ and $\operatorname{argmin} f_{y,\varepsilon} = y + K$. Moreover, if $f(y) \leq \inf f + \varepsilon$ then $f - 2\varepsilon \leq f_{y,\varepsilon} \leq f$, and so $g := f_{y,\varepsilon} + \varepsilon \in \Pi_f$ and $d(f, g) \leq \varepsilon$.*

Proof. Consider $\alpha := \inf f - \varepsilon < \inf f$. Denote $A := \overline{\operatorname{co}}(\operatorname{epi} f \cup \{(y, \alpha)\})$ and let $(x, t) \in A$ and $s > t$. We want to show that $t \geq \alpha$ and $(x, s) \in A$. If these happen then $\operatorname{epi} \varphi_A = A$ [φ_A being defined in Theorem 2.1.3(iv)], $f_{y,\varepsilon} := \varphi_A \in \Gamma(X)$, $f_{y,\varepsilon}(y) = \alpha = \inf f_{y,\varepsilon}$ and $f_{y,\varepsilon} \leq f$. Moreover, if $f(y) \leq \inf f + \varepsilon$, then $f(y) - 2\varepsilon \leq \alpha$, which shows that $(y, \alpha) \in \operatorname{epi}(f - 2\varepsilon)$. As $f - 2\varepsilon \leq f$, we obtain that $A \subset \operatorname{epi}(f - 2\varepsilon)$, and so $f - 2\varepsilon \leq f_{y,\varepsilon}$.

Indeed, there exist the nets $(\lambda_i)_{i\in I} \subset [0, 1]$ and $((x_i, t_i))_{i\in I} \subset \operatorname{epi} f$ such that $(x, t) = \lim_{i\in I} (\lambda_i(x_i, t_i) + (1 - \lambda_i)(y, \alpha))$, and so $x = \lim_{i\in I} \left(\lambda_i x_i + (1 - \lambda_i)y\right)$ and $t = \lim_{i\in I} (\lambda_i t_i + (1 - \lambda_i)\alpha)$. Since $[0, 1]$ is a compact set, we may suppose that $(\lambda_i)_{i\in I} \to \lambda \in [0, 1]$. Of course, $\lambda_i t_i + (1 - \lambda_i)\alpha \geq \lambda_i \inf f + (1 - \lambda_i)\alpha \geq \alpha$ for every $i \in I$, and so $t \geq \alpha$. If $(x, t) = (y, \alpha)$ then $\lim_{n\to\infty} \left(\frac{1}{n}(x_0, t_0 + ns - n\alpha) + \frac{n-1}{n}(y, \alpha)\right) = (x, s)$, and so $(x, s) \in A$, where (x_0, t_0) is a fixed element of epi f. In the contrary case $\lambda_i > 0$ for $i \succeq i_0$ for some $i_0 \in I$. Then $\lim_{i \succeq i_0} \left(\lambda_i(x_i, t_i + \lambda_i^{-1}(s-t)) + (1-\lambda_i)(y,\alpha)\right) = (x, s)$, and so $(x, s) \in A$.

Let $u \in K$; then $(u, 0) \in (\text{epi}\, f)_\infty$. Fixing $(x_0, t_0) \in \text{epi}\, f$, we have that $(x_0 + nu, t_0) \in \text{epi}\, f$ for every $n \in \mathbb{N}$, and so $\frac{1}{n}(x_0 + nu, t_0) + \frac{n-1}{n}(y, \alpha) \in A$ for every $n \in \mathbb{N}$. Taking the limit we obtain that $(y + u, \alpha) \in A$, whence $f_{y,\varepsilon}(y+u) \leq \alpha$. Hence $y + u \in \text{argmin}\, f_{y,\varepsilon}$. Conversely, let $z \in \text{argmin}\, f_{y,\varepsilon}$. Assume that $u := z - y \neq 0$. It follows that $(z, \alpha) \in A$, and so there exist the nets $(\lambda_i)_{i \in I} \subset [0,1]$ and $((x_i, t_i))_{i \in I} \subset \text{epi}\, f$ such that $(z, \alpha) = \lim_{i \in I} (\lambda_i(x_i, t_i) + (1 - \lambda_i)(y, \alpha))$. Because $z \neq y$, $\lambda_i > 0$ for $i \succeq i_0$ (for some $i_0 \in I$). As above, we may assume that $(\lambda_i)_{i \in I} \to \lambda \in [0,1]$. If $\lambda \neq 0$ then $((x_i, t_i))_{i \in I} \to \left(\lambda^{-1}z + (1 - \lambda^{-1})y, \alpha\right)$, a contradiction because $t_i \geq \inf f > \alpha$. Hence $\lambda = 0$. It follows that $\lim_{i \in I} \lambda_i (x_i, t_i) = (u, 0)$. Let $(x, t) \in \text{epi}\, f$. Then $\lambda_i (x_i, t_i) + (1 - \lambda_i)(x, t) \in \text{epi}\, f$, and so, taking the limit, we obtain that $(x, t) + (u, 0) \in \text{epi}\, f$. This shows that $(u, 0) \in \text{rec}(\text{epi}\, f) = (\text{epi}\, f)_\infty$. Therefore $u \in K$. Hence $\text{argmin}\, f_{y,\varepsilon} = y + K$. □

Theorem 2.5.5 *Let X be a reflexive Banach space and $f \in \Gamma(X)$ be such that $K \cap -K = \{0\}$, where $K := \{u \in X \mid f_\infty(u) \leq 0\}$. Then the following statements are equivalent:*

(i) *f is not coercive,*

(ii) *there exists $g \in \Pi_f$ such that* $\text{argmin}\, g = \emptyset$,

(iii) $\{g \in \Pi_f \mid \text{argmin}\, g = \emptyset\}$ *is a dense G_δ set (see page 34).*

Proof. It is obvious that (iii) ⇒ (ii).

(ii) ⇒ (i) Let $g \in \Pi_f$ be such that $\text{argmin}\, g = \emptyset$. From Theorem 2.5.1(ii) we have that g is not coercive. Since $g \in \Pi_f$, there exists $\gamma > 0$ such that $g - \gamma \leq f \leq g + \gamma$, which implies that f is not coercive, too.

(i) ⇒ (iii) First of all note that for any $g \in \Pi_f$, g is not coercive and $f_\infty = g_\infty$ (because $f - \gamma \leq g \leq f + \gamma$ for some $\gamma \in \mathbb{R}_+$).

If $\inf f = -\infty$ then $\mathcal{G} := \{g \in \Pi_f \mid \text{argmin}\, g = \emptyset\} = \Pi_f$, and so the conclusion holds.

Let $\inf f \in \mathbb{R}$. Without loss of generality we assume that $\inf f = 0$. For every $n \in \mathbb{N}$ consider the set

$$\mathcal{G}_n := \{g \in \Pi_f \mid \min_{x \in nU_X} g(x) > \inf g = 0\}.$$

The use of min instead of inf is possible because g is w-lsc and nU_X is w-compact. Also note that $\mathcal{G}_n$ is open in (Π_f, d); just take $g \in \mathcal{G}_n$ and

$0 < \delta < \inf_{x \in nU_X} g(x)$; then the ball $B(g, \delta) \subset \mathcal{G}_n$. As $\mathcal{G} = \bigcap_{n \in \mathbb{N}} \mathcal{G}_n$, it is sufficient to show that $\mathcal{G}_n$ is dense for any $n \geq 1$. For this fix $n \geq 1$, $\varepsilon > 0$ and $h \in \Pi_f$. We may assume that $h(0) \leq \varepsilon$ (otherwise replace h by $h(x_0 + \cdot)$, where x_0 is taken such that $h(x_0) \leq \varepsilon$).

There exists $y \in X$ such that $h(y) \leq \varepsilon$ and $(y+K) \cap nU_X = \emptyset$. Indeed, when $K = \{0\}$ take $y \in X \setminus nU_X$ such that $h(y) \leq \varepsilon$; this is possible because the set $[h \leq \varepsilon]$ is not bounded (see Exercise 2.41). Assume now that $K \neq \{0\}$. Then there exist $z \in K$ and $r > 0$ such that $(z + K) \cap rU_X = \emptyset$, or equivalently $z \notin rU_X - K$. Otherwise $K \subset \bigcap_{r>0} (rU_X - K) \subset -K$, a contradiction. The last inclusion is obtained as follows: consider $z \in \bigcap_{r>0} (rU_X - K)$; then $z = \frac{1}{n} u_n - k_n$ with $u_n \in U_X$ and $k_n \in K$, for every $n \in \mathbb{N}$, and so $(k_n) \to -z \in K$.

Consider $g := h_{y,\varepsilon} + \varepsilon$. By Lemma 2.5.4 we have that $g \in \Pi_f$, $d(h, g) \leq \varepsilon$ and $\operatorname{argmin} h = y + K$. Since $(y + K) \cap nU_X = \emptyset$, we have that $g \in \mathcal{G}_n$. Therefore $\mathcal{G}_n$ is dense in Π_f. □

Note that the reflexivity of the space was used in the proof of the preceding theorem only to ensure that the infimum of g on nU_X is attained; so the preceding result remain valid when working on the dual of a normed space, the considered convex functions being w^*-lsc. Also note that the condition $X_0 := K \cap -K = \{0\}$ in the preceding theorem is not essential; if $X_0 \neq \{0\}$ one obtains a similar result taking into account the constructions in Exercise 2.24. For a similar result when X is not a normed space see Exercise 2.25.

Another important problem in optimization theory is the uniqueness of the solution when it exists. The following result gives an answer to this problem.

Proposition 2.5.6 *Let $f \in \Lambda(X)$. Then $S(f, X)$ is a convex set. Furthermore, if f is strictly convex then $S(f, X)$ has at most one element.*

Proof. Let $\overline{x} \in S(f, X)$; then $S(f, X) = [f \leq f(\overline{x})]$, whence $S(f, X)$ is convex.

Let now f be strictly convex, and suppose that $S(f, X)$ contains (at least) two distinct elements x_1 and x_2; since f is proper, $S(f, X) \subset \operatorname{dom} f$. Then we obtain the contradiction

$$v(f, X) \leq f\left(\tfrac{1}{2}x_1 + \tfrac{1}{2}x_2\right) < \tfrac{1}{2}f(x_1) + \tfrac{1}{2}f(x_2) = v(f, X).$$

Therefore $S(f, X)$ has at most one element. □

As we already know, the practical method for determining extremum points of a function is to determine the points which verify the necessary conditions then to retain those which verify the sufficient conditions. In convex programming we have a very simple necessary and sufficient condition for optimal solutions.

Theorem 2.5.7 *If $f \in \Lambda(X)$, then $\overline{x} \in \operatorname{dom} f$ is a minimum point for f if and only if $0 \in \partial f(\overline{x})$.*

Proof. Indeed, $f(\overline{x}) \leq f(x)$ for every $x \in X$ if and only if $\overline{x} \in \operatorname{dom} f$ and $0 \leq f(x) - f(\overline{x})$ for every $x \in X$, which means that $0 \in \partial f(\overline{x})$. □

Therefore, in convex programming, the minimum necessary condition is also a sufficient condition.

In optimization theory local optimal solutions also play an important role; if $g : X \to \overline{\mathbb{R}}$, we say that $\overline{x} \in X$ is a *local minimum* (resp. *local maximum*) point if there exists $V \in \mathcal{N}_X(\overline{x})$ such that $f(\overline{x}) \leq f(x)$ (resp. $f(\overline{x}) \geq f(x)$) for every $x \in V$. The convex programming problems present a particularity.

Proposition 2.5.8 *Let $f : X \to \overline{\mathbb{R}}$ be a convex function.*

(i) *If $\overline{x} \in \operatorname{dom} f$ is a local minimum point for f, then $\overline{x}$ is also a global minimum point;*

(ii) *if $\overline{x} \in \operatorname{dom} f$ is a local maximum point for f, then $\overline{x}$ is a global minimum point for f.*

Proof. (i) By hypothesis there exists $V \in \mathcal{N}_X(\overline{x})$ such that $f(\overline{x}) \leq f(x)$ for every $x \in V$. Suppose that there exists $x \in X$ such that $f(x) < f(\overline{x})$; therefore $f(\overline{x}) \in \mathbb{R}$. Since $V \in \mathcal{N}_X(\overline{x})$, there exists $\lambda \in]0,1[$ such that $y := (1-\lambda)\overline{x} + \lambda x \in V$. Therefore

$$f(\overline{x}) \leq f(y) = f((1-\lambda)\overline{x} + \lambda x) \leq (1-\lambda) f(\overline{x}) + \lambda f(x),$$

whence the contradiction $f(\overline{x}) \leq f(x)$. Therefore $\overline{x}$ is a global minimum point for f.

(ii) By hypothesis there exists $U \in \mathcal{N}_X^c$ such that $f(\overline{x}+x) \leq f(\overline{x})$ for every $x \in U$. If $f(\overline{x}) = -\infty$, it is clear that $\overline{x}$ is a global minimum point of f. Suppose that $f(\overline{x}) \in \mathbb{R}$ (hence f is proper because $\overline{x} \in \operatorname{int}(\operatorname{dom} f)$). Then

$$\forall x \in U \;:\; f(\overline{x}) = f\left(\tfrac{1}{2}(\overline{x}+x) + \tfrac{1}{2}(\overline{x}-x)\right) \leq \tfrac{1}{2} f(\overline{x}+x) + \tfrac{1}{2} f(\overline{x}-x) \leq f(\overline{x}).$$

Therefore $f(x) = f(\overline{x}) \geq f(\overline{x})$ for every $x \in \overline{x} + U \in \mathcal{N}_X(\overline{x})$. Hence $\overline{x}$ is a global minimum point of f. □

This result explains why in a convex programming problem we look only for global minimum points.

In practical problems, solved numerically on computers, frequently it is not possible to determine the exact optimal solutions (because one works with approximate values). A simple example in this sense is the problem of minimizing the function $f : \mathbb{R} \to \mathbb{R}$, $f(t) := (t - \pi)^2$. Taking into account this fact, the notion of approximate solution is proved to be useful. More precisely, if $\varepsilon \in \mathbb{R}_+$, we say that $\overline{x} \in C$ is an **ε-(optimal) solution** of problem (P) if $f(\overline{x}) \leq f(x) + \varepsilon$ for every $x \in C$; we denote by $S_\varepsilon(P)$ or $S_\varepsilon(f, C)$ the set of ε-solutions of problem (P). It is obvious that when $C \cap \operatorname{dom} f \neq \emptyset$ we have that $S_\varepsilon(f, C) = S_\varepsilon(\tilde{f}, X)$; moreover, if f is proper and $S_\varepsilon(f, C) \neq \emptyset$, then $v(f, C) \in \mathbb{R}$ and $S_\varepsilon(f, C) = \{x \in C \mid f(x) \leq v(f, C) + \varepsilon\}$. Related to ε-solutions we have the following result.

Proposition 2.5.9 *Let $f \in \Lambda(X)$, $\overline{x} \in \operatorname{dom} f$ and $\varepsilon \in \mathbb{P}$. Then $S_\varepsilon(f, X)$ is convex; the set $S_\varepsilon(f, X)$ is nonempty if f is bounded from below. Furthermore, $\overline{x} \in S_\varepsilon(f, X)$ if and only if $0 \in \partial_\varepsilon f(\overline{x})$.* □

2.6 Perturbed Problems

We shall see (especially) in the following two sections that it is very useful to embed a minimization problem

(P) $\quad$ $\min\ f(x),\ x \in X,$

in a *family* of minimization problems.

In this section X, Y are separated locally convex spaces if not stated explicitly otherwise and $f : X \to \overline{\mathbb{R}}$.

Let us consider a function $\Phi : X \times Y \to \overline{\mathbb{R}}$ having the property that $f(x) = \Phi(x, 0)$ for every $x \in X$; Φ is called a **perturbation function**. For every $y \in Y$ consider the problem

(P_y) $\quad$ $\min\ \Phi(x, y),\ x \in X.$

It is obvious that problem (P) coincides with problem (P_0).

To remain in the convex framework, we assume that Φ is a convex function. Of course, when $f : X \to \overline{\mathbb{R}}$ is convex we can find a function Φ having the demanded property: $\Phi(x, y) := f(x)$ for all $(x, y) \in X \times Y$. For obtaining useful results one chooses adequate perturbation functions, as we shall see in the sequel.

Let $\Phi : X \times Y \to \overline{\mathbb{R}}$ be a convex function and $h : Y \to \overline{\mathbb{R}}$, $h(y) = v(P_y)$, its associated marginal function (see page 43); h is also called the **value** or **performance function** associated to problems (P_y). As noted in Theorem 2.1.3(v), h is convex, while from Eq. (2.8) we have that $\operatorname{dom} h = \Pr_Y(\operatorname{dom}\Phi)$.

The problem

$$(P) \qquad \min \ \Phi(x, 0), \ x \in X,$$

is called the **primal problem**; we associate to it, in a natural way, the following **dual problem**

$$(D) \qquad \max \ (-\Phi^*(0, y^*)), \ y^* \in Y^*.$$

It is obvious that (D) is equivalent to the convex programming problem

$$(D') \qquad \min \ \Phi^*(0, y^*), \ y^* \in Y^*.$$

The equivalence has to be understood in the sense that the problems (D) and (D') have the same (ε-)solutions; moreover $v(D') = -v(D)$ (of course, for a maximization problem the notions of (ε-)solution, local solution and value are defined dually to those for minimization problems). It is nice to observe that (D') and (P) are of the same type. In the following results we establish some properties which connect the problems (P), (D) and the function h.

Theorem 2.6.1 *Let $\Phi : X \times Y \to \overline{\mathbb{R}}$ and $h : Y \to \overline{\mathbb{R}}$ be the marginal function associated to Φ. Then:*

(i) *$h^*(y^*) = \Phi^*(0, y^*)$ for every $y^* \in Y^*$.*

(ii) *Let $(\overline{x}, \overline{y}) \in X \times Y$ be such that $\Phi(\overline{x}, \overline{y}) \in \mathbb{R}$. Then*

$$(0, y^*) \in \partial\Phi(\overline{x}, \overline{y}) \ \Leftrightarrow \ h(\overline{y}) = \Phi(\overline{x}, \overline{y}) \text{ and } y^* \in \partial h(\overline{y}).$$

(iii) *$v(P) = h(0)$ and $v(D) = h^{**}(0)$. Therefore $v(P) \geq v(D)$; in this case we say that one has **weak duality**.*

(iv) *Suppose that Φ is proper, $\overline{x} \in X$ and $\overline{y}^* \in Y^*$. Then $(0, \overline{y}^*) \in \partial\Phi(\overline{x}, 0)$ if and only if $\overline{x}$ is a solution of problem (P), $\overline{y}^*$ is a solution of (D) and $v(P) = v(D) \in \mathbb{R}$.*

Assume, moreover, that Φ is convex.

(v) *$h(0) \in \mathbb{R}$ and h is lsc at $0 \Leftrightarrow v(P) = v(D) \in \mathbb{R}$; in this case one says that we have **strong duality**.*

(vi) *$h(0) \in \mathbb{R}$ and $\partial h(0) \neq \emptyset \Leftrightarrow v(P) = v(D) \in \mathbb{R}$ and (D) has optimal solutions. In this situation $S(D) = \partial h(0)$.*

(vii) *Suppose that Φ is proper. Then [$\overline{h}$ is proper] $\Leftrightarrow$ [h^* is proper] $\Leftrightarrow$ [h is minorized by an affine continuous functional] $\Leftrightarrow$*

$$\exists \overline{y}^* \in Y^*,\ \exists \alpha \in \mathbb{R},\ \forall (x,y) \in X \times Y \ :\ \Phi(x,y) \geq \langle y, \overline{y}^* \rangle + \alpha. \tag{2.52}$$

Proof. (i) We have

$$\begin{aligned} h^*(y^*) &= \sup_{y \in Y} (\langle y, y^* \rangle - h(y)) = \sup_{y \in Y} \left(\langle y, y^* \rangle - \inf_{x \in X} \Phi(x,y) \right) \\ &= \sup_{y \in Y} \sup_{x \in X} (\langle y, y^* \rangle - \Phi(x,y)) = \sup_{(x,y) \in X \times Y} (\langle x, 0 \rangle + \langle y, y^* \rangle - \Phi(x,y)) \\ &= \Phi^*(0, y^*). \end{aligned}$$

(ii) Assume that $\Phi(\overline{x}, \overline{y}) \in \mathbb{R}$. Let $(0, y^*) \in \partial \Phi(\overline{x}, \overline{y})$. Then by (i) and Theorem 2.4.2 (iii),

$$h(\overline{y}) \leq \Phi(\overline{x}, \overline{y}) = \langle \overline{x}, 0 \rangle + \langle \overline{y}, y^* \rangle - \Phi^*(0, y^*) = \langle \overline{y}, y^* \rangle - h^*(y^*) \leq h(\overline{y}),$$

and so $h(\overline{y}) = \Phi(\overline{x}, \overline{y})$ and $y^* \in \partial h(\overline{y})$. Conversely, if these two conditions hold then

$$\Phi(\overline{x}, \overline{y}) = h(\overline{y}) = \langle \overline{y}, y^* \rangle - h^*(y^*) = \langle \overline{x}, 0 \rangle + \langle \overline{y}, y^* \rangle - \Phi^*(0, y^*),$$

whence, again by Theorem 2.4.2 (iii), $(0, y^*) \in \partial \Phi(\overline{x}, \overline{y})$.

(iii) It is obvious that $v(P) = h(0)$; moreover,

$$v(D) = \sup_{y^* \in Y^*} (-\Phi(0, y^*)) = \sup_{y^* \in Y^*} (\langle 0, y^* \rangle - h^*(y^*)) = h^{**}(0).$$

Therefore $v(P) \geq v(D)$.

(iv) If $(0, \overline{y}^*) \in \partial \Phi(\overline{x}, 0)$ then

$$v(P) = h(0) \leq \Phi(\overline{x}, 0) = -\Phi^*(0, \overline{y}^*) \leq v(D) = x^{**}(0) \leq h(0),$$

whence $\overline{x}$ is a solution of (P), $\overline{y}^*$ is a solution of (D) and $v(P) = v(D) \in \mathbb{R}$. Conversely, if these last assertions are true, we have

$$-\Phi^*(0, \overline{y}^*) = h^{**}(0) = h(0) = \Phi(\overline{x}, 0) \in \mathbb{R},$$

whence

$$(\overline{x}, 0) \in \operatorname{dom} \Phi \quad \text{and} \quad \Phi(\overline{x}, 0) + \Phi^*(0, \overline{y}^*) = \langle \overline{x}, 0 \rangle + \langle 0, \overline{y}^* \rangle,$$

which shows that $(0, \overline{y}^*) \in \partial\Phi(\overline{x}, 0)$.

Assume now that Φ is convex.

(v) If h is lsc at 0 and $h(0) \in \mathbb{R}$ we have that $\overline{h}(0) \in \mathbb{R}$, whence $h^{**} = \overline{h}$. Therefore $v(D) = h^{**}(0) = h(0) = v(P) \in \mathbb{R}$. Conversely, if this last relation is true, then $\overline{h}(0) = h(0)$, *i.e.* h is lsc at 0. The conclusion holds.

(vi) Suppose that $h(0) \in \mathbb{R}$ and $\partial h(0) \neq \emptyset$; then h is lsc at 0, whence, by (iv), we have that $v(P) = v(D) \in \mathbb{R}$. Let $\overline{y}^* \in \partial h(0)$. Then $h(0) + h^*(\overline{y}^*) = 0$, whence

$$\forall y^* \in Y^* \ : \ v(D) = v(P) = h(0) = -h^*(\overline{y}^*) = -\Phi^*(0, \overline{y}^*) \geq -\Phi^*(0, y^*).$$

Therefore $\emptyset \neq \partial h(0) \subset S(D)$. Conversely, suppose that $v(P) = v(D) \in \mathbb{R}$ and that (D) has solutions. Let $\overline{y}^* \in S(D)$. Then $h(0) = h^{**}(0) = -h^*(\overline{y}^*) \in \mathbb{R}$, whence $\overline{y}^* \in \partial h(0)$. Thus we have that $\emptyset \neq S(D) \subset \partial h(0)$.

(vii) The mentioned equivalences are obvious since $\overline{h}$ is convex, and $\operatorname{dom} \overline{h} \neq \emptyset$. □

We say that the problem (P) is **normal** if $v(P) = v(D) \in \mathbb{R}$; (P) is **stable** if $v(P) = v(D) \in \mathbb{R}$ and (D) has optimal solutions. Theorem 2.6.1 above gives characterizations of these notions.

In the following result we establish formulas for $\partial_\varepsilon h(y)$ and $\partial_\varepsilon h^*(y^*)$ when h is proper.

Theorem 2.6.2 *Let $\Phi \in \Lambda(X \times Y)$ satisfy condition Eq. (2.52) and $\varepsilon \in \mathbb{R}_+$. Then:*

(i) *for every $y \in \operatorname{dom} h$*

$$\begin{aligned}\partial_\varepsilon h(y) &= \bigcap\nolimits_{\eta>0} \bigcup\nolimits_{x\in X} \{y^* \in Y^* \mid (0, y^*) \in \partial_{\varepsilon+\eta}\Phi(x, y)\} \\ &= \bigcap_{\eta>0} \ \bigcap_{\Phi(x,y)\leq h(y)+\eta} \{y^* \in Y^* \mid (0, y^*) \in \partial_{\varepsilon+\eta}\Phi(x, y)\};\end{aligned}$$

(ii) *for every $y^* \in \operatorname{dom} h^*$,*

$$\partial_\varepsilon h^*(y^*) = \bigcap\nolimits_{\eta>0} \operatorname{cl}\{y \in Y \mid \exists x \in X \ : \ (0, y^*) \in \partial_{\varepsilon+\eta}\Phi(x, y)\}.$$

Proof. (i) It is obvious that

$$\bigcap_{\eta>0}\ \bigcap_{\Phi(x,y)\leq h(y)+\eta} \{y^* \mid (0,y^*) \in \partial_{\varepsilon+\eta}\Phi(x,y)\}$$
$$\subset \bigcap\nolimits_{\eta>0}\bigcup\nolimits_{x\in X}\{y^* \mid (0,y^*) \in \partial_{\varepsilon+\eta}\Phi(x,y)\}$$
$$\subset \partial_\varepsilon h(y).$$

Let $y^* \in \partial_\varepsilon h(y)$, *i.e.* $h(y) + h^*(y^*) \leq \langle y, y^*\rangle + \varepsilon$, and let $\eta > 0$ and $x \in X$ be such that $\Phi(x,y) \leq h(y) + \eta$. Then

$$\Phi(x,y) + \Phi^*(0,y^*) \leq h(y) + \eta + h^*(y^*) \leq \langle y, y^*\rangle + \varepsilon + \eta,$$

i.e. $(0,y^*) \in \partial_{\varepsilon+\eta}\Phi(x,y)$. Therefore

$$\partial_\varepsilon h(y) \subset \bigcap_{\eta>0}\ \bigcap_{\Phi(x,y)\leq h(y)+\eta} \{y^* \mid (0,y^*) \in \partial_{\varepsilon+\eta}\Phi(x,y)\},$$

and so the desired equalities hold.

(ii) Let $y \in \partial_\varepsilon h^*(y^*)$ and $\eta > 0$. Then

$$h^{**}(y) + h^*(y^*) = \overline{h}(y) + h^*(y^*) \leq \langle y, y^*\rangle + \varepsilon.$$

It follows that for every $V \in \mathcal{N}(y)$ there exists $y_V \in Y$ such that

$$h(y_V) + h^*(y^*) < \langle y_V, y^*\rangle + \varepsilon + \eta.$$

From the definition of h we get $x_V \in X$ such that

$$\Phi(x_V, y_V) + h^*(y^*) = \Phi(x_V, y_V) + \Phi^*(0, y^*) < \langle y_V, y^*\rangle + \varepsilon + \eta,$$

i.e. $(0,y^*) \in \partial_{\varepsilon+\eta}\Phi(x_V, y_V)$. Therefore

$$y \in \operatorname{cl}\{y \in Y \mid \exists x \in X \ :\ (0,y^*) \in \partial_{\varepsilon+\eta}\Phi(x,y)\}.$$

Taking into account that $\eta > 0$ is arbitrary, we obtain that

$$\partial_\varepsilon h^*(y^*) \subset \bigcap\nolimits_{\eta>0} \operatorname{cl}\{y \mid \exists x \in X \ :\ (0,y^*) \in \partial_{\varepsilon+\eta}\Phi(x,y)\}.$$

Let y be in the set from the right-hand side and $\eta > 0$. Then, using the continuity of y^*, there exists $V_0 \in \mathcal{N}(y)$ such that $\langle z, y^*\rangle < \langle y, y^*\rangle + \eta/2$ for every $z \in V_0$. On the other hand, for every $V \in \mathcal{N}(y)$ there exist

$(x_V, y_V) \in X \times Y$ such that $y_V \in V \cap V_0$ and $(0, y^*) \in \partial_{\varepsilon+\eta/2}\Phi(x_V, y_V)$; hence

$$h(y_V)+h^*(y^*) \leq \Phi(x_V, y_V)+\Phi^*(0, y^*) \leq \langle y_V, y^*\rangle+\varepsilon+\eta/2 < \langle y, v^*\rangle+\varepsilon+\eta.$$

Therefore

$$\forall V \in \mathcal{N}(y) \ : \ \inf_{y \in V} h(y) + h^*(y^*) \leq \langle y, y^*\rangle + \varepsilon + \eta,$$

i.e. $\overline{h}(y) + h^*(y^*) \leq \langle y, y^*\rangle + \varepsilon + \eta$. Since $\eta > 0$ is arbitrary, we obtain that

$$h^{**}(y) + h^*(y^*) = \overline{h}(y) + h^*(y^*) \leq \langle y, y^*\rangle + \varepsilon,$$

i.e. $y \in \partial_\varepsilon h^*(y^*)$. □

The preceding result is useful for deriving formulas for ε-subdifferentials for other functions.

Theorem 2.6.3 *Let* $\Phi \in \Gamma(X \times Y)$ *and* $\varphi : X \to \overline{\mathbb{R}}$, $\varphi(x) := F(x, 0)$. *Then for every* $\varepsilon \in \mathbb{R}_+$ *and every* $x \in \operatorname{dom} \varphi$ *we have*

$$\begin{aligned}\partial_\varepsilon\varphi(x) &= \bigcap\nolimits_{\eta>0} w^*\text{-cl}\{x^* \in X^* \mid \exists y^* \in Y^* \ : \ (x^*, y^*) \in \partial_{\varepsilon+\eta}\Phi(x, 0)\} \\ &= \bigcap\nolimits_{\eta>0} w^*\text{-cl}\{x^* \in X^* \mid \exists y^* \in Y^* \ : \ (x, 0) \in \partial_{\varepsilon+\eta}\Phi^*(x^*, y^*)\}.\end{aligned}$$

Proof. Let us consider the spaces X^*, Y^* endowed with their weak* topologies and the function

$$k : X^* \to \overline{\mathbb{R}}, \quad k(x^*) := \inf_{y^* \in Y^*} \Phi^*(x^*, y^*).$$

Then $k^* : X \to \overline{\mathbb{R}}$, $k^*(x) = \Phi^{**}(x, 0) = \Phi(x, 0) = \varphi(x)$. Using Theorem 2.6.2, we have

$$\begin{aligned}\partial_\varepsilon\varphi(x) = \partial_\varepsilon k^*(x) &= \bigcap\nolimits_{\eta>0} w^*\text{-cl}\{x^* \mid \exists y^* : (x, 0) \in \partial_{\varepsilon+\eta}\Phi^*(x^*, y^*)\} \\ &= \bigcap\nolimits_{\eta>0} w^*\text{-cl}\{x^* \mid \exists y^* \ : \ (x^*, y^*) \in \partial_{\varepsilon+\eta}\Phi(x, 0)\}.\end{aligned}$$

The proof is complete. □

Let us apply the results of Theorems 2.6.2 and 2.6.3 in some particular situations. Stronger conclusions, but under more restrictive conditions, will be established in Section 2.8.

Corollary 2.6.4 *Let $A \in \mathcal{L}(X,Y)$ and $f : X \to \overline{\mathbb{R}}$ be a convex function for which*

$$\exists \overline{y}^* \in Y^*,\ \exists \alpha \in \mathbb{R},\ \forall x \in X\ :\ f(x) \geq \langle x, A^*\overline{y}^*\rangle + \alpha.$$

If $(Af)(y) \in \mathbb{R}$ ($\Leftrightarrow y \in A(\operatorname{dom} f) = \operatorname{dom} Af$), $y^* \in \operatorname{dom}(f^* \circ A^*)$ and $\varepsilon \geq 0$, then

$$\begin{aligned}\partial_\varepsilon (Af)(y) &= \bigcap_{\eta>0} \bigcup_{Ax=y} {A^*}^{-1}(\partial_{\varepsilon+\eta} f(x)) \\ &= \bigcap_{\eta>0} \bigcap_{Ax=y, f(x)\leq (Af)(y)+\eta} {A^*}^{-1}(\partial_{\varepsilon+\eta} f(x)),\end{aligned}$$

and

$$\partial_\varepsilon (f^* \circ A^*)(y^*) = \bigcap_{\eta>0} \operatorname{cl}\{y \in Y \mid \exists x \in X\ :\ Ax = y,\ A^*y^* \in \partial_{\varepsilon+\eta} f(x)\}.$$

Proof. Let us consider

$$\Phi : X \times Y \to \overline{\mathbb{R}},\quad \Phi(x,y) := \begin{cases} f(x) & \text{if} \quad Ax = y, \\ \infty & \text{if} \quad Ax \neq y. \end{cases}$$

Then $\Phi^*(x^*, y^*) = f^*(x^* + A^*y^*)$ and

$$(0, y^*) \in \partial_\eta \Phi(x,y) \Leftrightarrow Ax = y,\ A^*y^* \in \partial_\eta f(x).$$

The result follows immediately from Theorem 2.6.2. □

Corollary 2.6.5 *Let $A \in \mathcal{L}(X,Y)$ and $f \in \Gamma(Y)$. Then*

$$\partial_\varepsilon (f \circ A)(x) = \bigcap\nolimits_{\eta>0} w^*\text{-}\operatorname{cl} A^*\big(\partial_{\varepsilon+\eta} f(Ax)\big)$$

for every $x \in A^{-1}(\operatorname{dom} f) = \operatorname{dom}(f \circ A)$ and every $\varepsilon \geq 0$.

Proof. Let us consider $\Phi : X \times Y \to \overline{\mathbb{R}}$, $\Phi(x,y) := f(Ax+y)$, and $\varphi(x) := \Phi(x,0) = f(Ax)$. Then $\Phi^*(x^*, y^*) = f^*(y^*)$ if $A^*y^* = x^*$, $\Phi^*(x^*, y^*) = \infty$ otherwise. Moreover

$$\begin{aligned}(x^*, y^*) \in \partial_\eta \Phi(x, 0) &\Leftrightarrow A^*y^* = x^* \ \text{ and } \ f(Ax) + f^*(y^*) \leq \langle Ax, y^*\rangle + \eta \\ &\Leftrightarrow A^*y^* = x^* \ \text{ and } \ y^* \in \partial_\eta f(Ax).\end{aligned}$$

The conclusion follows from Theorem 2.6.3. □

Corollary 2.6.6 *Let $f_1, f_2 \in \Lambda(X)$ for which*

$$\exists x^* \in X^*,\ \exists \alpha \in \mathbb{R},\ \forall x \in X,\ \forall i \in \{1,2\}\ :\ f_i(x) \geq \langle x, x^* \rangle + \alpha.$$

If $(f_1 \Box f_2)(x) \in \mathbb{R}$ and $\varepsilon \geq 0$, then:

$$\begin{aligned}\partial_\varepsilon (f_1 \Box f_2)(x) &= \bigcap_{\eta>0} \bigcup_{y\in X, \varepsilon_i \geq 0, \varepsilon+\eta=\varepsilon_1+\varepsilon_2} (\partial_{\varepsilon_1} f_1(x-y) \cap \partial_{\varepsilon_2} f_2(y)) \\ &= \bigcap_{\eta>0} \bigcap_{y \in S_\eta(x)} \bigcup_{\varepsilon_i \geq 0, \varepsilon+\eta=\varepsilon_1+\varepsilon_2} (\partial_{\varepsilon_1} f_1(x-y) \cap \partial_{\varepsilon_2} f_2(y)), \\ \partial (f_1 \Box f_2)(x) &= \bigcap\nolimits_{\eta>0} \bigcup\nolimits_{y\in X} (\partial_\eta f_1(x-y) \cap \partial_\eta f_2(y)),\end{aligned}$$

where $S_\eta(x) := \{y \in X \mid f_1(x-y) + f_2(y) \leq (f_1 \Box f_2)(x) + \eta\}$.

Proof. Let us consider $f : X \times X \to \overline{\mathbb{R}}$, $f(x_1, x_2) := f_1(x_1) + f_2(x_2)$ and $A \in \mathcal{L}(X \times X, X)$, $A(x_1, x_2) := x_1 + x_2$. The conclusion follows from Corollary 2.6.4. □

Corollary 2.6.7 *Let $f_1, f_2 \in \Gamma(X)$. If $x \in \operatorname{dom} f_1 \cap \operatorname{dom} f_2$ and $\varepsilon \geq 0$ then:*

$$\begin{aligned}\partial_\varepsilon (f_1 + f_2)(x) &= \bigcap_{\eta>0} w^*\text{-cl}\left(\bigcup_{\varepsilon_i \geq 0, \varepsilon+\eta=\varepsilon_1+\varepsilon_2} (\partial_{\varepsilon_1} f_1(x) + \partial_{\varepsilon_2} f_2(x)) \right), \\ \partial (f_1 + f_2)(x) &= \bigcap\nolimits_{\eta>0} w^*\text{-cl}(\partial_\eta f_1(x) + \partial_\eta f_2(x)).\end{aligned}$$

Proof. Let us consider $f : X \times X \to \overline{\mathbb{R}}$, $f(x_1, x_2) := f_1(x_1) + f_2(x_2)$ and $A \in \mathcal{L}(X, X \times X)$, $Ax := (x, x)$. The conclusion follows from Corollary 2.6.5. □

2.7 The Fundamental Duality Formula

The following theorem is very useful for obtaining important results in convex programming; this is the reason for calling formula (2.53) the *fundamental duality formula* of convex analysis.

Theorem 2.7.1 *Let $\Phi \in \Lambda(X \times Y)$ be such that $0 \in \operatorname{Pr}_Y(\operatorname{dom} \Phi)$. Consider $Y_0 := \operatorname{lin}(\operatorname{Pr}_Y(\operatorname{dom} \Phi))$. Suppose that one of the following conditions is satisfied:*

(i) *there exists* $\lambda_0 \in \mathbb{R}$ *such that* $V_0 := \{y \in Y \mid \exists x \in X,\ \Phi(x,y) \le \lambda_0\} \in \mathcal{N}_{Y_0}(0)$;

(ii) *there exist* $\lambda_0 \in \mathbb{R}$ *and* $x_0 \in X$ *such that*

$$\forall U \in \mathcal{N}_X \ : \ \{y \in Y \mid \exists x \in x_0 + U,\ \Phi(x,y) \le \lambda_0\} \in \mathcal{N}_{Y_0}(0);$$

(iii) *there exists* $x_0 \in X$ *such that* $(x_0, 0) \in \operatorname{dom}\Phi$ *and* $\Phi(x_0,\cdot)$ *is continuous at* 0;

(iv) X *and* Y *are metrizable,* $\operatorname{epi}\Phi$ *satisfies condition* (Hwx) *on page 14 and* $0 \in {}^{ib}(\operatorname{Pr}_Y(\operatorname{dom}\Phi))$;

(v) X *is a Fréchet space,* Y *is metrizable,* Φ *is a li-convex function and* $0 \in {}^{ib}(\operatorname{Pr}_Y(\operatorname{dom}\Phi))$;

(vi) X *is a Fréchet space,* Φ *is lsc and* $0 \in {}^{ib}(\operatorname{Pr}_Y(\operatorname{dom}\Phi))$;

(vii) X, Y *are Fréchet spaces,* Φ *is lsc and* $0 \in {}^{ic}(\operatorname{Pr}_Y(\operatorname{dom}\Phi))$;

(viii) $\dim Y_0 < \infty$ *and* $0 \in {}^{i}(\operatorname{Pr}_Y(\operatorname{dom}\Phi))$;

(ix) *there exists* $x_0 \in X$ *such that* $\Phi(x_0,\cdot)$ *is quasi-continuous and the sets* $\{0\}$, $\operatorname{Pr}_Y(\operatorname{dom}\Phi)$ *are united.*

Then either $h(0) = -\infty$ *or* $h(0) \in \mathbb{R}$ *and* $h|_{Y_0}$ *is continuous at* 0. *In both cases we have*

$$\inf_{x \in X} \Phi(x,0) = \max_{y^* \in Y^*} \big(- \Phi^*(0,y^*)\big). \tag{2.53}$$

Furthermore, $\overline{x} \in X$ *is a minimum point for* $\Phi(\cdot,0)$ *if and only if there exists* $\overline{y}^* \in Y^*$ *such that* $(0,\overline{y}^*) \in \partial\Phi(\overline{x},0)$.

Proof. Since $0 \in \operatorname{Pr}_Y(\operatorname{dom}\Phi)$, $h(0) < \infty$. If $h(0) = -\infty$ we have $h^*(y^*) = \infty$ for every $y^* \in Y^*$, whence $-\Phi^*(0,y^*) = -\infty = h(0)$ for every $y^* \in Y^*$. Therefore the conclusion is true in this case. Let us consider now the case $h(0) \in \mathbb{R}$.

Suppose that condition (i) is verified. Then $h|_{Y_0}$ is bounded above by λ_0 on V_0. Since h is convex, $h|_{Y_0}$ is continuous at 0. By Theorem 2.4.12 we have that $\partial h(0) \ne \emptyset$. Then relation (2.53) follows Theorem 2.6.1 (vi).

(ii) $\Rightarrow$ (i) This implication is obvious (just take $U = X$).

(iii) $\Rightarrow$ (ii) Suppose that (iii) holds. Since $\Phi(x_0,\cdot)$ is continuous at 0, the set $V_0 := \{y \mid \Phi(x_0,y) \le \Phi(x_0,0) + 1\}$ is a neighborhood of 0 (in particular $Y_0 = Y$). Taking $\lambda_0 := \Phi(x_0,0) + 1$, it is obvious that $V_0 \subset \{y \mid \exists x \in x_0 + U \ : \ \Phi(x,y) \le \lambda_0\}$ for every $U \in \mathcal{N}_X$. Therefore (ii) holds.

(iv) $\Rightarrow$ (ii) Suppose that (iv) holds. Let us consider the relation $\mathcal{R} : X \times \mathbb{R} \rightrightarrows Y$ whose graph is given by

$$\operatorname{gr} \mathcal{R} := \{(x,t,y) \mid (x,y,t) \in \operatorname{epi} \Phi\}.$$

From the hypothesis $\mathcal{R}$ satisfies condition (Hwx), whence, by Proposition 1.2.6(i), $\mathcal{R}$ satisfies condition (Hw(x,t)), too. Moreover $0 \in {}^{ib}(\operatorname{Im} \mathcal{R})$. Let $(x_0,t_0) \in X \times \mathbb{R}$ be such that $0 \in \mathcal{R}(x_0,t_0)$. Applying Theorem 1.3.5 we obtain that $\mathcal{R}((x_0+U)\times\,]-\infty,t_0+1]) \in \mathcal{N}_{Y_0}(0)$ for every $U \in \mathcal{N}_X$. This shows that (ii) holds with $\lambda_0 := t_0 + 1$.

(v) $\Rightarrow$ (ii) By Proposition 2.2.18, there exist a Fréchet space Z and a cs-closed function $F : Z \times X \times Y \to \overline{\mathbb{R}}$ such that $\Phi(x,y) = \inf_{z\in Z} F(z,x,y)$ for all $(x,y) \in X \times Y$. The conclusion follows like in the preceding case by replacing X by $Z \times X$, x by (z,x) and U by $Z \times U$.

(vi) $\Rightarrow$ (ii) The proof is the same as for (iv) $\Rightarrow$ (ii) with the exception that one uses Ursescu's theorem (Theorem 1.3.7) instead of Simons' theorem.

It is obvious that (vii) implies conditions (iv), (v) and (vi).

If (viii) is verified, the conclusion follows from Theorem 2.4.12.

(ix) $\Rightarrow$ (i) It is obvious that $\Phi(x_0,\cdot) \geq h$. By Proposition 2.2.15 we have that h is quasi-continuous. It follows that $\operatorname{rint}(\operatorname{dom} h) \neq \emptyset$. Using Proposition 1.2.8 we obtain that $0 \in \operatorname{rint}(\operatorname{dom} h)$. Therefore $h|_{Y_0}$ is continuous at 0, and so (i) holds.

Of course, if $\overline{x}$ is a minimum point for $\Phi(\cdot,0)$ then $\overline{x}$ is a solution of (P) (from p. 107); it follows that $\Phi(\overline{x},0) = v(P) = v(D) \in \mathbb{R}$. Let $\overline{y}^*$ be a solution of (D) (in our conditions a solution exists). Then, from Theorem 2.6.1(iv), $(0,\overline{y}^*) \in \partial\Phi(\overline{x},0)$. The converse implication follows from the same result. □

When the properness condition in the preceding theorem is violated relation (2.53) is automatically verified. Indeed, in this case there exists $(x_1,y_1) \in X \times Y$ such that $\Phi(x_1,y_1) = -\infty$, and so $h(y_1) = -\infty$. Because $0 \in {}^{i}(\operatorname{dom} h)$, by Proposition 2.2.5 we have that $h(0) = -\infty$.

In applications it is important to have conditions on Φ which also ensure that the functions $\widetilde{\Phi}$, $\widetilde{\Phi}(x,y) := \Phi(x,y) - \langle x,x^*\rangle$ with $x^* \in X^*$, satisfy them. Such conditions are conditions (ii)–(ix) of Theorem 2.7.1, but this is not always true for (i).

Other conditions of this type are:

$$\exists\,\lambda_0 \in \mathbb{R},\ \exists\, B \in \mathcal{B}_X\ :\ \{y \in Y \mid \exists\, x \in B,\ \Phi(x,y) \le \lambda_0\} \in \mathcal{N}_{Y_0}(0), \quad (2.54)$$

$$\forall U \in \mathcal{N}_X,\ \exists\,\lambda > 0\ :\ \{y \in Y \mid \exists\, x \in \lambda U,\ \Phi(x,y) \le \lambda\} \in \mathcal{N}_{Y_0}(0), \quad (2.55)$$

where $\mathcal{B}_X$ is the class of bounded subsets of X and, as in Theorem 2.7.1, $Y_0 = \operatorname{lin}(\operatorname{Pr}_Y(\operatorname{dom}\Phi))$.

Proposition 2.7.2 *Let* $\Phi \in \Lambda(X \times Y)$. *Conditions* (i) – (viii) *being those from Theorem 2.7.1, we have:* (iii) $\Rightarrow$ (2.54) $\Rightarrow$ (2.55) $\Leftrightarrow$ (ii) $\Rightarrow$ (i), (2.55) $\Rightarrow$ (2.54) *if X is a normed vector space,* (vii) $\Rightarrow$ (iv) $\wedge$ (v) $\wedge$ (vi), (iv) $\vee$ (v) $\vee$ (vi) $\Rightarrow$ (ii) *and* (viii) $\Rightarrow$ (2.54).

Moreover, taking $D = \operatorname{Pr}_Y(\operatorname{dom}\Phi)$, *one has: if* $\dim(\operatorname{lin} D) < \infty$ *then* ${}^{i}D = \operatorname{rint} D$; *if X,Y are metrizable,* $\operatorname{epi}\Phi$ *satisfies* $\mathrm{H}(x)$ *and* ${}^{ib}D \neq \emptyset$ *then* ${}^{ib}D = \operatorname{rint} D$; *similarly for the situations corresponding to conditions* (v) – (vii).

Proof. The implications (vii) $\Rightarrow$ (iv)$\wedge$(v)$\wedge$(vi), (iv)$\vee$(v)$\vee$(vi) $\Rightarrow$ (ii) $\Rightarrow$ (i) were already observed (or proved) during the proof of the preceding theorem.

The implication (iii) $\Rightarrow$ (2.54) is obvious; just take $B = \{x_0\}$.

(2.54) $\Rightarrow$ (2.55) Let $U \in \mathcal{N}_X$; there exists $\mu > 0$ such that $B \subset \mu U$. Taking $\lambda = \max\{\lambda_0, \mu\}$ we have that

$$\begin{aligned}\{y \in Y \mid \exists\, x \in B,\ \Phi(x,y) \le \lambda_0\} &\subset \{y \in Y \mid \exists\, x \in \mu U,\ \Phi(x,y) \le \lambda_0\}\\ &\subset \{y \in Y \mid \exists\, x \in \lambda U,\ \Phi(x,y) \le \lambda\}.\end{aligned}$$

The conclusion follows.

(ii) $\Rightarrow$ (2.55) Consider $\lambda_0 \in \mathbb{R}$ and $x_0 \in X$ given by (ii). Let $U \in \mathcal{N}_X$. There exists $\mu > 0$ such that $x_0 \in \mu U$. Let $V = \{y \in Y \mid \exists\, x \in x_0 + U,\ \Phi(x,y) \le \lambda_0\} \in \mathcal{N}_{Y_0}$. Taking $\lambda = \max\{\lambda_0, \mu + 1\}$ and $y \in V$, there exists $x \in x_0 + U$ such that $\Phi(x,y) \le \lambda_0$. As $x \in x_0 + U \subset \mu U + U = (\mu + 1)U \subset \lambda U$, the conclusion follows.

(2.55) $\Rightarrow$ (ii) It is obvious that there exists $x_0 \in X$ such that $\Phi(x_0, 0) < \infty$. Consider $\lambda_0 = \max\{\Phi(x_0,0), 0\} + 1$ and let $U \in \mathcal{N}_X$. There exists $U_0 \in \mathcal{N}_X$ such that $U_0 + U_0 \subset U$. There exists also $\lambda_1 > 0$ such that $x_0 \in \lambda_1 U_0$. By hypothesis, there exists $\lambda \ge \lambda_0 + \lambda_1$ such that $V_0 = \{y \in Y \mid \exists\, x \in \lambda U_0,\ \Phi(x,y) \le \lambda\} \in \mathcal{N}_{Y_0}(0)$. Let $V = \lambda^{-1} V_0$ and take $y \in V$. As $\lambda y \in V_0$, there exists $x' \in \lambda U_0$ such that $\Phi(x', \lambda y) \le \lambda$. It follows that

$$\Phi\left((1-\lambda^{-1})(x_0,0) + \lambda^{-1}(x', \lambda y)\right) \le (1-\lambda^{-1})\Phi(x_0,0) + \lambda^{-1}\Phi(x', \lambda y) \le \lambda_0,$$

whence $\Phi(x,y) \le \lambda_0$, where $x := x_0 + \lambda^{-1}(x' - x_0) \in x_0 + U_0 + U_0 \subset x_0 + U$.

(2.55) $\Rightarrow$ (2.54) when X is a normed vector space. Take $U = U_X = \{x \in X \mid \|x\| \le 1\}$. There exists $\lambda_0 > 0$ such that $\{y \in Y \mid \exists x \in \lambda_0 U,\ \Phi(x,y) \le \lambda_0\} \in \mathcal{N}_{Y_0}(0)$. As $\lambda_0 U$ is bounded, the conclusion follows.

(viii) $\Rightarrow$ (2.54) Suppose that $\dim Y_0 < \infty$ and $0 \in {}^i(\mathrm{Pr}_Y(\mathrm{dom}\,\Phi))$. It follows that there exist $y_1, \ldots, y_m \in \mathrm{Pr}_Y(\mathrm{dom}\,\Phi)$ such that $V_0 = \mathrm{co}\{y_1, \ldots, y_m\} \in \mathcal{N}_{Y_0}(0)$. For every $i \in \overline{1,m}$ there exists $x_i \in X$ such that $(x_i, y_i) \in \mathrm{dom}\,\Phi$. Let $\lambda_0 = \max\{\Phi(x_i, y_i) \mid 1 \le i \le m\}$ and $B = \mathrm{co}\{x_1, \ldots, x_m\}$. It is obvious that B is bounded and for $y \in V_0$ there exist $\lambda_1, \ldots, \lambda_m \ge 0$ with $\sum_{i=1}^m \lambda_i = 1$ such that $y = \sum_{i=1}^m \lambda_i y_i$. Then $x = \sum_{i=1}^m \lambda_i x_i \in B$ and $\Phi(x,y) \le \lambda_0$.

The fact that ${}^iD = \mathrm{rint}\,D$ if $\dim(\mathrm{lin}\,D) < \infty$ is obvious. Let X, Y be metrizable, $\mathrm{epi}\,\Phi$ satisfy (Hx), and consider $y_0 \in {}^{ib}D$. Taking Φ_0 defined by $\Phi_0(x,y) = \Phi(x, y + y_0)$, we have that $0 \in {}^{ib}\mathrm{Pr}_Y(\mathrm{dom}\,\Phi_0)$. It is easy to show that Φ_0 verifies condition (iv) of Theorem 2.7.1. As remarked above, condition (ii) holds for Φ_0, which implies that $0 \in \mathrm{rint}\,(\mathrm{Pr}_Y(\mathrm{dom}\,\Phi_0))$, *i.e.* $y_0 \in \mathrm{rint}\,D$. Similarly one obtains the other relations. □

Corollary 2.7.3 *Let $\Phi \in \Gamma(X \times Y)$. If one of the conditions* (ii)–(ix) *of Theorem 2.7.1,* (2.54) *or* (2.55) *holds, then*

$$(\Phi(\cdot,0))^*(x^*) = \min_{y^* \in Y^*} \Phi^*(x^*, y^*) =: \psi(x^*)$$

for every $x^* \in X^*$. In particular $\psi \in \Gamma^*(X^*)$.

Proof. It is obvious that $\Phi(\cdot,0) \in \Gamma(X)$ in our conditions. Let $x^* \in X^*$ and consider $\widetilde{\Phi} : X \times Y \to \overline{\mathbb{R}}$ defined by $\widetilde{\Phi}(x,y) := \Phi(x,y) - \langle x, x^* \rangle$. As observed above, the function $\widetilde{\Phi}$ satisfies the same condition as Φ among those mentioned in the statement of the corollary. Applying Theorem 2.7.1 (and eventually the preceding proposition), we have that $\inf_{x \in X} \widetilde{\Phi}(x,0) = \max_{y^* \in Y^*} \left(-\widetilde{\Phi}^*(0,y^*)\right)$. But $\widetilde{\Phi}^*(0,y^*) = \Phi^*(x^*,y^*)$, and so the conclusion follows. □

We state another duality formula which will be useful in the sequel.

Theorem 2.7.4 *Let $F \in \Lambda(X \times Y)$, $\mathcal{C} : X \rightrightarrows Y$ be a convex multifunction, and $D = \cup\{\mathcal{C}(x) - y \mid (x,y) \in \mathrm{dom}\,F\}$. Assume that $0 \in D$ and let $Y_0 = \mathrm{lin}\,D$. If one of the following conditions holds:*

(i) *for every $U \in \mathcal{N}_X$ there exist $\lambda > 0$ and $V \in \mathcal{N}_{Y_0}$ such that*

$$\{0\} \times V \subset \mathrm{gr}\,\mathcal{C} \cap (\lambda U \times Y) - [F \le \lambda];$$

(ii) *there exist* $\lambda_0 \in \mathbb{R}$, $B \in \mathcal{B}_X$ *and* $V_0 \in \mathcal{N}_{Y_0}$ *such that*

$$\{0\} \times V_0 \subset \operatorname{gr} \mathcal{C} \cap (B \times Y) - [F \leq \lambda_0];$$

(iii) *there exists* $(x_0, y_0) \in \operatorname{gr} \mathcal{C} \cap \operatorname{dom} F$ *such that* $F(x_0, \cdot)$ *is continuous at* y_0*;*

(iv) X, Y *are metrizable,* $0 \in {}^{ib}D$ *and either* F *is cs-complete and* $\mathcal{C}$ *is cs-closed, or* F *is cs-closed and* $\mathcal{C}$ *is cs-complete;*

(v) X, Y *are Fréchet spaces,* F *and* $\mathcal{C}$ *are li-convex, and* $0 \in {}^{ib}D$*;*

(vi) $\dim Y_0 < \infty$ *and* $0 \in {}^{i}D$*,*

then there exists $z^* \in Y^*$ *such that*

$$\inf\{F(x,y) \mid (x,y) \in \operatorname{gr} \mathcal{C}\} = \inf\{F(x,y) + \langle z, z^* \rangle \mid (x, y+z) \in \operatorname{gr} \mathcal{C}\}.$$

Proof. Let $Z := Y$ and

$$\Phi : (X \times Z) \times Y \to \overline{\mathbb{R}}, \quad \Phi(x,z;y) := F(x,z) + \iota_{\operatorname{gr} \mathcal{C}}(x, y+z).$$

It follows easily that Φ is convex and $\operatorname{Pr}_Y(\operatorname{dom} \Phi) = D$. It is obvious that the conclusion of the theorem is equivalent to $\inf_{(x,z) \in X \times Z} \Phi(x,z;0) = \max_{y^* \in Y^*} -\Phi^*(0,0;y^*)$. So we have to show that if one of the conditions of the theorem is verified then a condition of Theorem 2.7.1 holds.

If (i) holds it is immediate that Φ verifies condition (i) of Theorem 2.7.1. The implication (ii) $\Rightarrow$ (i) is obvious. We also have that (iii) $\Rightarrow$ (ii); just take $B := \{x_0\}$, $\lambda_0 := F(x_0, y_0) + 1$ and $V_0 = \{y \in Y \mid F(x_0, y_0 + y) \leq \lambda_0\}$ (in this case $Y_0 = Y$). Similar to the proof in Proposition 2.7.2 we have that (vi) $\Rightarrow$ (ii).

(iv) $\Rightarrow$ (i) Let $\sum_{n \geq 1} \lambda_n (x_n, z_n, y_n, t_n)$ be a convex series with elements of $\operatorname{epi} \Phi$ such that $\sum_{n \geq 1} \lambda_n x_n$ and $\sum_{n \geq 1} \lambda_n z_n$ are Cauchy, $\sum_{n \geq 1} \lambda_n y_n = y \in Y$ and $\sum_{n \geq 1} \lambda_n t_n = t \in \mathbb{R}$. Then $(x_n, z_n, t_n) \in \operatorname{epi} F$ and $(x_n, z_n + y_n) \in \operatorname{gr} \mathcal{C}$ for every $n \geq 1$. If F is cs-complete it follows that $\sum_{n \geq 1} \lambda_n x_n$ and $\sum_{n \geq 1} \lambda_n z_n$ are convergent with sums $x \in X$ and $z \in Z$, respectively; moreover, $(x, z, t) \in \operatorname{epi} F$, whence $(x, z + y) \in \operatorname{gr} \mathcal{C}$ since $\operatorname{gr} \mathcal{C}$ is cs-closed. The same conclusion holds in the other case. Therefore Φ verifies condition ($\mathrm{H}(x,z)$) in our hypotheses. Hence condition (iv) of Theorem 2.7.1 is verified. So, by Proposition 2.7.2, relation (2.55) holds. Therefore for $U \times Y \in \mathcal{N}_{X \times Z}$ there exists $\lambda > 0$ such that

$$\begin{aligned} &\{y \in Y \mid \exists\, (x,z) \in \lambda U \times Y,\ \Phi(x,z;y) \leq \lambda\} \\ &\quad = \{y \in Y \mid \exists\, x \in \lambda U,\ \exists\, z \in Z \ :\ (x, y+z) \in \mathcal{C},\ F(x,z) \leq \lambda\} \in \mathcal{N}_{Y_0}(0), \end{aligned}$$

and so (i) holds.

(v) ⇒ (i) Consider the set

$$\begin{aligned} A :&= \{(x,z,y) \in X \times Z \times Y \mid y + z \in \mathcal{C}(x)\} \\ &= \{(x, y' - y, y) \mid (x,y') \in \operatorname{gr}\mathcal{C},\ y \in Y\} \\ &= \operatorname{gr}\mathcal{C} \times \{0\} + \{0\} \times \{(-y,y) \mid y \in Y\}. \end{aligned}$$

Using Propositions 1.2.4 (ii) and 1.2.5 (ii) we obtain that A is li-convex (as sum of two li-convex subsets of a Fréchet space). Since $\Phi(x,z;y) = F(x,z) + \iota_A(x,z,y)$ and the functions F and ι_A are li-convex, Φ is li-convex, too. Thus Φ satisfies condition (v) of Theorem 2.7.1; the other conditions being obviously satisfied, as in (iv) ⇒ (i), we get that (i) holds, too. □

Note that every condition of the preceding theorem is verified by $\widetilde{F}$, $\widetilde{F}(x,y) = F(x,y) - \langle x, x^* \rangle$, where $x^* \in X^*$, when the same condition is verified by F.

Taking $\operatorname{gr}\mathcal{C} = X \times \{0\}$, the conclusion of the preceding theorem is just the conclusion of Theorem 2.7.1. Conditions (i), (ii), (iii) and (vi) become conditions (2.55), (2.54), (iii) and (viii) of Theorem 2.7.1, respectively; to conditions (iv) and (v) correspond slightly stronger forms of conditions (iv) and (v) of Theorem 2.7.1, respectively.

Remark 2.7.1 If $F(x,y) = f(x) + g(y)$ with $f \in \Lambda(X)$, $g \in \Lambda(Y)$, for condition (i) of Theorem 2.7.4 it is sufficient (and necessary if f, g have proper conjugates) to have

$$\forall U \in \mathcal{N}_X,\ \exists \lambda > 0,\ \exists V \in \mathcal{N}_{Y_0} \ :\ V \subset [g \le \lambda] - \mathcal{C}(\lambda U \cap [f \le \lambda]),$$

while for condition (ii) of Theorem 2.7.4 it is sufficient (and necessary if f, g have proper conjugates) to have

$$\exists \lambda_0 \in \mathbb{R},\ B \in \mathcal{B}_X,\ V_0 \in \mathcal{N}_{Y_0} \ :\ V_0 \subset [g \le \lambda_0] - \mathcal{C}(B \cap [f \le \lambda_0]).$$

Of course, these two conditions are equivalent if X is a normed space.

Corollary 2.7.5 *Let $f \in \Lambda(X)$ and $A \in \mathcal{L}(X,Y)$. Suppose that one of the following conditions is verified:*

(i) *f is continuous on* $\operatorname{int}(\operatorname{dom} f)$, *assumed to be nonempty, and A is* relatively open, *i.e. $A(D)$ is open in* $\operatorname{Im} A$ *for every open subset $D \subset X$;*

(ii) *X and Y are metrizable, either* (a) *f is cs-complete or* (b) *f is cs-closed and* $\operatorname{gr} A$ *satisfies condition* (Hx), *and* ${}^{ib}A(\operatorname{dom} f) \neq \emptyset$;

(iii) *X is a Fréchet space, Y is metrizable, f is a li-convex function and* ${}^{ib}A(\operatorname{dom} f) \neq \emptyset$;

(iv) *X is a Fréchet space, f is lower semicontinuous and* ${}^{ib}A(\operatorname{dom} f) \neq \emptyset$;

(v) $\dim\left(\operatorname{lin} A(\operatorname{dom} f)\right) < \infty$.

Then, either Af is $-\infty$ on ${}^{i}\left(A(\operatorname{dom} f)\right)$ or Af is proper and $(Af)|_{Y_0}$ is continuous on ${}^{i}\left(A(\operatorname{dom} f)\right)$. Moreover

$$\forall y \in {}^{i}\left(A(\operatorname{dom} f)\right) \ : \ (Af)(y) = \max\{\langle y, y^*\rangle - f^*(A^*y^*) \mid y^* \in Y^*\}.$$

Proof. Let us consider $y_0 \in {}^{i}\left(A(\operatorname{dom} f)\right)$ and

$$\Phi : X \times Y \to \overline{\mathbb{R}}, \quad \Phi(x,y) := f(x) + \iota_{\operatorname{gr} A}(x, y_0 + y).$$

We have that $\operatorname{Pr}_Y(\operatorname{dom}\Phi) = A(\operatorname{dom} f) - y_0$. If condition (ii), (iii), (iv) or (v) is verified, then Φ satisfies condition (iv), (v), (vi) or (viii) of Theorem 2.7.1, respectively. Suppose that condition (i) is satisfied. (Obviously, it is impossible that condition (iii) of Theorem 2.7.1 be verified in this case: $F(x_0,\cdot) = \iota_{\{Ax_0\}}$.) Since A is relatively open we have that ${}^{i}\left(A(\operatorname{dom} f)\right) = \operatorname{int}_{\operatorname{Im} A}\left(A(\operatorname{dom} f)\right) = A(\operatorname{int}(\operatorname{dom} f))$ (Exercise!), whence $y_0 = Ax_0$ for some $x_0 \in \operatorname{int}(\operatorname{dom} f)$. It follows that f is bounded above on a neighborhood V_0 of x_0, and so h (the marginal function associated to Φ) is bounded above by the same constant on $A(V_0) - y_0$, which is a neighborhood of 0. Therefore condition (i) of Theorem 2.7.1 is verified. So, under each of the five conditions, we have that

$$(Af)(y_0) = \inf_{x \in X} \Phi(x,0) = \max_{y^* \in Y^*}\left(-\Phi^*(0,y^*)\right) = \max_{y^* \in Y^*}\left(\langle y_0, y^*\rangle - f^*(A^*y^*)\right),$$

doing similar calculations to those from Theorem 2.3.1(ix). $\square$

Corollary 2.7.6 *Let $f_1, \dots, f_n \in \Lambda(X)$ and take $f := f_1 \square \cdots \square f_n$. Suppose that one of the following conditions is verified:*

(i) *f_1 is continuous at some point in* $\operatorname{dom} f_1$,

(ii) *X is a Fréchet space, the functions $f_1, \dots, f_n$ are li-convex and* ${}^{ib}(\operatorname{dom} f) \neq \emptyset$,

(iii) $\dim X < \infty$.

Then either f is $-\infty$ on ${}^{i}(\operatorname{dom} f)$ or $f|_{\operatorname{aff}(\operatorname{dom} f)}$ is finite and continuous on ${}^{i}(\operatorname{dom} f)$, in which case f is subdifferentiable on this set. Furthermore, for every $x \in {}^{i}(\operatorname{dom} f)$ we have that

$$f(x) = \max_{x^* \in X^*}\left(\langle x, x^*\rangle - f_1^*(x^*) - \cdots - f_n^*(x^*)\right) = (f_1^* + \cdots + f_n^*)^*(x).$$

Proof. Recall that $\operatorname{dom} f = \operatorname{dom} f_1 + \cdots + \operatorname{dom} f_n$. In the cases (ii) and (iii) the result is an immediate consequence of Corollary 2.7.5 taking Φ and A as in Corollary 2.4.7, while for (i) one does the proof by induction. □

2.8 Formulas for Conjugates and ε-Subdifferentials, Duality Relations and Optimality Conditions

In the preceding sections we have considered only situations in which it was simple to compute conjugates and ε-subdifferentials: sum of functions with separated variables, convolution of convex functions, and functions of type Af. For the other types of functions, generally, it is more difficult to compute the conjugate functions or the ε-subdifferentials. In this section, we intend to establish sufficient conditions, as general as possible, in order to ensure the validity of such formulas. In this section the spaces $X, X_1, \ldots, X_n$ and Y are separated locally convex spaces if not stated explicitly otherwise.

We begin with the following result.

Theorem 2.8.1 *Let $F \in \Lambda(X \times Y)$, $A \in \mathcal{L}(X,Y)$ and $\varphi : X \to \overline{\mathbb{R}}$, $\varphi(x) := F(x, Ax)$. Assume that $0 \in D := \{Ax - y \mid (x,y) \in \operatorname{dom} F\}$ and take $Y_0 := \operatorname{lin} D$. Assume that one of the following conditions holds:*

(i) *there exist $\lambda_0 \in \mathbb{R}$, $V_0 \in \mathcal{N}_{Y_0}$ and $B \in \mathcal{B}_X$ such that*

$$\{0\} \times V_0 \subset \{(x, Ax) \mid x \in B\} - [F \le \lambda_0];$$

(ii) *for every $U \in \mathcal{N}_X$ there exist $\lambda > 0$ and $V \in \mathcal{N}_{Y_0}$ such that*

$$\{0\} \times V \subset \{(x, Ax) \mid x \in \lambda U\} - [F \le \lambda];$$

(iii) *there exists $x_0 \in X$ such that $(x_0, Ax_0) \in \operatorname{dom} F$ and $F(x_0, \cdot)$ is continuous at Ax_0;*

(iv) *X and Y are metrizable, $0 \in {}^{ib}D$ and either F is cs-closed and $\operatorname{gr} A$ is cs-complete or F is cs-complete;*

(v) *X is a Fréchet space, Y is metrizable, F is li-convex and $0 \in {}^{ib}D$;*

(vi) *X is a Fréchet space, F is lsc and $0 \in {}^{ib}D$;*

(vii) *X and Y are Fréchet spaces, F is lsc and $0 \in {}^{ic}D$;*

(viii) $\dim Y_0 < \infty$ *and $0 \in {}^{i}D$,*

(ix) *there exists* $x_0 \in X$ *such that* $F(x_0, \cdot)$ *is quasi-continuous and* $\{0\}$ *and* D *are united.*

Then for $x^* \in X^*$, $x \in \operatorname{dom} \varphi$ *and* $\varepsilon \geq 0$ *we have:*

$$\varphi^*(x^*) = \min\{F^*(x^* - A^*y^*, y^*) \mid y^* \in Y^*\}, \tag{2.56}$$

$$\partial_\varepsilon \varphi(x) = \{A^*y^* + x^* \mid (x^*, y^*) \in \partial_\varepsilon F(x, Ax)\}. \tag{2.57}$$

Proof. It is easy to verify that

$$\forall x^* \in X^* \ : \ \varphi^*(x^*) \leq \inf\{F^*(x^* - A^*y^*, y^*) \mid y^* \in Y^*\}$$

for every function F and every operator $A \in \mathfrak{L}(X, Y)$.

Consider $\Phi : X \times Y \to \overline{\mathbb{R}}$, $\Phi(x, y) = F(x, Ax - y)$. Then Φ satisfies one of the conditions (2.54) or (ii) – (ix) of Theorem 2.7.1 when F satisfies one of the conditions (i) – (ix), respectively. It follows that the function $\widetilde{\Phi}$, defined by $\widetilde{\Phi}(x, y) := \Phi(x, y) - \langle x, x^* \rangle$, where $x^* \in X^*$, satisfies one of the conditions of Theorem 2.7.1, too. But

$$-\varphi^*(x^*) = \inf\{F(x, Ax) - \langle x, x^* \rangle \mid x \in X\} = \inf\left\{\widetilde{\Phi}(x, 0) \;\middle|\; x \in X\right\}.$$

Applying Theorem 2.7.1, we obtain that

$$-\varphi^*(x^*) = \inf\left\{\widetilde{\Phi}(x, 0) \;\middle|\; x \in X\right\} = \max\left\{-\widetilde{\Phi}^*(0, -y^*) \;\middle|\; y^* \in Y^*\right\}. \tag{2.58}$$

But

$$\begin{aligned}
\widetilde{\Phi}^*(0, -y^*) &= \sup\{\langle x, x^* \rangle + \langle y, -y^* \rangle - F(x, Ax - y) \mid (x, y) \in X \times Y\} \\
&= \sup\{\langle x, x^* \rangle + \langle z - Ax, y^* \rangle - F(x, z) \mid (x, z) \in X \times Y\} \\
&= \sup\{\langle x, x^* - A^*y^* \rangle + \langle z, y^* \rangle - F(x, z) \mid (x, z) \in X \times Y\} \\
&= F^*(x^* - A^*y^*, y^*).
\end{aligned}$$

From Eq. (2.58) we get immediately Eq. (2.56).

Note that the inclusion "$\supset$" in Eq. (2.57) is true for every function F and every operator $A \in \mathfrak{L}(X, Y)$. Let $x \in \operatorname{dom} \varphi$ and $x^* \in \partial_\varepsilon \varphi(x)$. Then (see Theorem 2.4.2)

$$\varphi(x) + \varphi^*(x^*) \leq \langle x, x^* \rangle + \varepsilon.$$

By Eq. (2.56) there exists $y^* \in Y^*$ such that $\varphi^*(x^*) = F^*(x^* - A^*y^*, y^*)$, whence

$$F(x, Ax) + F^*(x^* - A^*y^*, y^*) \leq \langle x, x^* - A^*y^* \rangle + \langle Ax, y^* \rangle + \varepsilon,$$

i.e. $(x^* - A^*y^*, y^*) =: (\overline{x}^*, \overline{y}^*) \in \partial_\varepsilon F(x, Ax)$. Therefore $x^* = \overline{x}^* + A^*\overline{y}^*$, which proves that the inclusion "$\subset$" in Eq. (2.57) holds, too. □

Note that (viii) $\vee$ (iii) $\Rightarrow$ (i) $\Rightarrow$ (ii), (iv) $\vee$ (v) $\vee$ (vi) $\Rightarrow$ (ii), (vii) $\Rightarrow$ (iv) $\wedge$ (v) $\wedge$ (vi), and (ii) $\Rightarrow$ (i) if X is a normed space.

Note also that similar properties to those stated in the second part of Proposition 2.7.2 can be given for the situations of Theorem 2.8.1.

Corollary 2.8.2 *Under the conditions of Theorem 2.8.1 we have that*

$$\inf_{x\in X} F(x, Ax) = \max_{y^*\in Y^*} \big(- F^*(-A^*y^*, y^*)\big). \tag{2.59}$$

Furthermore, $\overline{x}$ is minimum point for φ if and only if there exists $\overline{y}^* \in Y^*$ such that $(-A^*\overline{y}^*, \overline{y}^*) \in \partial F(\overline{x}, A\overline{x})$.

Proof. The relation (2.59) follows from relation (2.56) taking $x^* = 0$. Moreover we have that $\overline{x}$ is minimum point for φ if and only if $0 \in \partial\varphi(\overline{x})$, *i.e.*, using Eq. (2.57), if and only if there exists $(\overline{x}^*, \overline{y}^*) \in \partial F(\overline{x}, A\overline{x})$ such that $0 = \overline{x}^* + A^*\overline{y}^*$. Therefore the conclusion holds. □

We note that the result of Corollary 2.8.2 can be used to obtain the relation (2.56), and so obtain Theorem 2.8.1.

When the function F has separated variables, to Theorem 2.8.1 corresponds the next result.

Theorem 2.8.3 *Let $f \in \Lambda(X)$, $g \in \Lambda(Y)$ and $A \in \mathcal{L}(X,Y)$. Assume that* $\operatorname{dom} f \cap A^{-1}(\operatorname{dom} g) \neq \emptyset$ *and let* $Y_0 := \operatorname{lin}\big(A(\operatorname{dom} f) - \operatorname{dom} g\big)$. *Consider $\varphi \in \Lambda(X)$, $\varphi(x) := f(x) + g(Ax)$. Assume that one of the following conditions holds:*

(i) *there exist $\lambda_0 \in \mathbb{R}$, $B \in \mathcal{B}_X$ and $V_0 \in \mathcal{N}_{Y_0}$ such that*

$$V_0 \subset A\left([f \le \lambda_0] \cap B\right) - [g \le \lambda_0];$$

(ii) *for every $U \in \mathcal{N}_X$ there exist $\lambda > 0$ and $V \in \mathcal{N}_{Y_0}$ such that*

$$V \subset A\left([f \le \lambda] \cap \lambda U\right) - [g \le \lambda];$$

(iii) *there exists $x_0 \in \operatorname{dom} f \cap A^{-1}(\operatorname{dom} g)$ such that g is continuous at Ax_0;*

(iv) *X, Y are metrizable, $0 \in {}^{ib}\big(A(\operatorname{dom} f) - \operatorname{dom} g\big)$, f and g have proper conjugates, either f, g are cs-closed and* $\operatorname{gr} A$ *is cs-complete, or f, g are cs-complete;*

(v) *X is a Fréchet space, Y is metrizable, f, g are li-convex functions and* $0 \in {}^{ib}(A(\operatorname{dom} f) - \operatorname{dom} g)$*;*

(vi) *X is a Fréchet space, f, g are lsc and* $0 \in {}^{ib}(A(\operatorname{dom} f) - \operatorname{dom} g)$*;*

(vii) *X, Y are Fréchet spaces, f, g are lsc and* $0 \in {}^{ic}(A(\operatorname{dom} f) - \operatorname{dom} g)$*;*

(viii) $\dim Y_0 < \infty$ *and* $0 \in {}^{i}(A(\operatorname{dom} f) - \operatorname{dom} g)$*;*

(ix) *g is quasi-continuous and $A(\operatorname{dom} f)$ and $\operatorname{dom} g$ are united;*

(x) $Y = \mathbb{R}^n$, $\operatorname{qri}(\operatorname{dom} f) \neq \emptyset$ *and* $A(\operatorname{qri}(\operatorname{dom} f)) \cap {}^{i}\operatorname{dom} g \neq \emptyset$.

Then for every $x^* \in X^*$, $x \in \operatorname{dom} \varphi$ and $\varepsilon \geq 0$ we have:

$$\begin{aligned}
&\varphi^*(x^*) = \min\{f^*(x^* - A^*y^*) + g^*(y^*) \mid y^* \in Y^*\},\\
&\partial_\varepsilon \varphi(x) = \bigcup\{\partial_{\varepsilon_1} f(x) + \partial_{\varepsilon_2} g(Ax) \mid \varepsilon_1, \varepsilon_2 \geq 0,\ \varepsilon_1 + \varepsilon_2 = \varepsilon\},\\
&\partial\varphi(x) = \partial f(x) + A^*(\partial g(Ax)).
\end{aligned} \tag{2.60}$$

Proof. We apply Theorem 2.8.1 to A and $F : X \times Y \to \overline{\mathbb{R}}$ defined by $F(x, y) := f(x) + g(y)$. It is easy to see that if one of the conditions (i)-(ix) holds, then the corresponding condition of Theorem 2.8.1 is verified. If (x) holds, using the properties of the algebraic relative interior in finite dimensional spaces (p. 3) and Proposition 1.2.7, we have

$${}^{i}(A(\operatorname{dom} f) - \operatorname{dom} g) = {}^{i}(A(\operatorname{dom} f)) - {}^{i}(\operatorname{dom} g) = A(\operatorname{qri}(\operatorname{dom} f)) - {}^{i}(\operatorname{dom} g),$$

and so (viii) holds, too. The conclusion follows then applying Theorems 2.8.1, 2.3.1 (viii) and Corollary 2.4.5. □

Note that, as in Theorem 2.8.1, we have that (viii) ∨ (iii) ⇒ (i) ⇒ (ii), (iv) ∨ (v) ∨ (vi) ⇒ (ii), (vii) ⇒ (iv) ∧ (v) ∧ (vi), (ii) ⇒ (i) if X is a normed space, and of course, as mentioned in the proof, (ix) ⇒ (viii).

Applying the preceding result we obtain a formula for normal cones.

Corollary 2.8.4 *Let $A \in \mathfrak{L}(X, Y)$ and $L \subset X$, $M \subset Y$ be convex sets. Suppose that one of the following conditions is verified:*

(i) *there exists $x_0 \in L$ such that $Ax_0 \in \operatorname{int} M$,*

(ii) *X, Y are of Fréchet spaces, L, M are li-convex and* $0 \in {}^{ib}(A(L) - M)$,

(iii) $\dim Y < \infty$ *and* $0 \in {}^{i}(A(L) - M)$.

Then for every $x \in L \cap A^{-1}(M)$

$$N\left(L \cap A^{-1}(M); x\right) = N(L; x) + A^*(N(M; Ax)). \tag{2.61}$$

Proof. Using the preceding theorem for $f := \iota_L$, $g := \iota_M$ and A, formula (2.61) follows from formula (2.60). □

Corollary 2.8.5 *Under the conditions of Theorem 2.8.3 we have the following relation, called the* ***Fenchel–Rockafellar duality formula,***

$$\inf_{x\in X}\big(f(x)+g(Ax)\big)=\max_{y^*\in Y^*}\big(-f^*(-A^*y^*)-g^*(y^*)\big).$$

Furthermore, $\overline{x}$ is a minimum point for $f+g\circ A$ if and only if there exists $\overline{y}^*\in Y^*$ such that $-A^*\overline{y}^*\in\partial f(\overline{x})$ and $\overline{y}^*\in\partial g(A\overline{x})$.

Proof. We proceed as in Corollary 2.8.2 (or apply this corollary). □

Two particular cases of Theorem 2.8.3 are important in applications: $f=0$ and $A=\mathrm{Id}_X$. The next theorem is stated even for A replaced by a convex process $\mathcal{C}$.

Theorem 2.8.6 *Let $g\in\Lambda(Y)$ and $\mathcal{C}: X\rightrightarrows Y$ be a convex process. Assume that $0\in D$, where $D:=\operatorname{Im}\mathcal{C}-\operatorname{dom}g$. Consider $Y_0:=\operatorname{lin}D$ and the function $\varphi: X\to\overline{\mathbb{R}}$, $\varphi(x)=\inf\{g(y)\mid y\in\mathcal{C}(x)\}$. Assume that one of the following conditions holds:*

(i) *for every $U\in\mathcal{N}_X$ there exist $\lambda>0$ and $V\in\mathcal{N}_{Y_0}$ such that $V\subset[g\le\lambda]-\mathcal{C}(\lambda U)$;*

(ii) *there exist $\lambda_0\in\mathbb{R}$, $B\in\mathcal{B}_X$ and $V_0\in\mathcal{N}_{Y_0}$ such that $V_0\subset[g\le\lambda]-\mathcal{C}(B)$;*

(iii) *there exists $y_0\in\operatorname{dom}g\cap\operatorname{Im}\mathcal{C}$ such that g is continuous at y_0;*

(iv) *X, Y are metrizable, g has proper conjugate, either g is cs-closed and $\mathcal{C}$ is cs-complete, or g is cs-complete and $\mathcal{C}$ verifies* (Hx), *and $0\in{}^{it}D$;*

(v) *X, Y are Fréchet spaces, g, $\mathcal{C}$ are li-convex and $0\in{}^{ib}D$,*

(vi) *$\dim Y_0<\infty$ and $0\in{}^{i}D$.*

Then

$$\forall x^*\in X^*\ :\ \varphi^*(x^*)=\min\{g^*(y^*)\mid x^*\in\mathcal{C}^*(y^*)\}.$$

Moreover, if $\overline{x}\in\operatorname{dom}\varphi=\mathcal{C}^{-1}(\operatorname{dom}g)$ is such that $\varphi(\overline{x})=g(\overline{y})$ with $\overline{y}\in\mathcal{A}(\overline{x})$, and $\varepsilon\ge0$ then $\partial_\varepsilon\varphi(\overline{x})\subset\mathcal{C}^*\left(\partial_\varepsilon g(\overline{y})\right)$ (with equality if $\operatorname{gr}\mathcal{C}$ is a linear subspace).

Proof. Let $x^*\in X^*$. Consider $F: X\times Y\to\overline{\mathbb{R}}$, $F(x,y):=g(y)-\langle x,x^*\rangle$. Then

$$\varphi^*(x^*)=\sup\{\langle x,x^*\rangle-\varphi(x)\mid x\in X\}=-\inf\{F(x,y)\mid(x,y)\in\operatorname{gr}\mathcal{C}\}.$$

If one of the conditions (i)–(vi) holds then the corresponding condition of Theorem 2.7.4 holds. (In fact, in case (iv), if g is cs-complete and $\mathcal{C}$ satisfies (Hx) one verifies directly that the function Φ from the proof of Theorem 2.7.4 satisfies (H(x,z)).) So, by Theorem 2.7.4, there exists $y^* \in Y^*$ such that

$$\begin{aligned}
\varphi^*(x^*) &= -\inf\{F(x,y) + \langle z, y^*\rangle \mid (x, y+z) \in \operatorname{gr}\mathcal{C}\} \\
&= \sup\{\langle x, x^*\rangle - g(y) - \langle y' - y, y^*\rangle \mid (x,y') \in \operatorname{gr}\mathcal{C},\ y \in Y\} \\
&= \sup\{\langle y, y^*\rangle - g(y) \mid y \in Y\} \\
&\quad + \sup\{\langle x, x^*\rangle + \langle z, -y^*\rangle - \iota_{\operatorname{gr}\mathcal{C}}(x,z) \mid x \in X, z \in Y\} \\
&= g^*(y^*) + \iota_{-(\operatorname{gr}\mathcal{C})^+}(x^*, -y^*) = g^*(y^*) + \iota_{\operatorname{gr}\mathcal{C}^*}(y^*, x^*).
\end{aligned}$$

Since

$$\forall\, z^* \in Y^* \ :\ \varphi^*(x^*) \le g^*(z^*) + \iota_{\operatorname{gr}\mathcal{C}^*}(z^*, x^*), \tag{2.62}$$

the conclusion follows. Let $\overline{x} \in \operatorname{dom}\varphi = \mathcal{A}^{-1}(\operatorname{dom} g)$ be such that $\varphi(\overline{x}) = g(\overline{y})$ with $\overline{y} \in \mathcal{C}(\overline{x})$ and $\varepsilon \ge 0$. Let $x^* \in \partial_\varepsilon\varphi(\overline{x})$. Since $\varphi(\overline{x}) + \varphi^*(x^*) \le \langle \overline{x}, x^*\rangle + \varepsilon$, there exists $y^* \in \mathcal{C}^{*-1}(x^*)$ such that $\varphi^*(x^*) = g^*(y^*)$. It follows that $g(\overline{y}) + g^*(y^*) \le \langle \overline{x}, x^*\rangle + \varepsilon \le \langle \overline{y}, y^*\rangle + \varepsilon$, whence $y^* \in \partial_\varepsilon g(\overline{y})$. Therefore $\partial_\varepsilon\varphi(\overline{x}) \subset \mathcal{C}^*(\partial_\varepsilon g(\overline{y}))$. Conversely, suppose that $\operatorname{gr}\mathcal{C}$ is a linear subspace and take $x^* \in \mathcal{C}^*(y^*)$ with $y^* \in \partial_\varepsilon g(\overline{y})$. Then $\langle \overline{x}, x^*\rangle = \langle \overline{y}, y^*\rangle$; using Eq. (2.62) we obtain that $\varphi(\overline{x}) + \varphi^*(x^*) \le \langle \overline{x}, x^*\rangle + \varepsilon$, *i.e.* $x^* \in \partial_\varepsilon\varphi(\overline{x})$. □

Note that (vi) ∨ (iii) ⇒ (ii) ⇒ (i) and (iv) ∨ (v) ⇒ (i).

Important situations when $\operatorname{gr}\mathcal{C}$ is a linear subspace are: $\mathcal{C} = A^{-1}$ with $A \in \mathcal{L}(X,Y)$ and $\mathcal{C} = A$ with A a densely defined closed operator (*i.e.* $A : D(A) \to Y$, $D(A)$ being a dense linear subspace of X, A being a linear operator and $\operatorname{gr} A$ a closed subset of $X \times Y$; see also Exercise 1.11).

Theorem 2.8.7 *Let* $f, g \in \Lambda(X)$. *Assume that* $\operatorname{dom} f \cap \operatorname{dom} g \neq \emptyset$ *and let* $X_0 := \operatorname{lin}(\operatorname{dom} f - \operatorname{dom} g)$. *Assume that one of the following conditions holds:*

(i) *there exist* $\lambda_0 \in \mathbb{R}$, $B \in \mathcal{B}_X$ *and* $V_0 \in \mathcal{N}_{X_0}$ *such that*

$$V_0 \subset [f \le \lambda_0] \cap B - [g \le \lambda_0];$$

(ii) *for every* $U \in \mathcal{N}_X$ *there exist* $\lambda > 0$ *and* $V \in \mathcal{N}_{X_0}$ *such that*

$$V \subset [f \le \lambda] \cap \lambda U - [g \le \lambda];$$

(iii) *there exists* $x_0 \in \operatorname{dom} f \cap \operatorname{dom} g$ *such that* g *is continuous at* x_0;

(iv) X *is metrizable,* f, g *have proper conjugates,* f *is cs-closed,* g *is cs-complete and* $0 \in {}^{ib}(\operatorname{dom} f - \operatorname{dom} g)$;

(v) X *is a Fréchet space,* f, g *are li-convex and* $0 \in {}^{ib}(\operatorname{dom} f - \operatorname{dom} g)$;

(vi) X *is a Fréchet space,* f, g *are lsc and* $0 \in {}^{ib}(\operatorname{dom} f - \operatorname{dom} g)$;

(vii) X *is a Fréchet space,* f, g *are lsc and* $0 \in {}^{ic}(\operatorname{dom} f - \operatorname{dom} g)$;

(viii) $\dim X_0 < \infty$ *and* $0 \in {}^{i}(\operatorname{dom} f - \operatorname{dom} g)$;

(ix) g *is quasi-continuous and* $\operatorname{dom} f$ *and* $\operatorname{dom} g$ *are united;*

(x) X *is a Fréchet space,* f, g *are li-convex and* $(0,0) \in {}^{ib}(\{(x,x) \mid x \in X\} - \operatorname{dom} f \times \operatorname{dom} g)$.

Then for $x^* \in X^*$, $x \in \operatorname{dom} f \cap \operatorname{dom} g$ *and* $\varepsilon \geq 0$ *we have:*

$$(f+g)^*(x^*) = \min\{f^*(x^* - y^*) + g^*(y^*) \mid v^* \in X^*\} = (f^* \square g^*)(x^*), \tag{2.63}$$

$$\partial_\varepsilon (f+g)(x) = \bigcup \{\partial_{\varepsilon_1} f(x) + \partial_{\varepsilon_2} g(x) \mid \varepsilon_1, \varepsilon_2 \geq 0,\ \varepsilon_1 + \varepsilon_2 = \varepsilon\},$$
$$\partial (f+g)(x) = \partial f(x) + \partial g(x).$$

Proof. Taking $A = \mathrm{Id}_X$, the conclusion follows from Theorem 2.8.3 under conditions (i) – (iii), (v) – (ix). If (iv) holds, taking $Y = X$ and $\Phi(x,y) := f(x) + g(x-y)$, condition (iv) of Theorem 2.7.1 is verified. If (x) holds consider $F : X \times X \to \overline{\mathbb{R}}$, $F(x,y) := f(x) + g(y)$ and $A : X \to X \times X$, $A(x) := (x,x)$. Then condition (v) of Theorem 2.8.6 holds; applying it we obtain the conclusion, taking into account that $A^*(x^*, y^*) = x^* + y^*$. □

The same implications as in Theorem 2.7.1 (mentioned in Proposition 2.7.2) hold.

Corollary 2.8.8 *Let* X *be a Fréchet space and* $f, g \in \Lambda(X)$. *If* f *and* g *are li-convex then for every* $x \in {}^{ib}(\operatorname{dom} f + \operatorname{dom} g)$ *we have*

$$(f \square g)(x) = (f^* + g^*)^*(x) = \max\{\langle x, x^*\rangle - f^*(x^*) - g^*(x^*) \mid x^* \in X^*\}. \tag{2.64}$$

Proof. Let $x_0 \in {}^{ib}(\operatorname{dom} f + \operatorname{dom} g)$ and consider $h \in \Lambda(X)$, $h(x) := g(x_0 - x)$. Then $\operatorname{dom} h = x_0 - \operatorname{dom} g$, and so $\operatorname{dom} f - \operatorname{dom} h = \operatorname{dom} f + \operatorname{dom} g - x_0$. Therefore $0 \in {}^{ib}(\operatorname{dom} f - \operatorname{dom} h)$, and so condition (v) of

Theorem 2.8.7 is satisfied. From formula (2.63) applied for $x^* = 0$ we get

$$\begin{aligned}(f\Box g)(x_0) &= \inf_{x\in X}(f(x)+h(x)) = -(f+h)^*(0)\\ &= -\min_{x^*\in X^*}(f^*(x^*)+h^*(-x^*)).\end{aligned}$$

But $h^*(-x^*) = \sup\{-\langle x,x^*\rangle - g(x_0-x) \mid x\in X\} = g^*(x^*) - \langle x_0,x^*\rangle$, and so (2.64) holds for $x = x_0$. □

Of course, one can obtain the conclusion of the preceding corollary also for other situations corresponding to conditions of Theorem 2.8.7. For example, if there exist $\lambda_0\in\mathbb{R}$, $B\in\mathcal{B}_X$, $V_0\in\mathcal{N}_{X_0}(x)$ such that $V_0\subset[f\leq\lambda_0]\cap B+[g\leq\lambda_0]$, where $X_0 = \operatorname{aff}(\operatorname{dom} f+\operatorname{dom} g)$, then Eq. (2.64) holds.

Proposition 2.8.9 *Let $f,g\in\Lambda(X)$ and take $D = \operatorname{dom} f - \operatorname{dom} g$. If $\dim(\operatorname{lin} D)<\infty$ then ${}^iD = \operatorname{rint} D$, while if X is metrizable, f,g have proper conjugates, f is cs-closed, g is cs-complete and ${}^{ib}D\neq\emptyset$ then ${}^{ib}D = \operatorname{rint} D$; similarly for the situations corresponding to conditions* (v)–(vii) *of Theorem 2.8.7.*

Suppose that X is a Banach space and f,g are li-convex. Then for every $x\in{}^{ib}D$ there exist $\eta,\lambda>0$ such that

$$(x+\eta U_X)\cap\operatorname{aff} D\subset[f\leq\lambda]\cap\lambda U_X-[g\leq\lambda]\cap\lambda U_X. \tag{2.65}$$

Proof. The first part follows from Proposition 2.7.2 taking $Y = X$ and $\Phi(x,y) = f(x)+g(x-y)$.

Suppose that X is a Banach space and f,g are li-convex; consider $x\in{}^{ib}D$. Replacing f by $\widetilde{f}$, $\widetilde{f}(u) = f(u+x)$, we may suppose that $x = 0$. It follows that condition (v) of Theorem 2.8.7 holds, and, as noted after its proof, condition (i) is verified. Therefore there exist $\eta>0$, $\lambda_0\in\mathbb{R}$ and $B\in\mathcal{B}_X$ such that

$$\eta U_X\cap\operatorname{aff} D\subset[f\leq\lambda_0]\cap B-[g\leq\lambda_0].$$

Taking $\lambda'>\max\{\lambda_0,0\}$ such that $B\subset\lambda' U_X$ and $\lambda = \lambda'+\eta$ we obtain that

$$\begin{aligned}\eta U_X\cap\operatorname{aff} D &\subset[f\leq\lambda_0]\cap B-[g\leq\lambda_0]\cap(B+\eta U_X)\\ &\subset[f\leq\lambda]\cap\lambda U_X-[f\leq\lambda]\cap\lambda U_X.\end{aligned}$$

The proof is complete. □

Another important result is the following.

Theorem 2.8.10 *Let Y be ordered by the convex cone Q, $f \in \Lambda(X)$, $H : X \to Y^\bullet$ be convex and $g \in \Lambda(Y)$ be Q-increasing on $H(\operatorname{dom} H) + Q$. Then $\varphi := f + g \circ H$ is convex. Assume that $0 \in D$, where $D := H(\operatorname{dom} H \cap \operatorname{dom} f) - \operatorname{dom} g + Q$, and consider $Y_0 := \operatorname{lin} D$. Assume that one of the following conditions holds:*

(i) *for every $U \in \mathcal{N}_X$ there exist $\lambda > 0$ and $V \in \mathcal{N}_{Y_0}$ such that*

$$V \subset H(\lambda U \cap [f \le \lambda] \cap \operatorname{dom} H) - [g \le \lambda] + Q;$$

(ii) *there exist $\lambda_0 \in \mathbb{R}$, $B \in \mathcal{B}_X$ and $V_0 \in \mathcal{N}_{Y_0}$ such that*

$$V_0 \subset H(B \cap [f \le \lambda_0] \cap \operatorname{dom} H) - [g \le \lambda_0] + Q;$$

(iii) *there exists $x_0 \in \operatorname{dom} f \cap H^{-1}(\operatorname{dom} g)$ such that g is continuous at $H(x_0)$;*

(iv) *X, Y are metrizable, f, g have proper conjugates, either f, g are cs-closed and $\operatorname{epi} H$ is cs-complete, or f, g are cs-complete and $\operatorname{epi} H$ is cs-closed, and $0 \in {}^{ib}D$;*

(v) *X, Y are Fréchet spaces, f, g, $\operatorname{epi} H$ are li-convex and $0 \in {}^{ib}D$;*

(vi) *$\dim Y_0 < \infty$ and $0 \in {}^{i}D$;*

(vii) *g is quasi-continuous, and $\operatorname{dom} g$ and $H(\operatorname{dom} H \cap \operatorname{dom} f) + Q$ are united.*

Then for $x^* \in X^*$, $x \in \operatorname{dom} \varphi = \operatorname{dom} f \cap H^{-1}(\operatorname{dom} g)$ and $\varepsilon \ge 0$, we have:

$$\varphi^*(x^*) = \min\{(f + y^* \circ H)^*(x^*) + g^*(y^*) \mid y^* \in Q^+\}, \tag{2.66}$$

$$\partial_\varepsilon \varphi(x) = \bigcup \left\{\partial_{\varepsilon_1}(f + y^* \circ H)(x) \mid y^* \in Q^+ \cap \partial_{\varepsilon_2} g(H(x)),\ \varepsilon_1 + \varepsilon_2 = \varepsilon\right\}. \tag{2.67}$$

Proof. First observe that for all $x^* \in X^*$, $y^* \in Q^+$ and $x \in \operatorname{dom}\varphi$ we have

$$\varphi^*(x^*) \le (f + y^* \circ H)^*(x^*) + g^*(y^*), \tag{2.68}$$

$$\partial_\varepsilon \varphi(x) \supset \bigcup \{\partial_{\varepsilon_1}(f + y^* \circ H)(x) \mid y^* \in Q^+ \cap \partial_{\varepsilon_2} g(H(x)),\ \varepsilon_1 + \varepsilon_2 = \varepsilon\}, \tag{2.69}$$

without any supplementary condition on f, g and H. Indeed, let $x \in \operatorname{dom}\varphi = \operatorname{dom} f \cap H^{-1}(\operatorname{dom} g)$, $x^* \in X^*$ and $y^* \in Q^+$. Then

$$f(x) + (y^* \circ H)(x) + (f + y^* \circ H)^*(x^*) \geq \langle x, x^* \rangle,$$
$$g(H(x)) + g^*(y^*) \geq \langle H(x), y^* \rangle,$$

whence, adding them side by side, we get $g^*(y^*) + (f + y^* \circ H)^*(x^*) \geq \langle x, x^* \rangle - \varphi(x)$. Thus Eq. (2.68) holds.

Let now $x \in \operatorname{dom}\varphi$, $y^* \in Q^+ \cap \partial_{\varepsilon_2} g(H(x))$, and $x^* \in \partial_{\varepsilon_1}(f + y^* \circ H)(x)$, where $\varepsilon_1, \varepsilon_2 \geq 0$, $\varepsilon = \varepsilon_1 + \varepsilon_2$. Then

$$f(x) + (y^* \circ H)(x) + (f + y^* \circ H)^*(x^*) \leq \langle x, x^* \rangle + \varepsilon_1,$$
$$g(H(x)) + g^*(y^*) \leq \langle H(x), y^* \rangle + \varepsilon_2.$$

Adding them side by side we get $\varphi(x) + g^*(y^*) + (f + y^* \circ H)^*(x^*) \leq \langle x, x^* \rangle + \varepsilon$. Using Eq. (2.68) we obtain that $x^* \in \partial_\varepsilon \varphi(x)$. Therefore Eq. (2.69) holds.

Let $F(x,y) := f(x) + g(y)$ and $\mathcal{C} : X \rightrightarrows Y$ with $\operatorname{gr}\mathcal{C} := \operatorname{epi} H$; then $F \in \Lambda(X \times Y)$ and $\mathcal{C}$ is convex. Since g is increasing on $H(\operatorname{dom} H) + Q$, it follows that $\varphi(x) = \inf\{F(x,y) + \iota_{\operatorname{epi} H}(x,y) \mid y \in Y\}$ for every $x \in X$. Hence φ is the marginal function associated to the convex function $F + \iota_{\operatorname{epi} H}$; hence φ is convex.

If one of the conditions (i) – (vi) is verified, then F and $\mathcal{C}$ satisfy the corresponding condition of Theorem 2.7.4. (When f and g are cs-complete and have proper conjugates, similarly to the proof of Proposition 2.2.17, we get that F is cs-complete.) As noticed after the proof of that theorem, also the perturbed function $\widetilde{F}$, $\widetilde{F}(x,y) = F(x,y) - \langle x, x^* \rangle$, satisfies the same condition. Therefore there exists $z^* \in Y^*$ such that

$$\begin{aligned}\alpha &:= \inf_{(x,y) \in \operatorname{epi} H} (f(x) + g(y) - \langle x, x^* \rangle) \\ &= \inf_{(x,y+z) \in \operatorname{epi} H} (f(x) + g(y) - \langle x, x^* \rangle + \langle z, z^* \rangle) =: \beta.\end{aligned}$$

Using again the fact that g is increasing on $H(\operatorname{dom} H) + Q$, we have

$$\begin{aligned}\alpha &= \inf_{x \in \operatorname{dom} H} \inf_{y \in H(x) + Q} (f(x) + g(y) - \langle x, x^* \rangle) \\ &= \inf_{x \in \operatorname{dom} H} (f(x) + g(H(x)) - \langle x, x^* \rangle) = -\varphi^*(x^*),\end{aligned}$$

and

$$\beta = \inf_{x \in \operatorname{dom} H} \inf_{q \in Q,\, y \in Y} \left(f(x) + g(y) - \langle x, x^* \rangle + \langle H(x) + q - y, z^* \rangle \right).$$

It follows that $\beta = -\infty$ if $z^* \notin Q^+$. If $z^* \in Q^+$ then

$$\begin{aligned} -\beta &= \sup_{x \in \operatorname{dom} H} \sup_{y \in Y} \left(\langle x, x^* \rangle - f(x) - \langle H(x), z^* \rangle + \langle y, z^* \rangle - g(y) \right) \\ &= (f + z^* \circ H)^*(x^*) + g^*(z^*). \end{aligned}$$

Taking into account Eq. (2.68), it is clear that Eq. (2.66) holds.

In the case (vii) consider $F : X \times Y \to \overline{\mathbb{R}}$, $F(x,y) := f(x) + g(H(x)+y) - \langle x, x^* \rangle$. Because $0 \in D$ and g is Q-increasing on $H(\operatorname{dom} H) + Q$, there exists $x_0 \in \operatorname{dom} f \cap H^{-1}(\operatorname{dom} g)$. It follows that $F(x_0, \cdot)$ is quasi-continuous and $\{0\}$ and $D = \Pr_Y(\operatorname{dom} F)$ are united. Applying Theorem 2.7.1(ix) we have that $\inf_{x \in X} F(x, 0) = \max_{y^* \in Y^*}(-F^*(0, y^*))$. With a similar calculation as above we obtain that Eq. (2.66) holds.

Let $x \in \operatorname{dom} \varphi$ and $x^* \in \partial_\varepsilon \varphi(x)$. Using Eq. (2.66), there exists $y^* \in Q^+$ such that $\varphi^*(x^*) = (f + y^* \circ H)^*(x^*) + g^*(y^*)$. It follows that

$$(f + y^* \circ H)(x) + (f + y^* \circ H)^*(x^*) - \langle x, x^* \rangle + g(H(x)) + g^*(y^*) - \langle H(x), y^* \rangle \le \varepsilon.$$

Taking $\varepsilon_1 := (f + y^* \circ H)(x) + (f + y^* \circ H)^*(x^*) - \langle x, x^* \rangle$ (≥ 0 by the Young–Fenchel inequality) and $\varepsilon_2 = \varepsilon - \varepsilon_1$, we have that $y^* \in \partial_{\varepsilon_2} g(H(x))$ and $x^* \in \partial_{\varepsilon_1}(f + y^* \circ H)(x)$. Therefore the inclusion $\subset$ holds in Eq. (2.67); taking into account Eq. (2.69), we have the desired equality. □

We use the preceding theorem to obtain formulas for conjugates and subdifferentials of functions of "max" type.

Corollary 2.8.11 *Let* $f_1, \ldots, f_n \in \Lambda(X)$ *and*

$$\varphi : X \to \overline{\mathbb{R}}, \quad \varphi(x) := f_1(x) \vee \cdots \vee f_n(x).$$

Suppose that $\operatorname{dom} \varphi = \bigcap_{i=1}^n \operatorname{dom} f_i \neq \emptyset$ *and* $\varepsilon \in \mathbb{R}_+$. *For* $x \in \operatorname{dom} \varphi$ *denote by* $I(x)$ *the set* $\{i \in \overline{1,n} \mid f_i(x) = \varphi(x)\}$. *Then, for all* $x^* \in X^*$,

$$\varphi^*(x^*) = \min\{(\lambda_1 f_1 + \cdots + \lambda_n f_n)^*(x^*) \mid (\lambda_1, \ldots, \lambda_n) \in \Delta_n\},$$

while for every $x \in \operatorname{dom}\varphi$,

$$\partial_\varepsilon\varphi(x) = \bigcup\Big\{\partial_\eta(\lambda_1 f_1 + \cdots + \lambda_n f_n)(x) \,\Big|\, (\lambda_1,\ldots,\lambda_n) \in \Delta_n,$$
$$\eta \in [0,\varepsilon],\ \sum\nolimits_{i=1}^{n} \lambda_i f_i(x) \geq \varphi(x) + \eta - \varepsilon_0\Big\},$$
$$\partial\varphi(x) = \bigcup\Big\{\partial\left(\sum\nolimits_{i=1}^{n} \lambda_i f_i\right)(x) \,\Big|\, (\lambda_1,\ldots,\lambda_n) \in \Delta_n,$$
$$\forall i \notin I(x)\ :\ \lambda_i = 0\Big\}.$$

Proof. Let us consider the functions

$$H : X \to (\mathbb{R}^{n\bullet}, \mathbb{R}^n_+),\ \ H(x) := \begin{cases} (f_1(x),\ldots,f_n(x)) & \text{if } x \in \bigcap_{i=1}^n \operatorname{dom} f_i, \\ \infty & \text{otherwise,} \end{cases}$$
$$g : \mathbb{R}^n \to \mathbb{R},\quad g(y) := y_1 \vee \cdots \vee y_n.$$

It is clear that condition (iii) of the preceding theorem is verified. Taking into account the expressions of g^* and $\partial_\varepsilon g(y)$ given in Corollary 2.4.17 we obtain immediately the relations from our statement. □

An application of the previous result is given in the following example.

Example 2.8.1 Let $f \in \Lambda(X)$ and consider $f_+ := f \vee 0$. Then

$$\partial f_+(x) = \begin{cases} \partial f(x) & \text{if} \quad f(x) > 0, \\ \bigcup\{\partial(\lambda f)(x) \mid \lambda \in [0,1]\} & \text{if} \quad f(x) = 0, \\ \partial(0f)(x) = \partial\iota_{\operatorname{dom} f}(x) & \text{if} \quad f(x) < 0. \end{cases}$$

A useful particular case of the result in Corollary 2.8.11, in which we can give explicit formulas for φ^* and $\partial_\varepsilon\varphi$ without supplementary conditions, is presented in the following corollary.

Corollary 2.8.12 *Let* $f_i \in \Lambda(X_i)$ *for* $i \in \overline{1,n}$ *and*

$$\varphi : X := \prod\nolimits_{i=1}^{n} X_i \to \overline{\mathbb{R}},\quad \varphi(x_1,\ldots,x_n) := f_1(x_1) \vee \cdots \vee f_n(x_n).$$

For $x = (x_1,\ldots,x_n) \in \operatorname{dom}\varphi = \prod_{i=1}^n \operatorname{dom} f_i$ *consider* $I(x) := \{i \in \overline{1,n} \mid f_i(x_i) = \varphi(x)\}$. *Let* $x^* = (x_1^*,\ldots,x_n^*) \in \prod_{i=1}^n X_i^* = X^*$. *Then*

$$\varphi^*(x^*) \;=\; \min\Big\{\sum\nolimits_{i=1}^{n} (\lambda_i f_i)^*(x_i^*) \,\Big|\, (\lambda_1,\ldots,\lambda_n) \in \Delta_n\Big\}. \qquad (2.70)$$

For $x = (x_1, \ldots, x_n) \in \operatorname{dom}\varphi$ *and* $\varepsilon \in \mathbb{P}$ *we have:*

$$\partial_\varepsilon \varphi(x) = \bigcup \Big\{ \prod\nolimits_{i=1}^{n} \partial_{\varepsilon_i}(\lambda_i f_i)(x_i) \,\Big|\, (\lambda_1, \ldots, \lambda_n) \in \Delta_n,\ \varepsilon_i \geq 0, \\ \sum\nolimits_{i=0}^{n} \varepsilon_i = \varepsilon,\ \sum\nolimits_{i=1}^{n} \lambda_i f_i(x_i) \geq \varphi(x) - \varepsilon_0 \Big\},$$

$$\partial \varphi(x) = \bigcup \Big\{ \prod\nolimits_{i=1}^{n} \partial(\lambda_i f_i)(x_i) \,\Big|\, (\lambda_1, \ldots, \lambda_n) \in \Delta_n,\ \forall i \notin I(x)\ : \lambda_i = 0 \Big\}.$$

Proof. We apply the preceding corollary to the functions $\tilde{f}_i : X \to \overline{\mathbb{R}}$, $\tilde{f}_i(x) := f_i(x_i)$, then we use Theorem 2.3.1 (viii) for arbitrary n and Corollary 2.4.5. □

Other situations when one has explicit formulas for φ^* and $\partial\varphi(x)$ in Corollary 2.8.11 are specified in the next result.

Corollary 2.8.13 *Let* $f, g \in \Lambda(X)$ *satisfy one of the conditions* (i)–(iii) *or* (v)–(x) *of Theorem 2.8.7 and consider* $\varphi := f \vee g$. *Then for all* $x^* \in X^*$ *and* $x \in \operatorname{dom}\varphi = \operatorname{dom} f \cap \operatorname{dom} g$,

$$\varphi^*(x^*) = \min \left\{ (\lambda f)^*(u^*) + (\mu g)^*(v^*) \mid (\lambda, \mu) \in \Delta_2,\ u^*, v^* \in X^*, \\ u^* + v^* = x^* \right\},$$

$$\partial\varphi(x) = \bigcup \left\{ \partial(\lambda f)(x) + \partial(\mu g)(x) \mid (\lambda, \mu) \in \Delta_2,\ \lambda f(x) + \mu g(x) = \varphi(x) \right\}.$$

Proof. When f and g verify one of the conditions (i)–(iii), (v), (viii)–(x) of Theorem 2.8.7 and $\lambda, \mu \geq 0$, then the functions λf and μg also verify the same condition. If condition (vi) or (vii) holds then, by the relations among the classes of convex functions on page 68, condition (v) holds. So, $(\lambda f + \mu g)^*(x^*) = \min\{(\lambda f)^*(u^*) + (\mu g)^*(v^*) \mid u^* + v^* = x^*\}$ and $\partial(\lambda f + \mu g)(x) = \partial(\lambda f)(x) + \partial(\mu g)(x)$. Using Corollary 2.8.11 one obtains the conclusion. □

Taking into account that for $x \in \operatorname{dom} f$, one has $\partial f(x) + N(\operatorname{dom} f, x) = \partial f(x)$ (see also Exercise 2.23), $\partial(\lambda f)(x) = \lambda\partial f(x)$ for $\lambda > 0$ and $\partial(0f)(x) = N(\operatorname{dom} f, x)$, we get for $f, g \in \Lambda(X)$ and $x \in X$ with $f(x) = g(x) \in \mathbb{R}$,

$$\bigcup \left\{ \partial(\lambda f)(x) + \partial(\mu g)(x) \mid (\lambda, \mu) \in \Delta_2 \right\} \\ = \operatorname{co}\left(\partial f(x) \cup \partial g(x)\right) + N(\operatorname{dom} f, x) + N(\operatorname{dom} g, x).$$

This formula can be extended easily to a finite number of functions.

Using Corollary 2.8.12 we get the following result concerning the max-convolution.

Corollary 2.8.14 *Let $f_1, f_2 \in \Lambda(X)$ and $\varepsilon \in \mathbb{R}_+$. For every $x^* \in X^*$ we have that*

$$(f_1 \Diamond f_2)^*(x^*) = \min\{(\lambda_1 f_1)^*(x^*) + (\lambda_2 f_2)^*(x^*) \mid (\lambda_1, \lambda_2) \in \Delta_2\}. \quad (2.71)$$

Suppose that $(f_1 \Diamond f_2)(\overline{x}) = \max\{f_1(\overline{x}_1), f_2(\overline{x}_2)\}$, where $\overline{x}_1 \in \operatorname{dom} f_1$, $\overline{x}_2 \in \operatorname{dom} f_2$ and $\overline{x} = \overline{x}_1 + \overline{x}_2$. Then

$$\partial_\varepsilon(f_1 \Diamond f_2)(\overline{x}) = \bigcup\{\partial_{\varepsilon_1}(\lambda_1 f_1)(\overline{x}_1) \cap \partial_{\varepsilon_2}(\lambda_2 f_2)(\overline{x}_2) \mid (\lambda_1, \lambda_2) \in \Delta_2,$$
$$\varepsilon_1, \varepsilon_2 \geq 0,\ \varepsilon_1 + \varepsilon_2 \leq \varepsilon + \lambda_1 f_1(\overline{x}_1) + \lambda_2 f_2(\overline{x}_2) - (f_1 \Diamond f_2)(\overline{x})\}. \quad (2.72)$$

Furthermore, if $f_1(\overline{x}_1) = f_2(\overline{x}_2)$ then Eq. (2.73) is verified, while if $f_1(\overline{x}_1) > f_2(\overline{x}_2)$ then Eq. (2.74) is verified, where

$$\partial(f_1 \Diamond f_2)(\overline{x}) = \bigcup\{\partial(\lambda_1 f_1)(\overline{x}_1) \cap \partial(\lambda_2 f_2)(\overline{x}_2) \mid (\lambda_1, \lambda_2) \in \Delta_2\}, \quad (2.73)$$
$$\partial(f_1 \Diamond f_2)(\overline{x}) = \partial f_1(\overline{x}_1) \cap N(\operatorname{dom} f_2; \overline{x}_2). \quad (2.74)$$

Proof. Let $x^* \in X^*$; then

$$\begin{aligned}(f_1 \Diamond f_2)^*(x^*) &= \sup_{x \in X}\left(\langle x, x^*\rangle - \inf\{f_1(x_1) \vee f_2(x_2) \mid x_1 + x_2 = x\}\right)\\ &= \sup\left\{\langle x_1, x^*\rangle + \langle x_2, x^*\rangle - f_1(x_1) \vee f_2(x_2) \mid x_1,\ x_2 \in X\right\}\\ &= \min\{(\lambda_1 f_1)^*(x^*) + (\lambda_2 f_2)^*(x^*) \mid (\lambda_1, \lambda_2) \in \Delta_2\},\end{aligned}$$

the last equality being obtained by using Eq. (2.70).

The inclusion "$\supset$" of Eq. (2.72) can be verified easily. Consider $x^* \in \partial_\varepsilon(f_1 \Diamond f_2)(\overline{x})$. By Eq. (2.71), there exist $(\lambda_1, \lambda_2) \in \Delta_2$ such that

$$(f_1 \Diamond f_2)^*(x^*) = (\lambda_1 f_1)^*(x^*) + (\lambda_2 f_2)^*(x^*). \quad (2.75)$$

Using the preceding relation we obtain that

$$\begin{aligned}0 &\leq [(\lambda_1 f_1)(\overline{x}_1) + (\lambda_1 f_1)^*(x^*) - \langle \overline{x}_1, x^*\rangle]\\ &\qquad + [(\lambda_2 f_2)(\overline{x}_2) + (\lambda_2 f_2)^*(x^*) - \langle \overline{x}_2, x^*\rangle]\\ &\leq (f_1 \Diamond f_2)(\overline{x}) + (f_1 \Diamond f_2)^*(x^*) - \langle \overline{x}, x^*\rangle \leq \varepsilon.\end{aligned}$$

Taking $\varepsilon_i := (\lambda_i f_i)(\overline{x}_i) + (\lambda_i f_i)^*(x^*) - \langle \overline{x}_i, x^*\rangle \geq 0,\ i \in \{1, 2\}$, using the preceding relation and Eq. (2.75), we get

$$(f_1 \Diamond f_2)(\overline{x}) + \varepsilon_1 - \lambda_1 f_1(\overline{x}_1) + \varepsilon_2 - \lambda_2 f_2(\overline{x}_2) \leq \varepsilon,$$

and so x^* belongs to the set on the right-hand side of relation (2.72).

Let us prove now the equalities (2.73) and (2.74). The inclusions "$\supset$" follow directly from Eq. (2.72). Let us prove the converse inclusions. Let $x^* \in \partial(f_1 \Diamond f_2)(\overline{x}) = \partial_0(f_1 \Diamond f_2)(\overline{x})$. By Eq. (2.72), there exist $(\lambda_1, \lambda_2) \in \Delta_2$ and $\varepsilon_1, \varepsilon_2 \geq 0$ such that

$$x^* \in \partial_{\varepsilon_1}(\lambda_1 f_1)(\overline{x}_1) \cap \partial_{\varepsilon_2}(\lambda_2 f_2)(\overline{x}_2),$$
$$\varepsilon_1 + \varepsilon_2 \leq \lambda_1 f_1(\overline{x}_1) + \lambda_2 f_2(\overline{x}_2) - (f_1 \Diamond f_2)(\overline{x}) \leq 0.$$

Therefore $\varepsilon_1 = \varepsilon_2 = 0$ and $\lambda_1 f_1(\overline{x}_1) + \lambda_2 f_2(\overline{x}_2) = (f_1 \Diamond f_2)(\overline{x})$; hence x^* belongs to the set on the right-hand side of Eq. (2.73) when $f_1(\overline{x}_1) = f_2(\overline{x}_2)$. If $f_1(\overline{x}_1) > f_2(\overline{x}_2)$, the preceding relation shows that $\lambda_2 = 0,\ \lambda_1 = 1$, and so

$$x^* \in \partial f_1(\overline{x}_1), \quad x^* \in \partial(0 \cdot f_2)(\overline{x}_2) = \partial \iota_{\operatorname{dom} f_2}(\overline{x}_2) = N(\operatorname{dom} f_2; \overline{x}_2).$$

This shows that x^* belongs to the set on the right-hand side of relation (2.74). □

Note that in Eq. (2.72) we can take $\varepsilon_1 + \varepsilon_2 = \varepsilon + \cdots$. Moreover, if f_1 is continuous at $\overline{x}_1$ and f_2 is continuous at $\overline{x}_2$, then in Eq. (2.74) we have $\partial(f_1 \Diamond f_2)(\overline{x}) = \{0\}$. Indeed, in this situation, $N(\operatorname{dom} f_2; \overline{x}_2) = \{0\}$ (since $\overline{x}_2 \in \operatorname{int}(\operatorname{dom} f_2)$); since $f_1 \Diamond f_2$ is continuous at $\overline{x}$, $\partial(f_1 \Diamond f_2)(\overline{x}) \neq \emptyset$. In particular, it follows that $\overline{x}_1$ is a minimum point of f_1. Furthermore, in this situation, $f_1 \Diamond f_2$ is even Gâteaux differentiable.

Introducing the convention that $0 \cdot \partial f(x) := N(\operatorname{dom} f; x)$ for $f \in \Lambda(X)$ and $x \in \operatorname{dom} f$, formula (2.75) may be written in the form

$$\partial(f_1 \Diamond f_2)(\overline{x}) = \partial f_1(x_1) \diamond \partial f_2(x_2),$$

where $A \diamond B$ represents the harmonic sum of the sets A and B, which generalizes the inverse sum of A and B.

In order to have more explicit formulas in Corollary 2.8.11 it is necessary to impose some supplementary conditions. Let us give such conditions and the corresponding formulas in three situations for $n = 2$.

Corollary 2.8.15 *Let $f_1, f_2 \in \Lambda(X)$, $\varepsilon \in \mathbb{R}_+$ and $\varphi : X \to \overline{\mathbb{R}}$, $\varphi := \max\{f_1, f_2\}$. Suppose that one of following conditions is verified:*

(i) *there exists $x_0 \in \operatorname{dom} f_1 \cap \operatorname{dom} f_2$ such that f_2 is continuous at x_0;*

(ii) *X is a Fréchet space, f_1, f_2 are li-convex and $0 \in {}^{ib}(\operatorname{dom} f_1 - \operatorname{dom} f_2)$;*

(iii) $\dim X < \infty$ *and* ${}^{i}(\operatorname{dom} f_1) \cap {}^{i}(\operatorname{dom} f_2) \neq \emptyset$.

Then, for all $x^* \in X^*$ *and* $x \in \operatorname{dom} f_1 \cap \operatorname{dom} f_2$ *we have:*

$$\varphi^*(x^*) = \min\{(\lambda_1 f_1)^*(x_1^*) + (\lambda_2 f_2)^*(x_2^*) \mid (\lambda_1, \lambda_2) \in \Delta_2,\ x_1^* + x_2^* = x^*\},$$

$$\partial_\varepsilon \varphi(x) = \bigcup \{\partial_{\varepsilon_1}(\lambda_1 f_1)(x) + \partial_{\varepsilon_2}(\lambda_2 f_2)(x) \mid (\lambda_1, \lambda_2) \in \Delta_2,\ \varepsilon_0, \varepsilon_1, \varepsilon_2 \geq 0,$$
$$\varepsilon_0 + \varepsilon_1 + \varepsilon_2 = \varepsilon,\ \lambda_1 f_1(x) + \lambda_2 f_2(x) \geq \varphi(x) - \varepsilon_0\},$$

$$\partial \varphi(x) = \bigcup \{\partial(\lambda_1 f_1)(x) + \partial(\lambda_2 f_2)(x) \mid (\lambda_1, \lambda_2) \in \Delta_2,$$
$$\lambda_1 f_1(x) + \lambda_2 f_2(x) = \varphi(x)\}. \qquad \square$$

Let $x^*, x_1^*, \ldots, x_n^* \in X^*$ and consider the functions $f_1, f_2 : X \to \mathbb{R}$ defined by

$$f_1(x) := \max\{\langle x, x_1^* \rangle, \ldots, \langle x, x_n^* \rangle\}, \quad f_2(x) := |\langle x, x^* \rangle|.$$

Applying the preceding corollary, we get the following formulas for every $x \in X$:

$$\partial f_1(x) = \left\{ \sum\nolimits_{i=1}^n \lambda_i x_i^* \,\middle|\, (\lambda_1, \ldots, \lambda_n) \in \Delta_n,\ \lambda_i = 0 \text{ if } \langle x, x_i^* \rangle < f_1(x) \right\},$$

$$\partial f_2(x) = \begin{cases} \{x^*\} & \text{if} \quad \langle x, x^* \rangle > 0, \\ \{\lambda x^* \mid \lambda \in [-1, +1]\} & \text{if} \quad \langle x, x^* \rangle = 0, \\ \{-x^*\} & \text{if} \quad \langle x, x^* \rangle < 0. \end{cases}$$

2.9 Convex Optimization with Constraints

Let us come back to the general problem of convex programming as considered in Section 2.5,

(P) $\quad \min\ f(x),\ x \in C,$

where $f \in \Lambda(X)$, $C \subset X$ is a convex set and $C \cap \operatorname{dom} f \neq \emptyset$; the spaces considered in the present section are separated locally convex spaces if not stated explicitly otherwise. Since $\overline{x}$ is solution of (P) exactly when it minimizes $f + \iota_C$, $\overline{x}$ is solution of (P) if and only if $0 \in \partial(f + \iota_C)(\overline{x})$. Taking into account Theorem 2.8.7, we have

Theorem 2.9.1 (Pshenichnyi–Rockafellar) *Let* $f \in \Lambda(X)$ *and* $C \subset X$ *be a convex set. Suppose that either* $\operatorname{dom} f \cap \operatorname{int} C \neq \emptyset$, *or there exists* $x_0 \in \operatorname{dom} f \cap C$, *where* f *is continuous. In these conditions* $\overline{x} \in C$ *is a solution of* (P) *if and only if* $\partial f(\overline{x}) \cap (-N(C; \overline{x})) \neq \emptyset$.

Proof. In the conditions of our statement we have that

$$\forall x \in C \cap \operatorname{dom} f \; : \; \partial(f + \iota_C)(x) = \partial f(x) + \partial\iota_C(x) = \partial f(x) + N(C; x),$$

whence the conclusion is obvious. □

Very often the set C from problem (P) is introduced as the set of solutions of a system of equalities and/or inequalities.

Let Y be ordered by a closed convex cone $Q \subset Y$ and $H : X \to Y^\bullet$ be a Q-convex operator; the set $C := \{x \in X \mid H(x) \leq_Q 0\}$ is a convex set. In this case the problem (P) takes the form:

(P_0) $\quad$ $\min\ f(x),\ H(x) \leq_Q 0.$

An element $x \in X$ for which $H(x) \leq_Q 0$ is called an **admissible solution** of problem (P_0). Of course, we assume that $\operatorname{dom} f \cap \{x \in X \mid H(x) \leq_Q 0\} \neq \emptyset$; in particular $\operatorname{dom} f \cap \operatorname{dom} H \neq \emptyset$.

The problem (P_0) may be embedded in a natural way into a family of minimization problems (P_y), $y \in Y$:

(P_y) $\quad$ $\min\ f(x),\ H(x) \leq_Q y.$

Let us consider the function

$$\Phi : X \times Y \to \overline{\mathbb{R}}, \quad \Phi(x, y) := \begin{cases} f(x) & \text{if } H(x) \leq_Q y, \\ \infty & \text{otherwise.} \end{cases} \tag{2.76}$$

The problem (P_y) becomes now

(P_y) $\quad$ $\min\ \Phi(x, y),\ x \in X.$

We obtain, without difficulty, that Φ is convex. Moreover

$$\begin{aligned} \Phi^*(x^*, -y^*) &= \sup\{\langle x, x^*\rangle + \langle y, -y^*\rangle - \Phi(x, y) \mid x \in X, y \in Y\} \\ &= \sup\{\langle x, x^*\rangle - \langle y, y^*\rangle - f(x) \mid x \in X, y \in Y, H(x) \leq_Q y\} \\ &= \sup\{\langle x, x^*\rangle - \langle H(x) + q, y^*\rangle - f(x) \mid x \in \operatorname{dom} H,\ q \in Q\} \\ &= \sup\{\langle x, x^*\rangle - \langle H(x), y^*\rangle - f(x) \mid x \in \operatorname{dom} H\} \\ &\qquad + \sup\{-\langle q, y^*\rangle \mid q \in Q\}. \end{aligned}$$

Hence

$$\Phi^*(x^*, -y^*) = \sup\{\langle x, x^*\rangle - f(x) - \langle H(x), y^*\rangle \mid x \in \operatorname{dom} H\}$$

if $y^* \in Q^+$ and $\Phi^*(x^*, -y^*) = \infty$ if $y^* \notin Q^+$.

The function

$$L : X \times Q^+ \to \overline{\mathbb{R}}, \quad L(x, y^*) := \begin{cases} f(x) + \langle H(x), y^* \rangle & \text{if } x \in \operatorname{dom} H, \\ \infty & \text{if } x \notin \operatorname{dom} H, \end{cases}$$

is called the **Lagrange function** (or *Lagrangian*) associated to problem (P_0). The above definition of L shows that $L(x, y^*) = f(x) + (y^* \circ H)(x)$ with the convention that $y^*(\infty) = \infty$ for $y^* \in Q^+$, convention which is used in the sequel. By what was shown above, we have that

$$\forall y^* \in Q^+ \; : \; \Phi^*(0, -y^*) = \sup_{x \in X} (-L(x, y^*)) = -\inf_{x \in X} L(x, y^*).$$

Therefore, the dual problem of problem (P_0) (see Section 2.6) is

(D_0) $\quad \max \big(-\Phi^*(0, y^*) \big), \; y^* \in Y^*,$

or equivalently,

(D_0) $\quad \max \big(\inf_{x \in X} L(x, y^*) \big), \; y^* \in Q^+.$

We have the following result.

Theorem 2.9.2 *Let $f \in \Lambda(X)$, $\overline{x} \in \operatorname{dom} f$ and $H : X \to (Y^\bullet, Q)$ be a Q-convex operator, where $Q \subset Y$ is a closed convex cone. Suppose that the following **Slater's condition** holds:*

(S) $\quad \exists \, x_0 \in \operatorname{dom} f \; : \; -H(x_0) \in \operatorname{int} Q.$

Then the problem (D_0) has optimal solutions and $v(P_0) = v(D_0)$, i.e. there exists $\overline{y}^* \in Q^+$ such that

$$\inf\{f(x) \mid H(x) \leq_Q 0\} = \inf\{L(x, \overline{y}^*) \mid x \in X\}.$$

Furthermore, the following statements are equivalent:

(i) *$\overline{x}$ is a solution of (P_0);*

(ii) *$H(\overline{x}) \leq_Q 0$ and there exists $\overline{y}^* \in Q^+$ such that*

$$0 \in \partial(f + \overline{y}^* \circ H)(\overline{x}) \quad \text{and} \quad \langle H(\overline{x}), \overline{y}^* \rangle = 0;$$

(iii) *there exists $\overline{y}^* \in Q^+$ such that $(\overline{x}, \overline{y}^*)$ is a **saddle point** for L, i.e.*

$$\forall x \in X, \; \forall y^* \in Q^+ \; : \; L(\overline{x}, y^*) \leq L(\overline{x}, \overline{y}^*) \leq L(x, \overline{y}^*). \tag{2.77}$$

Proof. The Slater condition (S) ensures that $(x_0, 0) \in \operatorname{dom} \Phi$ and $\Phi(x_0, \cdot)$ is continuous at 0, where Φ is the function defined by Eq. (2.76). Applying Theorem 2.7.1, there exists $\overline{y}^* \in Y^*$ such that $v(P_0) = -\Phi^*(0, -\overline{y}^*)$. If $v(P_0) = -\infty$, then $\Phi^*(0, y^*) = \infty$ for every $y^* \in Y^*$; in this case we may take $\overline{y}^* = 0$. If $v(P_0) > -\infty$, using the expression of Φ^*, we have that $\overline{y}^* \in Q^+$. Therefore

$$v(P_0) = \inf\{f(x) \mid H(x) \leq_Q 0\} = -\Phi^*(0, -\overline{y}^*) = \inf\{L(x, \overline{y}^*) \mid x \in X\} \\ = v(D_0).$$

(i) $\Rightarrow$ (ii) Of course, $\overline{x}$ being a solution for (P_0), we have that $H(\overline{x}) \leq_Q 0$. By what was proved above, there exists $\overline{y}^* \in Q^+$ such that

$$f(\overline{x}) = v(P_0) = \inf\{L(x, \overline{y}^*) \mid x \in X\},$$

whence

$$\forall x \in X \ : \ f(\overline{x}) + \langle H(\overline{x}), \overline{y}^*\rangle \leq f(\overline{x}) \leq f(x) + \langle H(x), \overline{y}^*\rangle. \tag{2.78}$$

Relation (2.78), without the term from its middle, says that $\overline{x}$ is a minimum point for $f + \overline{y}^* \circ H$, and so $0 \in \partial(f + \overline{y}^* \circ H)(\overline{x})$; taking $x = \overline{x}$ in relation (2.78) we obtain that $\langle H(\overline{x}), \overline{y}^*\rangle = 0$.

(ii) $\Rightarrow$ (iii) Since $0 \in \partial(f + \overline{y}^* \circ H)(\overline{x})$ we have

$$\forall x \in X \ : \ L(\overline{x}, \overline{y}^*) = f(\overline{x}) + \langle H(\overline{x}), \overline{y}^*\rangle \leq f(x) + \langle H(x), \overline{y}^*\rangle = L(x, \overline{y}^*).$$

Furthermore, for $y^* \in Q^+$ we have that

$$L(\overline{x}, y^*) = f(\overline{x}) + \langle H(\overline{x}), y^*\rangle \leq f(\overline{x}) = f(\overline{x}) + \langle H(\overline{x}), \overline{y}^*\rangle = L(\overline{x}, \overline{y}^*).$$

Therefore Eq. (2.77) holds, *i.e.* $(\overline{x}, \overline{y}^*)$ is a saddle point of L.

(iii) $\Rightarrow$ (i) Taking successively $y^* = 0$ and $y^* = 2\overline{y}^*$ on the left-hand side of Eq. (2.77) we obtain that $\langle H(\overline{x}), \overline{y}^*\rangle = 0$. Using again the left-hand side of Eq. (2.77), we obtain that $\langle H(\overline{x}), y^*\rangle \leq 0$ for every $y^* \in Q^+$; thus, using the bipolar theorem (Theorem 1.1.9), we have that $-H(\overline{x}) \in Q^{++} = Q$, *i.e.* $\overline{x}$ is an admissible solution of (P_0). From the right-hand side of relation (2.77) we get

$$\forall x \in X,\ H(x) \leq_Q 0 \ : \ f(\overline{x}) = f(\overline{x}) + \langle H(\overline{x}), \overline{y}^*\rangle \leq f(x) + \langle H(x), \overline{y}^*\rangle \leq f(x).$$

Therefore $\overline{x}$ is solution of problem (P_0). □

The element $\overline{y}^* \in Q^+$ obtained in Theorem 2.9.2 is called a **Lagrange multiplier** of problem (P).

Note that when H is finite-valued (*i.e.* $\operatorname{dom} H = X$) and continuous,

$$\forall x \in \operatorname{dom} f \ : \ \partial(f + \overline{y}^* \circ H)(x) = \partial f(x) + \partial(\overline{y}^* \circ H)(x);$$

the fact that Q is closed was used only for the implication (iii) $\Rightarrow$ (i) of the above theorem, to prove that $\overline{x}$ is an admissible solution.

An important particular case is when there are a finite number of constraints. Let $f, g_1, \ldots, g_n \in \Lambda(X)$ and consider the problem

(P_1) $\quad$ $\min\ f(x),\ g_i(x) \leq 0, 1 \leq i \leq n.$

The dual problem of (P_1) is

(D_1) $\quad$ $\max\ \inf_{x\in X}(f(x)+\lambda_1 g_1(x)+\cdots+\lambda_n g_n(x)),\ \lambda_1 \geq 0, \ldots, \lambda_n \geq 0.$

Then the following result holds.

Theorem 2.9.3 *Let $f, g_1, \ldots, g_n \in \Lambda(X)$. Suppose that the Slater condition holds, i.e.*

$$\exists\, x_0 \in \operatorname{dom} f,\ \forall i \in \overline{1,n}\ :\ g_i(x_0) < 0.$$

Then:

(i) *the dual problem (D_1) has optimal solutions and $v(P_1) = v(D_1)$, i.e. there exist* (Lagrange multipliers) $\overline{\lambda}_1, \ldots, \overline{\lambda}_n \in \mathbb{R}_+$ *such that*

$$\begin{aligned}\inf\{f(x) \mid g_1(x) \leq 0, \ldots, g_n(x) \leq 0\}\\ = \inf\left\{f(x) + \overline{\lambda}_1 g_1(x) + \cdots + \overline{\lambda}_n g_n(x) \mid x \in X\right\}.\end{aligned}$$

(ii) *Let $\overline{x} \in \operatorname{dom} f$; $\overline{x}$ is a solution of problem (P_1) if and only if $g_i(\overline{x}) \leq 0$ for every $i \in \overline{1,n}$ and there exist $\overline{\lambda}_1, \ldots, \overline{\lambda}_n \in \mathbb{R}_+$ such that $\overline{\lambda}_i g_i(\overline{x}) = 0$ for $i \in \overline{1,n}$ and*

$$0 \in \partial\left(f + \overline{\lambda}_1 g_1 + \ldots + \overline{\lambda}_n g_n\right)(\overline{x}).$$

If the functions $g_1, \ldots, g_n$ are continuous at $\overline{x}$, the last condition is equivalent to

$$0 \in \partial f(\overline{x}) + \overline{\lambda}_1 \partial g_1(\overline{x}) + \ldots + \overline{\lambda}_n \partial g_n(\overline{x}).$$

Proof. Let us consider $H : X \to (\mathbb{R}^{n\bullet}, \mathbb{R}^n_+)$, $H(x) := (g_1(x), \ldots, g_n(x))$ for $x \in \bigcap_{i=1}^n \operatorname{dom} g_i$, $H(x) = \infty$ otherwise; H is $\mathbb{R}^n_+$-convex. The result stated in the theorem is an immediate consequence of the preceding theorem. When g_i are continuous at $\overline{x}$ one has, by Theorem 2.8.7 (iii), that

$$\partial(f + \overline{\lambda}_1 g_1 + \ldots + \overline{\lambda}_n g_n)(\overline{x}) = \partial f(\overline{x}) + \partial(\overline{\lambda}_1 g_1)(\overline{x}) + \ldots + \partial(\overline{\lambda}_n g_n)(\overline{x}),$$

and $\partial(\overline{\lambda}_i g_i)(\overline{x}) = \overline{\lambda}_i \partial g_i(\overline{x})$. □

Corollary 2.9.4 *Let $f, g_1, \ldots, g_n : X \to \mathbb{R}$ be Gâteaux differentiable continuous convex functions. Suppose that there exists $x_0 \in X$ such that $g_i(x_0) < 0$ for every $i \in \overline{1,n}$. Then $\overline{x} \in X$ is solution of problem (P_1) if and only if $g_i(\overline{x}) \leq 0$ for every $i \in \overline{1,n}$ and there exist $\overline{\lambda}_1, \ldots, \overline{\lambda}_n \in \mathbb{R}_+$ such that*

$$-\nabla f(\overline{x}) = \overline{\lambda}_1 \nabla g_1(\overline{x}) + \ldots + \overline{\lambda}_n \nabla g_n(\overline{x}) \quad \textit{and} \quad \overline{\lambda}_i g_i(\overline{x}) = 0 \textit{ for } i \in \overline{1,n}. \quad \square$$

Corollary 2.9.5 *Let $g \in \Lambda(X)$, $\gamma \in]\inf g, \infty[$ and $\overline{x} \in [g \leq \gamma]$. Then*

$$N([g \leq \gamma]; \overline{x}) = \bigcup \{\partial(\lambda g)(\overline{x}) \mid \lambda \geq 0,\ \lambda(g(\overline{x}) - \gamma) = 0\}. \tag{2.79}$$

Proof. The inclusion "$\supset$" in Eq. (2.79) holds without any condition on g and γ. Indeed, let $x^* \in \partial(\lambda g)(\overline{x})$ for some $\lambda \geq 0$ with $\lambda(g(\overline{x}) - \gamma) = 0$. Then for $x \in [g \leq \gamma]$ we have that $\langle x - \overline{x}, x^* \rangle \leq \lambda g(x) - \lambda g(\overline{x}) = \lambda(g(x) - \gamma) \leq 0$, and so $x^* \in N([g \leq \gamma], \overline{x})$.

Let $x^* \in N([g \leq \gamma]; \overline{x})$. Then $\overline{x}$ is a solution of the problem

(P_1') $\qquad \min \langle x, -x^* \rangle,\ h(x) := g(x) - \gamma \leq 0,$

whence, by Theorem 2.9.3, there exists $\lambda \geq 0$ such that $\lambda h(\overline{x}) = \lambda(g(\overline{x}) - \gamma) = 0$ and $0 \in \partial(-x^* + \lambda h)(\overline{x})$, *i.e.* $x^* \in \partial(\lambda h)(\overline{x}) = \partial(\lambda g)(\overline{x})$. Therefore the inclusion "$\subset$" of Eq. (2.79) holds, too. $\square$

Note that $\partial(\lambda g)(\overline{x}) = \lambda \partial g(\overline{x})$ if $\lambda > 0$ and $\partial(0g)(\overline{x}) = \partial \iota_{\operatorname{dom} g}(\overline{x}) = N(\operatorname{dom} g; \overline{x})$.

In the case of normed vector spaces, taking $A = X$ in Proposition 3.10.16 from Section 3.10, we have another sufficient condition for the validity of formula (2.79).

Note that we can obtain the characterization of optimal solutions in Theorem 2.9.3, when the functions g_i are continuous, using Corollary 2.9.5 and the formula for the subdifferential of a sum. Indeed, $\overline{x} \in \operatorname{dom} f$ is a solution of (P_1) if and only if $\overline{x}$ is minimum point of the function $f + \iota_{C_1} + \cdots + \iota_{C_n}$, where $C_i := \{x \mid g_i(x) \leq 0\}$. By hypothesis $\operatorname{dom} f \cap \bigcap_{i=1}^n \operatorname{int} C_i \neq \emptyset$, and so, for every $x \in \operatorname{dom} f \cap \bigcap_{i=1}^n C_i$ we have

$$\partial \left(f + \iota_{C_1} + \ldots + \iota_{C_n}\right)(x) = \partial f(x) + N(C_1; x) + \ldots + N(C_n; x).$$

Using formula (2.79) we obtain the desired characterization.

Theorem 2.9.2 can be extended further to the case when there are also linear constraints. Let us consider the problem

(P_2) $\qquad \min\ f(x),\ H(x) \leq_Q 0,\ Tx = 0,$

where $T \in \mathcal{L}(X, Z)$. Consider the Lagrange function associated to problem (P_2):

$$L_2 : X \times (Q^+ \times Z^*) \to \overline{\mathbb{R}}, \quad L_2(x, y^*, z^*) := f(x) + \langle H(x), y^* \rangle + \langle Tx, z^* \rangle.$$

The dual problem associated to (P_2) is

(D_2) $\quad \max \left(\inf_{x \in X} L_2(x, y^*, z^*)\right), \ y^* \in Q^+, \ z^* \in Z^*.$

Then the following result holds.

Theorem 2.9.6 *Let $f \in \Lambda(X)$, $H : X \to (Y, Q)$ be a Q-convex operator, where $Q \subset Y$ is a closed convex cone, $T \in \mathcal{L}(X, Z)$ be a relatively open operator and $\overline{x} \in \operatorname{dom} f$. Suppose that there exists $x_0 \in \operatorname{dom} f$ such that f, H are continuous at x_0, $-H(x_0) \in \operatorname{int} Q$ and $Tx_0 = 0$. Then the problem (D_2) has optimal solutions and $v(P_2) = v(D_2)$, i.e. there exists $\overline{y}^* \in Q^+$, $\overline{z}^* \in Z^*$ such that*

$$\inf\{f(x) \mid H(x) \leq_Q 0, \ Tx = 0\} = \inf\{L_2(x, \overline{y}^*, \overline{z}^*) \mid x \in X\}.$$

Furthermore, the following statements are equivalent:

(i) *$\overline{x}$ is solution of problem (P_2);*

(ii) *$H(\overline{x}) \leq_Q 0$, $T(\overline{x}) = 0$ and there exists $\overline{y}^* \in Q^+$, $\overline{z}^* \in Z^*$ such that*

$$T^*\overline{z}^* \in \partial f(\overline{x}) + \partial(\overline{y}^* \circ H)(\overline{x}) \quad \text{and} \quad \langle H(\overline{x}), \overline{y}^* \rangle = 0;$$

(iii) *there exists $(\overline{y}^*, \overline{z}^*) \in Q^+ \times Z^*$ such that $(\overline{x}, (\overline{y}^*, \overline{z}^*))$ is a saddle point of L_2, i.e.*

$$L_2(\overline{x}, y^*, z^*) \leq L_2(\overline{x}, \overline{y}^*, \overline{z}^*) \leq L_2(x, \overline{y}^*, \overline{z}^*)$$

for all $x \in X$ and all $y^* \in Q^+$, $z^* \in Z^*$.

Proof. Let us consider the perturbation function

$$\Phi : X \times (Y \times Z) \to \overline{\mathbb{R}}, \quad \Phi(x, y, z) := \begin{cases} f(x) & \text{if } H(x) \leq_Q y, \ T(x) = z, \\ \infty & \text{otherwise.} \end{cases}$$

We intend to prove that condition (i) of Theorem 2.7.1 is verified. Note first that

$$\Pr_{Y \times Z}(\operatorname{dom} \Phi) = \{(H(x) + q, Tx) \mid x \in \operatorname{dom} f, \ q \in Q\} \subset Y \times \operatorname{Im} T.$$

Taking into account that f is continuous at $x_0 \in \operatorname{dom} f$, there exists $U_0 \in \mathcal{N}_X(x_0)$ such that

$$\forall\, x \in U_0 \ : \ f(x) \leq M := f(x_0) + 1.$$

Since $-H(x_0) \in \operatorname{int} Q$, there exists $V_0 \in \mathcal{N}_Y$ such that $-H(x_0) + V_0 \subset Q$. There exists $V \in \mathcal{N}_Y$ such that $V + V \subset V_0$. Since H is continuous at x_0, there exists $U \in \mathcal{N}_X(x_0)$, $U \subset U_0$, such that $H(x) \in H(x_0) - V$ for every $x \in U$. Since T is relatively open, there exists $W \in \mathcal{N}_{\operatorname{Im} T}$ such that $W \subset T(U)$. Of course, $V \times W \in \mathcal{N}_{Y \times \operatorname{Im} T}(0,0)$; let $(y, z) \in V \times W$. There exists $x \in U$ such that $Tx = z$. Since $x \in U$, we have that $H(x) = H(x_0) - y'$ for some $y' \in V$. Then

$$y - H(x) = y + y' - H(x_0) \in -H(x_0) + V + V \subset -H(x_0) + V_0 \subset Q;$$

hence $H(x) \leq_Q y$. Since $x \in U \subset U_0$, we have that $\Phi(x, y, z) \leq M$, *i.e.* condition (i) of Theorem 2.7.1 is verified. Therefore there exists $(\overline{y}^*, \overline{z}^*) \in Y^* \times Z^*$ such that $v(P_2) = -\Phi^*(0, -\overline{y}^*, -\overline{z}^*)$. But

$$\begin{aligned}
\Phi^*(0, -y^*, -z^*) &= \sup\nolimits_{(x,y,z) \in X \times Y \times Z}(\langle y, -y^* \rangle + \langle z, -z^* \rangle - \Phi(x, y, z)) \\
&= \sup\{-\langle y, y^* \rangle - \langle z, z^* \rangle - f(x) \mid H(x) \leq_Q y, Tx = z\} \\
&= \sup\{-\langle H(x) + q, y^* \rangle - \langle Tx, z^* \rangle - f(x) \mid x \in X, q \in Q\} \\
&= -\inf\{f(x) + \langle H(x), y^* \rangle + \langle Tx, z^* \rangle \mid x \in X\} \\
&\qquad + \sup\{-\langle q, y^* \rangle \mid q \in Q\}.
\end{aligned}$$

Thus

$$\Phi^*(0, -y^*, -z^*) = \begin{cases} -\inf_{x \in X} L_2(x, y^*, z^*) & \text{if} \quad y^* \in Q^+, \\ \infty & \text{if} \quad y^* \notin Q^+. \end{cases}$$

Therefore $\overline{y}^* \in Q^+$ if $v(P_2) \in \mathbb{R}$ and we can take $\overline{y}^* = 0$ if $v(P_2) = -\infty$.

The proof of the second part is completely similar to that of Theorem 2.9.2. We note only that, f and H being continuous at x_0, we have $\partial(f + \overline{y}^* \circ H)(x) = \partial f(x) + \partial(\overline{y}^* \circ H)(x)$ for every $x \in \operatorname{dom} f$. □

2.10 A Minimax Theorem

In the preceding section we have obtained some results which are stated using saddle points. In the sequel we establish a quite general result on the existence of saddle points as well as a minimax theorem. Let A and B be two nonempty sets and $f : A \times B \to \overline{\mathbb{R}}$. It is obvious that

$$\sup_{x\in A} \inf_{y\in B} f(x,y) \leq \inf_{y\in B} \sup_{x\in A} f(x,y). \tag{2.80}$$

The results which ensure equality in the preceding inequality are called "minimax theorems," the common value being called **saddle value**. Note that if f has a saddle point, *i.e.* there exists $(\overline{x},\overline{y}) \in A \times B$ such that

$$\forall x \in A,\ \forall y \in B\ :\ f(x,\overline{y}) \leq f(\overline{x},\overline{y}) \leq f(\overline{x},y), \tag{2.81}$$

then

$$\max_{x\in A} \inf_{y\in B} f(x,y) = \min_{y\in B} \sup_{x\in A} f(x,y), \tag{2.82}$$

where max (min) means, as usual, an attained supremum (infimum).

Indeed, let $(\overline{x},\overline{y}) \in A \times B$ satisfy Eq. (2.81). It follows that

$$\inf_{y\in B} \sup_{x\in A} f(x,y) \leq \sup_{x\in A} f(x,\overline{y}) \leq f(\overline{x},\overline{y}) \leq \inf_{y\in B} (\overline{x},y) \leq \sup_{x\in A} \inf_{y\in B} f(x,y).$$

Using Eq. (2.80) we obtain that all the terms are equal in the preceding relation, and so (2.82) holds (the maximum being attained at $\overline{x}$ and the minimum at $\overline{y}$). Conversely, if Eq. (2.82) holds then f has saddle points. Indeed, let $\overline{x} \in A$ and $\overline{y} \in B$ be such that

$$\inf_{y\in B} (\overline{x},y) = \sup_{x\in A} \inf_{y\in B} f(x,y) = \inf_{y\in B} \sup_{x\in A} f(x,y) = \sup_{x\in A} f(x,\overline{y}).$$

Since $\inf_{y\in B}(\overline{x},y) \leq f(\overline{x},\overline{y}) \leq \sup_{x\in A} f(x,\overline{y})$, from the above relation it follows that all the terms are equal in these inequalities; this shows that $(\overline{x},\overline{y})$ is a saddle point of f. So we have obtained the following result.

Theorem 2.10.1 *Let A and B be two nonempty sets and $f : A\times B \to \overline{\mathbb{R}}$. Then f has saddle points if and only if condition (2.82) holds.* □

The following result is an enough general minimax theorem (frequently utilized) which gives also a sufficient condition for the existence of saddle points.

Theorem 2.10.2 *Let X be a locally convex space, Y be a linear space, $A \subset X$ be a nonempty convex compact set and $B \subset Y$ be a nonempty convex set. Let also $f : A \times B \to \mathbb{R}$ be a function with the property that $f(\cdot, y)$*

is concave and upper semicontinuous for every $y \in B$ and $f(x,\cdot)$ is convex for every $x \in A$. Then

$$\max_{x\in A}\inf_{y\in B} f(x,y) = \inf_{y\in B}\max_{x\in A} f(x,y). \tag{2.83}$$

If moreover Y is a locally convex space, B is compact and $f(x,\cdot)$ is lower semicontinuous for every $x \in A$, then

$$\max_{x\in A}\min_{y\in B} f(x,y) = \min_{y\in B}\max_{x\in A} f(x,y); \tag{2.84}$$

in particular f has saddle points.

Proof. Let $\alpha \in \mathbb{R}$ be such that $\alpha > \max_{x\in A}\inf_{y\in B} f(x,y)$. For every $x \in A$ there exists $y_x \in B$ such that $f(x,y_x) < \alpha$. Since $f(\cdot,y_x)$ is upper semicontinuous at x, there exists an open neighborhood V_x of x such that $f(u,y_x) < \alpha$ for every $u \in V_x$. Since A is compact and $A \subset \bigcup_{x\in A} V_x$, there exist $x_1,\dots,x_p \in A$ such that $A \subset \bigcup_{i=1}^{p} V_{x_i}$. Let $y_i := y_{x_i}$. Consider the sets

$$C_1 := \operatorname{co}\{(f(x,y_1),\dots,f(x,y_p)) \mid x \in A\} \subset \mathbb{R}^p,$$
$$C_2 := \left\{(u_1,\dots,u_p) \mid u_i \geq \alpha\ \forall i \in \overline{1,p}\right\}.$$

It is obvious that C_1 and C_2 are nonempty convex sets, and C_2 has nonempty interior. Moreover $C_1 \cap C_2 = \emptyset$. In the contrary case there exist $q \in \mathbb{N}$, $(\lambda_1,\dots,\lambda_q) \in \Delta_q$ and $x_1,\dots,x_q \in A$ such that

$$\forall i \in \overline{1,p}\ :\ \alpha \leq z_i := \sum\nolimits_{j=1}^{q} \lambda_j f(x_j,y_i).$$

Since $f(\cdot,y_i)$ is concave, taking $x_0 = \sum_{j=1}^{q} \lambda_j x_j$ we have that $x_0 \in A$ and

$$\forall i \in \overline{1,p}\ :\ \alpha \leq z_i \leq f(x_0,y_i).$$

There exists $i_0 \in \overline{1,p}$ such that $x_0 \in V_{x_{i_0}}$. Therefore $f(x_0,y_{i_0}) < \alpha$, a contradiction.

Applying Theorem 1.1.3 there exists $(\mu_1,\dots,\mu_p) \in \mathbb{R}^p \setminus \{0\}$ such that

$$\forall x \in A,\ \forall u = (u_1,\dots,u_p) \in C_2\ :\ \sum\nolimits_{i=1}^{p} \mu_i f(x,y_i) \leq \sum\nolimits_{i=1}^{p} \mu_i u_i.$$

Letting $u_i \to \infty$ we have that $\mu_i \geq 0$ for every i, and so we can suppose that $(\mu_1,\dots,\mu_p) \in \Delta_p$. Taking $u = (\alpha,\dots,\alpha)$ we obtain that

$$\forall x \in A\ :\ \sum\nolimits_{i=1}^{p} \mu_i f(x,y_i) \leq \alpha.$$

Let $y_0 := \sum_{i=1}^{p} \mu_i y_i \in B$. Since $f(x, \cdot)$ is convex, we obtain that $f(x, y_0) \leq \alpha$ for every $x \in A$. Therefore $\alpha \geq \inf_{y \in B} \max_{x \in A} f(x, y)$, and so (2.83) holds. In Eq. (2.83) the suprema with respect to x are attained because the functions $f(\cdot, y)$, $y \in B$, and $\inf_{y \in B} f(\cdot, y)$ are upper semicontinuous and A is compact.

When B is compact and the functions $f(x, \cdot)$ are lsc for all $x \in A$ we obtain that the infima with respect to $y \in B$ are attained in Eq. (2.83), *i.e.* Eq. (2.84) holds. □

Note that the convexity assumptions in the preceding theorem can be weakened. More precisely, we can suppose that A is a nonempty compact subset of a topological space, B is nonempty, and f satisfies the following two conditions (similar to concavity and convexity, respectively):

$$\forall x_1, \ldots, x_p \in A,\ \forall (\lambda_1, \ldots, \lambda_p) \in \Delta_p,\ \exists x_0 \in A,\ \forall y \in B \ : \qquad f(x_0, y) \geq \sum_{i=1}^{p} \lambda_i f(x_i, y),$$

$$\forall y_1, \ldots, y_q \in B,\ \forall (\mu_1, \ldots, \mu_q) \in \Delta_q,\ \exists y_0 \in B,\ \forall x \in A \ : \qquad f(x, y_0) \leq \sum_{j=1}^{q} \mu_j f(x, y_j).$$

2.11 Exercises

Exercise 2.1 Let X be a linear space, $f \in \Lambda(X)$ and $x, u \in X$. Prove that the mapping $\psi :]0, \infty[\to \overline{\mathbb{R}}$ defined by $\psi(t) := t \cdot f(x + t^{-1}u)$ is convex.

Exercise 2.2 (a) Let $I \subset \mathbb{R}$ be an interval and $f : I \to \mathbb{R}$. Suppose that f is locally nondecreasing, *i.e.* for every $a \in I$ there exists $\varepsilon > 0$ such that the restriction of f to $I \cap [a - \varepsilon, a + \varepsilon]$ is nondecreasing. Prove that f is nondecreasing.

(b) Let $g : [a, b] \to \mathbb{R}$ ($a, b \in \mathbb{R}$, $a < b$) be a continuous function. 1) Assume that $g'_+(x) \in \mathbb{R}$ exists for every $x \in [a, b[$; show that there exists $x' \in [a, b[$ such that $g(b) - g(a) \leq g'_+(x')(b - a)$. 2) Assume that $g'_-(x) \in \mathbb{R}$ exists for every $x \in]a, b]$; show that there exists $x'' \in]a, b]$ such that $g(b) - g(a) \geq g'_-(x'')(b - a)$.

(c) Let X be a locally convex space and $A \subset X$ be an open convex set. If $f : A \to \mathbb{R}$ is locally convex, *i.e.* for every $a \in A$ there exists a neighborhood V of a such that $f|_{V \cap A}$ is convex, prove that f is convex.

Exercise 2.3 Let $f : \mathbb{R}^n \to \overline{\mathbb{R}}$ be a quasi-convex function and $\overline{x} \in \mathbb{R}^n$. For

$x \in \mathbb{R}^n$ consider the function $f_x : \mathbb{R}^n \to \overline{\mathbb{R}}$, $f_x(t) = f(\overline{x} + tx)$. Prove that

$$f \text{ is lsc at } \overline{x} \Leftrightarrow \forall x \in \mathbb{R}^n : f_x \text{ is lsc at } 0,$$
$$f \text{ is usc at } \overline{x} \Leftrightarrow \forall x \in \mathbb{R}^n : f_x \text{ is usc at } 0.$$

If $\mathbb{R}^n$ is replaced by an infinite dimensional normed space then the above properties do not hold, generally.

Exercise 2.4 Let X be a linear space, $f : X \to \mathbb{R}$ be a convex function and $\lambda > \inf_{x\in X} f(x)$. Prove that

$$\{x \in X \mid f(x) \leq \lambda\}^i = \{x \in X \mid f(x) < \lambda\}.$$

Moreover, if X is a topological vector space and f is continuous, then

$$\operatorname{int}\{x \in X \mid f(x) \leq \lambda\} = \{x \in X \mid f(x) < \lambda\}.$$

Exercise 2.5 Let X be a topological vector space and $f : X \to \overline{\mathbb{R}}$ a convex function. Prove that:

(a) $[f \leq t] = \operatorname{cl}[f < t]$ for every $t \in]\inf f, \infty[$ if and only if f is lsc;

(b) $[f < t] = \operatorname{int}[f \leq t]$ for every $t \in]\inf f, \infty[$ if and only if f is continuous on $\operatorname{dom} f$;

(c) $[f = t] = \operatorname{bd}[f \leq t]$ for every $t \in]\inf f, \infty[$ if and only if f is continuous (on X).

Exercise 2.6 Let $f : \mathbb{R}^n \to \overline{\mathbb{R}}$ be a proper convex function for which all the partial derivatives $\partial f/\partial x_i$ exists at $a \in \operatorname{int}(\operatorname{dom} f)$. Prove that f is Gâteaux differentiable at a (even Fréchet differentiable because f is Lipschitz on a neighborhood of a).

Exercise 2.7 Let $p \in [1, \infty[$ and $\varphi_p : \mathbb{R}_+ \to \mathbb{R}$ be defined by $\varphi_p(t) := |1-t|^p - |1+t|^p + 2pt$. Prove that φ_p is strictly convex and increasing for $p \in]1, 2[$, φ_p is strictly concave and decreasing for $p \in]2, \infty[$, φ_1 is convex and nondecreasing and φ_2 is constant.

Exercise 2.8 Take $\beta \in [1, \infty[$ and consider the function

$$f : \mathbb{R}^2 \to \overline{\mathbb{R}}, \quad f(x,y) := \begin{cases} \left(\arctan \frac{x}{y}\right)^\beta \sqrt{x^2+y^2} & \text{for } x, y \geq 0, \\ \infty & \text{otherwise,} \end{cases}$$

where, by convention, $\arctan \frac{x}{0} := \frac{\pi}{2}$ for $x \geq 0$. Prove that f is a lsc convex function, but not strictly convex.

Exercise 2.9 Consider the function

$$f : C[0,1] \to \mathbb{R}, \qquad f(x) := \int_0^1 \sqrt{1 + (x(t))^2}\, dt.$$

Prove that f is convex and Fréchet differentiable of order 2 on $C[0,1]$ with

$$\nabla f(x)(u) = \int_0^1 \frac{xu}{\sqrt{1+x^2}}\,dt, \qquad \nabla^2 f(x)(u,v) = \int_0^1 \frac{uv}{(1+x^2)\sqrt{1+x^2}}\,dt$$

for all $u, v \in C[0,1]$. Moreover, $\nabla^2 f$ is continuous on $C[0,1]$.

Exercise 2.10 Consider the function

$$f : L^1(0,1) \to \mathbb{R}, \qquad f(x) := \textstyle\int_0^1 \sqrt{1+(x(t))^2}\,dt,$$

where $L^1(0,1)$ is the Banach space of (classes of) Lebesgue integrable functions on the interval $[0,1]$. Prove that f is a Gâteaux differentiable convex function with

$$\forall x, u \in L^1(0,1) \ : \ \nabla f(x)(u) = \int_0^1 \frac{xu}{\sqrt{1+x^2}}\,dt,$$

but f is nowhere Fréchet differentiable.

Exercise 2.11 Using properties of convex functions, prove that

$$\forall a, b \in \mathbb{R}_+,\ \forall p, q \in]1,\infty[,\ \tfrac{1}{p}+\tfrac{1}{q}=1 \ : \ ab \le \tfrac{1}{p}a^p + \tfrac{1}{q}b^q,$$

the equality being valid if and only if $a^p = b^q$, and

$$\forall n \in \mathbb{N},\ \forall x_1,\ldots,x_n \in \mathbb{R}_+,\ \forall \alpha_1,\ldots,\alpha_n \in]0,1[,\ \alpha_1+\ldots+\alpha_n = 1 \ : \\ \alpha_1 x_1 + \ldots + \alpha_n x_n \ge x_1^{\alpha_1}\cdots x_n^{\alpha_n},$$

the equality being valid if and only if $x_1 = \cdots = x_n$. In particular,

$$\forall n \in \mathbb{N},\ \forall x_1,\ldots,x_n \in \mathbb{R}_+ \ : \ \tfrac{1}{n}(x_1+\ldots+x_n) \ge \sqrt[n]{x_1\cdots x_n}.$$

Exercise 2.12 Let $f : X \to \overline{\mathbb{R}}$ be a proper function. Prove that f is convex if and only if $f + x^*$ is quasi-convex for every $x^* \in X^*$.

Exercise 2.13 Let $\alpha_1,\ldots,\alpha_n > 0$ $(n \ge 1)$ be such that $\alpha_1+\cdots+\alpha_n \le 1$. Prove that the function $f : \mathbb{P}^n \to \mathbb{R}$, $f(x) := x_1^{\alpha_1}\cdot x_2^{\alpha_2}\cdots x_n^{\alpha_n}$, is concave; moreover, if $\alpha_1+\cdots+\alpha_n < 1$, then f is strictly concave.

Exercise 2.14 Consider $f : \mathbb{R}^n \to \mathbb{R}$, $f(x_1,\ldots,x_n) := \ln\left(\sum_{k=1}^n \exp x_k\right)$. Prove that f is convex.

Exercise 2.15 Consider $p \in \mathbb{R}\setminus\{0\}$ and the function

$$f : \mathbb{P}\times\mathbb{P}^n \to \mathbb{R}, \qquad f(t,x) := \frac{t^p}{\prod_{i=1}^n x_i}.$$

Prove that f is strictly convex for $p \in]-\infty,0[\,\cup\,]n+1,\infty[$ and convex if $p = n+1$.

Let $c \in \mathbb{R}^n$ and $\Delta := \{x \in \mathbb{P}^n \mid (x \,|\, c) \geq 0\}$. Prove that the function

$$g : \Delta \to \mathbb{R}, \qquad g(x) := \frac{(x \,|\, c)^p}{\prod_{i=1}^n x_i},$$

is strictly convex for $p > n+1$ (it is possible to prove that g is strictly convex for $p > n$). The function g has been used by Karmarkar for establishing his interior point algorithm.

Finally prove that the function

$$h : \mathbb{P}^n \to \mathbb{R}, \qquad h(x) := \left(\prod_{i=1}^n x_i\right)^{-1},$$

is strictly convex.

Exercise 2.16 Let $\psi : \mathbb{R}_+ \to \mathbb{R}_+$ be a continuous convex function such that $\psi(t) = 0 \Leftrightarrow t = 0$. Show that

$$\int_0^{\psi(\alpha)} \frac{dt}{\psi'_-\left(\psi^{-1}(t)\right)} = \int_0^{\psi(\alpha)} \frac{dt}{\psi'_+\left(\psi^{-1}(t)\right)} = \alpha$$

for every $\alpha > 0$.

Exercise 2.17 Let X, Y be normed spaces and $T \in \mathcal{L}(X,Y)$. Prove that the function

$$f : Y \to \overline{\mathbb{R}}, \qquad f(y) := \inf\{\|x\| \mid Tx = y\},$$

is a sublinear functional. Moreover, if T is an open operator, then $\operatorname{dom} f = X$ and f is continuous.

Exercise 2.18 Let $f : \mathbb{R}^k \to \overline{\mathbb{R}}$ be a strongly coercive proper lsc function. Prove that $\operatorname{co}(\operatorname{epi} f)$ is a closed set.

Exercise 2.19 Let X be a locally convex space, $f \in \Gamma(X)$, $g : X \to \overline{\mathbb{R}}$ be a proper function and $\alpha, \beta > 0$. Prove that $\overline{\operatorname{co}}\left(\overline{\operatorname{co}}(f + \alpha g) + \beta g\right) = \overline{\operatorname{co}}\left(f + (\alpha + \beta)g\right)$ and $\left((f + \alpha g)^{**} + \beta g\right)^* = \left(f + (\alpha + \beta)g\right)^*$.

Exercise 2.20 Let X be a separated locally convex space and $A, B, C \subset X$ be nonempty sets. If C is bounded and $A + C \subset B + C$ prove that $A \subset \overline{\operatorname{co}} B$.

Exercise 2.21 Let $(X, \|\cdot\|)$ be a normed space, $f \in \Gamma(X)$ and $L \geq 0$. Prove the equivalence of the following statements:

(i) $\operatorname{dom} f = X$ and $\forall\, x, y \in X \;:\; |f(x) - f(y)| \leq L\,\|x - y\|$;

(ii) $\exists\, \alpha \in \mathbb{R}, \forall\, x \in X \;:\; f(x) \leq L\,\|x\| + \alpha$;

(iii) $\forall\, u \in X \;:\; f_\infty(u) \leq L\,\|u\|$.

Exercise 2.22 Let X be a locally convex space and $f \in \Gamma(X)$. Prove that $f_\infty(u) = \sup\{f(x + u) - f(x) \mid x \in \operatorname{dom} f\}$ for every $u \in X$.

Exercise 2.23 Let X, Y be separated locally convex spaces, $f \in \Gamma(X)$, $F \in \Gamma(X \times Y)$, $x \in \operatorname{dom} f$ and $\lambda \in \mathbb{R}$, $\varepsilon \in \mathbb{R}_+$ be such that $[f \leq \lambda]$ and $\partial_\varepsilon f(x)$ are nonempty. Prove that: $[f \leq \lambda]_\infty = [f_\infty \leq 0]$, $(\partial_\varepsilon f(x))_\infty = N(\operatorname{dom} f; x)$, $(f^*)_\infty = s_{\operatorname{dom} f}$, $f_\infty = s_{\operatorname{dom} f^*}$ and $\{v^* \in Y^* \mid (F^*)_\infty(0, v^*) \leq 0\} = (\operatorname{Pr}_Y(\operatorname{dom} F))^-$. In particular, if $\{v^* \in Y^* \mid (F^*)_\infty(0, v^*) \leq 0\}$ is a linear subspace then $\{0\}$ and $\operatorname{Pr}_Y(\operatorname{dom} F)$ are united.

Exercise 2.24 Let X be a separated locally convex space and $f \in \Gamma(X)$. Consider $K := \{u \in X \mid f_\infty(u) \leq 0\}$ and $X_0 := K \cap (-K)$. Prove that:

(i) K is a closed convex cone, X_0 is a closed linear subspace and $f(x+u) = f(x)$ for all $x \in X$ and $u \in X_0$.

(ii) $\overline{\operatorname{lin}(\operatorname{dom} f^*)} = X_0^\perp$.

(iii) The function $\widehat{f} : X/X_0 \to \overline{\mathbb{R}}$, $\widehat{f}(\widehat{x}) := f(x)$, is well defined, $\widehat{f} \in \Gamma(X/X_0)$, $\widehat{f}_\infty(\widehat{u}) = f_\infty(u)$ for every $u \in X$, $\widehat{f}^* = f^*|_{X_0^\perp}$ and $\partial_\varepsilon \widehat{f}(\widehat{x}) = \partial_\varepsilon f(x)|_{X_0^\perp}$ for all $x \in X$ and $\varepsilon \in \mathbb{R}_+$, where $\widehat{x}$ represents the class of $x \in X$.

Exercise 2.25 Let X be a separated locally convex space and $f \in \Gamma(X)$. Assume that there exists $u \in X$ such that $f_\infty(u) \leq 0 < f_\infty(-u)$. Prove that for every $\varepsilon > 0$ there exists $f_\varepsilon \in \Gamma(X)$ such that $f(x) \leq f_\varepsilon(x) \leq f(x) + \varepsilon$ for every $x \in X$ and $\operatorname{argmin} f_\varepsilon = \emptyset$.

Exercise 2.26 Let X, Y be separated locally convex spaces and $F \in \Lambda(X \times Y)$. Assume that F satisfies one of the conditions (ii)–(viii) of Theorem 2.7.1. Prove that the marginal function $g : X^* \to \overline{\mathbb{R}}$, $g(x^*) := \inf_{y^* \in Y^*} F^*(x^*, y^*)$ is convex, w^*-lsc, the infimum is attained for every $x^* \in X^*$ and $g_\infty(u^*) = \min_{v^* \in Y^*} (F^*)_\infty(u^*, v^*)$. Moreover, $\{v^* \in Y^* \mid (F^*)_\infty(0, v^*) \leq 0\}$ is a linear subspace.

Exercise 2.27 (Toland–Singer duality formula) Let X be a separated locally convex space, $f : X \to \overline{\mathbb{R}}$ be a proper function and $g \in \Gamma(X)$. Then

$$\inf_{x \in X} (f(x) - g(x)) = \inf_{x^* \in X^*} (g^*(x^*) - f^*(x^*)).$$

Exercise 2.28 Let X be a separated locally convex space, $f \in \Gamma(X)$ be such that $0 \in \operatorname{dom} f$ and $p \in \mathbb{P}$. Consider the following assertions:

(i) the mapping $\mathbb{P} \ni t \mapsto t^{-p} f(tx)$ is nondecreasing for every $x \in \operatorname{dom} f$;

(ii) $f'(x, -x) + pf(x) \leq 0$ for every $x \in \operatorname{dom} f$;

(iii) $pf(x) \leq f'(x, x)$ for every $x \in \operatorname{dom} f$;

(iv) $\langle x, x^* \rangle \geq pf(x)$ for all $(x, x^*) \in \operatorname{gr} \partial f$;

(v) for every $x \in \operatorname{int}(\operatorname{dom} f)$ there exists $x^* \in \partial f(x)$ such that $\langle x, x^* \rangle \geq pf(x)$.

Prove that (i) $\Leftrightarrow$ (ii) $\Leftrightarrow$ (iii) $\Rightarrow$ (iv). Moreover, if f is continuous at 0 then all five assertions above are equivalent.

Exercise 2.29 Let $f : X \to \overline{\mathbb{R}}$ be a function such that $\overline{\text{co}}f$ is proper. Assume that for some $\overline{x} \in \text{dom}\,\overline{\text{co}}f$ we have that $\overline{x} = \sum_{i=1}^{k} \overline{\lambda}_i \overline{x}_i$ and $\overline{\text{co}}f(\overline{x}) = \sum_{i=1}^{k} \overline{\lambda}_i f(\overline{x}_i)$ with $\overline{x}_i \in \text{dom}\, f$, $\overline{\lambda}_i > 0$ for $i \in \overline{1,k}$ and $\sum_{i=1}^{k} \overline{\lambda}_i = 1$. Prove that $\partial \overline{\text{co}}f(\overline{x}) = \bigcap_{i=1}^{k} \partial f(\overline{x}_i)$.

Exercise 2.30 Let X be a separated locally convex space and $f \in \Lambda(X)$. Assume that $0 \in \text{dom}\, f$ and $f|_{X_0}$ is continuous at 0, where $X_0 := \text{aff}(\text{dom}\, f)$. Prove that

$$\sup\{\langle x, x^*\rangle \mid x^* \in \partial f(0)\} \begin{cases} = & f'(0,x) \quad \text{if} \quad x \in X_0, \\ < & f'(0,x) = \infty \quad \text{if} \quad x \in \overline{X_0} \setminus X_0, \\ = & f'(0,x) = \infty \quad \text{if} \quad x \in X \setminus \overline{X_0}, \end{cases}$$

the supremum being attained for $x \in X_0$. Moreover, X_0 is closed if and only if

$$\forall x \in X \ : \ f'(0,x) = \sup\{\langle x, x^*\rangle \mid x^* \in \partial f(0)\}.$$

Exercise 2.31 Let X be a separated locally convex space and $f \in \Lambda(X)$ be continuous on $\text{int}(\text{dom}\, f)$, supposed to be nonempty. Prove that for all $x, y \in \text{int}(\text{dom}\, f)$ there exist $z \in]x, y[$ and $z^* \in \partial f(z)$ such that $f(y) - f(x) = \langle y - x, z^*\rangle$.

Exercise 2.32 Let X be a separated locally convex space and $f \in \Gamma(X)$.

(i) Consider the conditions: a) ∂f is single valued on $\text{dom}\,\partial f$, b) $(\partial f)^{-1}(x_1^*) \cap (\partial f)^{-1}(x_2^*) = \emptyset$ for all distinct elements $x_1^*, x_2^* \in X^*$, c) f^* is strictly convex on every segment $[x_1^*, x_2^*] \subset \text{Im}\,\partial f$, d) $\text{dom}\,\partial f = \text{int}(\text{dom}\, f)$. Prove that a) $\Leftrightarrow$ b) $\Leftrightarrow$ c); moreover, if $\text{int}(\text{dom}\, f) \neq \emptyset$ and f is continuous on $\text{int}(\text{dom}\, f)$ then a) $\Rightarrow$ d).

(ii) If f is continuous on $\text{int}(\text{dom}\, f)$ and $(\partial f)^{-1}$ is single-valued on $\text{Im}\,\partial f$, show that f is strictly convex on $\text{int}(\text{dom}\, f)$.

Exercise 2.33 Let X, Y be separated locally convex spaces and $f \in \Lambda(X \times Y)$. Prove that if f is continuous at $(x_0, y_0) \in \text{dom}\, f$, then

$$\text{Pr}_{X^*}\big(\partial f(x_0, y_0)\big) = \partial f(\cdot, y_0)(x_0) \quad \text{and} \quad \text{Pr}_{Y^*}\big(\partial f(x_0, y_0)\big) = \partial f(x_0, \cdot)(y_0).$$

Exercise 2.34 Let X be a separated locally convex space and $f_0, f_1 \in \Lambda(X)$. Consider

$$v := \inf\{f_0(x) \mid f_1(x) \le 0\}, \quad v^* := \sup_{\lambda \ge 0} \inf_{x \in X} (f_0(x) + \lambda f_1(x)),$$

with the convention $0 \cdot \infty = \infty$. Suppose that $v \in \mathbb{R}$. Prove that

$$v^* = v \Leftrightarrow [\forall \varepsilon > 0 \ : \ \inf\{f_1(x) \mid f_0(x) \le y - \varepsilon\} > 0].$$

Exercise 2.35 Let $(X, \|\cdot\|)$ be a normed linear space and $f : X \to \overline{\mathbb{R}}$ be a proper convex function. Prove that there exists $x^* \in X^*$ with $x^* \le f$ if and only if there exists $M \ge 0$ such that $f(x) + M\|x\| \ge 0$ for all $x \in X$.

Exercise 2.36 Let $(X, \|\cdot\|)$ be a normed linear space and $f_1, f_2 : X \to \overline{\mathbb{R}}$ be proper convex functions. Prove that there exist $x^* \in X^*$ and $\alpha \in \mathbb{R}$ such that $-f_1 \le x^* + \alpha \le f_2$ if and only if there exists $M \ge 0$ such that $f_1(x_1) + f_2(x_2) + M\|x_1 - x_2\| \ge 0$ for all $x_1, x_2 \in X$.

Exercise 2.37 Let X be a linear space, $(Y, \|\cdot\|)$ be a normed linear space, $T : X \to Y$ be a linear operator, $y_0 \in Y$ and $f : X \to \overline{\mathbb{R}}$ be a proper convex function. Prove that $f(x) + \|Tx + y_0\|^2 \ge 0$ for all $x \in X$ if and only if there exists $y^* \in Y^*$ such that $f(x) - 2\langle Tx + y_0, y^*\rangle - \|y^*\|^2 \ge 0$ for all $x \in X$.

Exercise 2.38 Let $(X, \|\cdot\|)$ be a normed vector space, $C, D \subset X$ be closed convex cones and $\overline{x} \in X$, $\overline{x}^* \in X^*$. Prove that

$$d(\overline{x}, C) := \inf\{\|\overline{x} - x\| \mid x \in C\} = \max\{-\langle \overline{x}, x^*\rangle \mid x^* \in U_{X^*} \cap C^+\},$$
$$d(\overline{x}^*, C^+) = \min\{\|\overline{x}^* - x^*\| \mid x^* \in C^+\} = \sup\{-\langle x, \overline{x}^*\rangle \mid x \in U_X \cap C\},$$

and $\sup_{x \in U_X \cap C} d(x, D) = \sup_{x^* \in U_{X^*} \cap D^+} d(x^*, C^+)$.

Exercise 2.39 Let $f \in \Lambda(X)$ and $\overline{x} \in \operatorname{dom} f$ be such that $f(\overline{x}) > \inf f$. Consider the sets

$$\partial^{<} f(\overline{x}) := \{x^* \in X^* \mid \langle x - \overline{x}, x^*\rangle \le f(x) - f(\overline{x})\ \forall x \in [f < f(\overline{x})]\},$$
$$\partial^{\le} f(\overline{x}) := \{x^* \in X^* \mid \langle x - \overline{x}, x^*\rangle \le f(x) - f(\overline{x})\ \forall x \in [f \le f(\overline{x})]\}.$$

Prove that $\partial^{<} f(\overline{x}) = \partial^{\le} f(\overline{x}) = [1, \infty[\cdot \partial f(\overline{x})$. The set $\partial^{<} f(\overline{x})$ is Plastria's subdifferential of f at $\overline{x}$.

Exercise 2.40 Let X be a normed space, $(a_n)_{n \ge 1} \subset X$ and $(\lambda_n)_{n \ge 1} \subset [0, \infty[$ be such that $\sum_{n=1}^{\infty} \lambda_n = 1$. Consider the function

$$f : X \to \overline{\mathbb{R}}, \qquad f(x) = \sum_{n=1}^{\infty} \lambda_n \|x - a_n\|^2.$$

1) Prove that the following statements are equivalent: (a) $\sum_{n=1}^{\infty} \lambda_n \|a_n\|^2 < \infty$ (*i.e.* $0 \in \operatorname{dom} f$), (b) $\operatorname{dom} f \ne \emptyset$, (c) $\operatorname{dom} f = X$.

2) Prove that f is finite, convex and continuous when $\sum_{n=1}^{\infty} \lambda_n \|a_n\|^2 < \infty$.

3) Suppose that $\sum_{n=1}^{\infty} \lambda_n \|a_n\|^2 < \infty$. Prove that for every $x \in X$ one has

$$\partial f(x) = \left\{ 2\sum_{n=1}^{\infty} \lambda_n \|x - a_n\| x_n^* \;\middle|\; x_n^* \in \partial \|\cdot\|(x - a_n)\ \forall n \in \mathbb{N} \right\}.$$

Exercise 2.41 Let X be a normed space and $f \in \Gamma(X)$. Prove that the following statements are equivalent:

(a) $\lim_{\|x\| \to \infty} f(x) = \infty$;

(b) $[f \le \lambda]$ is a bounded set for every $\lambda > \inf_{x \in X} f(x)$;

(c) there exists $\lambda_0 > \inf_{x \in X} f(x)$ such that $[f \le \lambda_0]$ is bounded;

(d) there exists $\alpha, \beta \in \mathbb{R}$, $\alpha > 0$, such that $f(x) \geq \alpha\|x\| + \beta$ for every $x \in X$;

(e) $\liminf_{\|x\|\to\infty} f(x)/\|x\| > 0$;

(f) $0 \in \operatorname{int}(\operatorname{dom} f^*)$.

Moreover, if $\dim X < \infty$ then the conditions above are equivalent to

(c′) $[f \leq \inf f]$ is nonempty and bounded.

Furthermore, if $p, q \in]1, \infty[$ are such that $1/p - 1/q = 1$, the following statements are equivalent:

(g) $\liminf_{\|x\|\to\infty} f(x)/\|x\|^p > 0$;

(h) $\limsup_{\|x^*\|\to\infty} f^*(x^*)/\|x^*\|^q < \infty$;

(i) there exists $\alpha, \beta \in \mathbb{R}$, $\alpha > 0$, such that $f(x) \geq \alpha\|x\|^p + \beta$ for every $x \in X$;

(j) there exists $\alpha, \beta \in \mathbb{R}$, $\alpha > 0$, such that $f^*(x^*) \leq \alpha\|x^*\|^q + \beta$ for every $x^* \in X^*$.

Exercise 2.42 Let $f, f_n : \mathbb{R}^m \to \mathbb{R}$ $(n \in \mathbb{N})$ be convex functions such that $(f_n(x)) \to f(x)$ for every $x \in \mathbb{R}^m$. Assume that f is coercive. Prove that there exist $\alpha, \beta \in \mathbb{R}$ with $\alpha > 0$, $n_0 \in \mathbb{N}$ such that $f_n(x) \geq \alpha\,\|x\| + \beta$ for all $x \in \mathbb{R}^m$ and $n \geq n_0$.

Exercise 2.43 Let $(X, \|\cdot\|)$ be a n.v.s. and $C \subset X$ be a nonempty closed convex set. Consider the following assertions:

(i) there exists $x_0^* \in X^*$ such that $\langle x, x_0^* \rangle \geq \|x\|$ for every $x \in C$;

(ii) there exist $x_0 \in X$, $x_0^* \in X^*$ such that $\langle x - x_0, x_0^* \rangle \geq \|x - x_0\|$ for every $x \in C$;

(iii) $\operatorname{int}(\operatorname{dom} s_C) \neq \emptyset$;

(iv) a) there exists $x_0^* \in X^*$ such that $\langle u, x_0^* \rangle > 0$ for every $u \in C_\infty \setminus \{0\}$ and b) for every sequence $(x_n) \subset C$ with $(\|x_n\|) \to \infty$ and $\left(\|x_n\|^{-1} x_n\right) \xrightarrow{w} u$ one has $u \neq 0$.

Prove that (i) $\Rightarrow$ (ii) $\Rightarrow$ (iii) $\Rightarrow$ (iv). If $0 \notin C$ then (ii) $\Rightarrow$ (i). Moreover, if X is a reflexive Banach space then (iv) $\Rightarrow$ (ii).

Exercise 2.44 Let X be a normed space and $f \in \Lambda(X)$ be lower bounded. Consider $\lambda > \inf_{x\in X} f(x) =: \inf f$ and $\rho > 0$. We envisage the conditions:

(a) $[f \leq \lambda] \subset \rho U$;

(b) $\forall x \in X \;:\; f(x) \geq \inf f + \dfrac{\lambda - \inf f}{2\rho} \cdot \max\{0, \|x\| - \rho\}$;

(c) $\forall x^* \in \dfrac{\lambda - \inf f}{2\rho} \cdot U_{X^*} \;:\; f^*(x^*) \leq f^*(0) + \rho\|x^*\|$.

Prove that (a) $\Rightarrow$ (b) $\Leftrightarrow$ (c).

Exercise 2.45 Let X be a normed space and $f \in \Gamma(X)$. Prove the following equivalences:

(a) $\liminf_{\|x\|\to\infty} f(x)/\|x\| > 0 \Leftrightarrow 0 \in \operatorname{int}(\operatorname{dom} f^*)$; in this case

$$\liminf_{\|x\|\to\infty} \frac{f(x)}{\|x\|} = \sup\{\mu > 0 \mid f^* \text{ is upper bounded on } \mu U^*\}.$$

(b) $\liminf_{\|x\|\to\infty} f(x)/\|x\| = 0 \Leftrightarrow 0 \in \operatorname{Bd}(\operatorname{dom} f^*)$.

(c) $\liminf_{\|x\|\to\infty} f(x)/\|x\| < 0 \Leftrightarrow 0 \notin \operatorname{cl}(\operatorname{dom} f^*)$; in this case

$$\liminf_{\|x\|\to\infty} \frac{f(x)}{\|x\|} = -d_{\operatorname{dom} f^*}(0).$$

Exercise 2.46 Let $p \in]1, \infty[$ and

$$f : \ell^p \to \overline{\mathbb{R}}, \qquad f\big((x_n)_{n\geq 1}\big) := \sum\nolimits_{n=1}^{\infty} n|x_n|^n.$$

Prove that f is finite, convex, continuous and $\lim_{\|y\|\to\infty} f^*(y)/\|y\| = 1$.

Exercise 2.47 Let X, Y be Hilbert spaces, $C \subset X$ be a nonempty closed convex set and $A \in \mathcal{L}(X, Y)$ be a surjective operator. Consider the problem

(P) $\quad \min \frac{1}{2}\|Ax\|^2,\ x \in C.$

A solution $\overline{x}$ of problem (P) is called a *spline function* in C associated to A.

(a) Prove that (P) has optimal solutions if $C + \ker A$ is closed.

(b) Prove that $\overline{x} \in C$ is an optimal solution of (P) if and only if $\overline{y} := A^*(A\overline{x})$ satisfies the relation $(\overline{x} \mid \overline{y}) = \min\{(x \mid \overline{y}) \mid x \in C\}$.

Exercise 2.48 Let X, Y be Hilbert spaces, $C := \{x \in X \mid \forall i, 1 \leq i \leq k : (x \mid a_i) \leq \beta_i\}$, where $a_1, \ldots, a_k \in X$, $\beta_1, \ldots, \beta_k \in \mathbb{R}$ and $A \in \mathcal{L}(X, Y)$ be a surjective operator. Consider the problem

(P) $\quad \min \frac{1}{2}\|Ax\|^2,\ x \in C.$

Prove that (P) has optimal solutions. Suppose, moreover, that there exists $\tilde{x} \in X$ such that $(\tilde{x} \mid a_i) < \beta_i$ for every i, $1 \leq i \leq k$. Prove that $\overline{x}$ is a solution of (P) if and only if there exists $(\lambda_i)_{1\leq i\leq k} \subset \mathbb{R}_+$ such that

$$A^*(A\overline{x}) = \lambda_1 a_1 + \cdots + \lambda_k a_k \ \text{ and } \forall i, 1 \leq i \leq k \ : \ \lambda_i((\overline{x} \mid a_i) - \beta_i) = 0.$$

Exercise 2.49 Let X be a Hilbert space and a_1, a_2, a_3 be three non colinear elements of X (*i.e.* they are not situated on the same straight-line). Prove that there exists a unique element $\overline{x} \in X$ such that

$$\forall x \in X \ : \ \|\overline{x} - a_1\| + \|\overline{x} - a_2\| + \|\overline{x} - a_3\| \leq \|x - a_1\| + \|x - a_2\| + \|x - a_3\|.$$

Moreover, prove that $\overline{x} \in \operatorname{co}\{a_1, a_2, a_3\}$. The point $\overline{x}$ is called the *Toricelli point* of the triangle of vertices a_1, a_2, a_3.

Exercise 2.50 Determine the optimal solutions (when they exist) and the value of the problem

$$(P_i^\mu) \qquad \min \int_0^1 \left(tx(t) + \mu\sqrt{1 + (u(t))^2}\right) dt, \; x \in X_i,$$

for every $\mu \in \mathbb{R}_+$ and every $i \in \{1, 2, 3, 4\}$, where

$$X_1 := C[0,1], \qquad X_2 := L^1(0,1),$$

$$X_3 := \left\{x \in C[0,1] \;\middle|\; \int_0^1 x(t)\, dt = 0\right\}, \quad X_4 := \left\{x \in L^1(0,1) \;\middle|\; \int_0^1 x(t)\, dt = 0\right\}.$$

Exercise 2.51 Consider the (convex) optimization problem

$$(P) \qquad \max \int_0^1 x(t)\, dt, \; x \in X, \; x(0) = x(1) = 0, \; \int_0^1 \sqrt{1 + (x'(t))^2}\, dt \le L,$$

where $L > 0$ and $X = C^1[0,1] := \{x : [0,1] \to \mathbb{R} \mid x \text{ derivable, } x' \text{ continuous on } [0,1]\}$ or $X = AC[0,1] := \{x : [0,1] \to \mathbb{R} \mid x \text{ absolutely continuous}\}$. Determine the optimal solutions of problem (P), when they exist, and its value (using, eventually, the dual problem).

2.12 Bibliographical Notes

Many results of this chapter are well-known and can be found in several books treating convex analysis: [Rockafellar (1970); Hiriart-Urruty and Lemaréchal (1993)] (for finite dimensional spaces), [Ekeland and Temam (1974); Ioffe and Tikhomirov (1974); Barbu and Precupanu (1986); Castaing and Valadier (1977); Phelps (1989); Azé (1997)]. In the sequel we point only those results that are not contained in these books but the last one.

Sections 2.1-2.5: The assertion (vii) of Theorem 2.1.5 was established in [Zălinescu (1983b)]. The characterization of convex functions using upper Dini directional derivatives in Theorem 2.1.16 as well as Lemma 2.1.15 are established by Luc and Swaminathan (1993) Theorem 2.2.2 was established by Crouzeix (1980) in finite dimensional spaces and Theorem 2.2.11 by Zălinescu (1978). The notion of quasi-continuous function was introduced by Joly and Laurent (1971); Proposition 2.2.15 and Corollary 2.2.16 are established by Moussaoui and Volle (1997). The notions of cs-convex, cs-closed and cs-complete functions are introduced by Simons (1990), while that of lcs-closed function by Amara and Ciligot-Travain (1999); all the results concerning lcs-closed functions are due to Amara and Ciligot-Travain (1999). Theorem 2.2.22 (for $I = \mathbb{N}$) and Proposition 2.2.24 can be found in [Kosmol and Müller-Wichards (2001)]; when $\dim X < \infty$ Theorem 2.2.22, Corollary 2.2.23 and Proposition 2.2.24 are stated in [Marti (1977)] under the weaker hypothesis that the pointwise boundedness or pointwise convergence holds on a dense subset of C. Proposition 2.2.17 is stated in [Zălinescu (1992b)]. Proposition 2.4.3 is established by Rockafellar (1966), Theorem 2.4.11 is proved in the book [Rockafellar (1970)] in $\mathbb{R}^n$ and stated by Hiriart-Urruty

(1982) in the general case; one can find another proof in [Azé (1997)]. The local boundedness of ∂f in Theorem 2.4.13 and Corollary 2.4.10 can be found in many books on convex analysis. The assertions (iv) and (vii) of Theorem 2.4.14 are established in [Zălinescu (1980)]; for (i) see also [Anger and Lembcke (1974)]. Theorem 2.4.18 is from the book [Ioffe and Tikhomirov (1974)]. Theorem 2.5.2 was proved by Polyak (1966) under the more stringent conditions that f is a quasi-convex function which is bounded and Lipschitz on bounded sets and attains its infimum on every closed convex subset of X. Theorem 2.5.5 and Lemma 2.5.3 are established by Borwein and Kortezov (2001), while Lemma 2.5.4 and other results on non-attaining convex functionals are established by Adly *et al.* (2001a).

Sections 2.6-2.9: The systematic use of perturbation functions for calculating conjugates and subdifferentials was done for the first time by Rockafellar (1974). The author of this book continued this approach in [Zălinescu (1983a); Zălinescu (1987); Zălinescu (1989); Zălinescu (1992a); Zălinescu (1992b); Zălinescu (1999)]; this permitted, for example, to give simpler proofs to several results stated in [Kutateladze (1977); Kutateladze (1979a); Kutateladze (1979b)]. Theorem 2.6.2(i) was established by Moussaoui and Seeger (1994), but Theorem 2.6.2(ii) and Theorem 2.6.3 are new. Corollaries 2.6.4–2.6.7 can be found in [Hiriart-Urruty and Phelps (1993)] and [Moussaoui and Seeger (1994)]; for other results in this direction see the survey paper [Hiriart-Urruty *et al.* (1995)].

The sufficient conditions for the fundamental duality formula and for the validity of the formulas for conjugates and ε-subdifferentials are, mainly, those from author's survey paper [Zălinescu (1999)], where one can find detailed historical notes; here we mention only the first use (to our knowledge) of them; actually, all these sufficient conditions are mentioned in that paper, excepting those which use li-convex or lcs-closed functions. So, conditions (iii) and (viii) of Theorem 2.7.1 and the corresponding ones in Theorems 2.8.1, 2.8.3, 2.8.7 and 2.8.10 are the classical ones and can be found in all the mentioned books which treat them. Condition (i) of Theorem 2.7.1 was used by Rockafellar (1974) when $Y_0 = Y$ and by Zălinescu (1998) in the present form (see also [Combari *et al.* (1999)]), (ii) and (vi) were introduced in [Zălinescu (1983a)] for $Y_0 = Y$ (and $\mathbb{R}$ replaced by a separated lcs ordered by a normal cone) and in [Zălinescu (1999)] in the present form, (iv) was introduced in [Zălinescu (1992b)] with (Hwx) replaced by (Hx), (v) is stated by Amara and Ciligot-Travain (1999) for X, Y Fréchet spaces, Φ a lcs-closed function and ^{ib} replaced by ^{ic}, (vii) was introduced by Rockafellar (1974) for X, Y Banach spaces (X even reflexive) and $Y_0 = Y$, and by Zălinescu (1987) in the present form, while (ix) was used by Joly and Laurent (1971) for Φ lsc and by Moussaoui and Volle (1997) for arbitrary Φ; note that condition (b) of [Cominetti (1994), Th. 2.2] implies condition (2.54). Theorem 2.7.4(iv) was established in [Zălinescu (1992b)]; the other conditions were introduced in [Zălinescu (1999)] (but in (v) it was assumed that F is lsc and $\mathcal{C}$ is closed).

Condition (iii) of Theorem 2.8.1 was introduced in [Ekeland and Temam (1974)], (vii) was introduced in [Zălinescu (1984)] (for $Y_0 = Y$), (iv) was introduced in [Zălinescu (1992b)], (i) and (viii) in [Zălinescu (1998)] for X, Y normed spaces, while conditions (ii) and (vi) were introduced in [Zălinescu (1999)].

Condition (vii) of Theorem 2.8.3 was introduced by Borwein (1983) for $Y_0 = Y$ (see also [Zălinescu (1986)]), (x) was introduced by Borwein and Lewis (1992), (ix) was introduced by Moussaoui and Volle (1997), (i) by Zălinescu (1998) for X, Y normed spaces, (ii) was introduced by Combari *et al.* (1999), (iv) was introduced by Zălinescu (1999) (a slightly stronger form was used in [Simons (1990)]) as well as conditions (v) and (vi).

For $\mathcal{C} \in \mathcal{L}(X, Y)$, generally, Theorem 2.8.6 is obtained under the corresponding conditions of Theorem 2.8.3 (taking $f = 0$). For $\mathcal{C}$ a densely defined closed linear operator, Rockafellar (1974) and Hiriart-Urruty Hiriart-Urruty (1982) obtained the results for g lsc, X, Y Banach spaces and $Y_0 = Y$ (in [Rockafellar (1974)] X is reflexive and $\varepsilon = 0$); Azé (1994) obtained the formula for the conjugate under condition (ii) in normed spaces. For general $\mathcal{C}$ condition (iv) was used by Zălinescu (1992b); the other conditions (excepting (v)) were used in [Zălinescu (1999)].

Condition (ix) of Theorem 2.8.7 was used by Joly and Laurent (1971), (vii) for X a Banach space was introduced by Attouch and Brézis (1986), (vi) was introduced by Zălinescu (1992b), (i) was introduced by Azé (1994) in an equivalent form in normed spaces, while (ii) was introduced by Combari *et al.* (1999).

Corollary 2.8.8 was obtained by Volle (1994) for X a Banach space, f, g lsc and ^{ib} replaced by ^{ic}. Formula (2.65) from Proposition 2.8.9 was obtained by Azé (1994) and Simons (1998b) for $x = 0 \in (\operatorname{dom} f - \operatorname{dom} g)^i$ when f, g are lsc.

The case $f = 0$ of Theorem 2.8.10 was considered by several authors; condition (iii) was used in [Hiriart-Urruty (1982); Lemaire (1985); Zălinescu (1983a); Zălinescu (1984)]; the algebraic case was considered in [Kutateladze (1979a); Kutateladze (1979b); Zălinescu (1983a)]; Zălinescu (1984) considered a stronger version of (v) (for lsc functions and $Y_0 = Y$); note that in these papers g is assumed to be Q-increasing on the entire space. The general case was considered in [Combari *et al.* (1994)] under (iii) and a condition stronger than (v) (f, g, H lsc and ^{ic} instead of ^{ib}) and in [Combari *et al.* (1999)] under condition (i) and a variant of (iv); it was also considered by Moussaoui and Volle (1997) under condition (vii).

The formula for $\partial\varphi(x)$ in Corollary 2.8.11 can be found in [Combari *et al.* (1994)]. The formula for $\partial\varphi(x)$ in Corollary 2.8.13 is stated in [Volle (1992); Volle (1993)] under condition (vii) of Theorem 2.8.7 and in [Combari *et al.* (1994)] for f, g lower semicontinuous (even for an arbitrary finite number of functions) and $(0,0) \in {}^{ic}(\{(x,x) \mid x \in X\} - \operatorname{dom} f \times \operatorname{dom} g)$. The formula for $(f_1 \Diamond f_2)^*$ in Corollary 2.8.14 is stated by Seeger and Volle (1995); the formulas for $\partial(f_1 \Diamond f_2)(\overline{x})$ are stated in this paper for f_1 and f_2 continuous at $\overline{x}_1$ and $\overline{x}_2$, respectively.

The results from Section 2.9 are classical. Note that the formula (2.79) for the normal cone of a level set is given generally for a finite valued continuous convex function; in the present form, for $\gamma = g(\overline{x})$, it can be found in [Penot and Zălinescu (2000)].

Section 2.10: The minimax theorem is also classical and can be found in the books [Barbu and Precupanu (1986); Simons (1998a)].

Exercises: Exercise 2.2 is from [Penot and Bougeard (1988)], Exercise 2.3 is from [Crouzeix (1981)], Exercise 2.6 is from [Marti (1977)], Exercise 2.15 is from [Crouzeix *et al.* (1992)] (there in a more complete form), Exercises 2.18 and 2.29 are from [Hiriart-Urruty and Lemaréchal (1993)], Exercise 2.19 is from [Lions and Rochet (1986)], Exercise 2.20 is the celebrated cancellation lemma from [Hörmander (1955)], Exercise 2.21 is from [Hiriart-Urruty (1998)], Exercise 2.22 is from [Jourani (2000)], the assertions in Exercise 2.23 are well-known (see also [Azé (1997)]), Exercise 2.25 can be found in [Borwein and Kortezov (2001)] (in normed spaces), the formula for g_∞ under condition (vii) of Theorem 2.8.1 in Exercise 2.26 is obtained in [Amara and Ciligot-Travain (1999)] and [Amara (1998)], Exercise 2.27 is the well-known Toland–Singer duality formula (see [Toland (1978)] and [Singer (1979)]), Exercise 2.30 is from [Zălinescu (1999)], the equivalence of a) and c) (for Banach spaces) in Exercise 2.32 can be found in [Bauschke *et al.* (2001)] (see also [Barbu and Da Prato (1985)]), Exercises 2.33 and 2.34 are from [Eremin and Astafiev (1976)], Exercises 2.35, 2.36 and 2.37 are from [Simons (1998a)], the last formula in Exercise 2.38 is from [Walkup and Wets (1967)], Exercise 2.39 is from [Penot (1998a)], the formula for the subdifferential of the function considered in Exercise 2.40 and its proof are from [Aussel *et al.* (1995)], the equivalences (e) $\Leftrightarrow$ (f) and (g) $\Leftrightarrow$ (h) of Exercise 2.41 are from [Zălinescu (1983b)], the equivalence of conditions (ii)-(iv) in Exercise 2.43 are established in [Adly *et al.* (2001b)] in reflexive Banach spaces, a weaker variant of the implication (a) $\Rightarrow$ (c) of Exercise 2.44 is stated in [Azé and Rahmouni (1996)], Exercise 2.45(c) is from [Borwein and Vanderwerff (1995)], Exercises 2.47, 2.48 are from [Laurent (1972)].

Chapter 3

Some Results and Applications of Convex Analysis in Normed Spaces

3.1 Further Fundamental Results in Convex Analysis

Throughout this chapter X, Y are normed spaces and X^*, Y^* are their duals endowed with the dual norms.

As an application of Ekeland's variational principle and of some results relative to convex functions, we state the following multi-propose generalization of the Brøndsted–Rockafellar theorem.

Theorem 3.1.1 (Borwein) *Let X be a Banach space and $f \in \Gamma(X)$. Consider $\varepsilon \in \mathbb{P}$, $x_0 \in \operatorname{dom} f$, $x_0^* \in \partial_\varepsilon f(x_0)$ and $\beta \in \mathbb{R}_+$. Then there exist $x_\varepsilon \in X$, $y_\varepsilon^* \in U_{X^*}$ and $\lambda_\varepsilon \in [-1, +1]$ such that*

$$\|x_\varepsilon - x_0\| + \beta \cdot |\langle x_\varepsilon - x_0, x_0^* \rangle| \leq \sqrt{\varepsilon}, \tag{3.1}$$

$$x_\varepsilon^* := x_0^* + \sqrt{\varepsilon}(y_\varepsilon^* + \beta \lambda_\varepsilon x_0^*) \in \partial f(x_\varepsilon). \tag{3.2}$$

Moreover

$$\|x_\varepsilon - x_0\| \leq \sqrt{\varepsilon}, \quad \|x_\varepsilon^* - x_0^*\| \leq \sqrt{\varepsilon}(1 + \beta\|x_0^*\|), \tag{3.3}$$

$$|\langle x_\varepsilon - x_0, x_\varepsilon^* \rangle| \leq \varepsilon + \sqrt{\varepsilon}/\beta, \tag{3.4}$$

$$x_\varepsilon^* \in \partial_{2\varepsilon} f(x_0), \quad |f(x_\varepsilon) - f(x_0)| \leq \varepsilon + \sqrt{\varepsilon}/\beta, \tag{3.5}$$

with the convention $1/0 = \infty$.

Proof. The function

$$\|\cdot\|_0 : X \to \mathbb{R}, \quad \|x\|_0 := \|x\| + \beta \cdot |\langle x, x_0^* \rangle|,$$

is, obviously, a norm on X, equivalent to the initial norm. Therefore $(X, \|\cdot\|_0)$ is a Banach space. Consider the function $g := f - x_0^*$. It is obvious that $x_0 \in \operatorname{dom} g = \operatorname{dom} f = \operatorname{dom} \partial_\varepsilon f$ and g is lsc and lower bounded:

$$\forall x \in X \;:\; g(x) \geq g(x_0) - \varepsilon \quad [\Leftrightarrow x_0^* \in \partial_\varepsilon f(x_0)\,].$$

We apply Ekeland's theorem (Theorem 1.4.1) to g, $\sqrt{\varepsilon}$ and the metric d given by $d(x,y) := \|x-y\|_0$. So there exists $x_\varepsilon \in \operatorname{dom} g$ such that

$$g(x_\varepsilon) + \sqrt{\varepsilon} \cdot \|x_\varepsilon - x_0\|_0 \leq g(x_0), \tag{3.6}$$
$$\forall x \in X, x \neq x_\varepsilon \;:\; g(x_\varepsilon) < g(x) + \sqrt{\varepsilon} \cdot \|x - x_\varepsilon\|_0. \tag{3.7}$$

From Eq. (3.6) we obtain that

$$g(x_\varepsilon) + \sqrt{\varepsilon}(\|x_\varepsilon - x_0\| + \beta \cdot |\langle x_\varepsilon - x_0, x_0^*\rangle|) \leq g(x_0) \leq g(x_\varepsilon) + \varepsilon,$$

whence Eq. (3.1) follows immediately.

Let us consider the function $h : X \to \overline{\mathbb{R}}$,

$$h(x) := g(x) + \sqrt{\varepsilon} \cdot \|x - x_\varepsilon\|_0 = f(x) - \langle x, x_0^*\rangle + \sqrt{\varepsilon} \cdot \|x - x_\varepsilon\| + \beta\sqrt{\varepsilon} \cdot |\langle x - x_\varepsilon, x_0^*\rangle|;$$

by Eq. (3.7) x_ε is a minimum point of h. Therefore $0 \in \partial h(x_\varepsilon)$. Since h is the sum of four convex functions, three of them being continuous, and taking into account the expression of the subdifferential of a norm (Corollary 2.4.16) and of the absolute value of a linear functional (at the end of Section 2.8), we have that

$$0 \in \partial h(x_\varepsilon) = \partial f(x_\varepsilon) - x_0^* + \sqrt{\varepsilon} \cdot U_{X^*} + \beta\sqrt{\varepsilon} \cdot [-1, +1] \cdot x_0^*.$$

Therefore there exists $x_\varepsilon^* \in \partial f(x_\varepsilon)$, $y_\varepsilon^* \in U_{X^*}$ and $\lambda_\varepsilon \in [-1, 1]$ such that Eq. (3.2) is verified.

The estimations from Eq. (3.3) follow immediately from Eqs. (3.1) and (3.2). Using again Eqs. (3.1) and (3.2) we obtain that

$$\begin{aligned} |\langle x_\varepsilon - x_0, x_\varepsilon^* - x_0^*\rangle| &= \sqrt{\varepsilon} \cdot |\langle x_\varepsilon - x_0, y_\varepsilon^* + \beta\lambda_\varepsilon x_0^*\rangle| \\ &\leq \sqrt{\varepsilon}(\|x_\varepsilon - x_0\| \cdot \|y_\varepsilon^*\| + \beta|\lambda_\varepsilon| \cdot |\langle x_\varepsilon - x_0, x_0^*\rangle|) \\ &\leq \sqrt{\varepsilon}(\|x_\varepsilon - x_0\| + \beta \cdot |\langle x_\varepsilon - x_0, x_0^*\rangle|) \\ &\leq \sqrt{\varepsilon} \cdot \sqrt{\varepsilon} = \varepsilon. \end{aligned} \tag{3.8}$$

From Eqs. (3.1) and (3.8) we get

$$|\langle x_\varepsilon - x_0, x_\varepsilon^*\rangle| \leq |\langle x_\varepsilon - x_0, x_\varepsilon^* - x_0^*\rangle| + |\langle x_\varepsilon - x_0, x_0^*\rangle| \leq \varepsilon + \sqrt{\varepsilon}/\beta,$$

i.e. the estimation in Eq. (3.4) holds, too. Since $x_0^* \in \partial_\varepsilon f(x_0)$ and $x_\varepsilon^* \in \partial f(x_\varepsilon)$, we get

$$\begin{aligned}\langle x_0 - x_\varepsilon, x_\varepsilon^* - x_0^*\rangle + \langle x_0 - x_\varepsilon, x_0^*\rangle = \langle x_0 - x_\varepsilon, x_\varepsilon^*\rangle &\le f(x_0) - f(x_\varepsilon)\\ &\le \langle x_0 - x_\varepsilon, x_0^*\rangle + \varepsilon.\end{aligned}$$

Using relations (3.1) and (3.8), we obtain that

$$|f(x_0) - f(x_\varepsilon)| \le |\langle x_0 - x_\varepsilon, x_0^*\rangle| + \varepsilon \le \varepsilon + \sqrt{\varepsilon}/\beta,$$

i.e. the second relation in (3.5) holds. Using Eq. (3.8) and the fact that $x_0^* \in \partial_\varepsilon f(x_0)$, $x_\varepsilon^* \in \partial f(x_\varepsilon)$, we obtain that

$$\begin{aligned}\langle x - x_0, x_\varepsilon^*\rangle &= \langle x - x_\varepsilon, x_\varepsilon^*\rangle + \langle x_\varepsilon - x_0, x_\varepsilon^* - x_0^*\rangle + \langle x_\varepsilon - x_0, x_0^*\rangle\\ &\le f(x) - f(x_\varepsilon) + \varepsilon + f(x_\varepsilon) - f(x_0) + \varepsilon\\ &= f(x) - f(x_0) + 2\varepsilon\end{aligned}$$

for every $x \in X$, *i.e.* $x_\varepsilon \in \partial_{2\varepsilon} f(x_0)$; hence the first relation in Eq. (3.5) holds, too. □

The next result is well-known. The first part is an immediate consequence of Borwein's theorem, while the density part, which follows easily from the first one, will be reinforced in Theorem 3.1.4 below.

Theorem 3.1.2 (Brøndsted–Rockafellar) *Let X be a Banach space and $f \in \Gamma(X)$. Consider $\varepsilon \ge 0$ and $(x_0, x_0^*) \in \operatorname{gr} \partial_\varepsilon f$. Then there exists $(x_\varepsilon, x_\varepsilon^*) \in \operatorname{gr} \partial f$ such that $\|x_\varepsilon - x_0\| \le \sqrt{\epsilon}$ and $\|x_\varepsilon^* - x_0^*\| \le \sqrt{\epsilon}$. In particular* $\operatorname{dom} f \subset \operatorname{cl}(\operatorname{dom} \partial f)$ *and* $\operatorname{dom} f^* \subset \operatorname{cl}(\operatorname{Im} \partial f)$. □

Another consequence of Borwein's theorem is the following result which will be completed in Proposition 3.1.10 below.

Proposition 3.1.3 *Let X be a Banach space and $f \in \Gamma(X)$. Then*

$$\begin{aligned}f(x) &= \sup\{\langle x - z, z^*\rangle + f(z) \mid (z, z^*) \in \operatorname{gr} \partial f\}\\ &= \sup\{\langle x, z^*\rangle - f^*(z^*) \mid z^* \in \operatorname{Im}(\partial f)\}\end{aligned} \tag{3.9}$$

for every $x \in \operatorname{dom} f$.

Proof. The second equality in Eq. (3.9) is obvious because $f(z)+f^*(z^*) = \langle z, z^*\rangle$ for any $(z, z^*) \in \operatorname{gr} \partial f$. Let $x \in \operatorname{dom} f$ be fixed. When $(z, z^*) \in \operatorname{gr} \partial f$ we have $\langle x - z, z^*\rangle + f(z) \le f(x)$ for every $x \in X$; hence $f(x) \ge \sup\{\langle x - z, z^*\rangle + f(z) \mid (z, z^*) \in \operatorname{gr} \partial f\}$. Because f is lsc, by Theorem 2.4.4 (iii), there exists $x^* \in \partial_{\varepsilon/2} f(x)$. Applying Borwein's theorem we get $(x_\varepsilon, x_\varepsilon^*) \in \operatorname{gr} \partial f$ such that $x_\varepsilon^* \in \partial_\varepsilon f(x)$. Therefore $\langle x_\varepsilon - x, x_\varepsilon^* \rangle \le f(x_\varepsilon) - f(x) + \varepsilon$, whence $f(x) - \varepsilon \le \langle x - x_\varepsilon, x_\varepsilon^* \rangle + f(x_\varepsilon)$. Hence relation (3.9) holds. □

As announced before, the density results in Theorem 3.1.2 can be stated in a stronger form. In the next result $(x_n) \to_f x$ means that $(x_n) \to x$ and $(f(x_n)) \to f(x)$, and similarly for $(x_n^*) \to_{f^*} x^*$.

Theorem 3.1.4 *Let X be a Banach space and $f \in \Gamma(X)$. Then*

(i) $\forall x \in \operatorname{dom} f,\ \exists\, ((x_n, x_n^*)) \subset \operatorname{gr} \partial f \ :\ (x_n) \to_f x$;

(ii) $\forall x^* \in \operatorname{dom} f^*,\ \exists\, ((x_n, x_n^*)) \subset \operatorname{gr} \partial f \ :\ (x_n^*) \to_{f^*} x^*$.

Proof. (i) Consider $x \in \operatorname{dom} f$. Because f is lsc, by Theorem 2.4.4 (iii), for every $n \in \mathbb{N}$ there exists $\overline{x}_n^* \in \partial_{n^{-2}} f(x)$. Applying Borwein's theorem for $(x, \overline{x}_n^*)$, $\varepsilon = n^{-2}$ and $\beta = 1$, we get $(x_n, x_n^*) \in \operatorname{gr} \partial f$ such that $\|x_n - x\| \le n^{-1}$, $|f(x_n) - f(x)| \le n^{-2} + n^{-1}$. Hence $(x_n) \to_f x$.

(ii) Because f^* is lsc, it is sufficient to show that for every $\varepsilon > 0$ there exists $(y, y^*) \in \partial f$ such that $\|y^* - x^*\| \le \varepsilon$ and $f^*(y^*) \le f^*(x^*) + \varepsilon$. Fix $\overline{x} \in \operatorname{dom} f$. Let $\varepsilon > 0$ and take $r := \|\overline{x}\| + 2\varepsilon^{-1} (\varepsilon + f(\overline{x}) + f^*(x^*) - \langle \overline{x}, x^* \rangle) > 0$. Consider also the function $g := f \Box (x^* + \iota_{rU_X}) \in \Lambda(X)$. Then $g(x) \ge (f \Box x^*)(x) = \langle x, x^* \rangle - f^*(x^*)$ for every $x \in X$. It follows that $\overline{g} \in \Gamma(X)$ and $f(\overline{x}) + \langle -\overline{x}, x^* \rangle \ge \overline{g}(0) \ge -f^*(x^*)$. By Proposition 3.1.3 there exists $(z, z^*) \in \partial \overline{g}$ such that $-g^*(z^*) = \overline{g}(z) - \langle z, z^* \rangle > -f^*(x^*) - \varepsilon/2$. But $g^*(z^*) = f^*(z^*) + (x^* + \iota_{rU_X})^*(z^*) = f^*(z^*) + r\|x^* - z^*\|$. Therefore

$$f^*(z^*) + r\|x^* - z^*\| < f^*(x^*) + \varepsilon/2.$$

Because $f^*(z^*) \ge \langle \overline{x}, z^* \rangle - f(\overline{x}) \ge -\|\overline{x}\| \cdot \|z^* - x^*\| + \langle \overline{x}, x^* \rangle - f(\overline{x})$, from the preceding inequality we obtain that $(r - \|\overline{x}\|)\|z^* - x^*\| < \varepsilon + f(\overline{x}) + f^*(x^*) - \langle \overline{x}, x^* \rangle$, and so $\|z^* - x^*\| < \varepsilon/2$. The inequality above shows also that $f^*(z^*) < f^*(x^*) + \varepsilon/2$. On the other hand, because $(z, \overline{g}(z)) \in \operatorname{cl}(\operatorname{epi} g)$, there exists $(z_n) \subset \operatorname{dom} f$ and $(u_n) \subset rU_X$ such that $(z_n + u_n) \to z$ and

$$\begin{aligned} \limsup \left(f(z_n) + \langle u_n, x^* \rangle\right) &\le \overline{g}(z) = \langle z, z^* \rangle - g^*(z^*) \\ &= \langle z, z^* \rangle - f^*(z^*) - r\|x^* - z^*\|. \end{aligned} \tag{3.10}$$

Setting $\varepsilon_n := f(z_n) + f^*(z^*) - \langle z_n, z^* \rangle$, we have that

$$0 \le \varepsilon_n = f(z_n) + \langle u_n, x^* \rangle - \langle z_n + u_n, z^* \rangle + \langle u_n, z^* - x^* \rangle + f^*(z^*).$$

Taking into account that $\langle u_n, z^* - x^* \rangle \le r\|x^* - z^*\|$, from Eq. (3.10) we obtain that $\limsup \varepsilon_n \le 0$. Hence $\lim \varepsilon_n = 0$. Because (u_n) is bounded, so is (z_n). Therefore there exists $r' > 0$ such that $(z_n) \subset r'U_X$. Take $n \in \mathbb{N}$ so that $\delta := \varepsilon_n \le 1$ and $\sqrt{\delta}(r' + 2)(1 + \|z^*\|) < \varepsilon/2$; set $z := z_n$. Of course, we have that $z^* \in \partial_\delta f(z)$. By Borwein's theorem (for $\beta = 1$) there exists $(y, y^*) \in \partial f$ such that $\|z - y\| \le \sqrt{\delta}$, $\|z^* - y^*\| \le \sqrt{\delta}(1 + \|z^*\|)$ and $|f(z) - f(y)| \le \delta + \sqrt{\delta}$. Hence $\|y^* - x^*\| \le \sqrt{\delta}(1 + \|z^*\|) + \varepsilon/2 < \varepsilon$ and

$$\begin{aligned} f^*(y^*) &= \langle y, y^* \rangle - f(y) \\ &= \langle y, y^* - z^* \rangle + \langle y - z, z^* \rangle - \delta + f^*(z^*) + f(z) - f(y) \\ &\le \sqrt{\delta}(\sqrt{\delta} + r')(1 + \|z^*\|) + \sqrt{\delta}\,\|z^*\| - \delta + f^*(x^*) + \varepsilon/2 + \delta + \sqrt{\delta} \\ &\le f^*(x^*) + \varepsilon/2 + \sqrt{\delta}(r' + 2)(1 + \|z^*\|) < f^*(x^*) + \varepsilon. \end{aligned}$$

The proof is complete. □

Using Brøndsted–Rockafellar's theorem we can add other sufficient conditions for the validity of the conclusions of Theorems 2.8.1 and 2.8.7.

Proposition 3.1.5 *Let X, Y be Banach spaces, $F \in \Gamma(X \times Y)$, $A \in \mathcal{L}(X, Y)$, $D = \{Ax - y \mid (x, y) \in \operatorname{dom} F\}$ and $E = \{Ax - y \mid (x, y) \in \operatorname{dom} \partial F\}$. Then*

$$^{ic}E = \operatorname{ri} E = {}^{ic}(\operatorname{co} E) = \operatorname{ri}(\operatorname{co} E) = {}^{ic}D = \operatorname{ri} D.$$

Moreover, if one of the above sets is nonempty then $\overline{^{ic}E} = \overline{E} = \overline{D}$; in particular the sets ^{ic}E and $\overline{E}$ are convex.

Proof. Consider the linear operator $T : X \times Y \to Y$ defined by $T(x, y) := Ax - y$. Since $\operatorname{dom} \partial F \subset \operatorname{dom} F$ and $\operatorname{dom} F$ is convex, we have that

$$E = T(\operatorname{dom} \partial F) \subset \operatorname{co} E = T\big(\operatorname{co}(\operatorname{dom} \partial F)\big) \subset D = T(\operatorname{dom} F).$$

By Theorem 3.1.2 we have that $\operatorname{dom} F \subset \overline{\operatorname{dom} \partial F}$. Hence, using the continuity of T, we have that

$$E \subset D \subset T\big(\overline{\operatorname{dom} \partial F}\big) \subset \overline{T(\operatorname{dom} \partial F)} = \overline{E}.$$

Using the properties of the affine hull (see p. 2), it follows that $\operatorname{aff} E \subset \operatorname{aff} D \subset \overline{\operatorname{aff} E}$. Therefore, if $\operatorname{aff} E$ is closed then $\operatorname{aff} D = \operatorname{aff} E$; the above relation shows that $^{ic}E \subset {}^{ic}D$. Let us show the converse inclusion. Let $y_0 \in {}^{ic}D$ and consider the function $F_0 \in \Gamma(X \times Y)$, $F_0(x, y) := F(x, y - y_0)$. Of course, $\operatorname{dom} F_0 = \operatorname{dom} F + (0, y_0)$ and $\operatorname{dom} \partial F_0 = \operatorname{dom} \partial F + (0, y_0)$. Therefore $0 \in {}^{ic}\big(\operatorname{Pr}_Y(\operatorname{dom} F_0)\big)$. Taking $\varphi_0(x) := F_0(x, Ax)$, $\varphi_0 \in \Gamma(X)$. Hence $\operatorname{dom} \partial\varphi_0 \ne \emptyset$, which shows, by Theorem 2.8.1, that there exists x

such that $\partial F_0(x, Ax) = \partial F(x, Ax - y_0) \neq \emptyset$. Therefore $y_0 \in E$. So we obtained that ${}^{i}D = {}^{ic}D \subset E \subset D$, which shows that $\operatorname{aff} E = \operatorname{aff} D$ and ${}^{i}D \subset {}^{i}E$. It follows that ${}^{ic}E = {}^{ic}D = {}^{ic}(\operatorname{co} E)$. As observed after the proof of Theorem 2.8.1, in our situation, if ${}^{ic}D$ is nonempty we have that ${}^{ic}D = \operatorname{rint} D$.

Suppose that ${}^{ic}D \neq \emptyset$ (for example). From what was proved above, we have that

$$\operatorname{rint} D = {}^{i}D = {}^{ic}D = \operatorname{rint} E = {}^{i}E = {}^{ic}E \subset E \subset D \subset \overline{E}.$$

Hence $\overline{E} = \overline{{}^{ic}E} = \overline{D}$. The conclusion follows. □

Remark 3.1.1 Taking into account the preceding proposition, for X, Y Banach spaces and $F \in \Gamma(X \times Y)$, we may add to the sufficient conditions in Theorem 2.8.1 the following conditions:

$$0 \in \operatorname{ri}\{Ax - y \mid (x, y) \in \operatorname{dom} \partial F\},$$
$$0 \in {}^{ic}\{Ax - y \mid (x, y) \in \operatorname{dom} \partial F\},$$
$$0 \in {}^{ic}\{Ax - y \mid (x, y) \in \operatorname{co}(\operatorname{dom} \partial F)\},$$
$$Y_0 = \operatorname{cone}\{Ax - y \mid (x, y) \in \operatorname{dom} \partial F\} \text{ is a closed linear space.}$$

Indeed, the first three conditions are, evidently (using the preceding proposition), equivalent to $0 \in {}^{ic}\{Ax - y \mid (x, y) \in \operatorname{dom} F\}$. In the fourth case $Y_0 = \operatorname{aff}\{Ax - y \mid (x, y) \in \operatorname{co}(\operatorname{dom} \partial F)\}$, and so $0 \in {}^{ic}\{Ax - y \mid (x, y) \in \operatorname{dom} F\}$.

An important particular case of the preceding proposition is when $X = Y$, $A = \operatorname{Id}_X$ and $F(x, y) = f(x) + g(y)$ with $f, g \in \Gamma(X)$; in this case $D = \operatorname{dom} f - \operatorname{dom} g$ and $E = \operatorname{dom} \partial f - \operatorname{dom} \partial g$.

Using the Borwein theorem one obtains a formula for the subdifferential of a composition $g \circ A$ of a lsc convex function and a continuous linear operator without qualification conditions.

Theorem 3.1.6 *Let X, Y be Banach spaces, $A \in \mathcal{L}(X, Y)$, $g \in \Gamma(Y)$, $f = g \circ A$ and $x \in \operatorname{dom} f$. Then $x^* \in \partial f(x)$ if and only if there exists a net $\left((y_i, y_i^*)\right)_{i \in I} \subset \operatorname{gr} \partial g$ such that $(y_i) \to y := Ax$, $\left(g(y_i)\right) \to g(y)$, $\left(\langle y_i - y, y_i^* \rangle\right) \to 0$ and $(A^* y_i^*) \xrightarrow{w^*} x^*$.*

If X is reflexive one can take sequences instead of nets and impose norm convergence instead of w^*-convergence.

Proof. Let $((y_i, y_i^*))_{i\in I} \subset \operatorname{gr} \partial g$ be such that $(y_i) \to y$, $(\langle y_i - y, y_i^* \rangle) \to 0$, $(g(y_i)) \to g(y)$ and $(A^* y_i^*) \xrightarrow{w^*} x^*$. Then $\langle y - y_i, y_i^* \rangle \le g(y) - g(y_i)$ for all $i \in I$ and $y \in Y$. It follows that

$$\langle x' - x, A^* y_i^* \rangle + \langle y - y_i, y_i^* \rangle = \langle Ax' - y_i, y_i^* \rangle \le f(x') - g(y_i)$$

for all $i \in I$ and $x' \in X$. Taking the limit we obtain that $\langle x' - x, x^* \rangle \le f(x') - f(x)$ for all $x' \in X$, and so $x^* \in \partial f(x)$.

Let now $x^* \in \partial f(x)$ and consider $(\varepsilon_n)_{n\in\mathbb{N}} \downarrow 0$. Using Corollary 2.6.5 we have that $x^* \in w^*\text{-cl}\, A^*(\partial_{\varepsilon_n} g(y))$ for every $n \in \mathbb{N}$. Let $\mathcal{N}$ be a base of w^*-neighborhoods of x^* and consider $I = \mathbb{N}\times\mathcal{N}$ with $(n, V) \succeq (n', V')$ iff $n \ge n'$ and $V \subset V'$. Then for all $i = (n, V) \in I$ there exists $z_i^* \in \partial_{\varepsilon_n} g(y)$ such that $A^* z_i^* \in V$. Taking $\beta = 1$ and $\varepsilon = \varepsilon_n^2$ in Theorem 3.1.1, there exists $(y_i, y_i^*) \in \partial g$ such that $\|y_i - y\| \le \varepsilon_n$, $\|y_i^* - z_i^*\| \le \varepsilon_n$, $|g(y_i) - g(y)| \le \varepsilon_n(\varepsilon_n + 1)$ and $|\langle y_i - y, y_i^* \rangle| \le \varepsilon_n(\varepsilon_n + 1)$. Because $\|A^* z_i^* - A^* y_i^*\| \le \|A^*\| \cdot \|y_i^* - z_i^*\| \le \varepsilon_n \|A^*\|$ and $(A^* z_i^*)_{i\in I} \xrightarrow{w^*} x^*$, we obtain that $(A^* y_i^*)_{i\in I} \xrightarrow{w^*} x^*$.

If X is reflexive then $w^*\text{-cl}\, A^*(\partial_{\varepsilon_n} g(y)) = \text{cl}\, A^*(\partial_{\varepsilon_n} g(y))$, and so, for every $n \in \mathbb{N}$, we can take $z_n^* \in \partial_{\varepsilon_n} g(y)$ such that $\|A^* z_n^* - x^*\| \le \varepsilon_n$. The conclusion follows. □

Using the preceding result one obtains a similar formula for the subdifferential of the sum of two lsc proper convex functions.

Theorem 3.1.7 *Let X be a Banach space, $f_1, f_2 \in \Gamma(X)$ and $x \in \operatorname{dom} f_1 \cap \operatorname{dom} f_2$. Then $x^* \in \partial(f_1 + f_2)(x)$ if and only if there exist two nets $((x_{k,i}, x_{k,i}^*))_{i\in I} \subset \operatorname{gr} \partial f_k$, $k = 1, 2$, such that $(x_{k,i})_{i\in I} \to x$, $(f_k(x_{k,i}))_{i\in I} \to f_k(x)$, $((\langle x_{k,i} - x, x_{k,i}^* \rangle))_{i\in I} \to 0$ for $k = 1, 2$ and $(x_{1,i}^* + x_{2,i}^*)_{i\in I} \xrightarrow{w^*} x^*$.*

If X is reflexive one can take sequences instead of nets and impose norm convergence instead of w^*-convergence.

Proof. We apply the preceding result for $Y := X\times X$, $A : X \to Y$ defined by $Ax := (x, x)$ and $g : Y \to \overline{\mathbb{R}}$ defined by $g(x_1, x_2) := f_1(x_1) + f_2(x_2)$. Then $f := f_1 + f_2 = g\circ A$. The sufficiency is immediate by taking $y = (x, x) = Ax$, $y_i = (x_{1,i}, x_{2,i})$ and $y_i^* = (x_{1,i}^*, x_{2,i}^*)$ for $i \in I$. Let $x^* \in \partial f(x)$. By Theorem 3.1.6 there exists a net $((y_i, y_i^*))_{i\in I} \subset \operatorname{gr} \partial g$ verifying the conditions of the theorem with $y := (x, x)$. Taking $y_i = (x_{1,i}, x_{2,i})$ and $y_i^* = (x_{1,i}^*, x_{2,i}^*)$ and taking into account that $\partial g(x_1, x_2) = \partial f_1(x_1) \times \partial f_2(x_2)$, we have that $((x_{k,i}, x_{k,i}^*))_{i\in I} \subset \operatorname{gr} \partial f_k$ and $(x_{k,i})_{i\in I} \to x$ for $k = 1, 2$. Assume that $(f_1(x_{1,i})) \not\to f_1(x)$. Because f_1 is lsc at x, $\limsup_{i\in I} f_1(x_{1,i}) > f_1(x)$. Thus there exists $\delta > 0$ such that $J := \{i \in I \mid f_1(x_{1,i}) - f_1(x) > \delta\}$ is cofinal. It follows that $g(y_i) - g(y) \geq \delta + \big(f_2(x_{2,i}) - f_2(x)\big)$ for all $i \in J$. Taking the limit inferior, we obtain the contradiction $0 \geq \delta$. Therefore $\big(f_k(x_{k,i})\big)_{i\in I} \to f_k(x)$ for $k = 1, 2$. The inequalities

$$f_1(x_{1,i}) - f_1(x) \leq \langle x_{1,i} - x, x_{1,i}^*\rangle \leq \langle y_i - y, y_i^*\rangle - (f_2(x_{2,i}) - f_2(x))$$

for $i \in I$ imply that $\big(\langle x_{1,i} - x, x_{1,i}^*\rangle\big) \to 0$. □

Using Brøndsted–Rockafellar theorem we obtain another famous result.

Theorem 3.1.8 (Bishop–Phelps) *Let X be a Banach space and $C \subset X$, $C \neq X$, be a nonempty closed convex set. Then*

(i) *The set of support points of C is dense in* $\operatorname{Bd} C$.

(ii) *The set of support functionals of C is dense in the set of continuous linear functionals which are bounded above on C. Moreover, if C is bounded, then the set of support functionals of C is dense in X^*.*

Proof. (i) Let $x_0 \in \operatorname{Bd} C$ and $\varepsilon \in]0,1[$. Taking into account that $C \neq X$, there exists $x_1 \in X \setminus C$ such that $\|x_1 - x_0\| \leq \varepsilon^2$. Applying a separation theorem, there exists $x_0^* \in S_{X^*}$ such that $\sup_{x\in C}\langle x, x_0^*\rangle < \langle x_1, x_0^*\rangle$. So, for every $x \in C$,

$$\langle x - x_0, x_0^*\rangle = \langle x - x_1, x_0^*\rangle + \langle x_1 - x_0, x_0^*\rangle \leq \langle x - x_1, x_0^*\rangle + \|x_1 - x_0\| \leq \varepsilon^2,$$

whence $x_0^* \in \partial_{\varepsilon^2}\iota_C(x_0)$. Applying the Brøndsted–Rockafellar theorem we have that

$$\exists\, x_\varepsilon \in C,\ \exists\, x_\varepsilon^* \in \partial\iota_C(x_\varepsilon)\ :\ \|x_\varepsilon - x_0\| \leq \varepsilon,\ \|x_\varepsilon^* - x_0^*\| \leq \varepsilon < 1.$$

Since $\|x_0^*\| = 1$, we have that $x_\varepsilon^* \neq 0$, and so x_ε is a support point of C with $\|x_\varepsilon - x_0\| \leq \varepsilon$. The conclusion holds.

(ii) Using the second part of Theorem 3.1.2 we have that

$$\operatorname{dom}(\iota_C)^* \subset \operatorname{cl}(\operatorname{Im}\partial\iota_C) = \operatorname{cl}(\operatorname{Im}\partial\iota_C \setminus \{0\}),$$

which shows that the conclusion of the theorem is true. □

The following theorem will be used for a simple proof of the Rockafellar theorem.

Theorem 3.1.9 (Simons) *Let X be a Banach space, $f \in \Gamma(X)$ and $\overline{x} \in X$, $\eta \in \mathbb{R}$ be such that* $\inf f < \eta < f(\overline{x})$. *Consider*

$$L := \sup_{x \in X \setminus \{\overline{x}\}} \frac{\eta - f(x)}{\|\overline{x} - x\|},$$

and $d_{\overline{x}} : X \to \mathbb{R}$, $d_{\overline{x}}(x) := \|x - \overline{x}\|$. *Then:*

(i) $0 < L < \infty$ *and* $\inf(f + Ld_{\overline{x}}) \geq \eta$;

(ii) $\forall \varepsilon \in]0,1[,\ \exists y \in X\ :\ (f + Ld_{\overline{x}})(y) < \inf(f + Ld_{\overline{x}}) + \varepsilon L\|\overline{x} - y\|$;

(iii) $\forall \varepsilon \in]0,1[,\ \exists (z, z^*) \in \operatorname{gr} \partial f\ :\ \langle \overline{x} - z, z^* \rangle \geq (1 - \varepsilon)L\|\overline{x} - z\| > 0$ *and* $\|z^*\| \leq (1 + \varepsilon)L$;

(iv) $\forall \varepsilon \in]0,1[,\ \exists (z, z^*) \in \operatorname{gr} \partial f\ :\ \langle \overline{x} - z, z^* \rangle + f(z) > \eta$ *and* $L \leq \|z^*\| \leq (1 + \varepsilon)L$.

Proof. (i) Because $\eta > \inf f$, it is clear that $L > 0$. Since $\eta < f(\overline{x})$ and f is lsc at $\overline{x}$, there exists $\rho > 0$ such that $f(x) > \eta$ for every $x \in B(\overline{x}, \rho)$. Furthermore, since $f \in \Gamma(X)$, there exist $x^* \in X^*$ and $\alpha \in \mathbb{R}$ such that $f(x) \geq \langle x, x^* \rangle - \alpha$ for every $x \in X$. Therefore

$$\eta - f(x) \leq \eta - \langle x, x^* \rangle + \alpha \leq \eta + \|x - \overline{x}\| \cdot \|x^*\| - \langle \overline{x}, x^* \rangle + \alpha$$

for every $x \in X$ with $\|x - \overline{x}\| \geq \rho$. Let $\gamma := \max\{0, \eta + \alpha - \langle \overline{x}, x^* \rangle\}$. So,

$$\forall x \in X,\ \|x - \overline{x}\| \geq \rho\ :\ \frac{\eta - f(x)}{\|\overline{x} - x\|} \leq \|x^*\| + \frac{\gamma}{\|\overline{x} - x\|} \leq \|x^*\| + \frac{\gamma}{\rho}.$$

This relation and the choice of ρ show that $L < \infty$. From the expression of L we obtain immediately that $\inf(f + Ld_{\overline{x}}) \geq \eta$.

(ii) Let $\varepsilon \in]0,1[$. Since $(1 - \varepsilon)L < L$, from the definition of L there exists $y \in X$, $y \neq \overline{x}$, such that $(\eta - f(y))/\|\overline{x} - y\| > (1 - \varepsilon)L$, and so

$$(f + Ld_{\overline{x}})(y) < \eta + \varepsilon L d_{\overline{x}}(y) \leq \inf(f + Ld_{\overline{x}}) + \varepsilon L\|y - \overline{x}\|.$$

(iii) Let $\varepsilon \in]0,1[$ be fixed. The function $f + Ld_{\overline{x}}$ is proper, lsc and bounded from below, while the element y from (ii) is in $\operatorname{dom}(f + Ld_{\overline{x}}) = \operatorname{dom} f$. Taking the metric d on X defined by $d(x_1, x_2) := L\|x_1 - x_2\|$, (X, d) is a complete metric space. Using Ekeland's variational principle we get the existence of $z \in X$ such that

$$(f + Ld_{\overline{x}})(z) + \varepsilon L\|z - y\| \leq (f + Ld_{\overline{x}})(y)$$

and

$$(f + Ld_{\overline{x}})(z) \leq (f + Ld_{\overline{x}})(x) + \varepsilon L\|x - z\| \quad \forall x \in X.$$

The first relation and (ii) give $\|z - y\| < \|\overline{x} - y\|$, and so $z \neq \overline{x}$. The second relation says that z is a minimum point of the function $f + Ld_{\overline{x}} + \varepsilon Ld_z$. Taking into account that $d_{\overline{x}}$ and d_z are continuous convex functions, it follows that

$$0 \in \partial(f + Ld_{\overline{x}} + \varepsilon Ld_z)(z) = \partial f(z) + L\partial d_{\overline{x}}(z) + \varepsilon L\partial d_z(z).$$

But $\partial d_z(z) = \partial\|\cdot\|(0) = U_{X^*}$ and $\partial d_{\overline{x}}(z) = \{x^* \in U_{X^*} \mid \langle z - \overline{x}, x^*\rangle = \|z - \overline{x}\|\}$. Hence there exist $z^* \in \partial f(z)$ and x^*, $y^* \in U_{X^*}$ such that $z^* = Lx^* + \varepsilon Ly^*$ and $\langle \overline{x} - z, x^*\rangle = \|\overline{x} - z\|$. Therefore $\|z^*\| \leq (1 + \varepsilon)L$ and

$$\begin{aligned}\langle \overline{x} - z, z^*\rangle &= \langle \overline{x} - z, Lx^* + \varepsilon Ly^*\rangle = L\|\overline{x} - z\| + \varepsilon L\langle \overline{x} - z, y^*\rangle \\ &\geq L\|\overline{x} - z\| - \varepsilon L\|\overline{x} - z\| = L(1 - \varepsilon)\|\overline{x} - z\| > 0.\end{aligned}$$

(iv) Let $\varepsilon \in]0, 1[$ be fixed and consider $\varepsilon' := \varepsilon/3$. Let us take $M := (1 + 2\varepsilon')L$. Since $f + Md_{\overline{x}} \geq f + Ld_{\overline{x}} \geq \eta$, we can apply Ekeland's theorem for $f + Md_{\overline{x}}$, an element x_0 of dom f, $\varepsilon' > 0$ and the metric defined at (iii). We get so the existence of $z \in X$ such that

$$\forall x \in X \; : \; (f + Md_{\overline{x}})(z) \leq (f + Md_{\overline{x}})(x) + \varepsilon' L\|x - z\|.$$

As in the proof of (iii), there exist $z^* \in \partial f(z)$, x^*, $y^* \in U_{X^*}$ such that $z^* = Mx^* + \varepsilon' Ly^*$ and $\langle \overline{x} - z, x^*\rangle = \|\overline{x} - z\|$. Thus $\|z^*\| \leq (1 + \varepsilon)L$ and

$$\langle \overline{x} - z, z^*\rangle \geq (M - \varepsilon' L)\|\overline{x} - z\| = (1 + \varepsilon' L)\|\overline{x} - z\|.$$

So, for $\overline{x} = z$ we have that $\langle \overline{x} - z, z^*\rangle + f(z) = f(\overline{x}) > \eta$, while for $\overline{x} \neq z$ we have

$$\langle \overline{x} - z, z^*\rangle + f(z) \geq (1 + \varepsilon')L\|\overline{x} - z\| + f(z) = (f + Ld_{\overline{x}})(z) + \varepsilon' L\|\overline{x} - z\| > \eta.$$

Therefore $\langle \overline{x} - z, z^*\rangle + f(z) > \eta$. Moreover

$$\begin{aligned}\|\overline{x} - x\| \cdot \|z^*\| \geq \langle \overline{x} - x, z^*\rangle &= [\langle \overline{x} - z, z^*\rangle + f(z)] - [\langle x - z, z^*\rangle + f(z)] \\ &> \eta - f(x)\end{aligned}$$

for every $x \in X$; the last inequality holds because $z^* \in \partial f(z)$. Dividing by $\|\overline{x} - x\|$ for $x \neq \overline{x}$, we obtain that $\|z^*\| \geq L$. □

Using the preceding theorem we reinforce slightly Proposition 3.1.3.

Proposition 3.1.10 *Let X be a Banach space and $f \in \Gamma(X)$. Then relation (3.9) holds for every $x \in X$.*

Proof. Taking into account Proposition 3.1.3 we have to show that $\infty = \sup\{\langle x - z, z^* \rangle + f(z) \mid (z, z^*) \in \operatorname{gr} \partial f\}$ when $x \notin \operatorname{dom} f$. So, let $x \in X \setminus \operatorname{dom} f$ and take $\inf f < \eta < \infty$. Using Theorem 3.1.9 (iv), there exists $(z, z^*) \in \operatorname{gr} \partial f$ such that $\langle x - z, z^* \rangle + f(z) > \eta$. The conclusion follows. □

The maximal monotone operators are of a great importance in the theory of evolution equations. A significant example of such operators is the subdifferential of a proper lsc convex function on a Banach space.

Theorem 3.1.11 (Rockafellar) *Let X be a Banach space and $f \in \Gamma(X)$. Then ∂f is a maximal monotone operator.*

Proof. Let $(\overline{x}, \overline{x}^*) \in X \times X^* \setminus \operatorname{gr} \partial f$. Then $\overline{x}^* \notin \partial f(\overline{x})$, and so $0 \notin \partial \tilde{f}(\overline{x})$, where $\tilde{f} := f - \overline{x}^*$. It follows that $\inf \tilde{f} < \tilde{f}(\overline{x})$. Applying assertion (iii) of Simons' theorem, there exists $(z, \tilde{z}^*) \in \operatorname{gr} \partial \tilde{f}$ such that $\langle \overline{x} - z, \tilde{z}^* \rangle > 0$. Consider $z^* := \tilde{z}^* + \overline{x}^*$; we have that $(z, z^*) \in \partial f$ and $\langle z - \overline{x}, z^* - \overline{x}^* \rangle < 0$, which shows that the set $\partial f \cup \{(\overline{x}, \overline{x}^*)\}$ is not monotone. Therefore ∂f is a maximal monotone operator. □

3.2 Convexity and Monotonicity of Subdifferentials

The aim of this section is to show that the monotonicity of an abstract subdifferential of a lower semicontinuous function ensures its convexity; among these abstract subdifferentials one can mention the Clarke subdifferential we introduce below.

Throughout this section $(X, \|.\|)$ is a normed vector space. Consider $\emptyset \neq M \subset X$ and $\overline{x} \in \operatorname{cl} M$. In the sequel we shall denote by $(x_n) \to_M \overline{x}$ a sequence $(x_n) \subset M$ with $(x_n) \to \overline{x}$; more generally, $x \to_M \overline{x}$ will mean $x \in M$ and $x \to \overline{x}$. The **Clarke's tangent cone of M at** $\overline{x}$ is defined by

$$T_C(M, \overline{x}) := \{u \in X \mid \forall (t_n) \downarrow 0,\ \forall (x_n) \to_M \overline{x},\ \exists (u_n) \to u,\ \forall n \in \mathbb{N} : x_n + t_n u_n \in M\}.$$

Recall another cone introduced in Section 2.3:

$$\mathcal{C}(M, \overline{x}) := \operatorname{cl}\left(\bigcup\nolimits_{t \geq 0} t \cdot (M - \overline{x})\right) = \overline{\operatorname{cone}}(M - \overline{x}).$$

Because $\operatorname{cl} M - \overline{x} \subset \mathcal{C}(M,\overline{x}) \subset \mathcal{C}(\operatorname{cl} M,\overline{x})$, we have $\mathcal{C}(M,\overline{x}) = \mathcal{C}(\operatorname{cl} M,\overline{x})$. Note (Exercise!) that

$$0 \in T_C(M,\overline{x}) \subset \mathcal{C}(M,\overline{x}). \tag{3.11}$$

Several properties of the tangent cone in the sense of Clarke are collected in the following proposition.

Proposition 3.2.1 *Let $\emptyset \neq M \subset X$ and $\overline{x} \in \operatorname{cl} M$. Then:*

(i) *$T_C(M,\overline{x})$ is a nonempty closed convex cone;*

(ii) *$T_C(M,\overline{x}) = T_C(\operatorname{cl} M,\overline{x}) = T_C(M \cap V,\overline{x})$ for all $V \in \mathcal{N}(\overline{x})$;*

(iii) *if M is a convex set then $T_C(M,\overline{x}) = \mathcal{C}(M,\overline{x})$.*

Proof. (i) It is obvious that $\lambda u \in T_C(M,\overline{x})$ for $\lambda > 0$ and $u \in T_C(M,\overline{x})$; hence $T_C(M,\overline{x})$ is a cone. Let $u, v \in T_C(M,\overline{x})$ and consider $(t_n) \downarrow 0$ and $(x_n) \to_M \overline{x}$. Then there exists $(u_n) \to u$ such that $x_n' := x_n + t_n u_n \in M$ for every $n \in \mathbb{N}$. Of course $(x_n') \to \overline{x}$. Hence there exists $(v_n) \to v$ such that $x_n + t_n(u_n + v_n) = x_n' + t_n v_n \in M$ for every n. Therefore $u + v \in T_C(M,\overline{x})$.

Consider now $(u^k)_{k\in\mathbb{N}} \subset T_C(M,\overline{x})$ with $(u^k) \to u \in X$. Let us show that $u \in T_C(M,\overline{x})$. For this take $(t_n) \downarrow 0$ and $(x_n) \to_M \overline{x}$. Because $u^k \in T_C(M,\overline{x})$, there exists $(u_n^k) \to u^k$ such that $x_n + t_n u_n^k \in M$ for every $n \in \mathbb{N}$. Hence $x_n + t_n u_n^k \in M$ for all $k, n \in \mathbb{N}$. For every $n \in \mathbb{N}$ there exists $k_n' \in \mathbb{N}$ such that $\|u_n^k - u^k\| \leq n^{-1}$ for every $k \geq k_n'$. Let $(k_n) \subset \mathbb{N}$ be an increasing sequence such that $k_n \geq k_n'$; then $\|u_n^k - u^k\| \leq n^{-1}$ for all $k, n \in \mathbb{N}$ with $k \geq k_n$. Let $u_n := u_n^{k_n}$. Of course, $x_n + t_n u_n = x_n + t_n u_n^{k_n} \in M$ for every n. As $\|u_n - u\| = \|u_n^{k_n} - u\| \leq \|u_n^{k_n} - u^{k_n}\| + \|u^{k_n} - u\| \leq n^{-1} + \|u^{k_n} - u\|$, we have that $(u_n) \to u$. Hence $u \in T_C(M,\overline{x})$. Therefore $T_C(M,\overline{x})$ is a closed convex cone.

(ii) Let $u \in T_C(M,\overline{x})$ and take $(t_n) \downarrow 0$ and $(\overline{x}_n) \to_{\operatorname{cl} M} \overline{x}$. For every $n \in \mathbb{N}$ there exists $x_n \in M$ such that $\|\overline{x}_n - x_n\| \leq t_n/n$, *i.e.* $x_n = \overline{x}_n + t_n n^{-1} u_n'$ with $u_n' \in U_X$. Hence $(x_n) \to_M \overline{x}$. Therefore there exists $(u_n) \to u$ with $x_n + t_n u_n = \overline{x}_n + t_n(n^{-1}u_n' + u_n) \in M \subset \operatorname{cl} M$ for every $n \in \mathbb{N}$. As $(n^{-1}u_n' + u_n) \to u$, we have that $u \in T_C(\operatorname{cl} M,\overline{x})$.

Conversely, let $u \in T_C(\operatorname{cl} M,\overline{x})$ and take $(t_n) \downarrow 0$ and $(x_n) \to_{\overline{M}} \overline{x}$. There exists $(u_n) \to u$ with $x_n + t_n u_n \in \operatorname{cl} M$ for every n. Like above, for every $n \in \mathbb{N}$ there exists $u_n' \in U_X$ such that $x_n + t_n u_n + t_n n^{-1} u_n' \in M$. Because $(u_n + n^{-1}u_n') \to u$, we have that $u \in T_C(M,\overline{x})$. Hence $T_C(M,\overline{x}) = T_C(\operatorname{cl} M,\overline{x})$.

The equality $T_C(M,\overline{x}) = T_C(M \cap V, \overline{x})$ is obvious when V is a neighborhood of $\overline{x}$.

(iii) Let M be convex. Taking into consideration (ii) we may assume that M is closed. Consider $x \in M$ and take $(t_n) \downarrow 0$ and $(x_n) \to_M \overline{x}$. There exists $n_0 \in \mathbb{N}$ such that $t_n \leq 1/2$ for $n \geq n_0$. Take $u_n := \overline{x} + x - 2x_n$ for $n \geq n_0$ and $u_n = 0$ otherwise. Of course, $(u_n) \to x - \overline{x}$. As $x_n + t_n u_n = (1-2t_n)x_n + t_n\overline{x} + t_n x \in M$ for $n \geq n_0$, we have that $x - \overline{x} \in T_C(M,\overline{x})$. It follows that $C(M,\overline{x}) \subset T_C(M,\overline{x})$. Using Eq. (3.11) we obtain the conclusion. □

From (ii) and Eq. (3.11) we obtain that

$$T_C(M,\overline{x}) \subset \bigcap\nolimits_{V \in \mathcal{N}(\overline{x})} C(M \cap V, \overline{x}) =: T_B(M,\overline{x});$$

$T_B(M,\overline{x})$ is the well known *tangent cone in the sense of Bouligand of M at $\overline{x} \in \operatorname{cl} M$.*

Let now $f : X \to \overline{\mathbb{R}}$ be a proper function and $\overline{x} \in \operatorname{dom} f$. It is natural to consider the tangent cone $T_C(\operatorname{epi} f, (\overline{x}, f(\overline{x})))$. This cone is related to the **Clarke–Rockafellar directional derivative** of f at $\overline{x}$ introduced as follows:

$$\begin{aligned} f^{\uparrow}(\overline{x}, u) &:= \sup_{\varepsilon > 0} \limsup_{t \downarrow 0,\, (x,\alpha) \to_{\operatorname{epi} f} (\overline{x}, f(\overline{x}))} \inf_{\|v-u\| \leq \varepsilon} \frac{f(x+tv) - \alpha}{t} \\ &:= \sup_{\varepsilon > 0} \inf_{\delta > 0} \sup_{0 < t \leq \delta,\, \|x - \overline{x}\| \leq \delta,\, f(x) \leq \alpha \leq f(\overline{x}) + \delta} \inf_{\|v-u\| \leq \varepsilon} \frac{f(x+tv) - \alpha}{t}. \end{aligned}$$

It follows that

$$f^{\uparrow}(\overline{x}, u) = \sup_{\varepsilon > 0} \inf_{\delta > 0} \sup_{0 < t \leq \delta,\, \|x - \overline{x}\| \leq \delta,\, f(x) \leq f(\overline{x}) + \delta} \inf_{\|v-u\| \leq \varepsilon} \frac{f(x+tv) - f(x)}{t},$$

and even

$$f^{\uparrow}(\overline{x}, u) = \sup_{\varepsilon > 0} \limsup_{t \downarrow 0,\, x \to_f \overline{x}} \inf_{\|v-u\| \leq \varepsilon} \frac{f(x+tv) - f(x)}{t} \tag{3.12}$$

if f is lower semicontinuous at $\overline{x}$, where $x \to_f \overline{x}$ means $x \to \overline{x}$ and $f(x) \to f(\overline{x})$. It is obvious that $f^{\uparrow}(\overline{x}, 0) \leq 0$ and $f^{\uparrow}(\overline{x}; \lambda u) = \lambda f^{\uparrow}(\overline{x}, u)$ for $\lambda > 0$ and $u \in X$. We may have $f^{\uparrow}(\overline{x}, 0) = -\infty$ even if f is continuous at $\overline{x}$. Take for example $f : \mathbb{R} \to \mathbb{R}$, $f(x) := -\sqrt{|x|}$, and $\overline{x} = 0$ (Exercise!).

The next result shows the connection between the preceding two notions.

Proposition 3.2.2 *Let $f : X \to \overline{\mathbb{R}}$ be a proper function and $\overline{x} \in \operatorname{dom} f$. Then* $\operatorname{epi} f^{\uparrow}(\overline{x}, \cdot) = T_C\big(\operatorname{epi} f, (\overline{x}, f(\overline{x}))\big)$. *In particular $f^{\uparrow}(\overline{x}, \cdot)$ is a lsc convex function; $f^{\uparrow}(\overline{x}, \cdot)$ is a lsc sublinear function if and only if $f^{\uparrow}(\overline{x}, 0) = 0$. Moreover, if f is convex then* $\operatorname{epi} f^{\uparrow}(\overline{x}, \cdot) = \operatorname{cl}(\operatorname{epi} f'(\overline{x}, \cdot))$.

Proof. Let $(u, \lambda) \in T_C\big(\operatorname{epi} f, (\overline{x}, f(\overline{x}))\big)$; assume that $f^{\uparrow}(\overline{x}, u) > \lambda$, and take $f^{\uparrow}(\overline{x}, u) > \lambda' > \lambda$. Then there exist $\varepsilon_0 > 0$, $\big((x_n, \alpha_n)\big) \to_{\operatorname{epi} f} (\overline{x}, f(\overline{x}))$ and $(t_n) \downarrow 0$ such that $\inf_{v \in D(u, \varepsilon_0)} t_n^{-1}\big(f(x_n + t_n v) - \alpha_n\big) > \lambda'$. Because $(u, \lambda) \in T_C\big(\operatorname{epi} f, (\overline{x}, f(\overline{x}))\big)$, there exists the sequence $\big((u_n, \lambda_n)\big)$ converging to (u, λ) such that $(x_n, \alpha_n) + t_n(u_n, \lambda_n) \in \operatorname{epi} f$, *i.e.* $f(x_n + t_n u_n) \le \alpha_n + t_n \lambda_n$ for all $n \in \mathbb{N}$. Since $(u_n) \to u$, there exists n_0 such that $u_n \in D(u, \varepsilon_0)$ for $n \ge n_0$. Hence

$$\lambda' < \inf_{\|v-u\| \le \varepsilon_0} \frac{f(x_n + t_n v) - \alpha_n}{t_n} \le \frac{f(x_n + t_n u_n) - \alpha_n}{t_n} \le \lambda_n \quad \forall n \ge n_0.$$

Taking the limit we get the contradiction $\lambda' \le \lambda$.

Let now $f^{\uparrow}(\overline{x}, u) < \lambda$ and take $\big((x_n, \alpha_n)\big) \to_{\operatorname{epi} f} (\overline{x}, f(\overline{x}))$ and $(t_n) \downarrow 0$. Let us fix $(\varepsilon_k) \downarrow 0$. Because

$$\inf_{\delta > 0} \sup_{0 < t \le \delta,\, (x, \alpha) \in \operatorname{epi} f,\, \|x - \overline{x}\| \le \delta,\, |\alpha - f(\overline{x})| \le \delta} \inf_{\|v-u\| \le \varepsilon_k} \frac{f(x + tv) - \alpha}{t} < \lambda$$

for every $k \in \mathbb{N}$, there exists $\delta_k > 0$ such that

$$\sup_{0 < t \le \delta_k,\, \|x - \overline{x}\| \le \delta_k,\, f(x) \le \alpha \le f(\overline{x}) + \delta_k} \inf_{\|v-u\| \le \varepsilon_k} \frac{f(x + tv) - \alpha}{t} < \lambda.$$

There exists $n'_k \in \mathbb{N}$ such that $0 < t_n \le \delta_k$, $\|x_n - \overline{x}\| \le \delta_k$, $|\alpha_n - f(\overline{x})| \le \delta_k$ for all $n \ge n'_k$. Therefore $\inf_{v \in D(u, \varepsilon_k)} t_n^{-1}\big(f(x_n + t_n v) - \alpha_n\big) < \lambda$, which shows that for every k and every $n \ge n'_k$ there exists $u_n^k \in D(u, \varepsilon_k)$ such that $f(x_n + t_n u_n^k) \le \alpha_n + \lambda t_n$. Consider an increasing sequence $(n_k) \subset \mathbb{N}$ such that $n_k \ge n'_k$ for every $k \in \mathbb{N}$. Taking $u_n := u_n^k$ for $n_k \le n < n_{k+1}$ we have that $\big((u_n, \lambda)\big) \to (u, \lambda)$ and $(x_n, \alpha_n) + t_n(u_n, \lambda) \in \operatorname{epi} f$ for every $n \in \mathbb{N}$. Hence $(u, \lambda) \in T_C\big(\operatorname{epi} f, (\overline{x}, f(\overline{x}))\big)$. Because $T_C\big(\operatorname{epi} f, (\overline{x}, f(\overline{x}))\big)$ is closed, we have that $\operatorname{epi} f^{\uparrow}(\overline{x}, \cdot) \subset T_C(\operatorname{epi} f, (\overline{x}, f(\overline{x})))$. Therefore $\operatorname{epi} f^{\uparrow}(\overline{x}, \cdot) = T_C\big(\operatorname{epi} f, (\overline{x}, f(\overline{x}))\big)$. From this relation and Proposition 3.2.1 (i) we have that $f^{\uparrow}(\overline{x}, \cdot)$ is lsc and convex. The other statement follows from Proposition 2.2.7.

Assume now that f is convex. Because

$$\operatorname{cone}\big(\operatorname{epi} f - (\overline{x}, f(\overline{x}))\big) \subset \operatorname{epi} f'(\overline{x}, \cdot) \subset \overline{\operatorname{cone}}\big(\operatorname{epi} f - (\overline{x}, f(\overline{x}))\big),$$

we have that $\operatorname{epi} f^{\uparrow}(\overline{x},\cdot) = \mathcal{C}(\operatorname{epi} f, (\overline{x}, f(\overline{x}))) = \operatorname{cl}(\operatorname{epi} f'(\overline{x},\cdot))$. $\square$

Note that when f is not lsc at $\overline{x}$, $f^{\uparrow}(\overline{x},\cdot)$ and $\overline{f}^{\uparrow}(\overline{x},\cdot)$ may be different, where, as usual, $\overline{f}$ is the lower semicontinuous envelope of f. Take for example $f : \mathbb{R} \to \mathbb{R}$, $f(x) = 0$ for $x \neq 0$, $f(0) = 1$; $f^{\uparrow}(0,\cdot) = -\infty$, but $\overline{f}^{\uparrow}(0,\cdot) = 0$. When f is lsc at $\overline{x}$ then $f^{\uparrow}(\overline{x},\cdot)$ and $\overline{f}^{\uparrow}(\overline{x},\cdot)$ coincide.

When f has additional properties, $f^{\uparrow}$ has a simpler expression.

Proposition 3.2.3 *Let $f : X \to \overline{\mathbb{R}}$ be a proper function and $\overline{x} \in \operatorname{dom} f$. Then for every $u \in X$ we have:*

(i) $\limsup_{x \to_f \overline{x}} f^{\uparrow}(x,u) \le f^{\uparrow}(\overline{x},u)$; *moreover, if f is lsc at $\overline{x}$ then*

$$f^{\uparrow}(\overline{x},u) \le \sup_{\varepsilon>0} \inf_{\delta>0} \sup_{\|x-\overline{x}\|\le\delta,\, 0<t\le\delta} \inf_{\|v-u\|\le\varepsilon} \frac{f(x+tv)-f(x)}{t}$$
$$\le \inf_{\delta>0} \sup_{\|x-\overline{x}\|\le\delta,\, 0<t\le\delta} \frac{f(x+tu)-f(x)}{t}. \tag{3.13}$$

(ii) *If f is continuous at $\overline{x}$ then*

$$f^{\uparrow}(\overline{x},u) = \sup_{\varepsilon>0} \inf_{\delta>0} \sup_{\|x-\overline{x}\|\le\delta,\, 0<t\le\delta} \inf_{\|v-u\|\le\varepsilon} \frac{f(y+tv)-f(x)}{t}.$$

(iii) *If f is finite and L-Lipschitz on $B(\overline{x},r)$ for some $r>0$ and $L \ge 0$, then*

$$f^{\uparrow}(\overline{x},u) = \inf_{\delta>0} \sup_{\|x-\overline{x}\|\le\delta,\, 0<t\le\delta} \frac{f(x+tu)-f(x)}{t}$$
$$= \limsup_{x\to\overline{x},\, t\downarrow 0} \frac{f(x+tu)-f(x)}{t} \le L\cdot\|u\|. \tag{3.14}$$

(iv) *If f is finite and Gâteaux differentiable on $B(\overline{x},r)$, and ∇f is continuous at $\overline{x}$ then $f^{\uparrow}(\overline{x},\cdot) = \nabla f(\overline{x})$.*

Proof. Throughout the proof $u \in X$ is a fixed element.

(i) Let $f^{\uparrow}(\overline{x},u) < \lambda$ and consider $\varepsilon > 0$; there exists $\delta > 0$ such that

$$\sup_{\|x-\overline{x}\|\le\delta,\, 0<t\le\delta,\, f(x)\le f(\overline{x})+\delta} \inf_{\|v-u\|\le\varepsilon} \frac{f(x+tv)-f(x)}{t} < \lambda.$$

Take $\delta' := \delta/2$. Let $\overline{x}' \in B(\overline{x},\delta')$ with $|f(\overline{x}')-f(\overline{x})| \le \delta'$ and x, t be such that $\|x-\overline{x}'\| \le \delta'$, $0 < t \le \delta'$ and $f(x) \le f(\overline{x}')+\delta'$. Then $\|x-\overline{x}\| \le \delta$,

$0 < t \le \delta$ and $f(x) \le f(\overline{x}) + \delta$. It follows that

$$\begin{aligned} \sup\big\{\inf\nolimits_{v\in D(u,\varepsilon)} t^{-1}\big(f(x+tv)-f(x)\big) \mid t\in\,]0,\delta'],\ x\in D(\overline{x}',\delta'),\\ f(x)\le f(\overline{x}')+\delta'\big\}\\ \le \sup\big\{\inf\nolimits_{v\in D(u,\varepsilon)} t^{-1}\big(f(x+tv)-f(x)\big) \mid t\in\,]0,\delta],\ x\in D(\overline{x},\delta),\\ f(x)\le f(\overline{x})+\delta\big\} < \lambda \end{aligned}$$

for all $\overline{x}' \in B(\overline{x},\delta')$ with $|f(\overline{x}') - f(\overline{x})| \le \delta'$. Therefore $f^{\uparrow}(\overline{x}',u) < \lambda$ for such $\overline{x}'$, and so $\limsup_{x\to_f \overline{x}} f^{\uparrow}(x,u) \le f^{\uparrow}(\overline{x},u)$.

The first inequality in Eq. (3.13) follows immediately from Eq. (3.12), while the second is obvious.

(ii) Taking into account Eq. (3.13), one must show the inequality $\ge$. For this take $\varepsilon > 0$ and $\delta > 0$. There exists $\delta' \in\,]0,\delta]$ such that $f(x) \le f(\overline{x}) + \delta$ for every $x \in D(\overline{x},\delta')$. Then

$$\sup_{\substack{0<t\le\delta\\ \|x-\overline{x}\|\le\delta\\ f(x)\le f(\overline{x})+\delta}} \inf_{\|v-u\|\le\varepsilon} \frac{f(y+tv)-f(y)}{t} \ge \sup_{\substack{0<t\le\delta'\\ \|x-\overline{x}\|\le\delta'}} \inf_{\|v-u\|\le\varepsilon} \frac{f(y+tv)-f(y)}{t},$$

whence the conclusion follows.

(iii) The inequality $\le$ is proved in Eq. (3.13). Assume that f is Lipschitz on $B(x,r)$ with Lipschitz constant L. Take $\lambda > \limsup_{x\to\overline{x},\, t\downarrow 0} \frac{f(x+tu)-f(x)}{t}$ and fix $\varepsilon > 0$. Consider $\delta_0 > 0$ such that $\sup_{\|x-\overline{x}\|\le\delta_0,\, 0<t\le\delta_0} \frac{f(x+tu)-f(x)}{t} < \lambda$. Let $\delta := \min\{\delta_0, r/(1+\varepsilon+\|u\|)\}$. Then for all $x \in B(\overline{x},\delta)$, $t \in\,]0,\delta]$ and $v \in D(u,\varepsilon)$ we have that $\|x+tv-\overline{x}\| < \delta + \delta\,\|v\| \le \delta(1+\varepsilon+\|u\|) \le r$. Therefore for such x, t, v we have

$$\begin{aligned} \frac{f(x+tv)-f(x)}{t} &\le \frac{f(x+tv)-f(x+tu)}{t} + \frac{f(x+tu)-f(x)}{t}\\ &\le L\,\|v-u\| + \frac{f(x+tu)-f(x)}{t} \le \frac{f(x+tu)-f(x)}{t} + L\varepsilon. \end{aligned}$$

Hence

$$\sup_{\substack{\|x-\overline{x}\|\le\delta\\ 0<t\le\delta}} \inf_{\|v-u\|\le\varepsilon} \frac{f(x+tv)-f(x)}{t} \le \sup_{\substack{\|x-\overline{x}\|\le\delta\\ 0<t\le\delta}} \frac{f(x+tu)-f(x)}{t} + L\varepsilon,$$

and so

$$\inf_{\delta>0} \sup_{\|x-\overline{x}\|\le\delta,\, 0<t\le\delta} \inf_{\|v-u\|\le\varepsilon} \frac{f(x+tv)-f(x)}{t}$$

$$\le \sup_{\|x-\overline{x}\|\le\delta_0,\, 0<t\le\delta_0} \frac{f(x+tu)-f(x)}{t} + L\varepsilon \le \lambda + L\varepsilon.$$

Taking the limit for $\varepsilon \to 0$, we obtain

$$\lambda \ge \lim_{\varepsilon\downarrow 0} \inf_{\delta>0} \sup_{\|x-\overline{x}\|\le\delta,\, 0<t\le\delta} \inf_{\|v-u\|\le\varepsilon} \frac{f(x+tv)-f(x)}{t} = f^{\uparrow}(\overline{x}, u).$$

Taking into account (ii), we get Eq. (3.14). Taking $\delta = r/(1+\|u\|)$, for $x \in B(\overline{x}, \delta)$ and $t \in]0, \delta]$ we have that $x + tu, x \in B(\overline{x}, r)$, and so $f(x + tu) - f(x) \le tL\|u\|$, whence $f^{\uparrow}(\overline{x}, u) \le L\|u\|$.

(iv) Let $r > 0$ be such that f is Gâteaux differentiable on $B(\overline{x}, r)$. Since ∇f is continuous at $\overline{x}$, ∇f is bounded on a neighborhood of $\overline{x}$. So we may assume that $\|\nabla f(x)\| \le L$ for $x \in B(\overline{x}, r)$. Using the mean-value theorem we obtain that f is L-Lipschitz on $B(\overline{x}, r)$, and so Eq. (3.14) holds. Let $(t_n) \downarrow 0$ and $(x_n) \to \overline{x}$ be such that $f^{\uparrow}(\overline{x}, u) = \lim t_n^{-1}\left(f(x_n+t_nu) - f(x_n)\right)$. But $x_n + t_nu, x_n \in B(\overline{x}, r)$ for $n \ge n_0$ (for some $n_0 \in \mathbb{N}$); applying the mean value theorem, there exists $\theta_n \in]0, t_n[$ such that $f(x_n + t_nu) - f(x_n) = \nabla f(x_n + \theta_n u)(t_n u)$ for every $n \ge n_0$. Because ∇f is continuous at $\overline{x}$, we obtain that $f^{\uparrow}(\overline{x}, u) = \nabla f(\overline{x})(u)$. □

Let us introduce now the **Clarke subdifferential** of $f : X \to \overline{\mathbb{R}}$ at $\overline{x} \in X$ with $f(\overline{x}) \in \mathbb{R}$. This is

$$\partial_C f(\overline{x}) = \{x^* \in X^* \mid \langle u, x^*\rangle \le f^{\uparrow}(\overline{x}, u)\ \forall u \in X\};$$

if $f(\overline{x}) \notin \mathbb{R}$ we consider that $\partial_C f(\overline{x}) = \emptyset$.

Taking into account Theorem 2.4.14, $\partial_C f(\overline{x}) \ne \emptyset$ if and only if $f^{\uparrow}(\overline{x}, 0) = 0$. Moreover, $\partial_C f(\overline{x})$ is a nonempty w^*-compact subset of X^* if f is Lipschitz on a neighborhood of $\overline{x}$. Other properties of ∂_C are collected in the following result.

Theorem 3.2.4 *Let $f, g : X \to \overline{\mathbb{R}}$ be proper functions and $\overline{x} \in \operatorname{dom} f \cap \operatorname{dom} g$.*

(i) *If f and g coincide on a neighborhood of $\overline{x}$ then $\partial_C f(\overline{x}) = \partial_C g(\overline{x})$.*

(ii) *If f is convex then $\partial_C f(\overline{x}) = \partial f(\overline{x})$.*

(iii) *If $\overline{x}$ is a local minimum of f then $0 \in \partial_C f(\overline{x})$.*

(iv) $\limsup_{x\to\overline{x}} \partial_C f(x) \subset \partial_C f(\overline{x})$, *where* $\limsup_{x\to\overline{x}} \partial_C f(x)$ *is the set*

$$\left\{x^* \in X^* \;\middle|\; \exists\,(x_n) \to_f \overline{x},\ \exists\,(x_n^*) \xrightarrow{w^*} x^*,\ \forall n \in \mathbb{N} \,:\, x_n^* \in \partial_C f(x_n)\right\}. \tag{3.15}$$

(v) *If g is finite and Lipschitz on a neighborhood of $\overline{x}$ then*

$$(f+g)^{\uparrow}(\overline{x},\cdot) \le f^{\uparrow}(\overline{x},\cdot) + g^{\uparrow}(\overline{x},\cdot), \quad \partial_C(f+g)(\overline{x}) \subset \partial_C f(\overline{x}) + \partial_C g(\overline{x}).$$

(vi) *If g is Gâteaux differentiable on a neighborhood of $\overline{x}$ and ∇g is continuous at $\overline{x}$ then*

$$(f+g)^{\uparrow}(\overline{x},\cdot) = f^{\uparrow}(\overline{x},\cdot) + \nabla g(\overline{x}), \quad \partial_C(f+g)(\overline{x}) = \partial_C f(\overline{x}) + \nabla g(\overline{x}).$$

Proof. (i) and (iii) are obvious because $f^{\uparrow}(\overline{x},\cdot) = (f|_V)^{\uparrow}(\overline{x},\cdot)$ for every $V \in \mathcal{N}(\overline{x})$, while (ii) follows from the second part of Proposition 3.2.2.

(iv) Let $x^* \in \limsup_{x\to_f\overline{x}} \partial_C f(y)$; then there exist $(x_n) \to_f \overline{x}$ and $(x_n^*) \xrightarrow{w^*} x^*$ such that $x_n^* \in \partial_C f(x_n)$ for all $n \in \mathbb{N}$. Let $u \in X$ be fixed. Then $\langle u, x_n^*\rangle \le f^{\uparrow}(x_n, u)$ for every $n \in \mathbb{N}$, whence, by Proposition 3.2.3 (i), $\langle u, x^*\rangle \le f^{\uparrow}(x,u)$, and so $x^* \in \partial_C f(x)$.

(v) Let $r > 0$ be such that g is L-Lipschitz on $B(\overline{x}, r)$; we may assume that $L \ge 1$. Let $u \in X$, $\varepsilon > 0$ and $\delta > 0$ be fixed and take $\delta' = \min\{\delta/(2L), r/(1+\varepsilon+\|u\|)\} > 0$. Let v, x and t be such that $\|v-u\| \le \varepsilon$, $\|x-\overline{x}\| \le \delta'$, $0 < t \le \delta'$ and $f(x)+g(x) \le f(\overline{x})+g(\overline{x})+\delta'$. It follows that $f(x) \le f(\overline{x}) + (g(\overline{x}) - g(x)) + \delta' \le f(\overline{x}) + L\delta' + \delta' \le f(\overline{x}) + \delta$. Furthermore

$$\frac{(f+g)(x+tv)-(f+g)(x)}{t} \le \frac{f(x+tv)-f(x)}{t} + \frac{g(x+tu)-g(x)}{t} + L\varepsilon,$$

whence, taking first the supremum with respect to $v \in D(u,\varepsilon)$, we obtain that $\alpha \le A_1(\delta) + A_2(\delta) + L\varepsilon$, where

$$\alpha := \sup_{\substack{\|x-\overline{x}\|\le\delta',\, 0<t\le\delta' \\ (f+g)(x)\le(f+g)(\overline{x})+\delta'}} \ \inf_{\|v-u\|\le\varepsilon} \frac{(f+g)(x+tv)-(f+g)(x)}{t},$$

$$A_1(\delta) := \sup_{\|x-\overline{x}\|\le\delta,\, 0<t\le\delta,\, f(x)\le f(\overline{x})+\delta} \ \inf_{\|v-u\|\le\varepsilon} \frac{f(x+tv)-f(x)}{t},$$

$$A_2(\delta) := \sup\left\{t^{-1}\left(g(x+tu)-g(x)\right) \mid x \in D(\overline{x},\delta)\right\}.$$

Therefore $\beta \leq A_1(\delta) + A_2(\delta) + L\varepsilon$ for every $\delta > 0$, where

$$\beta := \inf_{\delta'>0} \sup_{\substack{\|x-\overline{x}\|\leq\delta',\, 0<t\leq\delta' \\ (f+g)(x)\leq(f+g)(\overline{x})+\delta'}} \inf_{\|v-u\|\leq\varepsilon} \frac{(f+g)(x+tv)-(f+g)(x)}{t}.$$

Taking the limit of $A_1(\delta) + A_2(\delta)$ for $\delta \to 0$ (taking into account that A_1 and A_2 are nondecreasing for $\delta > 0$), we get

$$\inf_{\delta>0} \sup_{\|x-\overline{x}\|\leq\delta,\, 0<t\leq\delta,\, (f+g)(x)\leq(f+g)(\overline{x})+\delta} \inf_{\|v-u\|\leq\varepsilon} \frac{(f+g)(x+tv)-(f+g)(x)}{t}$$

$$\leq \inf_{\delta>0} \sup_{\|x-\overline{x}\|\leq\delta,\, 0<t\leq\delta,\, f(x)\leq f(\overline{x})+\delta} \inf_{\|v-u\|\leq\varepsilon} \frac{f(x+tv)-f(x)}{t}$$

$$+ \inf_{\delta>0} \sup_{\|x-\overline{x}\|\leq\delta} \frac{g(x+tu)-g(x)}{t} + L\varepsilon.$$

Taking now the limit for $\varepsilon \to 0$, we get the desired conclusion. The relation for the subdifferentials follows from the preceding inequality and the fact that $g^{\uparrow}(\overline{x},\cdot)$ is continuous.

(vi) As mentioned in the proof of Proposition 3.2.3 (iv), g is Lipschitz on a neighborhood of $\overline{x}$ and so the conclusion of (iv) holds with $g^{\uparrow}(\overline{x},\cdot) = \nabla g(\overline{x})$. Applying again (iv) for $f+g$ and $-g$ (we may assume that g is finite on X; otherwise take $g = 0$ outside a neighborhood of $\overline{x}$), we obtain that $f^{\uparrow}(\overline{x},\cdot) \leq (f+g)^{\uparrow}(\overline{x},\cdot) - \nabla g(\overline{x})$. The conclusion is now obvious. □

In the rest of this section we use an abstract subdifferential. Before introducing this notion let us consider the following class of finite-valued convex functions:

$$\mathfrak{C}(X) := \{g : X \to \mathbb{R} \mid g \text{ is convex and Lipschitz}\};$$

of course, $\mathfrak{C}(X)$ is a convex cone in the vector space $\mathbb{R}^X$ of all functions from X into $\mathbb{R}$. Let $\mathfrak{C}_1(X) \subset \mathbb{R}^X$ be the cone generated by $X^* \cup \{d_{[a,b]} \mid a,b \in X\}$ $(\subset \mathfrak{C}(X))$ and $\mathfrak{C}_2(X) \subset \mathbb{R}^X$ be the cone generated by

$$X^* \cup \left\{d^2_{[a,b]} \mid a,b \in X\right\} \cup \mathfrak{D}_0,$$

where

$$\mathfrak{D}_0 := \left\{\sum\nolimits_{k\geq 0} \mu_k d^2_{u_k} \,\middle|\, (\mu_k) \subset \mathbb{R}_+,\ \sum\nolimits_{k\geq 0} \mu_k = 1,\ (u_k)_{k\geq 0} \text{ convergent}\right\}.$$

These cones will be used below.

We call an **abstract subdifferential** on the nonempty class $\mathcal{F} \subset \overline{\mathbb{R}}^X$ a multifunction $\overline{\partial} : X \times \mathcal{F} \rightrightarrows X^*$, which associates to (x, f) a set denoted by $\overline{\partial} f(x)$, satisfying condition (P1) below:

(P1) $0 \in \limsup_{y \to x} \overline{\partial} f(y) + \partial g(x)$ if $f \in \mathcal{F}$, $g \in \mathfrak{C}(X)$, $f(x) \in \mathbb{R}$ and x is a local minimum of $f + g$, where $\limsup_{y \to x} \overline{\partial} f(y)$ is defined as in Eq. (3.15) and $\partial g(x)$ is the Fenchel subdifferential of g at x.

A stronger form of (P1) is

(P2) $0 \in \overline{\partial} f(x) + \partial g(x)$ if $f \in \mathcal{F}$, $g \in \mathfrak{C}(X)$, $f(x) \in \mathbb{R}$ and x is a local minimum of $f + g$.

Sometimes one asks

(P3) $\overline{\partial} f(x) = \partial f(x)$ if $f \in \mathcal{F} \cap \mathfrak{C}(X)$.

Of course, (P2) holds if (P4) and (P5) below are satisfied:

(P4) $0 \in \overline{\partial} f(x)$ if $f \in \mathcal{F}$, $f(x) \in \mathbb{R}$ and x is a local minimum of f,
(P5) $\mathcal{F} + \mathfrak{C}(X) \subset \mathcal{F}$ and $\overline{\partial}(f + g)(x) \subset \overline{\partial} f(x) + \partial g(x)$ when $f \in \mathcal{F}$ and $g \in \mathfrak{C}(X)$.

Note that

$$\overline{\mathbb{R}}^X + \mathfrak{C}(X) = \overline{\mathbb{R}}^X, \quad \Lambda(X) + \mathfrak{C}(X) = \Lambda(X), \quad \Gamma(X) + \mathfrak{C}(X) = \Gamma(X).$$

Remark 3.2.1 From the preceding theorem we observe that Clarke's subdifferential ∂_C and the Fenchel subdifferential ∂ are abstract subdifferentials on $\overline{\mathbb{R}}^X$ and $\Lambda(X)$, respectively; in fact they satisfy conditions (P1)–(P5).

There are many other subdifferentials which satisfy condition (P1).

Remark 3.2.2 If the abstract subdifferential $\overline{\partial}$ on $\mathcal{F}$ satisfies the stronger condition (P2) then for any proper function $f \in \mathcal{F}$ one has $\partial f(x) \subset \overline{\partial} f(x)$ for all $x \in \operatorname{dom} f$.

Indeed, if $x^* \in \partial f(x)$ then x is a (local) minimum point of $f + (-x^*)$, and so, by (P2), $0 \in \overline{\partial} f(x) + \partial(-x^*)(x) = \overline{\partial} f(x) - x^*$. Hence $x^* \in \overline{\partial} f(x)$.
The following approximate mean value theorem proves to be useful in nonconvex analysis.

Theorem 3.2.5 (Zagrodny) *Let $(X, \|\cdot\|)$ be a Banach space and $\overline{\partial}$ be an abstract subdifferential on $\mathcal{F} \subset \overline{\mathbb{R}}^X$. Let $f \in \mathcal{F}$ be lsc, $a, b \in X$ with $a \in \operatorname{dom} f$ and $a \neq b$, and $r \in \mathbb{R}$ with $r \leq f(b)$. Then there exist $(x_n) \to_f c \in [a, b[$ and $x_n^* \in \partial f(x_n)$ for every $n \in \mathbb{N}$ such that*

(i) $r - f(a) \leq \liminf \langle b - a, x_n^* \rangle,$

(ii) $0 \leq \liminf \langle c - x_n, x_n^* \rangle,$

(iii) $\frac{\|b-c\|}{\|b-a\|} (r - f(a)) \leq \liminf \langle b - x_n, x_n^* \rangle,$

(iv) $\|b - a\| (f(c) - f(a)) \leq \|c - a\| (r - f(a)).$

Proof. There exists $x^* \in X^*$ such that $\langle b - a, x^* \rangle = r - f(a)$. Consider $h := f - x^*$; then $h(a) \leq h(b)$. Because h is lsc, there exists $c \in [a, b[$ such that $h(c) \leq h(x)$ for all $x \in [a, b]$. Therefore $c = (1 - \mu)a + \mu b$ for some $\mu \in [0, 1[$. It follows that $c - a = \mu(b - a)$ and $\|c - a\| = \mu \|b - a\|$. Therefore $f(c) - f(a) = h(c) - h(a) + \langle c - a, x^* \rangle \leq \mu (r - f(a))$, whence (iv) follows.

Let $\gamma < h(c)$; then there exists $r > 0$ such that $\gamma < h(x)$ for every $x \in [a, b] + rU_X$. Otherwise, for every $n \in \mathbb{N}$ there exists $x_n \in [a, b] + n^{-1}U_X$ such that $h(x_n) \leq \gamma$. Hence $x_n = d_n + y_n$ with $d_n \in [a, b]$ and $y_n \in n^{-1}U_X$. As the segment $[a, b]$ is a compact set, there exists a subsequence (d_{n_k}) converging to $d \in [a, b]$. It follows that $(x_{n_k}) \to d$, and so, by the lower semicontinuity of h, we get the contradiction $h(d) \leq \gamma$.

Let $r > 0$ correspond to $\gamma := h(c) - 1$, and take $U := [a, b] + rU_X$; of course, U is closed. Like above, for any $n \in \mathbb{N}$ there exists $r_n \in]0, r[$ such that $h(x) \geq h(c) - n^{-2}$ for $x \in [a, b] + r_n U_X$; choose $t_n \geq n$ such that $\gamma + t_n r_n \geq h(c) - n^{-2}$. Then one has

$$\forall x \in U \ : \ h(c) \leq h(x) + t_n d_{[a,b]}(x) + n^{-2}. \tag{3.16}$$

Indeed, the inequality is obvious for $x \in [a, b] + r_n U_X$. If $x \in U \setminus ([a, b] + r_n U_X)$ then $d_{[a,b]}(x) > r_n$, and so $h(x) + t_n d_{[a,b]}(x) \geq \gamma + t_n r_n \geq h(c) - n^{-2}$. Consider $H_n := h + \iota_U + t_n d_{[a,b]}$. From Eq. (3.16) we have that $H_n(c) \leq \inf_X H_n + n^{-2}$; moreover, H_n is lsc and bounded from below. Applying Corollary 1.4.2 for H_n, c, $\varepsilon := n^{-2}$ and $\lambda := n^{-1}$, we get $u_n \in X$ such that

$$H_n(u_n) \leq H_n(c), \quad \|c - u_n\| \leq n^{-1}, \tag{3.17}$$

$$H_n(u_n) \leq H_n(x) + n^{-1} \|x - u_n\| \quad \forall x \in X. \tag{3.18}$$

Since $(u_n) \to c \in [a, b[$, we may assume that $u_n \in \operatorname{int} U$ for every n.

Relation (3.18) can be written as

$$\left(f + t_n d_{[a,b]} + n^{-1} d_{u_n} - x^*\right)(u_n) \leq \left(f + t_n d_{[a,b]} + n^{-1} d_{u_n} - x^*\right)(x)$$

for every $x \in U$, where d_a denotes $d_{[a,a]} = d_{\{a\}}$. Therefore u_n is a local minimum of $f + \left(t_n d_{[a,b]} + n^{-1} d_{u_n} - x^*\right)$; the function between the parentheses being convex and Lipschitz, by (P1) we have that

$$0 \in \limsup_{y \to u_n} \overline{\partial} f(y) + \partial\left(t_n d_{[a,b]} + n^{-1} d_{u_n} - x^*\right)(u_n).$$

The formula for the Fenchel subdifferential of a sum of (continuous) convex functions, we obtain the existence of $u_n^* \in \limsup_{y \to_f u_n} \overline{\partial} f(y)$, $v_n^* \in \partial d_{[a,b]}(u_n)$ and $b_n^* \in U_{X^*}$ such that

$$u_n^* = x^* - (t_n v_n^* + n^{-1} b_n^*).$$

Let $y_n \in [a,b]$ be such that $d_{[a,b]}(u_n) = \|u_n - y_n\|$ (such an element exists because $[a,b]$ is compact). Because $d_{[a,b]}$ is continuous we have that $(u_n - y_n) \to 0$, and so $(y_n) \to c$. Since $c \neq b$, we may suppose that $y_n \neq b$ for every $n \in \mathbb{N}$. Hence $y_n = (1-\lambda_n)a + \lambda_n b$ with $\lambda_n \in [0,1[$. From Proposition 3.8.3(ii) on page 239, $\partial d_{[a,b]}(u_n) = \partial \|\cdot\| (u_n - y_n) \cap N([a,b], y_n)$. Therefore $\langle (1-\lambda)a + \lambda b - y_n, v_n^* \rangle = (\lambda - \lambda_n) \langle b - a, v_n^* \rangle \leq 0$ for every $\lambda \in [0,1[$. Hence $\langle b - a, v_n^* \rangle \leq 0$ for every $n \in \mathbb{N}$. Moreover,

$$\langle c - u_n, v_n^* \rangle = \langle c - y_n, v_n^* \rangle - \langle u_n - y_n, v_n^* \rangle \leq -\|u_n - y_n\| \leq 0.$$

It follows that $\langle b - a, u_n^* \rangle \geq \langle b - a, x^* \rangle - n^{-1} \langle b - a, b_n^* \rangle \geq r - f(a) - n^{-1} \|b - a\|$, whence

$$\liminf \langle b - a, u_n^* \rangle \geq r - f(a).$$

Similarly, $\langle c - u_n, u_n^* \rangle \geq \langle c - u_n, x^* \rangle - n^{-1} \|c - u_n\|$, whence

$$\liminf \langle c - u_n, u_n^* \rangle \geq 0.$$

From Eq. (3.17) we get $f(u_n) - \langle u_n, x^* \rangle \leq f(c) - \langle c, x^* \rangle$, and so $f(c) \leq \liminf f(u_n) \leq \limsup f(u_n) \leq f(c)$; hence $\lim f(u_n) = f(c)$ which shows that $(u_n) \to_f c$.

Because $u_k^* \in \limsup_{u \to_f u_k} \overline{\partial} f(u)$, for every $k \in \mathbb{N}$ there exist $(u_{n,k}) \to_f u_k$ and $(u_{n,k}^*) \xrightarrow{w^*} u_k^*$ (for $n \to \infty$) such that $u_{n,k}^* \in \overline{\partial} f(u_{n,k})$ for all $n, k \in \mathbb{N}$.

It follows that $(\langle c - u_{n,k}, u^*_{n,k}\rangle) \to \langle c - u_k, u^*_k\rangle$ for $n \to \infty$. Therefore for every $k \in \mathbb{N}$ there exists $n'_k \in \mathbb{N}$ such that

$$\|u_{n,k} - u_k\| \le k^{-1}, \quad |f(u_{n,k}) - f(u_k)| \le k^{-1}, \quad |\langle b - a, u^*_{n,k} - u^*_k\rangle| \le k^{-1},$$
$$\langle c - u_{n,k}, u^*_{n,k}\rangle - \langle c - u_k, u^*_k\rangle \,| \le k^{-1}$$

for $n \ge n'_k$. Consider $(n_k) \subset \mathbb{N}$ an increasing sequence such that $n_k \ge n'_k$ for all $k \in \mathbb{N}$. Let $x_n := u_{n,k}$ and $x^*_n := u^*_{n,k}$ for $n_k \le n < n_{k+1}$. Since for $n_k \le n < n_{k+1}$ we have

$$\begin{aligned}
\|x_n - c\| &\le \|u_{n,k} - u_k\| + \|u_k - c\| \le k^{-1} + \|u_k - c\|, \\
|f(x_n) - f(c)| &\le k^{-1} + |f(u_k) - f(c)|, \\
\langle b - a, x^*_n\rangle &= \langle b - a, u^*_k\rangle + \langle b - a, u^*_{n,k} - u^*_k\rangle \ge \langle b - a, u^*_k\rangle - k^{-1} \\
\langle c - x_n, x^*_n\rangle &= \langle c - u_{n,k}, u^*_{n,k}\rangle \ge \langle c - u_k, u^*_k\rangle - k^{-1},
\end{aligned}$$

we obtain that $(x_n) \to_f c$,

$$\liminf_{n\to\infty} \langle b - a, x^*_n\rangle \ge \liminf_{k\to\infty} \langle b - a, u^*_k\rangle \ge r - f(a)$$

and

$$\liminf_{n\to\infty} \langle c - x_n, x^*_n\rangle \ge \liminf_{k\to\infty} \langle c - u_k, u^*_k\rangle \ge 0.$$

Taking into account that $b - c = (1 - \mu)(b - a)$ and $\|b - c\| = (1 - \mu)\,\|b - a\|$, from the preceding two relations we get

$$\begin{aligned}
\liminf \langle b - x_n, x^*_n\rangle &= \liminf \left((1 - \mu)\,\langle b - a, x^*_n\rangle + \langle c - x_n, x^*_n\rangle\right) \\
&\ge (1 - \mu) \liminf \langle b - a, x^*_n\rangle + \liminf \langle c - x_n, x^*_n\rangle \\
&\ge \tfrac{\|b-c\|}{\|b-a\|}\,(r - f(a))\,.
\end{aligned}$$

The proof is complete. □

Remark 3.2.3 The conclusion of Theorem 3.2.5 remains valid if (P1) holds only for functions g in the cone $\mathfrak{C}_1(X)$ or in the cone $\mathfrak{C}_2(X)$ defined on page 177.

Indeed, for the first part just note that in the proof of Theorem 3.2.5 we used (P1) only for $g = t_n d_{[a,b]} + n^{-1} d_{u_n} - x^* \in \mathfrak{C}_1(X)$.

For the second part one must notice that one can find $t_n > 0$ such that $\gamma + t_n r_n^2 > h(c)$, and so Eq. (3.16) holds with $d_{[a,b]}$ replaced by $d^2_{[a,b]}$. Applying the Borwein–Preiss variational principle for $p = 2$, H_n,

c, $\varepsilon := 2n^{-2}$ and $\lambda := \sqrt{2/n}$, we find $u_n \in B(c, \sqrt{2/n})$ and a function $\Theta_{2,n}$ generated by the sequences $(\mu_{n,k})_{k\geq 0} \subset \mathbb{R}_+$ with $\sum_{k\geq 0} \mu_{n,k} = 1$ and $(u_{n,k})_{k\geq 0} \subset B(c, \sqrt{2})$ (see Eq. (1.11) on page 31) such that u_n is a local minimum of $f + \left(t_n d^2_{[a,b]} + n^{-1}\Theta_{2,n} - x^*\right)$. Then, in the expression of u_n^* we have $t_n \|u_n - y_n\|$ instead of t_n and $b_n^* \in 2^{3/2}U_{X^*}$ instead of $b_n^* \in U_{X^*}$ (see also Exercise 2.40). The rest of the proof is the same.

Corollary 3.2.6 *Let X be a Banach space and $\overline{\partial}$ be an abstract subdifferential on $\mathcal{F} \subset \overline{\mathbb{R}}^X$. If $f \in \mathcal{F}$ is a lsc proper function then for every $x \in \operatorname{dom} f$ there exists $((x_n, x_n^*)) \subset \operatorname{gr}\overline{\partial} f$ such that $(x_n) \to_f x$ and* $\liminf \langle x - x_n, x_n^*\rangle \geq 0$. *In particular* $\operatorname{dom} f \subset \operatorname{cl}(\operatorname{dom}\overline{\partial} f)$.

Proof. Let $x \in \operatorname{dom} f$. If x is a local minimum of f then 0 belongs to $\limsup_{y\to_f x} \overline{\partial} f(y)$ by (P2) (just take $g = 0$). Therefore there exists $(x_n) \to_f x$ and $(x_n^*) \xrightarrow{w^*} 0$ such that $x_n^* \in \overline{\partial} f(x_n)$ for every $n \in \mathbb{N}$. Of course, $\lim \langle x - x_n, x_n^*\rangle = 0$ in this case, and so the conclusion holds.

Assume that x is not a local minimum of f. Then there exists $(\overline{x}_n) \to x$ such that $f(\overline{x}_n) < f(x)$ for every $n \in \mathbb{N}$. Since f is lsc, we get $(\overline{x}_n) \to_f x$. Applying the preceding theorem for $\overline{x}_n$, x and $r = f(x)$, there exists $\left((x_{k,n}, x_{k,n}^*)\right)_{k\in\mathbb{N}} \subset \operatorname{dom}\overline{\partial} f$ such that $(x_{k,n}) \to_f y_n \in [\overline{x}_n, x[$ for $k \to \infty$ and

$$0 < \tfrac{\|x-y_n\|}{\|x-\overline{x}_n\|}\left(f(x) - f(\overline{x}_n)\right) \leq \liminf\nolimits_{k\to\infty} \langle x - x_{k,n}, x_{k,n}^*\rangle,$$
$$f(y_n) \leq f(\overline{x}_n) + \tfrac{\|y_n-\overline{x}_n\|}{\|x-\overline{x}_n\|}\left(f(x) - f(\overline{x}_n)\right) \leq f(x)$$

for every $n \in \mathbb{N}$. Since $\|y_n - x\| \leq \|\overline{x}_n - x\|$ we have that $(y_n) \to x$, which, together with the preceding inequality and the lower semicontinuity of f, yields $(y_n) \to_f x$. Then, for every $n \in \mathbb{N}$ there exists $k_n' \in \mathbb{N}$ such that $\|x_{k,n} - y_n\| \leq n^{-1}$, $|f(x_{k,n}) - f(y_n)| \leq n^{-1}$ and $0 \leq \langle x - x_{k,n}, x_{k,n}^*\rangle$ for every $k \geq k_n'$. Let $(k_n) \subset \mathbb{N}$ be an increasing sequence such that $k_n \geq k_n'$ for $n \in \mathbb{N}$. Then

$$\|x_{k_n,n} - x\| \leq \|x_{k_n,n} - y_n\| + \|y_n - x\| \leq n^{-1} + \|y_n - x\|,$$
$$|f(x_{k_n,n}) - f(x)| \leq |f(x_{k_n,n}) - f(y_n)| + |f(y_n) - f(x)|$$
$$\leq n^{-1} + |f(y_n) - f(x)|,$$
$$0 \leq \left\langle x - x_{k_n,n}, x_{k_n,n}^*\right\rangle$$

for all $n \in \mathbb{N}$. Taking $x_n := x_{k_n,n}$ and $x_n^* := x_{k_n,n}^*$, we have the desired conclusion. □

When $\mathcal{F} = \Lambda(X)$ and $\overline{\partial}$ is the Fenchel subdifferential, the statement of the preceding corollary is very close to assertion (i) of Theorem 3.1.2.

Theorem 3.2.7 *Let X be a Banach space, $\overline{\partial}$ be an abstract subdifferential on $\mathcal{F} \subset \overline{\mathbb{R}}^X$ and $f \in \mathcal{F}$ be a lsc proper function. If $\overline{\partial} f$ is a monotone multifunction then f is convex.*

Proof. We prove the theorem in three steps: 1) $a \in \operatorname{dom} f$, $b \in \operatorname{dom} \overline{\partial} f$ $\Rightarrow [a, b] \subset \operatorname{dom} f$, 2) $\operatorname{gr} \overline{\partial} f \subset \operatorname{gr} \partial f$, 3) f is convex.

1) Assume that there exists $b' \in [a, b] \setminus \operatorname{dom} f$; so $b' = \lambda a + (1-\lambda) b$ with $\lambda \in]0,1[$. Fix $y^* \in \overline{\partial} f(b)$ and take $r \in \mathbb{R}$,

$$r > f(a) + \lambda^{-1}(1-\lambda)\,\|b-a\| \cdot \|y^*\|.$$

By Theorem 3.2.5, there exist $(x_n) \to_f c \in [a, b'[$ and $x_n^* \in \overline{\partial} f(x_n)$ for every $n \in \mathbb{N}$ such that

$$\liminf \langle b' - x_n, x_n^* \rangle \geq 0, \quad \liminf \langle b' - a, x_n^* \rangle \geq r - f(a).$$

Since $\overline{\partial} f$ is monotone we get the contradiction

$$\begin{aligned} \|b-a\| \cdot \|y^*\| &\geq \|b-c\| \cdot \|y^*\| \geq \langle b-c, y^* \rangle \\ &= \liminf \langle b - x_n, y^* \rangle \geq \liminf \langle b - x_n, x_n^* \rangle \\ &= \liminf \left(\lambda(1-\lambda)^{-1} \langle b' - a, x_n^* \rangle + \langle b' - x_n, x_n^* \rangle \right) \\ &\geq \lambda(1-\lambda)^{-1} \liminf \langle b - a, x_n^* \rangle + \liminf \langle b' - x_n, x_n^* \rangle \\ &\geq \lambda(1-\lambda)^{-1}(r - f(a)). \end{aligned}$$

2) Let $x^* \in \overline{\partial} f(b)$ with $b \in X$. Consider $x \in \operatorname{dom} f$. Applying again Theorem 3.2.5 for $r = f(b)$, there exist $(x_n) \to_f c \in [x, b[$ and $x_n^* \in \overline{\partial} f(x_n)$ for every $n \in \mathbb{N}$ such that

$$\liminf \langle b - x_n, x_n^* \rangle \geq \tfrac{\|b-c\|}{\|b-x\|} \left(f(b) - f(x) \right).$$

Since $\overline{\partial} f$ is monotone and $c = \lambda x + (1-\lambda) b$ with $\lambda \in]0,1]$, we have that

$$\begin{aligned} \lambda \langle b - x, x^* \rangle &= \langle b - c, x^* \rangle = \liminf \langle b - x_n, x^* \rangle \geq \liminf \langle b - x_n, x_n^* \rangle \\ &\geq \lambda \left(f(b) - f(x) \right), \end{aligned}$$

and so $\langle x - b, x^* \rangle \leq f(x) - f(b)$ for every $x \in \operatorname{dom} f$. Therefore $x^* \in \partial f(x)$.

3) Let $x, y \in \operatorname{dom} f$ and $z := (1-\lambda)x + \lambda y$ with $\lambda \in]0,1[$. By Corollary 3.2.6, there exists $(y_n) \subset \operatorname{dom} \overline{\partial} f$ such that $(y_n) \to_f y$. Let $z_n := (1-\lambda)x + \lambda y_n$ for $n \in \mathbb{N}$; by 1) we have that $z_n \in \operatorname{dom} f$ for every n. Using again

Corollary 3.2.6, there exists $((z_{n,k}, z_{n,k}^*)) \subset \operatorname{gr} \overline{\partial} f$ such that $(z_{n,k}) \to_f z_n$ for $k \to \infty$ and $\liminf_{k\to\infty} \langle z_n - z_{n,k}, z_{n,k}^* \rangle \geq 0$. By 2) we have that $z_{n,k}^* \in \partial f(z_{n,k})$, and so

$$\langle x - z_{n,k}, z_{n,k}^* \rangle + f(z_{n,k}) \leq f(x), \quad \langle y_n - z_{n,k}, z_{n,k}^* \rangle + f(z_{n,k}) \leq f(y_n)$$

for all $n, k \in \mathbb{N}$. Multiplying the first inequality by $(1 - \lambda) > 0$ and the second one by $\lambda > 0$ we get

$$\langle z_n - z_{n,k}, z_{n,k}^* \rangle + f(z_{n,k}) \leq (1 - \lambda) f(x) + \lambda f(y_n) \quad \forall n, k \in \mathbb{N}.$$

Taking the limit inferior for $k \to \infty$ we get $f(z_n) \leq (1 - \lambda) f(x) + \lambda f(y_n)$ for every n. Passing to the limit inferior for $n \to \infty$ and taking into account the lower semicontinuity of f and the fact that $(y_n) \to_f y$ we get $f(z) \leq (1 - \lambda) f(x) + \lambda f(y)$, and so f is convex. □

We give now another short proof for the Rockafellar theorem on the maximal monotonicity of the subdifferential of a lsc convex function.

Theorem 3.2.8 *Let X be a Banach space and $f \in \Gamma(X)$. Then ∂f is maximal monotone.*

Proof. Let $x \in X$ and $x^* \notin \partial f(x)$. Then x is not a minimum point for $g := f - x^*$, and so there exists $z \in X$ such that $g(z) < g(x)$. Let $r \in \mathbb{R}$ be between these numbers. By Theorem 3.2.5 applied for ∂ on $\Gamma(X)$, there exists $((y_n, y_n^*)) \subset \operatorname{gr} \partial g$ such that $(y_n) \to_g y \in [z, x[$ and $\liminf \langle x - y_n, y_n^* \rangle \geq \frac{\|x-y\|}{\|x-z\|} (r - g(z)) > 0$. Therefore $\langle x - y_n, y_n^* \rangle > 0$ for $n \geq n_0$, for some $n_0 \in \mathbb{N}$. It follows that $y^* := y_{n_0}^* + x^* \in \partial f(y)$, with $y := y_{n_0}$, and $\langle x - y, x^* - y^* \rangle < 0$. Therefore ∂f is maximal monotone. □

Theorem 3.2.5 can also be used for proving that two convex functions with the same subdifferential differ by an additive constant. In fact one obtains even more.

Theorem 3.2.9 *Let C be an open convex subset of the Banach space X and $\overline{\partial}$ an abstract subdifferential on $\mathcal{F} \subset \overline{\mathbb{R}}^X$. Let $f \in \mathcal{F}$ be proper and lower semicontinuous and $g \in \Gamma(X)$. Assume there exists $\varepsilon \in \mathbb{R}_+$ such that*

$$\overline{\partial} f(x) \subset \partial g(x) + \varepsilon U_{X^*} \quad \forall x \in C. \tag{3.19}$$

Then $C \cap \operatorname{dom} f = C \cap \operatorname{dom} g$ and

$$g(x) - g(y) - \varepsilon \|x - y\| \leq f(x) - f(y) \leq g(x) - g(y) + \varepsilon \|x - y\| \tag{3.20}$$

for all $x \in C$ *and* $y \in C \cap \operatorname{dom} g$.

Proof. Consider $\gamma > \varepsilon$. By Corollary 3.2.6 we have that $\operatorname{dom} f \subset \operatorname{cl}(\operatorname{dom} \overline{\partial} f)$, and so $C \cap \operatorname{dom} \overline{\partial} f \neq \emptyset$. Let $y \in C \cap \operatorname{dom} \overline{\partial} f$ and $x \in C$. From our assumption we have that $y \in \operatorname{dom} \partial g$, and so $y \in \operatorname{dom} g$. We are going to prove that

$$f(x) - f(y) \leq g(x) - g(y) + \varepsilon \|x - y\|. \tag{3.21}$$

The inequality is obvious if $x \notin \operatorname{dom} g$ or $x = y$, and so assume that $x \in \operatorname{dom} g$ and $x \neq y$. Because $\partial g(y) \neq \emptyset$, we have that $g'(y, u) > -\infty$ for all $u \in X$. Consider $\varphi : \mathbb{R} \to \overline{\mathbb{R}}$ defined by $\varphi(t) := g(y + tu)$, where $u := \|x - y\|^{-1}(x - y)$. Of course $\varphi \in \Gamma(\mathbb{R})$ and φ is continuous on $\operatorname{dom} \varphi \supset [0, \|x - y\|]$. It is obvious that $\varphi'(t) = g'(y + tu, u)$ for every $t \in \operatorname{dom} \varphi$. By Theorem 2.1.5(v) we have that

$$\lim_{t \downarrow 0} g'(y + tu, u) = \lim_{t \downarrow 0} \varphi'(t) = \varphi'(0) = g'(y, u),$$

and this quantity is finite. Therefore there exists $t_0 \in]0, \|x - y\|[$ such that $g'(y + t_0 u, u) \leq g'_+(y, u) + \frac{1}{2}(\gamma - \varepsilon)$. Let $x_0 := y + t_0 u \in]y, x[$. Hence

$$g'(x_0, u) \leq \frac{g(y + t_0 u) - g(y)}{t_0} + \tfrac{1}{2}(\gamma - \varepsilon) = \frac{g(x_0) - g(y)}{\|x_0 - y\|} + \tfrac{1}{2}(\gamma - \varepsilon).$$

We want to show that

$$f(x_0) - f(y) \leq g(x_0) - g(y) + \gamma \|x_0 - y\|. \tag{3.22}$$

Assume that Eq. (3.22) does not hold, *i.e.*

$$f(x_0) - f(y) > g(x_0) - g(y) + \gamma \|x_0 - y\|.$$

Choose $r \in \mathbb{R}$ such that $r \leq f(x_0)$ and

$$r - f(y) > g(x_0) - g(y) + \gamma \|x_0 - y\|. \tag{3.23}$$

By Theorem 3.2.5, there exist $(x_n) \to_f z \in [y, x_0[$ and $x_n^* \in \overline{\partial} f(x_n)$ for every $n \in \mathbb{N}$ such that

$$\liminf \langle x_0 - x_n, x_n^* \rangle \geq \frac{\|x_0 - z\|}{\|x_0 - y\|}(r - f(y)).$$

Because $(x_n) \to z$ we have that $(\|x_0 - x_n\|) \to \|x_0 - z\|$, and so

$$\liminf \left\langle \frac{x_0 - x_n}{\|x_0 - x_n\|}, x_n^* \right\rangle \geq \frac{r - f(y}{\|x_0 - y\|}.$$

Taking into account Eq. (3.23), there exists $n_0 \in \mathbb{N}$ such that

$$\left\langle \frac{x_0 - x_n}{\|x_0 - x_n\|}, x_n^* \right\rangle > \frac{g(x_0) - g(y)}{\|x_0 - y\|} + \gamma \quad \forall n \geq n_0.$$

Since $(x_n) \to z \in [y, x_0[\subset C$, we may assume that $x_n \in C$ for $n \geq n_0$. By Eq. (3.19) we have that $x_n^* = z_n^* + \varepsilon u_n^*$ with $z_n^* \in \partial g(x_n)$ and $u_n^* \in U_{X^*}$ for $n \geq n_0$. Using also the convexity of g, it follows that

$$\frac{g(x_0) - g(x_n)}{\|x_0 - x_n\|} \geq \left\langle \frac{x_0 - x_n}{\|x_0 - x_n\|}, z_n^* \right\rangle > \frac{g(x_0) - g(y)}{\|x_0 - y\|} + \gamma - \varepsilon \quad \forall n \geq n_0.$$

Taking into account the lower semicontinuity of g at z and Eq. (3.19) we get

$$\frac{g(x_0) - g(z)}{\|x_0 - z\|} \geq \frac{g(x_0) - g(y)}{\|x_0 - y\|} + \gamma - \varepsilon \geq g'(x_0, u) + \tfrac{1}{2}(\gamma - \varepsilon) > g'(x_0, u).$$

Since $z \in [y, x_0[$, $z = y + su$ for some $s \in [0, t_0[$. Using the convexity of φ we get the contradiction

$$\frac{g(x_0) - g(z)}{\|x_0 - z\|} = \frac{\varphi(t_0) - \varphi(s)}{t_0 - s} \leq \varphi'_-(t_0) \leq \varphi'_+(t_0) = g'(x_0, u).$$

Therefore Eq. (3.22) holds. Consider

$$T := \{t \in]0, \|x - y\|] \mid f(y + tu) - f(y) \leq g(y + tu) - g(y) + \gamma t\}.$$

From Eq. (3.22) we have that $t_0 \in T$, and so $\bar{t} := \sup T \in]0, \|x - y\|]$. Because φ is continuous on $[0, \|x - y\|]$ and f is lsc, we obtain that $\bar{t} \in T$. Assume that $\bar{t} < \|x - y\|$ and take $\bar{y} := y + \bar{t}u$. Of course, $u = \|x - \bar{y}\|^{-1}(x - \bar{y})$ and

$$\frac{g(x) - g(\bar{y})}{\|x - \bar{y}\|} = \frac{\varphi(\|x - y\|) - \varphi(\bar{t})}{\|x - y\| - \bar{t}} \geq g'(\bar{y}, u) = \varphi'(\bar{t}) \geq \varphi'(0) = g'_+(y, 0),$$

whence $g'(\bar{y}, u) \in \mathbb{R}$. Moreover, $\bar{y} \in \operatorname{dom} f \cap \operatorname{dom} g$. Proceeding as above with y replaced by $\bar{y}$ (noting that we used only that $y \in \operatorname{dom} f \cap \operatorname{dom} g$ and $g'(y, u) \in \mathbb{R}$), we obtain some $s_0 \in]0, \|x - \bar{y}\|]$ such that

$$f(\bar{y} + s_0 u) - f(\bar{y}) \leq g(\bar{y} + s_0 u) - g(\bar{y}) + \gamma s_0.$$

As $f(\bar{y}) - f(y) \le g(\bar{y}) - g(y) + \gamma\bar{t}$, we obtain by addition that

$$f(y + (\bar{t} + s_0)u) - f(y) \le g(y + (\bar{t} + s_0)u) - g(y) + \gamma(\bar{t} + s_0),$$

contradicting our choice of $\bar{t}$. Hence $\bar{t} = \|x - y\|$, and so $f(x) - f(y) \le g(x) - g(y) + \gamma \|x - y\|$. Because $\gamma > \varepsilon$ is arbitrary, we obtain that Eq. (3.21) holds.

Taking $x \in C \cap \operatorname{dom} g$ and a fixed $y \in C \cap \operatorname{dom} \overline{\partial} f$, from Eq. (3.21) we obtain that $C \cap \operatorname{dom} g \subset C \cap \operatorname{dom} f$. Let now $y \in C \cap \operatorname{dom} f$. From Corollary 3.2.6 we get a sequence $(y_n) \subset \operatorname{dom} \overline{\partial} f$ with $(y_n) \to_f y$. Because $y \in C$, we may assume that $y_n \in C$ for every $n \in \mathbb{N}$. Let $x \in C \cap \operatorname{dom} g$ be fixed. From Eq. (3.21) we have that

$$f(x) - f(y_n) \le g(x) - g(y_n) + \varepsilon \|x - y_n\| \quad \forall n \in \mathbb{N}.$$

Since g is lsc at y, we get

$$f(x) - f(y) \le g(x) - g(y) + \varepsilon \|x - y\|, \tag{3.24}$$

and so $y \in \operatorname{dom} g$. Hence $C \cap \operatorname{dom} f = C \cap \operatorname{dom} g$. Let $x, y \in C \cap \operatorname{dom} f = C \cap \operatorname{dom} g$; then Eq. (3.24) holds. Interchanging x and y in Eq. (3.24), we obtain also that

$$f(x) - f(y) \ge g(x) - g(y) - \varepsilon \|x - y\|,$$

which proves that Eq. (3.20) holds for all $x, y \in C \cap \operatorname{dom} f$. Equation (3.20) being obvious for $x \in C \setminus \operatorname{dom} f = C \setminus \operatorname{dom} g$ and $y \in C \cap \operatorname{dom} g$, the conclusion follows. □

A more precise result is the following.

Corollary 3.2.10 *Let X be a Banach space and $\overline{\partial}$ an abstract subdifferential on $\mathcal{F} \subset \overline{\mathbb{R}}^X$. Consider a proper and lower semicontinuous function $f \in \mathcal{F}$ and $g \in \Gamma(X)$. If $\overline{\partial} f(x) \subset \partial g(x)$ for every $x \in X$ then there exists $k \in \mathbb{R}$ such that $f = g + k$.*

Proof. Take $C = X$ and $\varepsilon = 0$ in Theorem 3.2.9. Fixing $x_0 \in \operatorname{dom} f = \operatorname{dom} g$, we obtain that $f(x) = g(x) + f(x_0) - g(x_0)$ for every $x \in X$. □

Corollary 3.2.11 *Let X be a Banach space and $f, g \in \Gamma(X)$. If $\partial f(x) \subset \partial g(x)$ for every $x \in X$ then $f = g + k$ for some $k \in \mathbb{R}$.*

Proof. Apply the preceding corollary on $\Gamma(X)$ and the Fenchel subdifferential. □

The next result completes Proposition 2.4.3.

Corollary 3.2.12 *Let X be a Banach space and $T : X \rightrightarrows X^*$ be a maximal cyclically monotone multifunction. Then there exists $f \in \Gamma(X)$, unique up to an additive constant, such that $T = \partial f$.*

Proof. Because T is maximal monotone, $\operatorname{gr} T \neq \emptyset$. Fix $(x_0, x_0^*) \in \operatorname{gr} T$ and consider f_T defined by (2.37) on page 85. Then, by Proposition 2.4.3 we have that $f_T \in \Gamma(X)$ and $\operatorname{gr} T \subset \operatorname{gr} \partial f_T$. Since T is maximal monotone we even have $\operatorname{gr} T = \operatorname{gr} \partial f_T$, and so $T = \partial f$. The uniqueness follows from Corollary 3.2.11. □

3.3 Some Classes of Functions of a Real Variable and Differentiability of Convex Functions

Consider the following useful classes of functions:

$$\begin{aligned}
\mathcal{A} &:= \{\varphi : \mathbb{R}_+ \to \overline{\mathbb{R}}_+ \mid \varphi(0) = 0\},\\
\mathcal{A}_0 &:= \{\varphi \in \mathcal{A} \mid \varphi(t) = 0 \Leftrightarrow t = 0\},\\
N_k &:= \{\varphi \in \mathcal{A} \mid \mathbb{P} \ni t \mapsto t^{-k}\varphi(t) \in \overline{\mathbb{R}}_+ \text{ is nondecreasing}\},\ k \in \{0,1,2\},\\
\Omega_k &:= \{\varphi \in \mathcal{A} \mid \lim_{t\downarrow 0} t^{-k}\varphi(t) = 0\},\ k \in \{0,1\},\\
\Gamma &:= \{\varphi \in \mathcal{A} \mid \varphi \text{ is lsc and convex}\},\\
\Gamma_0 &:= \Gamma \cap \mathcal{A}_0, \quad \Gamma_0^2 := \{\varphi \in \Gamma_0 \mid \liminf_{t\to\infty} t^{-2}\varphi(t) > 0\},\\
\Sigma_1 &:= \Gamma \cap \Omega_1, \quad \Sigma_1^2 := \{\varphi \in \Sigma_1 \mid \limsup_{t\to\infty} t^{-2}\varphi(t) < \infty\}.
\end{aligned}$$

Of course, $N_2 \subset N_1 \subset N_0$, $\Sigma_1^2 \subset \Sigma_1 \subset \Omega_1 \subset \Omega_0$ and $\Gamma_0^2 \subset \Gamma_0 \subset \Gamma \subset N_1$. Moreover, if $\varphi \in N_1$ and $\varphi(t_0) < \infty$ for some $t_0 > 0$ then $\varphi \in \Omega_0$.

Let $\varphi \in N_0$ and consider the *lowest quasi-inverse* $\varphi^e : \mathbb{R}_+ \to \overline{\mathbb{R}}_+$ and the *greatest quasi-inverse* $\varphi^h : \mathbb{R}_+ \to \overline{\mathbb{R}}_+$ defined by

$$\varphi^e(s) := \inf\{t \geq 0 \mid \varphi(t) \geq s\}, \quad \varphi^h(s) := \sup\{t \geq 0 \mid \varphi(t) \leq s\}; \tag{3.25}$$

it is obvious that $\varphi^e \leq \varphi^h$, $\varphi^e \in N_0$, φ^h is nondecreasing and $\varphi^h \in N_0 \Leftrightarrow \varphi \in \mathcal{A}_0$.

For $\varphi \in \mathcal{A}$ and $s \in \mathbb{R}_+$ consider

$$\varphi^\#(s) := \sup\{st - \varphi(t) \mid t \geq 0\} \in \overline{\mathbb{R}};$$

it is obvious that $\varphi^\#(0) = 0$ and $\varphi^\#(s) \geq 0$ for $s \geq 0$, and so we get a function $\varphi^\# \in \mathcal{A}$; even more, $\varphi^\# \in \Gamma$.

Furthermore, one can extend φ to $\mathbb{R}$; such useful extensions can be defined for $t < 0$ by: $\varphi_0(t) := 0$, $\varphi_1(t) := \alpha\,|t|$ with $\alpha > 0$, $\varphi_2(t) := \frac{1}{2}t^2$, $\varphi_3(t) := \infty$. It is easy to show (Exercise!) that $\varphi_k^*(s) = \varphi^\#(s)$ for $s \geq 0$, while for $s < 0$, $\varphi_0^*(s) = \infty$, $\varphi_1^*(s) = \iota_{[-\alpha,0]}(s)$, $\varphi_2^*(s) = \frac{1}{2}s^2$, $\varphi_3^*(s) = 0$. Eventually using such extensions, we see that the most part of the properties of the conjugation are also valid for this operation on $\mathcal{A}$. It is obvious that for $\varphi \in \mathcal{A}$, $\operatorname{epi}\varphi \subset \mathbb{R}_+ \times \mathbb{R}_+$, and so $\overline{\operatorname{co}}(\operatorname{epi}\varphi) \subset \mathbb{R}_+ \times \mathbb{R}_+$, too; therefore the function whose epigraph is $\overline{\operatorname{co}}(\operatorname{epi}\varphi)$, denoted as usual by $\overline{\operatorname{co}}\varphi$, is in $\mathcal{A}$.

In the next auxiliary result we collect some properties of the quasi-inverse and conjugation operations on N_0 and $\mathcal{A}$, respectively.

Lemma 3.3.1 (i) *If $\varphi \in \mathcal{A}_0 \cap N_0$ then $\varphi^e, \varphi^h \in \Omega_0$; if $\varphi \in \Omega_0 \cap N_0$ then $\varphi^e \in \mathcal{A}_0$.*

(ii) *Let $\varphi \in \mathcal{A}$; then $\varphi^{\#\#} = \overline{\operatorname{co}}\varphi$. In particular $\varphi = \varphi^{\#\#}$ if and only if $\varphi \in \Gamma$.*

(iii) *If $\varphi \in \Omega_1$ then $\varphi^\# \in \Gamma_0$; if $\varphi \in \mathcal{A}_0 \cap N_1$ then $\varphi^\# \in \Omega_1$. In particular, if $\varphi \in \mathcal{A}_0 \cap N_1$ then $\overline{\operatorname{co}}\varphi \in \mathcal{A}_0$. Let $\psi \in \Gamma$; then $\psi \in \mathcal{A}_0 \Leftrightarrow \psi^\# \in \Omega_1$, $\psi \in \Gamma_0 \Leftrightarrow \psi^\# \in \Sigma_1$ and $\psi \in \Gamma_0^2 \Leftrightarrow \psi^\# \in \Sigma_1^2$.*

(iv) *Let $p, q \in]1, \infty[$ be such that $1/p + 1/q = 1$ and $\varphi \in \mathcal{A}$. If the mapping $t \mapsto t^{-p}\varphi(t)$ is nondecreasing (nonincreasing) on $\mathbb{P}$ then the mapping $s \mapsto s^{-q}\varphi^\#(s)$ is nonincreasing (nondecreasing) on $\mathbb{P}$.*

(v) *Let $(X, \|\cdot\|)$ be a normed space, $\varphi \in \mathcal{A}$ and $f : X \to \overline{\mathbb{R}}$, $f(x) := \varphi(\|x\|)$. Then $f^*(x^*) = \varphi^\#(\|x^*\|)$ for every $x^* \in X^*$ and $f^{**}(x) = \varphi^{\#\#}(\|x\|)$ for every $x \in X$.*

Proof. (i) Let $\varphi \in \mathcal{A}_0 \cap N_0$ and assume that $\varphi^h \notin \Omega_0$. Then there exist $(s_n) \downarrow 0$ and $\alpha > 0$ such that $\varphi^h(s_n) > \alpha$ for every $n \in \mathbb{N}$. So, for $n \in \mathbb{N}$ there exists $t_n \geq \alpha$ such that $\varphi(t_n) \leq s_n$. Since $\varphi \in N_0$, we have that $\varphi(\alpha) \leq s_n$, whence the contradiction $\varphi(\alpha) = 0$. As $\varphi^e \leq \varphi^h$, $\varphi^e \in \Omega_0$, too. Assume now that $\varphi \in \Omega_0 \cap N_0$ but $\varphi^e \notin \mathcal{A}_0$. Then there exists $\alpha > 0$ such that $\varphi^e(\alpha) = 0$. So there exists $(t_n) \downarrow 0$ such that $\varphi(t_n) \geq \alpha$ for every $n \in \mathbb{N}$, which contradicts $\varphi \in \Omega_0$.

(ii) Consider (for example) the extension φ_2 of φ defined above. By what was already noted, we have that $\varphi_2^{**}(t) = \varphi_2(t)$ for $t < 0$ and $\varphi_2^{**}(t) = \varphi^{\#\#}(t)$ for $t \geq 0$. Since $\varphi_2^{**} = \overline{\operatorname{co}}\varphi_2$, we obtain that $\varphi^{\#\#} = \overline{\operatorname{co}}\varphi$.

(iii) Let $\varphi \in \Omega_1$ and consider $s > 0$; then there exists $\varepsilon > 0$ such that $\varphi(t) \leq st/2$ for every $t \in [0, \varepsilon]$. Consider $\psi \in \mathcal{A}$ such that $\psi(t) := st/2$ for $t \in [0, \varepsilon]$ and $\psi(t) := \infty$ for $t > \varepsilon$. Then $\varphi \leq \psi$, and so $\varphi^\# \geq \psi^\#$. In particular $\varphi^{\#}(s) \geq \sup\{st - st/2 \mid t \in [0,\varepsilon]\} = \varepsilon s/2 > 0$. Hence $\varphi^{\#} \in \Gamma_0$.

Suppose now that $\varphi \in \mathcal{A}_0 \cap N_1$ and take $\varepsilon > 0$. For $0 \leq s \leq \varepsilon^{-1}\varphi(\varepsilon)$ we have that

$$\begin{aligned}\varphi^{\#}(t) &= \sup\{ts - \varphi(t) \mid t \geq 0\} \\ &= \max\{\sup\{ts - \varphi(t) \mid t \in [0,\varepsilon]\}, \sup\{ts - \varphi(t) \mid t \geq \varepsilon\}\} \\ &\leq \max\{s\varepsilon, \sup\{t(s - \varepsilon^{-1}\varphi(\varepsilon)) \mid t \geq \varepsilon\} = s\varepsilon,\end{aligned}$$

and so $\lim_{s\downarrow 0} s^{-1}\varphi^{\#}(s) = 0$. This means that $\varphi^{\#} \in \Omega_1$. From the first part we obtain that $\overline{\text{co}}\varphi = \varphi^{\#\#} \in \mathcal{A}_0$.

Let now $\psi \in \Gamma$; the first two equivalences are immediate from the first part, while for the last one can use Exercise 2.41.

(iv) Assume that $t \mapsto t^{-p}\varphi(t)$ is nondecreasing on $\mathbb{P}$; this means that $\varphi(ct) \geq c^p\varphi(t)$ for all $t \geq 0$ and $c \geq 1$. Let $d \geq 1$ and $s \geq 0$. Then

$$\begin{aligned}d^q\varphi^{\#}(s) &= \sup\{d^q st - (d^{q/p})^p\varphi(t) \mid t \geq 0\} \\ &\geq \sup\{(ds)(d^{q/p}t) - \varphi(d^{q/p}t) \mid t \geq 0\} = \varphi^{\#}(ds).\end{aligned}$$

Therefore $s \mapsto s^{-q}\varphi^{\#}(s)$ is nonincreasing on $\mathbb{P}$. The proof for the other statement is similar.

(v) For $x^* \in X^*$ we have that

$$\begin{aligned}f^*(x^*) &= \sup\{\langle x, x^*\rangle - \varphi(\|x\|) \mid x \in X\} \\ &= \sup\{\langle x, x^*\rangle - \varphi(x) \mid x \in X,\ t \geq 0,\ \|x\| = t\} \\ &= \sup_{t\geq 0}\ \sup_{\|x\|=t} (\langle x, x^*\rangle - \varphi(x)) = \sup_{t\geq 0}(t\|x^*\| - \varphi(x)) = \varphi^{\#}(\|x^*\|).\end{aligned}$$

The equality $f^{**}(x) = \varphi^{\#\#}(\|x)\|$ is obtained similarly. □

As an illustration of the use of the above classes of functions let us give the following interesting result about the differentiability of a convex function. First recall that a **bornology** on X is a family $\mathcal{B}$ of nonempty bounded subsets of X having the properties: 1) every set B of $\mathcal{B}$ is symmetric, 2) $\mathcal{B}$ is directed upwards, *i.e.* for all $B_1, B_2 \in \mathcal{B}$ there exists $B_3 \in \mathcal{B}$ such that $B_1 \cup B_2 \subset B_3$ and 3) $\cup\{B \mid B \in \mathcal{B}\} = X$. The topology induced by the bornology $\mathcal{B}$ on X^*, denoted by $\tau_{\mathcal{B}}$, is the topology of uniform convergence on every set B of $\mathcal{B}$. The most common bornologies on X are the *Gâteaux*, the *Hadamard* and the *Fréchet bornologies* which are obtained taking as $\mathcal{B}$ the classes of finite symmetric subsets, the weakly compact symmetric subsets and the bounded symmetric subsets of X, respectively. To the Gâteaux bornology corresponds the weak* topology on X^*, while to the Fréchet bornology corresponds the norm topology on X^*. It is clear that the topology $\tau_{\mathcal{B}}$ which corresponds to the bornology $\mathcal{B}$ is finer than the weak* topology and coarser than the norm topology. The function $g : X \to \overline{\mathbb{R}}$ is said to be **$\mathcal{B}$-differentiable** at $\overline{x}$ if f is finite on a neighborhood of $\overline{x}$ and there exists $x^* \in X^*$ such that

$$\lim_{t\to 0} \frac{f(\overline{x}+tu)-f(\overline{x})}{t} = \langle u, x^*\rangle \quad \forall\, u \in X, \tag{3.26}$$

the limit being uniform w.r.t. $u \in B$, for every $B \in \mathcal{B}$; x^* is denoted by $\nabla f(\overline{x})$ and is called the $\mathcal{B}$-differential or $\mathcal{B}$-derivative of f at $\overline{x}$. Of course, for the Gâteaux, Hadamard and Fréchet bornologies one obtains the Gâteaux, Hadamard and Fréchet derivatives, respectively. Condition 3) implies that f is Gâteaux differentiable at $\overline{x}$ whenever f is $\mathcal{B}$-differentiable at $\overline{x}$ for the bornology $\mathcal{B}$; in particular the $\mathcal{B}$-derivative is unique when it exists.

Theorem 3.3.2 *Let $f \in \Lambda(X)$ be continuous at $\overline{x} \in \operatorname{dom} f$ and $\mathcal{B}$ be a bornology on X. Consider the following statements:*

(i) *f is $\mathcal{B}$-differentiable at $\overline{x}$,*

(ii) *f is **$\mathcal{B}$-smooth** at $\overline{x}$, i.e. $\lim_{t\downarrow 0} t^{-1}\sigma_{\mathcal{B}}(t) = 0$, where, for $t \geq 0$,*

$$\sigma_{\mathcal{B}}(t) := \sup\{f(\overline{x}+ty) + f(\overline{x}-ty) - 2f(\overline{x}) \mid y \in B\},$$

(iii) *every selection of ∂f is norm to $\tau_{\mathcal{B}}$ continuous at $\overline{x}$,*

(iv) *for every $x^* \in \partial f(\overline{x})$ there exists a selection γ of ∂f which is norm to $\tau_{\mathcal{B}}$ continuous at $\overline{x}$ and $\gamma(\overline{x}) = x^*$,*

(v) *there exists a selection of ∂f which is norm to $\tau_{\mathcal{B}}$ continuous at $\overline{x}$,*

(vi) *$(x_n^*) \xrightarrow{\tau_{\mathcal{B}}} x^*$ whenever $x^* \in \partial f(\overline{x})$, $x_n^* \in \partial f(x_n)$ for $n \in \mathbb{N}$, and $(x_n) \to \overline{x}$,*

(vii) *$(x_n^*) \xrightarrow{\tau_{\mathcal{B}}} x^*$ whenever $x^* \in \partial f(\overline{x})$, $0 \leq \varepsilon_n$, $x_n^* \in \partial_{\varepsilon_n} f(x_n)$ for $n \in \mathbb{N}$, and $(\varepsilon_n) \to 0$, $(x_n) \to \overline{x}$,*

(viii) *$(x_n^*) \xrightarrow{\tau_{\mathcal{B}}} x^*$ whenever $x^* \in \partial f(\overline{x})$, $0 \leq \varepsilon_n$, $x_n^* \in \partial_{\varepsilon_n} f(\overline{x})$ for $n \in \mathbb{N}$, and $(\varepsilon_n) \to 0$.*

Then (i) $\Leftrightarrow$ (ii) $\Leftrightarrow$ (iii) $\Leftrightarrow$ (iv) $\Leftrightarrow$ (v) $\Leftrightarrow$ (vi) $\Leftarrow$ (vii) $\Leftrightarrow$ (viii). *Moreover, if X is a Banach space and f is lsc then* (vi) $\Rightarrow$ (vii).

Proof. (i) $\Rightarrow$ (ii) Assume that f is $\mathcal{B}$-differentiable at $\overline{x}$. Let $u \in X$. Then

$$\begin{aligned}\lim_{t\downarrow 0} & \frac{f(\overline{x}+tu)+f(\overline{x}-tu)-2f(\overline{x})}{t} \\ &= \lim_{t\downarrow 0}\frac{f(\overline{x}+tu)-f(\overline{x})}{t}+\lim_{t\downarrow 0}\frac{f(\overline{x}-tu)-f(\overline{x})}{t} \\ &= \langle u, \nabla f(\overline{x})\rangle + \langle -u, \nabla f(\overline{x})\rangle = 0.\end{aligned}$$

As the limits in the right-hand side of the first equality above is uniform w.r.t. $u \in B$, for every $B \in \mathcal{B}$, we obtain that f is $\mathcal{B}$-smooth at $\overline{x}$.

(ii) $\Rightarrow$ (i) Take $x^* \in \partial f(\overline{x})$; such an element exists by Theorem 2.4.9. Consider $B \in \mathcal{B}$ and take $u \in B$; one has

$$\sigma_B(t) \geq f(\overline{x}+tu)+f(\overline{x}-tu)-2f(\overline{x}) \geq f(\overline{x}+tu)-f(\overline{x})-t\langle u, x^*\rangle \geq 0,$$

and so

$$0 \leq \frac{f(\overline{x}+tu)-f(\overline{x})}{t} - \langle u, x^*\rangle \leq \frac{\sigma_B(t)}{t} \quad \forall u \in B.$$

Hence Eq. (3.26) holds, the limit being uniform w.r.t. $u \in B$.

(i) $\Rightarrow$ (iii) Let $\gamma : D \to X^*$ be a selection of ∂f, where $D := \operatorname{dom} \partial f$; of course, $\operatorname{int}(\operatorname{dom} f) \subset D$ and $\gamma(\overline{x}) = \nabla f(\overline{x})$. Let $B \in \mathcal{B}$. Consider

$$\overline{\sigma}_B : \mathbb{R}_+ \to \overline{\mathbb{R}}_+, \quad \overline{\sigma}_B(t) := \sup\{f(\overline{x}+tu)-f(\overline{x})-t\langle u, \gamma(\overline{x})\rangle \mid u \in B\}. \tag{3.27}$$

Because f is $\mathcal{B}$-differentiable at $\overline{x}$, $\lim_{t\downarrow 0} t^{-1}\overline{\sigma}_B(t) = 0$, *i.e.* $\overline{\sigma} \in \Omega_1$. Because f is convex, $\overline{\sigma}_B$ is convex, too. For $x \in D$, $u \in B$ and $t \geq 0$ we have that

$$\langle \overline{x}+tu-x, \gamma(x)\rangle \leq f(\overline{x}+tu)-f(x) \leq f(\overline{x})+t\langle u, \gamma(\overline{x})\rangle + \overline{\sigma}_B(t) - f(x),$$

whence

$$t\langle u, \gamma(x)-\gamma(\overline{x})\rangle - \overline{\sigma}_B(t) \leq f(\overline{x}) - f(x) + \langle x-\overline{x}, \gamma(x)\rangle.$$

Taking first the supremum with respect to $u \in B$, then with respect to $t \geq 0$ on the left-hand side of the above inequality we obtain that

$$\begin{aligned}(\overline{\sigma}_B)^{\#}\left(\sup\nolimits_{u\in B}|\langle u, \gamma(x)-\gamma(\overline{x})\rangle|\right) &\leq f(\overline{x})-f(x)+\langle x-\overline{x}, \gamma(x)\rangle \\ &\leq f(\overline{x})-f(x)+\|x-\overline{x}\|\cdot\|\gamma(x)\| \qquad \forall x \in D.\end{aligned}$$

Since f is continuous at $\overline{x}$, by Theorem 2.4.13, ∂f is bounded on a neighborhood of $\overline{x}$, and so γ is bounded on a neighborhood of $\overline{x}$. From the above inequality we obtain that $\lim_{x\to\overline{x}}(\overline{\sigma}_B)^{\#}\left(\sup_{u\in B}|\langle u,\gamma(x)-\gamma(\overline{x})\rangle|\right)=0$. Because, by Lemma 3.3.1, $\sigma^{\#}\in\Gamma_0$, we get $\lim_{x\to\overline{x}}\sup_{u\in B}|\langle u,\gamma(x)-\gamma(\overline{x})\rangle|=0$. Therefore γ is norm to $\tau_{\mathcal{B}}$ continuous at $\overline{x}$.

(iii) $\Rightarrow$ (iv) and (iv) $\Rightarrow$ (v) are obvious.

(v) $\Rightarrow$ (i) Let γ be a selection of ∂f which is norm to $\tau_{\mathcal{B}}$ continuous at $\overline{x}$. Consider $B\in\mathcal{B}$ and $\varepsilon>0$. Then there exists $\delta>0$ such that $D(\overline{x},\delta)\subset D$ and $\sup_{u\in B}|\langle u,\gamma(x)-\gamma(\overline{x})\rangle|\le\varepsilon$ for all $x\in D(\overline{x},\delta)$. Because B is bounded, $B\subset\beta U_X$ for some $\beta>0$.

Fix $u\in B$ and take $\varphi:\mathbb{R}\to\overline{\mathbb{R}}$, $\varphi(t):=f(\overline{x}+tu)-f(\overline{x})-t\langle u,\gamma(\overline{x})\rangle$. It is obvious that φ is proper, convex and $[0,\delta']\subset\operatorname{int}(\operatorname{dom}\varphi)$, where $\delta':=\beta^{-1}\delta>0$. Moreover,

$$\varphi'_+(t)=f'(\overline{x}+tu,u)-\langle u,\gamma(\overline{x})\rangle,\quad \varphi'_-(t)=-f'(\overline{x}+tu,-u)-\langle u,\gamma(\overline{x})\rangle,$$

and so $\varphi'_-(t)\le\langle u,\gamma(\overline{x}+tu)-\gamma(\overline{x})\rangle\le\varphi'_+(t)$ for every $t\in\operatorname{int}(\operatorname{dom}\varphi)$. Using Corollary 2.1.8, we obtain that

$$f(\overline{x}+tu)-f(\overline{x})-t\langle u,\gamma(\overline{x})\rangle=\varphi(t)-\varphi(0)=\int_0^t\langle u,\gamma(\overline{x}+su)-\gamma(\overline{x})\rangle\,ds\le t\varepsilon$$

for all $t\in[0,\delta']$. Hence $\lim_{t\downarrow 0}t^{-1}\overline{\sigma}_B(t)=0$, where $\overline{\sigma}_B$ is defined by Eq. (3.27). It follows that f is $\mathcal{B}$-differentiable at $\overline{x}$ and $\nabla f(\overline{x})=\gamma(\overline{x})$.

(iii) $\Rightarrow$ (vi) Let $x^*\in\partial f(\overline{x})$ and $((x_n,x_n^*))\subset\operatorname{gr}\partial f$ be such that $(x_n)\to\overline{x}$. Assume that x^* is not the limit of the sequence (x_n^*) for the topology $\tau_{\mathcal{B}}$. It follows that there exist $\varepsilon_0>0$, $B\in\mathcal{B}$ and an infinite subset P of $\mathbb{N}$ such that $\sup_{u\in B}|\langle u,x_n^*-x^*\rangle|\ge\varepsilon_0$ for every $n\in P$. From (iii) $\Rightarrow$ (i) we have that $\partial f(\overline{x})=\{x^*\}$, and so $x_n\ne x$ for $n\in P$. Since $(x_n)\to\overline{x}$, we may assume that $x_n\ne x_m$ for distinct $n,m\in P$. Let γ be a selection of ∂f such that $\gamma(x_n):=x_n^*$ for $n\in P$, $\gamma(\overline{x}):=x^*$ and $\gamma(x)$ an (arbitrary) element of $\partial f(x)$ otherwise. From (iii) we obtain that γ is norm to $\tau_{\mathcal{B}}$ continuous at $\overline{x}$, and so $(x_n^*)_{n\in P}\overset{\tau_{\mathcal{B}}}{\to}x^*$, contradicting our choice of P.

(vi) $\Rightarrow$ (iii) Let γ be a selection of ∂f, and assume that γ is not norm to $\tau_{\mathcal{B}}$ continuous at $\overline{x}$. Then (since $\overline{x}$ has a countable base of neighborhoods) there exists a sequence $(x_n)\subset D$ converging to $\overline{x}$ such that $(\gamma(x_n))$ does not converge to $\gamma(\overline{x})$ w.r.t. $\tau_{\mathcal{B}}$; this contradicts (vi).

(vii) $\Rightarrow$ (vi) is obvious; just take $\varepsilon_n=0$ for $n\in\mathbb{N}$.

(vii) $\Rightarrow$ (viii) is obvious; just take $x_n=\overline{x}$ for $n\in\mathbb{N}$.

(viii) $\Rightarrow$ (vii) Let $\overline{x}^*\in\partial f(\overline{x})$, $0\le\varepsilon_n$ and $x_n^*\in\partial_{\varepsilon_n}f(x_n)$ be such that

$(\varepsilon_n) \to 0$ and $(x_n) \to \overline{x}$. Consider $\overline{\varepsilon} := \sup\{\varepsilon_n \mid n \in \mathbb{N}\}$. Since f is continuous at $\overline{x}$, by Corollary 2.2.12, f is Lipschitz on a neighborhood of $\overline{x}$, while from Theorem 2.4.13 we have that $\partial_{\overline{\varepsilon}} f$ is bounded on a neighborhood of $\overline{x}$. Hence there exist $r, m > 0$ such that $|f(x) - f(y)| \le m \|x - y\|$ for $x, y \in D(\overline{x}, r)$ and $\partial_{\overline{\varepsilon}} f(D(\overline{x}, r)) \subset mU_{X^*}$. There exists $n_0 \in \mathbb{N}$ such that $x_n \in D(\overline{x}, r)$ for $n \ge n_0$. Let $x^* \in \partial_{\varepsilon_n} f(x)$. Then

$$\begin{aligned}\langle y - \overline{x}, x^*\rangle &= \langle y - x_n, x^*\rangle + \langle x_n - \overline{x}, x^*\rangle \\ &\le f(y) - f(x_n) + \varepsilon_n + \|x_n - \overline{x}\| \cdot \|x^*\| \\ &\le f(y) - f(\overline{x}) + m\|\overline{x} - x_n\| + \varepsilon_n + m \cdot \|x_n - \overline{x}\|\end{aligned}$$

for all $y \in X$, and so

$$\partial_{\varepsilon_n} f(x_n) \subset \partial_{\varepsilon_n + 2m\|x_n - \overline{x}\|} f(\overline{x}) \quad \forall n \ge n_0.$$

Because $(\varepsilon_n + 2m\|x_n - \overline{x}\|) \to 0$, from our hypothesis we obtain that $(x_n^*)_{n \in P} \overset{\tau_{\mathcal{B}}}{\to} x^*$.

Assume now that X is a Banach space.

(vi) $\Rightarrow$ (vii) Let $x^* \in \partial f(\overline{x})$, $0 \le \varepsilon_n$, $x_n^* \in \partial_{\varepsilon_n} f(x_n)$ be such that $(\varepsilon_n) \to 0$ and $(x_n) \to \overline{x}$. Because $x_n \in \operatorname{int}(\operatorname{dom} f)$ for large n and f and $\overline{f}$ coincide on $\operatorname{int}(\operatorname{dom} f)$, we may suppose that f is lsc. By the Borwein theorem, for every $n \in \mathbb{N}$ there exists $(x_n', x_n'^*) \in \operatorname{gr} \partial f$ such that $\|x_n - x_n'\| \le \sqrt{\varepsilon_n}$ and $\|x_n^* - x_n'^*\| \le \sqrt{\varepsilon_n}$. Therefore $(x_n') \to \overline{x}$. By (vi) we obtain that $(x_n'^*)_{n \in \mathbb{N}} \overset{\tau_{\mathcal{B}}}{\to} x^*$. As $(\|x_n^* - x_n'^*\|) \to 0$ and $\tau_{\mathcal{B}}$ is coarser than the norm topology, we obtain that $(x_n^*)_{n \in \mathbb{N}} \overset{\tau_{\mathcal{B}}}{\to} x^*$. □

Because any selection of ∂f coincides with the gradient of f at the points x where $\partial f(x)$ is a singleton, from the preceding theorem we get the following.

Corollary 3.3.3 *If $f \in \Lambda(X)$ is continuous and Gâteaux differentiable on the open set $D \subset \operatorname{dom} f$ then f is $\mathcal{B}$-differentiable on D if and only if ∇f is norm to $\tau_{\mathcal{B}}$ continuous on D.* □

As a consequence of Theorems 3.3.2 and 3.1.2 we get the following interesting result.

Corollary 3.3.4 *Let X be a Banach space and $f \in \Gamma(X)$. If f^* is Fréchet differentiable at $\overline{x}^* \in \operatorname{int}(\operatorname{dom} f^*)$ then $\nabla f^*(\overline{x}^*) \in X$.*

Proof. By the Brøndsted–Rockafellar theorem there exists $((x_n, x_n^*)) \subset \operatorname{gr} \partial f$ such that $(x_n^*) \to \overline{x}^*$. Then $x_n \in \partial f^*(\overline{x}_n^*)$ for the duality (X^*, X^{**}).

Taking the Fréchet bornology, from the implication (i) $\Rightarrow$ (vi) of the preceding theorem we get $(x_n) \to \nabla f^*(\overline{x}^*)$. Because X is closed in X^{**} we obtain that $\nabla f^*(\overline{x}^*) \in X$. □

A similar result holds for Gâteaux differentiability of f^* under supplementary conditions on X.

Corollary 3.3.5 *Let X be a weakly sequentially complete Banach space and $f \in \Gamma(X)$. If f^* is Gâteaux differentiable at $\overline{x}^* \in \operatorname{int}(\operatorname{dom} f^*)$ then $\nabla f^*(\overline{x}^*) \in X$.*

Proof. As in the proof of the preceding corollary, there exists a sequence $((x_n, x_n^*))$ in $\operatorname{gr} \partial f$ such that $(x_n^*) \to \overline{x}^*$. We have again that $x_n \in \partial f^*(\overline{x}_n^*)$ for the duality (X^*, X^{**}). Taking now the Gâteaux bornology, from the implication (i) $\Rightarrow$ (vi) of the preceding theorem we get $(x_n) \to \nabla f^*(\overline{x}^*)$ for the weak* topology of X^{**}, *i.e.* $(\langle x^*, x_n\rangle) \to \langle x^*, \nabla f^*(\overline{x}^*)\rangle$ for all $x^* \in X^*$. In particular (x_n) is a weakly Cauchy sequence, and so, by hypothesis, (x_n) converges weakly to some $\overline{x} \in X$. It follows that $\nabla f^*(\overline{x}^*) = \overline{x} \in X$. □

Note that every reflexive Banach space is weakly sequentially complete.

The next result shows that for convex functions the usual sufficient condition for Fréchet differentiability is also necessary.

Corollary 3.3.6 *Let $f \in \Gamma(X)$ be continuous and Gâteaux differentiable on a neighborhood of $\overline{x} \in \operatorname{dom} f$. Then f is Fréchet differentiable at $\overline{x}$ if and only if ∇f is continuous at $\overline{x}$.*

Proof. We may assume that f is continuous and Gâteaux differentiable on the ball $B(\overline{x}, r) \subset \operatorname{dom} f$ with $r > 0$. Because ∇f can be prolonged to a selection of ∂f, the conclusion follows from Theorem 3.3.2. □

Note that for convex functions defined on finite dimensional normed vector spaces Gâteaux and Fréchet differentiability coincide (use for example Theorem 3.3.2 and take into account the fact that the norm and weak topologies coincide). This is not true for infinite dimensional normed spaces as the functions in Exercises 2.10 and 3.1 show.

3.4 Well Conditioned Functions

Let $f \in \Lambda(X)$ be such that $\operatorname{argmin} f := \{x \in X \mid f(x) = \inf f\} \neq \emptyset$, $\psi \in \mathcal{A}$ and $S \subset X$ be a nonempty closed convex set. We say that f is

ψ-conditioned with respect to S if

$$\forall x \in X \ : \ f(x) \geq \inf f + \psi(d_S(x))$$

(see Eq. (3.62) on page 237 for the definition of d_S). If there exists $\eta > 0$ such that the above inequality holds for every $x \in X$ with $d_S(x) \leq \eta$, we say that f is *locally ψ-conditioned with respect to S*.

Let us note that if f is ψ-conditioned with respect to the subset S and $\psi \in \mathcal{A}_0$, then $\operatorname{argmin} f \subset S$. Taking into account this fact, we say that f is **ψ-conditioned** if f is ψ-conditioned with respect to $\operatorname{argmin} f$.

Assume that $S := \operatorname{argmin} f \neq \emptyset$ and consider the *conditioning gage* of f defined by

$$\psi_f : \mathbb{R}_+ \to \overline{\mathbb{R}}_+, \quad \psi_f(t) := \inf\{f(x) - \inf f \mid x \in X, \ d_S(x) = t\}. \quad (3.28)$$

It is clear that $\psi_f \in \mathcal{A}$ and f is ψ_f-conditioned. The following result holds.

Proposition 3.4.1 *Let $f \in \Lambda(X)$ be such that $S := \operatorname{argmin} f \neq \emptyset$. Then $\psi_f \in N_1$.*

Proof. Let $t > 0$ and $c > 1$; we must show that $\psi_f(ct) \geq c\psi_f(t)$. There exists a sequence $(x_n) \subset X$ such that $d_S(x_n) = ct$ and $\big(f(x_n) - \inf f\big) \to \psi_f(ct)$. Let $(\varepsilon_n) \downarrow 0$. For every n there exists $a_n \in S$ such that $\|x_n - a_n\| < ct+\varepsilon_n$. Let $\lambda_n := (\|x_n - a_n\| - ct + t) \,/\, \|x_n - a_n\| \in [c^{-1}, 1[$. We have that

$$d_S\left((1-\lambda_n)a_n + \lambda_n x_n\right) \geq d_S(x_n) - (1-\lambda_n)\|x_n - a_n\| = ct - ct + t = t.$$

Because $\lambda \mapsto d_S\left((1-\lambda)a_n + \lambda x_n\right)$ is a continuous function, there exists $\mu_n \in]0, \lambda_n]$ such that $d_S\left((1-\mu_n)a_n + \mu_n x_n\right) = t$. We obtain so that

$$\begin{aligned}\psi_f(t) &\leq f\left((1-\mu_n)a_n + \mu_n x_n\right) - \inf f \leq (1-\mu_n)f(a_n) + \mu_n f(x_n) - \inf f \\ &= \mu_n \left(f(x_n) - \inf f\right) \leq \lambda_n \left(f(x_n) - \inf f\right).\end{aligned}$$

Because $(\lambda_n) \to c^{-1}$, we obtain that $\psi_f(t) \leq c^{-1} \cdot \psi_f(ct)$. □

Let us remark we used only that $f((1-\lambda)a + \lambda x) \leq (1-\lambda)f(a) + \lambda f(x)$ for every $a \in S$, $x \in X$ and $\lambda \in [0,1]$, in other words the fact that f is star-shaped at every $a \in S$; of course, this condition ensures the convexity of S.

Corollary 3.4.2 *Let f and S be as in Proposition 3.4.1. If $\psi : [0, \alpha] \to \overline{\mathbb{R}}_+$, where $\alpha > 0$, is such that*

$$\forall x \in X, \ d_S(x) \leq \alpha \ : \ f(x) \geq \inf f + \psi\left(d_S(x)\right),$$

then f is ψ_α-conditioned, where

$$\psi_\alpha : \mathbb{R}_+ \to \overline{\mathbb{R}}_+, \quad \psi_\alpha(t) = \begin{cases} \psi(t) & \text{for} \quad t \in [0,\alpha], \\ \frac{\psi(\alpha)}{\alpha}t & \text{for} \quad t \in]\alpha, \infty[. \end{cases}$$

Proof. From the definition of ψ_f we have that $\psi_f \geq \psi$ on $[0,\alpha]$. For $d_S(x) = \gamma > \alpha$ we have that

$$\begin{aligned} f(x) &\geq \inf f + \psi_f(\gamma) \geq \inf f + \frac{\gamma}{\alpha}\psi_f(\alpha) \geq \inf f + \frac{\gamma}{\alpha}\psi(\alpha) \\ &= \inf f + \frac{\psi(\alpha)}{\alpha} d_S(x) = \inf f + \psi_\alpha\left(d_S(x)\right), \end{aligned}$$

whence the conclusion. □

Proposition 3.4.1 shows that $\psi_f(t) > 0$ for every $t > t_0$ if $\psi_f(t_0) > 0$. It is important the case in which $\psi_f(t) > 0$ for every $t > 0$, *i.e.* $\psi_f \in \mathcal{A}_0$; in such a case we say that f is **well-conditioned**. The following result holds.

Theorem 3.4.3 *Let X be a Banach space and $f \in \Gamma(X)$ be such that $S := \operatorname{argmin} f \neq \emptyset$. The following statements are equivalent:*

(i) *there exist $\psi \in \mathcal{A}$ and $t_0 > 0$ such that $\psi(t) > 0$ for every $t \in]0, t_0[$ and f is ψ-conditioned;*

(ii) *for every sequence $(x_n) \subset X$ with $(f(x_n)) \to \inf f$ we have that $(d_S(x_n)) \to 0$;*

(iii) *there exists $\psi \in \Gamma_0$ such that f is ψ-conditioned;*

(iv) $\exists \sigma \in \Sigma_1,\ \forall x^* \in X^* \ :\ f^*(x^*) \leq f^*(0) + \iota_S^*(x^*) + \sigma(\|x^*\|)$;

(v) $\exists \psi \in \Gamma_0,\ \forall \overline{x} \in S,\ \forall (x,x^*) \in \operatorname{gr} \partial f \ :\ \langle x - \overline{x}, x^* \rangle \geq \psi\left(d_S(x)\right)$;

(vi) $\exists \varphi \in \mathcal{A}_0 \cap N_0,\ \forall (x,x^*) \in \operatorname{gr} \partial f \ :\ \varphi(d_S(x)) \leq \|x^*\|$;

(vii) $\exists \theta \in \Omega_0 \cap N_0,\ \forall (x,x^*) \in \operatorname{gr} \partial f \ :\ d_S(x) \leq \theta(\|x^*\|)$;

(viii) $\forall (x_n) \subset X \ :\ \left(d_{\partial f(x_n)}(0)\right) \to 0 \Rightarrow \left(d_S(x_n)\right) \to 0$.

Furthermore, in the equivalence (iii) $\Leftrightarrow$ (iv) *we can take $\sigma = \psi^\#$, $\psi = \sigma^\#$, in* (iii) $\Rightarrow$ (v) *we can take the same ψ, in* (v) $\Rightarrow$ (vi) *we can take $\varphi \in \mathcal{A}$, $\varphi(t) = t^{-1}\psi(t)$ for $t > 0$, in* (vi) $\Rightarrow$ (vii) *we can take $\theta = \varphi^h$, in* (vii) $\Rightarrow$ (vi) *we can take $\varphi = \theta^e$ and in* (vi) $\Rightarrow$ (iii) *we can take $\psi(t) = (1-\gamma)\int_0^t \varphi(\gamma s)\, ds$ with $\gamma \in]0,1[$ fixed.*

Proof. (i) $\Rightarrow$ (iii) It is clear that $\psi_f \geq \psi$, and so $\psi_f \in \mathcal{A}_0$ (since ψ_f is nondecreasing). Furthermore, since $\liminf_{t\to\infty} t^{-1}\psi_f(t) > 0$ (because

$\psi_f \in \mathcal{A}_0 \cap N_1$), we have that $\widetilde{\psi} := \overline{\mathrm{co}}\psi_f \in \Gamma_0$ (see Lemma 3.3.1(iii)). Of course, f is $\widetilde{\psi}$-conditioned.

(iii) $\Rightarrow$ (ii) is obvious, since for $\psi \in \Gamma_0$ we have that $\big(\psi(t_n)\big) \to 0 \Rightarrow (t_n) \to 0$.

(ii) $\Rightarrow$ (i) Suppose that there exists $t_0 > 0$ such that $\psi_f(t_0) = 0$. Then there exists $(x_n) \subset X$ such that $d_S(x_n) = t_0$ and $(f(x_n) - \inf f) \to 0$, which evidently contradicts the hypothesis. As f is ψ_f-conditioned, the conclusion holds.

(iii) $\Rightarrow$ (iv) Using Theorems 2.8.10, 2.3.1(ix) and Corollaries 2.4.7, 2.4.16 (see also Exercise 3.7) we have that

$$(\psi \circ d_S)^* = \iota_S^* + \psi^\# \circ \|\cdot\|, \quad \left(\iota_S^* + \psi^\# \circ \|\cdot\|\right)^* = \psi \circ d_S. \tag{3.29}$$

Using Eq. (3.29), for $x^* \in X^*$ we have that

$$\begin{aligned} f^*(x^*) &= \sup_{x\in X} \big(\langle x, x^*\rangle - f(x)\big) \le \sup_{x\in X} \big(\langle x, x^*\rangle - \inf f - \psi(d_S(x))\big) \\ &= f^*(0) + (\psi \circ d_S)^*(x^*) = f^*(0) + \iota_S^*(x^*) + \psi^\#(\|x^*\|). \end{aligned}$$

The conclusion follows with $\sigma := \psi^\#$; $\sigma \in \Sigma_1$ by Lemma 3.3.1(iii).

(iv) $\Rightarrow$ (iii) is obtained similarly, using Eq. (3.29), with $\psi := \sigma^\#$.

(iii) $\Rightarrow$ (v) Let $\overline{x} \in S$ and $(x, x^*) \in \operatorname{gr}\partial f$. Then

$$f(x) \ge f(\overline{x}) + \psi(d_S(x)) \quad \text{and} \quad f(\overline{x}) \ge f(x) + \langle \overline{x} - x, x^*\rangle .$$

Adding these two relations we obtain the conclusion with the same ψ.

(v) $\Rightarrow$ (vi) From (v) we obtain that

$$\forall \overline{x} \in S,\ \forall (x, x^*) \in \operatorname{gr}\partial f \ :\ \|x - \overline{x}\| \cdot \|x^*\| \ge \psi(d_S(x)),$$

whence $\|x^*\| \cdot d_S(x) \ge \psi(d_S(x))$ for every $(x, x^*) \in \operatorname{gr}\partial f$. The conclusion follows taking $\varphi \in \mathcal{A}_0$ defined by $\varphi(t) := t^{-1}\psi(t)$ for $t > 0$.

(vi) $\Rightarrow$ (vii) Let $\theta := \varphi^h$. By Lemma 3.3.1(i) we have that $\theta \in N_0 \cap \Omega_0$. Furthermore, from the definition of θ, we have that $\theta(\|x^*\|) \ge d_S(x)$ for every $(x, x^*) \in \partial f$.

(vii) $\Rightarrow$ (vi) Let $\varphi := \theta^{\,e}$. By Lemma 3.3.1(i) we have that $\varphi \in \mathcal{A}_0$. By the definition of φ we have that $\varphi(d_S(x)) \le \|x^*\|$ for every $(x, x^*) \in \operatorname{gr}\partial f$.

(vi) $\Rightarrow$ (iii) Let $\gamma \in]0, 1[$ and

$$\psi : \mathbb{R}_+ \to \overline{\mathbb{R}}_+, \quad \psi(t) = (1-\gamma)t \cdot \varphi(\gamma t).$$

Suppose that (iii) does not hold for this ψ. Then there exists $\overline{x} \in X$ such that $f(\overline{x}) < \inf f + \psi(d_S(\overline{x}))$. It follows that $\overline{x} \notin S$ and there exists $c \in]0,1[$ such that

$$f(\overline{x}) < \inf f + c \cdot \psi(d_S(\overline{x})).$$

We take $\varepsilon = c \cdot \varphi\big(\gamma(d_S(\overline{x}))\big) > 0$. Using Ekeland's variational principle, there exists $u \in X$ such that

$$f(u) + \varepsilon \cdot \|u - \overline{x}\| \leq f(\overline{x}) \quad \text{and} \quad f(u) \leq f(x) + \varepsilon \cdot \|u - x\| \ \forall x \in X.$$

The last relation being equivalent to $\partial f(u) \cap \varepsilon U_{X^*} \neq \emptyset$, there exists $u^* \in X^*$ such that $(u, u^*) \in \operatorname{gr} \partial f$ and $\|u^*\| \leq \varepsilon$. From the first relation, by the choice of $\overline{x}$, the definitions of ψ and ε, and the hypothesis, it follows that

$$\|u - \overline{x}\| \leq (1 - \gamma) d_S(\overline{x}) \quad \text{and} \quad \varphi(d_S(u)) \leq \|u^*\| \leq \varepsilon = c \cdot \varphi\big(\gamma(d_S(\overline{x}))\big).$$

But

$$d_S(u) \geq d_S(\overline{x}) - \|u - \overline{x}\| \geq d_S(\overline{x}) - (1 - \gamma) d_S(\overline{x}) = \gamma \cdot d_S(\overline{x}),$$

whence the contradiction

$$\varphi(d_S(u)) \geq \varphi(\gamma \cdot d_S(\overline{x})) > c \cdot \varphi(\gamma \cdot d_S(\overline{x})).$$

Hence (iii) holds for this ψ. Because $t \cdot \varphi(\gamma t) \geq \int_0^t \varphi(\gamma s)\, ds$, we can replace ψ by $\widetilde{\psi} \in \Gamma_0$, where $\widetilde{\psi}(t) := (1 - \gamma) \int_0^t \varphi(\gamma s)\, ds$.

(vii) $\Rightarrow$ (viii) is obvious.

(viii) $\Rightarrow$ (vii) Consider the function

$$\theta : \mathbb{R}_+ \to \overline{\mathbb{R}}_+, \quad \theta(t) = \sup\{d_S(x) \mid (x, x^*) \in \operatorname{gr} \partial f, \ \|x^*\| \leq t\}.$$

It is obvious that θ is nondecreasing, $\theta(0) = 0$ (*i.e.* $\theta \in N_0$) and $d_S(x) \leq \theta(\|x^*\|)$ for every $(x, x^*) \in \operatorname{gr} \partial f$. If $\lim_{t \downarrow 0} \theta(t) > 0$, there exists $\varepsilon_0 > 0$ such that $\theta(t) \geq \varepsilon_0$ for every $t > 0$. From the definition of θ, for every $n \in \mathbb{N}$ there exists $(x_n, x_n^*) \in \partial f$ such that $\|x_n^*\| \leq 1/n$ and $d_S(x_n) \geq \varepsilon_0$. Of course this contradicts the hypothesis. □

We remark that we used the convexity of ψ only for the implication (iv) $\Rightarrow$ (iii); for the other implications it is sufficient that $\psi \in N_1$. The function ψ_f verifies this condition. Also, we used that X is a Banach space only in the implication (vi) $\Rightarrow$ (iii). We also note that in the implications (iii) $\Leftrightarrow$ (iv), (v) $\Rightarrow$ (vi) $\Leftrightarrow$ (vii) $\Leftrightarrow$ (viii), (vi) $\Rightarrow$ (iii) $\Rightarrow$ (ii) $\Rightarrow$ (i) we can take $S \subset X$ a nonempty closed convex set, and in the implication (iii) $\Rightarrow$ (v) we used only the fact that $S \subset \operatorname{argmin} f$.

Corollary 3.4.4 *Let X be a Banach space, $f \in \Gamma(X)$ and $(\overline{x}, \overline{x}^*) \in \operatorname{gr} \partial f$. The following statements are equivalent:*

(i) $\exists \psi \in \Gamma_0,\ \forall x \in X \ :\ f(x) \geq f(\overline{x}) + \langle x - \overline{x}, \overline{x}^* \rangle + \psi(\|x - \overline{x}\|)$;

(ii) $\exists \sigma \in \Sigma_1,\ \forall x^* \in X^* \ :\ f^*(x^*) \leq f(\overline{x}^*) + \langle \overline{x}, x^* - \overline{x}^* \rangle + \sigma(\|x^* - \overline{x}^*\|)$;

(iii) $\exists \psi \in \Gamma_0,\ \forall (x, x^*) \in \operatorname{gr} \partial f \ :\ \langle x - \overline{x}, x^* - \overline{x}^* \rangle \geq \psi(\|x - \overline{x}\|)$;

(iv) $\exists \varphi \in \mathcal{A}_0 \cap N_0,\ \forall (x, x^*) \in \operatorname{gr} \partial f \ :\ \varphi(\|x - \overline{x}\|) \leq \|x^* - \overline{x}^*\|$;

(v) $\exists \theta \in \Omega_0 \cap N_0,\ \forall (x, x^*) \in \operatorname{gr} \partial f \ :\ \|x - \overline{x}\| \leq \theta(\|x^* - \overline{x}^*\|)$;

(vi) $\forall (x_n) \subset X \ :\ (d_{\partial f(x_n)}(\overline{x}^*)) \to 0 \Rightarrow (x_n) \to \overline{x}$;

(vii) $\overline{x}^* \in \operatorname{int}(\operatorname{dom} f^*)$ *and f^* is Fréchet differentiable at $\overline{x}^*$.*

Proof. Taking the auxiliary function $g : X \to \overline{\mathbb{R}}$ defined by $g(x) = f(x) - \langle x, x^* \rangle$, we may assume that $\overline{x}^* = 0$.

The equivalence of conditions (i)–(vi) follows from Theorem 3.4.3 by taking $S := \{\overline{x}\}$ (taking into account, eventually, the remarks done after the proof of that theorem).

(ii) $\Rightarrow$ (vii) Taking into account that $(\overline{x}, \overline{x}^*) \in \operatorname{gr} \partial f$, we have that

$$\forall x^* \in X^* \ :\ 0 \leq f^*(x^*) - f^*(\overline{x}^*) - \langle \overline{x}, x^* - \overline{x}^* \rangle \leq \sigma(\|x^* - \overline{x}^*\|),$$

whence $\overline{x}^* \in \operatorname{int}(\operatorname{dom} f^*)$ and

$$\limsup_{x^* \to \overline{x}^*} \left| \frac{f^*(x^*) - f^*(\overline{x}^*) - \langle \overline{x}, x^* - \overline{x}^* \rangle}{\|x^* - \overline{x}^*\|} \right| \leq \limsup_{t \downarrow 0} \frac{\sigma(t)}{t} = \lim_{t \downarrow 0} \frac{\sigma(t)}{t} = 0.$$

This shows that f^* is Fréchet differentiable at $\overline{x}^*$ and $\nabla f^*(\overline{x}^*) = \overline{x}$.

(vii) $\Rightarrow$ (ii) Because $(\overline{x}^*, \overline{x}) \in \operatorname{gr}(\partial f)^{-1} \subset \operatorname{gr} \partial f^*$ (the last one for the duality (X^*, X^{**})), we obtain that $\overline{x} \in \partial f^*(\overline{x}^*)$. Since f^* is Fréchet differentiable at $\overline{x}^*$, we have that $\nabla f^*(\overline{x}^*) = \overline{x}$. For $t \in \mathbb{R}_+$ consider

$$\begin{aligned}\sigma(t) &:= \sup\{f^*(x^*) - f^*(\overline{x}^*) - \langle \overline{x}, x^* - \overline{x}^* \rangle \mid \|x^* - \overline{x}^*\| = t\} \\ &= \sup\{f^*(\overline{x}^* + tu^*) - f^*(\overline{x}^*) - t\,\langle \overline{x}, u^* \rangle \mid \|u^*\| = 1\}.\end{aligned}$$

Because the function $t \mapsto f^*(\overline{x}^* + tu^*) - f^*(\overline{x}^*) - t\,\langle \overline{x}, u^* \rangle$ $(\in \overline{\mathbb{R}}_+)$ is convex and lsc, it follows that $\sigma \in \Gamma$. Furthermore, from the Fréchet differentiability of the function f^* it follows that $\lim_{t \downarrow 0} t^{-1}\sigma(t) = 0$, and so $\sigma \in \Sigma_1$. The proof is complete. □

A sufficient condition for (i) from the preceding corollary is that

$$f((1-\lambda)\overline{x}+\lambda x)+\lambda(1-\lambda)\psi(\|x-\overline{x}\|)\leq\ (1-\lambda)f(\overline{x})+\lambda f(x) \tag{3.30}$$

for all $x \in X$ and $\lambda \in]0,1[$. Indeed, if Eq. (3.30) holds and $\overline{x} \in \operatorname{dom} f$ then

$$\big(f(\overline{x}+\lambda(x-\overline{x}))-f(\overline{x})\big)/\lambda \leq f(x)-f(\overline{x})-(1-\lambda)\psi(\|x-\overline{x}\|)$$

for all $x \in X$ and all $\lambda \in]0,1[$, whence

$$\forall\, x \in X\ :\ f(x) \geq f(\overline{x})+f'(\overline{x},x-\overline{x})+\psi(\|x-\overline{x}\|), \tag{3.31}$$

and so

$$\forall\, x \in X,\ \forall x^* \in \partial f(\overline{x})\ :\ f(x) \geq f(\overline{x})+\langle x-\overline{x},x^*\rangle+\psi(\|x-\overline{x}\|). \tag{3.32}$$

A function $f \in \Lambda(X)$ for which there exists $\psi \in \mathcal{A}_0$ satisfying condition (3.30) is said to be **uniformly convex at** $\overline{x}$ ($\in \operatorname{dom} f$).

Let $f \in \Lambda(X)$ and $\overline{x} \in \operatorname{dom} f$. The **gage of uniform convexity of** f **at** $\overline{x}$ is related to relation (3.30), and is defined by

$$\begin{aligned}\rho_{f,\overline{x}}(t) &:= \inf\left\{\frac{(1-\lambda)f(\overline{x})+\lambda f(x)-f((1-\lambda)\overline{x}+\lambda x)}{\lambda(1-\lambda)}\ \middle|\ \lambda \in]0,1[,\right.\\ &\qquad\qquad\qquad\qquad\qquad\qquad\qquad\qquad\left. \|x-\overline{x}\|=t\right\}\\ &= \inf\left\{\frac{(1-\lambda)f(\overline{x})+\lambda f(\overline{x}+tu)-f(\overline{x}+\lambda tu)}{\lambda(1-\lambda)}\ \middle|\ u \in S_X,\ \lambda \in]0,1[\right\}\end{aligned} \tag{3.33}$$

(with the convention $\infty-\infty=\infty$). Using Theorem 2.1.5(vii) one obtains easily that $\rho_{f,\overline{x}} \in N_0$ (Exercise!).

We may introduce also a gage related to relation (3.31). More exactly consider

$$\begin{aligned}\vartheta_{f,\overline{x}}(t) &:= \inf\left\{f(x)-f(\overline{x})-f'(\overline{x},x-\overline{x}) \mid x \in \operatorname{dom} f,\ \|x-\overline{x}\|=t\right\}\\ &= \inf\left\{f(\overline{x}+tu)-f(\overline{x})-tf'(\overline{x},u) \mid u \in S_X \cap \operatorname{cone}(\operatorname{dom} f-\overline{x})\right\}.\end{aligned}$$

Because the mapping $\mathbb{R}_+ \ni t \mapsto f(\overline{x}+tu)-f(\overline{x})-tf'(\overline{x},u) \in \overline{\mathbb{R}}_+$ belongs to $\Gamma \subset N_1$, we obtain that $\vartheta_{f,\overline{x}} \in N_1$. As in the proof of (3.30) $\Rightarrow$ (3.31) we have that $\vartheta_{f,\overline{x}} \geq \rho_{f,\overline{x}}$. Thus, when f is uniformly convex at $\overline{x} \in \operatorname{dom} f$ (*i.e.* $\rho_{f,\overline{x}} \in \mathcal{A}_0$) then $\vartheta_{f,\overline{x}} \in \mathcal{A}_0$. One may ask if the converse implication holds. The answer is affirmative when f is Fréchet differentiable at $\overline{x} \in \operatorname{dom} f$.

Proposition 3.4.5 *Let $f \in \Gamma(X)$ be Fréchet differentiable at $\overline{x} \in \operatorname{dom} f$. If $\vartheta_{f,\overline{x}} \in \mathcal{A}_0$ then f is uniformly convex at $\overline{x}$.*

Proof. Let $t > 0$ be such that $D(\overline{x}, t) \subset \operatorname{int}(\operatorname{dom} f)$. By hypothesis

$$f(x) \geq f(\overline{x}) + f'(\overline{x}, x - \overline{x}) + \delta \quad \forall x \in X,\ \|x - \overline{x}\| = t/2,$$

where $\delta = \vartheta_{f,\overline{x}}(t/2) > 0$. On the other hand, because f is Fréchet differentiable at $\overline{x}$, a selection $\gamma : \operatorname{dom} \partial f \to X^*$ of ∂f exists which is continuous at $\overline{x}$ (by Theorem 3.3.2). Hence there exists $\eta > 0$ such that $\|\gamma(z) - \nabla f(\overline{x})\| \leq \delta/t$ for $\|z - \overline{x}\| \leq \eta$. Let $\alpha \in]0, 1[$ be such that $1 - \alpha \leq 2\eta/t$. Consider $x \in X$ arbitrary satisfying $\|x - \overline{x}\| = t$ and take $z := \alpha\overline{x} + (1-\alpha)\frac{\overline{x}+x}{2} = \overline{x} + \frac{1-\alpha}{2}(x - \overline{x}) \in \operatorname{int}(\operatorname{dom} f)$. Hence $\|z - \overline{x}\| = \frac{1-\alpha}{2}t \leq \eta$, and so $\|\gamma(z) - \nabla f(\overline{x})\| \leq \delta/t$. Of course,

$$f(\overline{x}) \geq f(z) + f'(z, \overline{x} - z) = f(z) + f'\left(z, \tfrac{1-\alpha}{2}(\overline{x} - x)\right). \tag{3.34}$$

As $\frac{1}{2}(\overline{x} + x) = \frac{\alpha}{\alpha+1}x + \frac{1}{\alpha+1}z$, we have that

$$\tfrac{\alpha}{\alpha+1} f(x) + \tfrac{1}{\alpha+1} f(z) \geq f\left(\tfrac{1}{2}(\overline{x} + x)\right). \tag{3.35}$$

Multiplying Eq. (3.35) with $\alpha + 1$ and adding the result to (3.34) we obtain

$$f(\overline{x}) + \alpha f(x) \geq (1 + \alpha) f\left(\tfrac{1}{2}(\overline{x} + x)\right) + f'\left(z, \tfrac{1-\alpha}{2}(\overline{x} - x)\right),$$

whence

$$\begin{aligned} f(\overline{x}) + f(x) - 2f\left(\tfrac{1}{2}(\overline{x} + x)\right) &\geq \tfrac{1-\alpha}{\alpha}\left(f\left(\tfrac{1}{2}(\overline{x} + x)\right) - f(\overline{x}) + \tfrac{1}{2}f'(z, \overline{x} - x)\right) \\ &\geq \tfrac{1-\alpha}{\alpha}\left(\delta + \tfrac{1}{2}f'(\overline{x}, x - \overline{x}) + \tfrac{1}{2}f'(z, \overline{x} - x)\right) \\ &\geq \tfrac{1-\alpha}{\alpha}\left(\delta + \tfrac{1}{2}\langle x - \overline{x}, \nabla f(\overline{x}) - \gamma(z)\rangle\right) \\ &\geq \tfrac{1-\alpha}{\alpha}\left(\delta - \tfrac{t}{2}\|\nabla f(\overline{x}) - \gamma(z)\|\right) \\ &\geq \tfrac{1-\alpha}{\alpha} \cdot \tfrac{\delta}{2} \;=:\; \delta_0 > 0. \end{aligned}$$

Therefore $f(\overline{x}) + f(x) - 2f\left(\frac{1}{2}(\overline{x} + x)\right) \geq \delta_0$ for every $x \in X$ with $\|x - \overline{x}\| = t$. The conclusion follows like in Exercise 2.24(iv). □

Without Fréchet differentiability of f at $\overline{x}$ the result in the preceding proposition may be false (see Exercise 3.4).

3.5 Uniformly Convex and Uniformly Smooth Convex Functions

We "globalize" the notion of uniformly convex function at a point as follows. We say that the proper function $f : X \to \overline{\mathbb{R}}$ is **ρ-convex**, where $\rho \in \mathcal{A}$, if

$$f((1-\lambda)x + \lambda y) + \lambda(1-\lambda)\rho(\|x-y\|) \leq (1-\lambda)f(x) + \lambda f(y).$$

for all $x, y \in X$ and all $\lambda \in]0,1[$. Of course, if f is ρ-convex then f is convex; we say that f is **uniformly convex** if there exists $\rho \in \mathcal{A}_0$ such that f is ρ-convex. We say that f is **strongly convex** (with rate c or *c-strongly convex*) if f is ρ_c-convex for some $c > 0$, where $\rho_c \in \mathcal{A}_0$, $\rho_c(t) := \frac{1}{2}ct^2$. It is clear that when f is uniformly convex then f is uniformly convex at every point of its domain and when f is uniformly convex at every point of its domain then f is strictly convex.

When $f \in \Lambda(X)$ and $A \subset X$ is a convex set we say that f is **uniformly convex on** A if $f + \iota_A$ is uniformly convex.

It is natural to introduce the **gage of uniform convexity** of the convex function f. Namely, it is the function $\rho_f : \mathbb{R}_+ \to \overline{\mathbb{R}}_+$ defined by

$$\rho_f(t) = \inf\left\{ \frac{(1-\lambda)f(x) + \lambda f(y) - f((1-\lambda)x + \lambda y)}{\lambda(1-\lambda)} \;\middle|\; \lambda \in]0,1[,\right.$$
$$\left. x, y \in \operatorname{dom} f,\ \|x-y\| = t\right\}.$$

Of course, $\rho_f \geq \rho$ when f is ρ-convex. Hence f is uniformly convex if, and only if, $\rho_f \in \mathcal{A}_0$.

The gage of uniform convexity of the convex function f has a remarkable property.

Proposition 3.5.1 *Let $f \in \Lambda(X)$. Then*

$$\forall t \geq 0,\ \forall c \geq 1 \ :\ \rho_f(ct) \geq c^2 \rho_f(t),$$

i.e. the function $t \mapsto t^{-2}\rho_f(t)$ is nondecreasing on $\mathbb{P}$.

Proof. Let $t > 0$ and $c \in]1,2[$ be such that $\rho_f(ct) < \infty$. Let us fix $\varepsilon > 0$. There exist $x, y \in \operatorname{dom} f$ and $\lambda \in]0, \frac{1}{2}]$ such that $\|y - x\| = ct$ and

$$\rho_f(ct) + \varepsilon > \frac{(1-\lambda)f(x) + \lambda f(y) - f((1-\lambda)x + \lambda y)}{\lambda(1-\lambda)}.$$

Consider $x_\lambda := (1-\lambda)x + \lambda y$ and $x_c := (1-c^{-1})x + c^{-1}y$. We have that $\|x_c - x\| = t$ and $x_\lambda = (1-c\lambda)x + c\lambda x_c$ (of course $c\lambda \in]0,1[$). Hence

$$f(x_c) \le (1-c^{-1})f(x) + c^{-1}f(y) - c^{-1}(1-c^{-1})\rho_f(ct),$$
$$f(x_\lambda) \le (1-c\lambda)f(x) + c\lambda f(x_c) - c\lambda(1-c\lambda)\rho_f(t),$$
$$f(x_\lambda) > (1-\lambda)f(x) + \lambda f(y) - \lambda(1-\lambda)\rho_f(ct) - \varepsilon\lambda(1-\lambda).$$

From these relations (multiplying the first with $c\lambda > 0$) we obtain that

$$c^2\rho_f(t) < \rho_f(ct) + \varepsilon c \cdot \tfrac{1-\lambda}{1-c\lambda} \le \rho_f(ct) + \varepsilon \cdot \tfrac{2c}{2-c}.$$

Letting $\varepsilon \to 0$, we obtain that $c^2\rho_f(t) \le \rho_f(ct)$ for all $c \in]1,2[$ and $t > 0$. By mathematical induction one obtains immediately that $c^{2n}\rho_f(t) \le \rho_f(c^n t)$ for $c \in]1,2[$, $n \in \mathbb{N}$ and $t > 0$, and so $c^2\rho_f(t) \le \rho_f(ct)$ for every $c > 1$ and $t > 0$. The conclusion holds. □

A dual notion (as we shall see) for uniform convexity is that of uniform smoothness.

Let $f \in \Lambda(X)$, $\sigma \in \mathcal{A}$ and $\emptyset \ne A \subset \operatorname{dom} f$. We say that f is **σ-smooth on** A if for all $x, y \in X$ and all $\lambda \in]0,1[$ such that $(1-\lambda)x + \lambda y \in A$ we have

$$f((1-\lambda)x + \lambda y) + \lambda(1-\lambda)\sigma(\|x-y\|) \ge (1-\lambda)f(x) + \lambda f(y), \qquad (3.36)$$

or equivalently

$$f(x) + \lambda(1-\lambda)\sigma(\|y\|) \ge (1-\lambda)f(x-\lambda y) + \lambda f(x + (1-\lambda)y)$$

for all $x \in A$, $y \in X$ and $\lambda \in]0,1[$; we say that f is **σ-smooth** if f is σ-smooth on $\operatorname{dom} f$. Having in view this definition, it is natural to consider the following **gage of uniform smoothness on** A of the function f:

$$\begin{aligned}\sigma_{f,A}(t) := \sup\Bigg\{ \frac{(1-\lambda)f(x) + \lambda f(y) - f((1-\lambda)x+\lambda y)}{\lambda(1-\lambda)} \,\Bigg|\, &\lambda \in]0,1[,\\ &x,y \in X,\ (1-\lambda)x + \lambda y \in A,\ \|x-y\| = t\Bigg\}\\ = \sup\Bigg\{ \frac{(1-\lambda)f(x-\lambda t y) + \lambda f(x+(1-\lambda)ty) - f(x)}{\lambda(1-\lambda)} \,\Bigg|\, &\lambda \in]0,1[,\\ &x \in A,\ y \in S_X\Bigg\};\end{aligned} \qquad (3.37)$$

$\sigma_{f,\mathrm{dom}\,f}$ is denoted by σ_f and is called the **gage of uniform smoothness** of f. It is obvious that $\sigma_{f,A} \le \sigma$ if f is σ-smooth on A. Sometimes it is useful to consider the following *midpoint gage of smoothness*:

$$\begin{aligned}\sigma^0_{f,A}(t) &:= \sup\left\{f(x)+f(y)-2f(\tfrac{1}{2}x+\tfrac{1}{2}y) \mid \tfrac{1}{2}x+\tfrac{1}{2}y \in A,\ \|x-y\|=2t\right\}\\ &= \sup\left\{f(x-ty)+f(x+ty)-2f(x) \mid x\in A,\ y\in S_X\right\}. \qquad (3.38)\end{aligned}$$

It is easy to obtain that $2\sigma^0_{f,A}(t/2) \le \sigma_{f,A}(t) \le \sigma^0_{f,A}(t)$ for $t \ge 0$. As above, $\sigma^0_{f,\mathrm{dom}\,f}$ will be denoted simply σ^0_f. It is obvious that $\sigma_{f,A}, \sigma^0_{f,A} \in \mathcal{A}$; moreover, since f is convex, by Eqs. (3.37) and (3.38), they are also convex as suprema of convex functions.

Proposition 3.5.2 *Let $f \in \Lambda(X)$, $\emptyset \ne A \subset \mathrm{dom}\,f$ and $t_0 > 0$. If $\sigma_{f,A}(t_0) < \infty$ then $A + t_0 B_X \subset \mathrm{dom}\,f$; if $\sigma_f(t_0) < \infty$ then $\mathrm{dom}\,f = X$ and $\sigma_f \in \Gamma$.*

Proof. Suppose first that $\sigma_f(t_0) < \infty$ and take $x \in A$. From Eq. (3.37) we obtain that $x - \lambda t_0 y \in \mathrm{dom}\,f$ for every $\lambda \in]0,1[$ and $y \in S_X$, and so $B(x, t_0) \subset \mathrm{dom}\,f$. Therefore $A + t_0 B_X \subset \mathrm{dom}\,f$. If $A = \mathrm{dom}\,f$ we obtain that $\mathrm{dom}\,f + nt_0 B_X \subset \mathrm{dom}\,f$ for every $n \in \mathbb{N}$, whence $\mathrm{dom}\,f = X$. In this case the functions $t \mapsto f(x - \lambda ty)$ and $t \mapsto f(x + (1-\lambda)ty)$ are convex and finite, and so continuous; from Eq. (3.37) we obtain that $\sigma_f \in \Gamma$. □

The relationships between uniform convexity and uniform smoothness are given in the following result.

Proposition 3.5.3 *Let $f \in \Lambda(X)$, $h \in \Lambda(X^*)$ have proper conjugates and $\rho, \sigma \in \mathcal{A}$.*

(i) *If f and h are ρ-convex then f^* and h^* are $\rho^\#$-smooth, respectively.*

(ii) *If f and h are σ-smooth then f^* and h^* are $\sigma^\#$-convex, respectively.*

Proof. (i) Suppose that f is ρ-convex. Let $t > 0$ be fixed and consider $x^* \in X^*$, $y^* \in S_{X^*}$ and $\lambda \in]0,1[$; take $x_0^* := x^* - \lambda ty^*$, $x_1^* := x^* + (1-\lambda)ty^*$. Consider also $\gamma_0 < f^*(x_0^*)$ and $\gamma_1 < f^*(x_1^*)$. From the definition of f^* there exist $x_0, x_1 \in X$ such that

$$\gamma_0 < \langle x_0, x_0^*\rangle - f(x_0), \quad \gamma_1 < \langle x_1, x_1^*\rangle - f(x_1).$$

Multiplying the first relation by $1-\lambda$ and the second one by λ, then adding them to both sides of the (Young–Fenchel) inequality

$$0 \le f(x_\lambda) + f^*(x^*) - \langle x_\lambda, x^*\rangle,$$

where $x_\lambda := (1-\lambda)x_0 + \lambda x_1$, we obtain that

$$\begin{aligned}(1-\lambda)\gamma_0 + \lambda\gamma_1 \\ &\le f(x_\lambda) + f^*(x^*) - (1-\lambda)f(x_0) - \lambda f(x_1) + \lambda(1-\lambda)t\,\langle x_1 - x_0, y^*\rangle \\ &\le f^*(x^*) + \lambda(1-\lambda)\left(t\,\|x_0 - x_1\| - \rho(\|x_0 - x_1\|)\right) \\ &\le f^*(x^*) + \lambda(1-\lambda)\rho^{\#}(t).\end{aligned}$$

Letting $\gamma_i \to f^*(x_i^*)$ for $i = 1, 2$, we obtain that f^* is $\rho^{\#}$-smooth.

The same calculation shows that $h^* : X \to \overline{\mathbb{R}}$ is $\rho^{\#}$-smooth if h is ρ -convex.

(ii) Suppose now that f is σ-smooth. Consider $x_0^*, x_1^* \in X^*$ and $\lambda \in]0,1[$. For $x, y \in X$, denoting $x_\lambda^* := (1-\lambda)x_0^* + \lambda x_1^*$, we have that

$$\begin{aligned}(1-\lambda)f^*(x_0^*) + \lambda f^*(x_1^*) \\ &\ge (1-\lambda)\left(\langle x - \lambda y, x_0^*\rangle - f(x - \lambda y)\right) \\ &\qquad + \lambda\big(\langle x + (1-\lambda)y, x_1^*\rangle - f(x + (1-\lambda)y)\big) \\ &\ge (1-\lambda)\,\langle x - \lambda y, x_0^*\rangle + \lambda\,\langle x + (1-\lambda)y, x_1^*\rangle - f(x) - \lambda(1-\lambda)\sigma(\|y\|) \\ &= \langle x, x_\lambda^*\rangle - f(x) + \lambda(1-\lambda)\left(\langle y, x_1^* - x_0^*\rangle - \sigma(\|y\|)\right).\end{aligned}$$

Since $x, y \in X$ are arbitrary, we have that

$$\begin{aligned}(1-\lambda)f^*(x_0^*) + \lambda f^*(x_1^*) \\ &\ge \sup\nolimits_{x\in X}\left(\langle x, x_\lambda^*\rangle - f(x)\right) + \lambda(1-\lambda)\sup\nolimits_{y\in X}\left(\langle y, x_1^* - x_0^*\rangle - \sigma(\|y\|)\right) \\ &= f^*(x_\lambda^*) + \lambda(1-\lambda)\sigma^{\#}(\|x_0^* - x_1^*\|),\end{aligned}$$

which means that f^* is $\sigma^{\#}$-convex.

The proof for the fact that h^* is $\sigma^{\#}$-convex when h is σ-smooth is similar. □

Corollary 3.5.4 *Let $f \in \Gamma(X)$. Then $\sigma_{f^*} = (\rho_f)^{\#}$ and $\sigma_f = (\rho_{f^*})^{\#}$. Furthermore $t \mapsto t^{-2}\sigma_f(t)$ is nonincreasing on $\mathbb{P}$. In particular, if there exists $t_0 > 0$ such that $\sigma_f(t_0) < \infty$ then $\sigma_f(t) < \infty$ for every $t \ge 0$.*

Proof. From the preceding proposition we have that f^* is $(\rho_f)^{\#}$-smooth and $(\sigma_f)^{\#}$-convex, while $f = f^{**}$ is $(\rho_{f^*})^{\#}$-smooth and $(\sigma_{f^*})^*$ -convex. Therefore we have that $\sigma_f \le (\rho_{f^*})^{\#}$ and $\rho_{f^*} \ge (\sigma_f)^{\#}$. Passing to the conjugates in the second inequality we obtain that $\sigma_f \le (\rho_{f^*})^{\#} \le (\sigma_f)^{\#\#}$. Therefore $\sigma_f = (\rho_{f^*})^{\#} = (\sigma_f)^{\#\#}$ (in particular we have obtained again that σ_f is convex and lower semicontinuous). In a similar way we obtain that $\sigma_{f^*} = (\rho_f)^\#$.

Taking into account Proposition 3.5.1, Lemma 3.3.1(iv) and the relation $\sigma_f = (\rho_{f^*})^\#$, we have that $t \mapsto t^{-2}\sigma_f(t)$ is nonincreasing on $\mathbb{P}$. The remainder of the conclusion is immediate. □

We say that $f \in \Lambda(X)$ is **uniformly smooth** (on the nonempty subset A of $\operatorname{dom} f$) if there exists $\sigma \in \Omega_1$ such that f is σ-smooth (on A). When A is a singleton $\{\overline{x}\} \subset \operatorname{dom} f$, f is uniformly smooth on $\{\overline{x}\}$ if and only if f is smooth at $\overline{x}$ in the sense of Theorem 3.3.2. Taking into account Proposition 3.5.2, if f is uniformly smooth on $A \subset X$ then there exists $\varepsilon_0 > 0$ such that $A + \varepsilon_0 B_X \subset \operatorname{dom} f$, while if f is uniformly smooth then $\operatorname{dom} f = X$.

Summarizing Propositions 3.5.1–3.5.3 (and using Lemma 3.3.1(iii)) we obtain the following important result.

Theorem 3.5.5 *Let $f \in \Gamma(X)$. Then*

(i) *f is uniformly convex $\Leftrightarrow$ f^* is uniformly smooth;*

(ii) *f is uniformly smooth $\Leftrightarrow$ f^* is uniformly convex.* □

It is quite natural to say that the proper function $f : X \to \overline{\mathbb{R}}$ is *uniformly Fréchet differentiable on $\emptyset \neq A \subset X$* if there exists $\varepsilon_0 > 0$ such that $A + \varepsilon_0 B_X \subset \operatorname{dom} f$, f is Fréchet differentiable at any point of A and

$$\lim_{t \to 0} \frac{f(x + tu) - f(x)}{t} = \nabla f(x)(u),$$

uniformly with respect to $x \in A$ and $u \in S_X$. Of course, when $A = \{a\}$ this means that f is Fréchet differentiable at a. Similarly to the proof of Theorem 3.3.2, one obtains easily that $f \in \Lambda(X)$ is uniformly Fréchet differentiable on $A \subset \operatorname{dom} f$ if and only if f is continuous at some point of its domain and is uniformly smooth on A. The following result gives other characterizations of uniformly smooth convex functions.

Theorem 3.5.6 *Let $f \in \Lambda(X)$. Consider the following statements:*

(i) *f is uniformly smooth;*

(ii) $(\operatorname{dom} f)^i \neq \emptyset$ *and there exists $\sigma_2 \in \Omega_1$ such that*

$$\forall x \in (\operatorname{dom} f)^i,\ \forall y \in X \ :\ f(y) \leq f(x) + f'(x, y - x) + \sigma_2(\|y - x\|); \quad (3.39)$$

(iii) $\operatorname{dom} f = X$, *$f'(x, \cdot)$ is linear for every $x \in X$ and there exists $\sigma_3 \in$*

Ω_1 *such that*

$$\forall\, x,\, y \in X \;:\; f'(x, x-y) + f'(y, y-x) \le \sigma_3(\|y-x\|); \tag{3.40}$$

(iv) $\operatorname{gr}\partial f \neq \emptyset$ *and there exists* $\sigma_4 \in \Omega_1$ *such that*

$$\forall\, (x, x^*) \in \operatorname{gr}\partial f,\ \forall\, y \in X \;:\; f(y) \le f(x) + \langle y-x, x^*\rangle + \sigma_4(\|y-x\|); \tag{3.41}$$

(v) $\operatorname{dom}\partial f = X$ *and there exists* $\sigma_5 \in \Omega_1$ *such that*

$$\forall\, (x, x^*),\, (y, y^*) \in \operatorname{gr}\partial f \;:\; \langle y-x, y^*-x^*\rangle \le \sigma_5(\|y-x\|); \tag{3.42}$$

(vi) $\operatorname{dom}\partial f = X$ *and there exists* $\psi_1 \in \Gamma_0$ *such that*

$$\forall\, (x, x^*),\, (y, y^*) \in \operatorname{gr}\partial f \;:\; f(y) \ge f(x) + \langle y-x, x^*\rangle + \psi_1(\|y^*-x^*\|); \tag{3.43}$$

(vii) $\operatorname{dom}\partial f = X$ *and there exists* $\psi_2 \in \Gamma_0$ *such that*

$$\forall\, (x, x^*),\, (y, y^*) \in \operatorname{gr}\partial f \;:\; \langle y-x, y^*-x^*\rangle \ge \psi_2(\|y^*-x^*\|); \tag{3.44}$$

(viii) $\operatorname{dom}\partial f = X$ *and there exists* $\varphi \in \mathcal{A}_0 \cap N_0$ *such that*

$$\forall\, (x, x^*),\, (y, y^*) \in \operatorname{gr}\partial f \;:\; \|y-x\| \ge \varphi(\|y^*-x^*\|); \tag{3.45}$$

(ix) $\operatorname{dom}\partial f = X$ *and there exists* $\theta \in \Omega_0 \cap N_0$ *such that*

$$\forall\, (x, x^*),\, (y, y^*) \in \operatorname{gr}\partial f \;:\; \theta(\|y-x\|) \ge \|y^*-x^*\|; \tag{3.46}$$

(x) $\operatorname{dom} f = X$, *f is Fréchet differentiable on X and there exists* $\sigma_6 \in \Omega_1$ *such that*

$$\forall\, x,\, y \in X \;:\; \langle x-y, \nabla f(x) - \nabla f(y)\rangle \le \sigma_6(\|y-x\|); \tag{3.47}$$

(xi) $\operatorname{dom} f = X$, *f is Fréchet differentiable and ∇f is uniformly continuous.*

Then conditions (i) – (iii) *and* (iv) – (xi) *are equivalent, respectively, and* (iv) $\Rightarrow$ (ii). *Furthermore, if f is continuous at some point of* $\operatorname{dom} f$ *(or X is Banach space and f is lower semicontinuous) then the eleven statements are equivalent; in such a case if one of conditions* (i) – (xi) *holds then* $\operatorname{dom} f = X$ *and f is continuous.*

Proof. (i) $\Rightarrow$ (ii) Let $\sigma \in \Omega_1$ satisfy Eq. (3.36); we already observed above that $\operatorname{dom} f = X$. From Eq. (3.36) we have that

$$\lambda^{-1}\big(f(x + \lambda(y-x)) - f(x)\big) + (1-\lambda)\sigma(\|y-x\|) \ge f(y) - f(x)$$

for all $x, y \in X$ and $\lambda \in]0,1[$; taking the limit for $\lambda \to 0$ we obtain Eq. (3.39) with $\sigma_2 := \sigma$.

(ii) $\Rightarrow$ (iii) Let $x_0 \in (\operatorname{dom} f)^i$. From Theorem 2.1.13 we have that $f'(x_0, u) \in \mathbb{R}$ for every $u \in X$. Let $t_0 > 0$ be such that $\sigma_2(t_0) < \infty$. From Eq. (3.39), it follows that $\{y \in X \mid \|y - x_0\| = t_0\} \subset \operatorname{dom} f$, and so $B(x_0, t_0) \subset \operatorname{dom} f$ (because $\operatorname{dom} f$ is a convex set). Therefore $(\operatorname{dom} f)^i + t_0 B_X \subset \operatorname{dom} f$, whence, as in the proof of Proposition 3.5.2 we obtain that $\operatorname{dom} f = X$.

Changing x and y in Eq. (3.39), then adding side by side the corresponding relations, we obtain

$$\forall\, x,\, y \in X \;:\; 0 \le f'(x, y - x) + f'(y, x - y) + 2\sigma_2(\|y - x\|). \tag{3.48}$$

Let $x \in X$ be fixed and $u \in X \setminus \{0\}$. Consider the function $\varphi : \mathbb{R} \to \mathbb{R}$, $\varphi(t) = f(x + tu)$; φ is convex and continuous. Furthermore

$$\forall\, t \in \mathbb{R} \;:\; \varphi'_+(t) = f'(x + tu, u), \quad \varphi'_-(t) = -f'(x + tu, -u).$$

Taking into account Theorem 2.1.5, for $-t < 0 < t$ we obtain that $\varphi'_+(-t) \le \varphi'_-(0) \le \varphi'_+(0) \le \varphi'_-(t)$, and so

$$\begin{aligned} f'(x, u) + f'(x, -u) &= \varphi'_+(0) - \varphi'_-(0) \le \varphi'_-(t) - \varphi'_+(-t) \\ &= -f'(x + tu, -u) - f'(x + tu, u) \\ &\le (2t)^{-1} 2\sigma_2(2t\,\|u\|) = (2t\,\|u\|)^{-1} 2\sigma_2(2t\,\|u\|) \cdot \|u\| \end{aligned}$$

for every $t > 0$. Letting $t \to 0$, we obtain that $f'(x, u) + f'(x, -u) \le 0$. Since the converse inequality holds ($f'(x, \cdot)$ being sublinear), we have that $f'(x, -u) = -f'(x, u)$, and so $f'(x, \cdot)$ is linear. In this situation Eq. (3.40) coincides with Eq. (3.48) (and so $\sigma_3 = 2\sigma_2$).

(iii) $\Rightarrow$ (ii) Let $\overline{\sigma}_3 : \mathbb{R}_+ \to \overline{\mathbb{R}}_+$ be defined by $\overline{\sigma}_3(t) = \sup\{\sigma_3(s) \mid s \in [0, t]\}$. It is obvious that $\overline{\sigma}_3 \in \Omega_1 \cap N_0$ and $\overline{\sigma}_3 \ge \sigma$. Consider the function φ defined in the proof of the preceding implication. In our situation φ is derivable and

$$\forall\, s, t \in \mathbb{R} \;:\; (t - s)(\varphi'(t) - \varphi'(s)) \le \sigma_3(|t - s| \cdot \|u\|) \le \overline{\sigma}_3(|t - s| \cdot \|u\|).$$

It follows that

$$\begin{aligned} f(x+u)-f(x) &= \varphi(1)-\varphi(0)=\int_0^1 \varphi'(t)\,dt \\ &\le \int_0^1 \left(\varphi'(0)+t^{-1}\overline{\sigma}_3(t\|u\|)\right)dt \\ &= \varphi'(0)+\int_0^{\|u\|} t^{-1}\overline{\sigma}_3(t)\,dt = f'(x,u)+\sigma_2(\|u\|), \end{aligned}$$

where $\sigma_2(t)=\int_0^t s^{-1}\overline{\sigma}_3(s)\,ds$; it is obvious that $\sigma_2\in\Omega_1$. Therefore (ii) holds.

(ii) $\Rightarrow$ (i) From (ii) $\Rightarrow$ (iii) we have that $\operatorname{dom} f = X$ and $f'(x,\cdot)$ is linear for every $x\in X$. Let $\widetilde{\sigma}_f:\mathbb{R}_+\to\overline{\mathbb{R}}_+$ be defined by

$$\begin{aligned} \widetilde{\sigma}_f(t) &= \sup\{f(y)-f(x)-f'(x,y-x)\mid x,y\in X,\ \|y-x\|=t\} \\ &= \sup\{f(x+tu)-f(x)-tf'(x,u)\mid x,u\in X,\ \|u\|=1\}. \end{aligned} \tag{3.49}$$

Since $t\mapsto f(x+tu)-f(x)-tf'(x,u)$ is convex and continuous for all $x,u\in X$, it follows that $\widetilde{\sigma}_f$ is convex and lower semicontinuous. From the obvious inequality $\widetilde{\sigma}_f\le\sigma_2$, we have that $\widetilde{\sigma}_f\in\Omega_1$.

Let $x,y\in X$ and $\lambda\in]0,1[$; consider $x_\lambda:=(1-\lambda)x+\lambda y$. From the definition of $\widetilde{\sigma}_f$ we have that

$$\begin{aligned} f(x) &\le f(x_\lambda)+f'(x_\lambda,x-x_\lambda)+\widetilde{\sigma}_f(\|x-x_\lambda\|), \\ f(y) &\le f(x_\lambda)+f'(x_\lambda,y-x_\lambda)+\widetilde{\sigma}_f(\|y-x_\lambda\|). \end{aligned}$$

Multiplying the first relation with $1-\lambda$ and the second one with λ, then adding them side by side, we obtain

$$\begin{aligned} &(1-\lambda)f(x)+\lambda f(y) \\ &\quad\le f(x_\lambda)+\lambda(1-\lambda)[f'(x_\lambda,x-y)+f'(x_\lambda,y-x)] \\ &\qquad\quad +(1-\lambda)\widetilde{\sigma}_f(\lambda\|y-x\|)+\lambda\widetilde{\sigma}_f((1-\lambda)\|y-x\|) \qquad (3.50)\\ &\quad\le f(x_\lambda)+2\lambda(1-\lambda)\widetilde{\sigma}_f(\|y-x\|) \\ &\quad\le f(x_\lambda)+2\lambda(1-\lambda)\sigma_2(\|y-x\|); \end{aligned}$$

we have used the convexity of $\widetilde{\sigma}_f$. Therefore f is uniformly smooth.

(iv) $\Rightarrow$ (i) Let $(x_0,x_0^*)\in\operatorname{gr}\partial f$ and take $t_0>0$ such that $\sigma_4(t)\le 1$ for $t\in[0,t_0]$. From Eq. (3.41) we obtain that $f(x)\le f(x_0)+t_0\|x_0^*\|+1$ for $x\in B(x_0,t_0)$, and so $B(x_0,t_0)\subset\operatorname{dom} f$ and f is continuous at every point of $B(x_0,t_0)$. Hence $\operatorname{dom}\partial f+t_0B_X\subset\operatorname{dom}\partial f\subset\operatorname{dom} f$, which implies that $\operatorname{dom}\partial f=\operatorname{dom} f=X$ and f is continuous on X.

Let $x, y \in X$ and $\lambda \in]0,1[$. With the notation from the proof of the implication (ii) $\Rightarrow$ (i), there exists $x_\lambda^* \in \partial f(x_\lambda)$. Replacing $f'(x,\cdot)$ by $x^* \in \partial f(x)$ and $f'(x_\lambda,\cdot)$ by x_λ^* in the proof of that implication, we obtain that f is $2\sigma_4$-smooth.

(iv) $\Rightarrow$ (v) In the proof of the implication (iv) $\Rightarrow$ (i) we obtained that $\operatorname{dom}\partial f = X$. Taking $(x,x^*),(y,y^*) \in \operatorname{gr}\partial f$, adding then side by side the Eq. (3.41) and that obtained from Eq. (3.41) interchanging (x,x^*) and (y,y^*), we get Eq. (3.42) with $\sigma_5 := 2\sigma_4$.

(v) $\Rightarrow$ (iv) Let $y \in X$ and $(x,x^*) \in \operatorname{gr}\partial f$; by hypothesis there exists an element $y^* \in \partial f(y)$. Since

$$f(y) \le f(x) + \langle y-x, y^*\rangle \quad \text{and} \quad 0 \le \langle y-x, x^*-y^*\rangle + \sigma_5(\|y-x\|),$$

Eq. (3.41) follows by adding side by side these relations, with $\sigma_4 := \sigma_5$.

(iv) $\Rightarrow$ (vi) From (iv) $\Rightarrow$ (i) we have that $\operatorname{dom} f = X$ and f is continuous. Let (x,x^*), $(y,y^*) \in \operatorname{gr}\partial f$ be fixed and take an arbitrary $z \in X$. From Eq. (3.41) we have that $f(y+z) \le f(y) + \langle z, y^*\rangle + \sigma_4(\|z\|)$, and so

$$\begin{aligned} 0 &\le f(y+z) - f(x) - \langle y+z-x, x^*\rangle \\ &\le f(y) - f(x) - \langle y-x, x^*\rangle - \langle z, x^*-y^*\rangle + \sigma_4(\|z\|). \end{aligned}$$

Therefore

$$\forall z \in X \;:\; \langle z, x^*-y^*\rangle - \sigma_4(\|z\|) \le f(y) - f(x) - \langle y-x, x^*\rangle.$$

Taking the supremum with respect to $z \in X$ and using Lemma 3.3.1(v) we obtain that Eq. (3.43) holds with $\psi_1 := \sigma_4^{\#}$. By Lemma 3.3.1(iii) we have that $\psi_1 \in \Gamma_0$.

(vi) $\Rightarrow$ (vii) Changing x and y in Eq. (3.43) and summing the obtained relations side by side, Eq. (3.44) holds with $\psi_2 := 2\psi_1$.

(vii) $\Rightarrow$ (viii) From the obvious inequality $\langle y-x, y^*-x^*\rangle \le \|y-x\| \cdot \|y^*-x^*\|$ we obtain immediately that Eq. (3.45) holds with $\varphi \in \mathcal{A}_0 \cap N_0$, $\varphi(t) := t^{-1}\psi_2(t)$ for $t > 0$.

(viii) $\Rightarrow$ (ix) and (ix) $\Rightarrow$ (viii) The first implication is obvious by taking $\theta := \varphi^h$ and the second one by taking $\varphi := \theta^e$ using Lemma 3.3.1(i).

(ix) $\Rightarrow$ (v) The implication is obvious by taking $\sigma_5 \in \Omega_1$, $\sigma_5(t) := t\cdot\theta(t)$ for $t \ge 0$.

From what was shown above we have that conditions (i)–(iii) and (iv)–(ix) are equivalent, respectively, and (iv) $\Rightarrow$ (i). Moreover $\operatorname{dom} f = X$ and f is continuous if (iv) holds.

(ix) $\Rightarrow$ (xi) From what was said above, f is continuous; from Eq. (3.46) we have that ∂f is single valued; by Corollary 2.4.10 we have that f is Gâteaux differentiable. Using again Eq. (3.46), ∇f is uniformly continuous. Using Corollary 3.3.3 we obtain that f is Fréchet differentiable.

(xi) $\Rightarrow$ (ix) This implication is obvious; just take θ the gage of uniform continuity of ∇f.

(v) $\Rightarrow$ (x) In our hypothesis, as shown above, f is Fréchet differentiable. Hence Eq. (3.47) holds with $\sigma_6 := \sigma_5$.

(x) $\Rightarrow$ (v) is obvious with $\sigma_5 := \sigma_6$.

(ii) $\Rightarrow$ (iv) when f is continuous at some point of its domain. From (ii) $\Rightarrow$ (iii) we have that $\operatorname{dom} f = X$, and so f is continuous on X, and $f'(x, \cdot) = \nabla f(x)$ for every $x \in X$. Therefore Eq. (3.41) holds with $\sigma_4 := \sigma_2$.

(ii) $\Rightarrow$ (iv) when f is lsc and X is a Banach space. From Theorem 2.2.20 we have that f is continuous on $(\operatorname{dom} f)^i \neq \emptyset$. The conclusion follows from the previous case. □

The proof of the preceding theorem shows that $\widetilde{\sigma}_f \leq \sigma_f \leq 2\widetilde{\sigma}_f$. Moreover, when f is continuous, one can take $\sigma_2 = \sigma_4$ and $\sigma_3 = \sigma_5 = \sigma_6$.

Remark 3.5.1 In Theorem 3.5.6 the implication (ix) $\Rightarrow$ (iv) holds with $\sigma_4(t) := \int_0^t \theta(s)\, ds$.

Indeed, if (ix) holds, by the preceding theorem $\operatorname{dom} f = X$ and f is Fréchet differentiable. Let $x, y \in X$. Taking

$$\varphi : \mathbb{R} \to \mathbb{R}, \quad \varphi(t) := f\left((1-t)x + ty\right) - f(x) - t \cdot \langle y - x, \nabla f(x)\rangle,$$

we have that $\varphi(0) = 0$, $\varphi(1) = f(y) - f(x) - \langle y - x, \nabla f(x)\rangle$, φ is derivable and

$$\varphi'(t) = \langle y - x, \nabla f\left((1-t)x + ty\right) - \nabla f(x)\rangle \leq \|y - x\| \cdot \theta\left(t\,\|y - x\|\right)$$

for every $t \in \mathbb{R}$. It follows that

$$\begin{aligned} f(y) - f(x) - \langle y - x, \nabla f(x)\rangle &= \textstyle\int_0^1 \varphi'(t)\, dt \leq \int_0^1 \|y - x\| \cdot \theta\left(t\,\|y - x\|\right) dt \\ &= \textstyle\int_0^{\|y-x\|} \theta(s)\, ds = \sigma_4(\|y - x\|). \end{aligned}$$

Corollary 3.5.7 *Let $f : X \to \mathbb{R}$ be a continuous convex function and $p, q \in \mathbb{R}$ be such that $1 \leq p \leq 2 \leq q$ and $p^{-1} + q^{-1} = 1$. The following statements are equivalent:*

(i) *there exists $L_1 > 0$ such that f is σ-uniformly smooth, where $\sigma(t) := \frac{L_1}{p} t^p$ for $t \geq 0$;*

(ii) $\exists L_2 > 0, \forall x, y \in X, \forall x^* \in \partial f(x)$:

$$f(y) \leq f(x) + \langle y - x, x^* \rangle + \tfrac{L_2}{p} \|x - y\|^p ;$$

(iii) $\exists L_3 > 0, \forall (x, x^*), (y, y^*) \in \operatorname{gr} \partial f$:

$$f(y) \geq f(x) + \langle y - x, x^* \rangle + \tfrac{1}{L_3 q} \|x^* - y^*\|^q ;$$

(iv) $\exists L_4 > 0, \forall (x, x^*), (y, y^*) \in \operatorname{gr} \partial f$:

$$\langle y - x, y^* - x^* \rangle \leq \tfrac{2L_4}{p} \|x - y\|^p ;$$

(v) $\exists L_5 > 0, \forall (x, x^*), (y, y^*) \in \operatorname{gr} \partial f$:

$$\langle y - x, y^* - x^* \rangle \geq \tfrac{2}{L_5 q} \|x^* - y^*\|^q ;$$

(vi) *f is Fréchet differentiable and there exists $L_6 > 0$ such that*

$$\|\nabla f(x) - \nabla f(y)\| \leq L_6 \|x - y\|^{p-1} \quad \forall x, y \in X.$$

Proof. The conditions (i)–(vi) of the corollary correspond to conditions (i), (iv) (or (ii)), (vi), (v) (or (iii)), (vii) and (ix) of the preceding theorem, respectively. So, we have that (i) $\Rightarrow$ (ii) with $L_2 := L_1$ and (ii) $\Rightarrow$ (i) with $L_1 := 2^{2-p} L_2$ (using directly Eq. (3.50)). Also, (ii) $\Rightarrow$ (iii) with $L_3 := L_2^{q-1}$, (iii) $\Rightarrow$ (v) with $L_5 := L_3$, (v) $\Rightarrow$ (vi) with $L_6 := (L_5 q/2)^{p-1}$, (vi) $\Rightarrow$ (ii) with $L_2 := L_6$ (by the preceding remark), (ii) $\Rightarrow$ (iv) with $L_4 := L_2$ and (iv) $\Rightarrow$ (ii) with $L_2 := 2L_4/p$ (for this one uses the proof of the implication (iii) $\Rightarrow$ (ii) of the theorem). □

Remark 3.5.2 For $p = 2$ we may take the same $L_k = L > 0$ for all $k \in \overline{1,6}$ in the preceding corollary, as the proof shows.

The next result shows that every uniformly convex function on a Banach space is coercive and its subdifferential is onto.

Proposition 3.5.8 *Let X be a Banach space and $f \in \Gamma(X)$ be uniformly convex. Then*

$$\liminf_{\|x\| \to \infty} \frac{f(x)}{\|x\|^2} > 0 \quad \text{and} \quad \operatorname{Im} \partial f = X^*.$$

Moreover, if $(x_n) \subset X$ is a minimizing sequence for f then (x_n) converges to the unique minimum point of f.

Proof. By the Brøndsted–Rockafellar theorem there exists $(x_0, x_0^*) \in \operatorname{gr} \partial f$, and so

$$\forall x \in X \ : \ f(x) \geq f(x_0) + \langle x - x_0, x_0^* \rangle . \tag{3.51}$$

It follows that

$$\begin{aligned} f(x_0) + \lambda \langle x - x_0, x_0^* \rangle &\leq f\left((1-\lambda)x_0 + \lambda x\right) \\ &\leq (1-\lambda) f(x_0) + \lambda f(x) - \lambda(1-\lambda)\rho_f(\|x - x_0\|) \end{aligned}$$

for all $x \in X$ and $\lambda \in]0,1[$, and so $f(x) \geq f(x_0) + \langle x - x_0, x_0^* \rangle + (1-\lambda)\rho_f(\|x - x_0\|)$ for $x \in X$ and $\lambda \in]0,1[$. Letting $\lambda \to 0$ we obtain that

$$f(x_0) + \langle x - x_0, x_0^* \rangle + \rho_f(\|x - x_0\|) \leq f(x) \tag{3.52}$$

for all $x \in X$. Dividing the last inequality by $\|x - x_0\|^2$, taking into account that $\langle x - x_0, x_0^* \rangle \geq -\|x - x_0\| \cdot \|x_0^*\|$ and Proposition 3.5.1, we obtain that $\liminf_{\|x\| \to \infty} f(x) / \|x\|^2 \geq \liminf_{t \to \infty} \rho_f(t)/t^2 > 0$. From Eq. (3.51) we obtain that f is bounded below on bounded sets, and so, f being coercive, f is bounded from below. Let $(x_n) \subset \operatorname{dom} f$ be a minimizing sequence of f, *i.e.* $(f(x_n)) \to \inf f \in \mathbb{R}$. Because f is ρ_f-convex we have

$$\begin{aligned} \tfrac{1}{4}\rho_f(\|x_n - x_m\|) &\leq \tfrac{1}{2} f(x_n) + \tfrac{1}{2} f(x_m) - f(\tfrac{1}{2}x_n + \tfrac{1}{2}x_m) \\ &\leq \tfrac{1}{2} f(x_n) + \tfrac{1}{2} f(x_m) - \inf f \end{aligned}$$

for all $n, m \in \mathbb{N}$. Because $(t_n) \to 0$ when $(t_n) \subset [0, \infty[$ and $(\rho_f(t_n)) \to 0$, we obtain that (x_n) is a Cauchy sequence. As X is a Banach space, there exists $\overline{x} \in X$ such that $(x_n) \to \overline{x}$. The function f being lsc, $\inf f \leq f(\overline{x}) \leq \lim f(x_n) = \inf f$. Therefore $f(\overline{x}) = \inf f$, and so $0 \in \partial f(\overline{x})$. Replacing (x_0, x_0^*) by $(\overline{x}, 0)$ in Eq. (3.52) we obtain that

$$\forall x \in X \ : \ \rho_f(\|x - \overline{x}\|) \leq f(x) - f(\overline{x}).$$

From this relation it follows immediately that every minimizing sequence of f converges to $\overline{x}$.

Let now $x^* \in X^*$. Since $f - x^*$ is uniformly convex, by what precedes, there exists $\widetilde{x} \in X$ such that $f(\widetilde{x}) - \langle \widetilde{x}, x^* \rangle \leq f(x) - \langle x, x^* \rangle$ for every $x \in X$, *i.e.* $x^* \in \partial f(\widetilde{x})$. Therefore $\operatorname{Im} \partial f = X^*$. □

The subdifferential of $f \in \Gamma(X)$ is onto even in slightly weaker conditions. We call $f : X \to \overline{\mathbb{R}}$ **strongly coercive** if $\lim_{\|x\| \to \infty} f(x)/\|x\| = \infty$.

Corollary 3.5.9 *Let X be a Banach space and $f \in \Gamma(X)$. Assume that f is strongly coercive and uniformly convex on every convex and bounded subset of X. Then* $\operatorname{Im}\partial f = X^*$.

Proof. Let us show first that f attains its infimum on X. Indeed, let $\lambda > \inf f$. Since f is coercive, there exists $r > 0$ such that $[f \le \lambda] \subset rU_X$. Since f is uniformly convex on $C := rU_X$, we have that $g := f + \iota_C$ is uniformly convex and lower semicontinuous. From the preceding result there exists $\overline{x} \in X$ such that $g(\overline{x}) = \inf g$. Since $\inf g = \inf f$, it follows that $f(\overline{x}) = \inf f$.

Let now $x^* \in X^*$ be arbitrary. Then $f - x^*$ satisfies the hypotheses of our statement, and so $f - x^*$ attains its infimum on X. Thus $0 \in \operatorname{Im}\partial(f - x^*)$, whence $x^* \in \operatorname{Im}\partial f$. □

In the next result we give several (characteristic) properties of uniformly convex functions.

Theorem 3.5.10 *Let X be a Banach space and $f \in \Gamma(X)$. The following statements are equivalent:*

(i) *f is uniformly convex;*

(ii) $\exists\, \psi_1 \in \Gamma_0,\ \forall\, x, y \in \operatorname{dom} f\ :\ f(y) \ge f(x) + f'(x, y - x) + \psi_1(\|y - x\|)$;

(iii) $\exists\, \psi_2 \in \Gamma_0,\ \forall\, (x, x^*) \in \operatorname{gr}\partial f,\ \forall\, y \in X\ :$

$$f(y) \ge f(x) + \langle y - x, x^* \rangle + \psi_2(\|y - x\|);$$

(iv) $\exists\, \psi_3 \in \Gamma_0,\ \forall\, (x, x^*), (y, y^*) \in \operatorname{gr}\partial f\ :$

$$f(y) \ge f(x) + \langle y - x, x^* \rangle + \psi_3(\|y - x\|);$$

(v) $\exists\, \psi_4 \in \Gamma_0,\ \forall\, (x, x^*),\ (y, y^*) \in \operatorname{gr}\partial f\ :$

$$\langle y - x, y^* - x^* \rangle \ge \psi_4(\|y - x\|);$$

(vi) $\exists\, \varphi \in \mathcal{A}_0 \cap N_0,\ \forall\, (x, x^*),\ (y, y^*) \in \operatorname{gr}\partial f\ :\ \|y^* - x^*\| \ge \varphi(\|y - x\|)$;

(vii) $\exists\, \theta \in N_0 \cap \Omega_0,\ \forall\, (x, x^*),\ (y, y^*) \in \operatorname{gr}\partial f\ :\ \|y - x\| \le \theta(\|y^* - x^*\|)$;

(viii) $(\partial f)^{-1}$ *is single-valued and uniformly continuous on* $\operatorname{Im}\partial f$;

(ix) *f^* is uniformly smooth;*

(x) $\exists\, \sigma_1 \in \Sigma_1,\ \forall\, (x, x^*) \in \operatorname{gr}\partial f,\ \forall\, y^* \in X^*\ :$

$$f^*(y^*) \le f^*(x^*) + \langle x, y^* - x^* \rangle + \sigma_1(\|y^* - x^*\|); \tag{3.53}$$

(xi) $\operatorname{dom} f^* = X^*$ *and* $\exists \sigma_2 \in \Sigma_1,\ \forall (x,x^*),(y,y^*) \in \operatorname{gr} \partial f$:

$$\langle y - x, y^* - x^* \rangle \le \sigma_2(\|y^* - x^*\|);$$

(xii) $\operatorname{dom} f^* = X^*$, f^* *is Fréchet differentiable and* ∇f^* *is uniformly continuous.*

In (ii)–(v) *one can take* ψ_k *in* Γ_0^2 *instead of* Γ_0, *while in* (x) *and* (xi) *one can take* σ_1, σ_2 *in* Σ_1^2 *instead of* Σ_1.

Proof. (i) $\Rightarrow$ (ii) $\Rightarrow$ (iii) follows as in the proof of the implications (3.30) $\Rightarrow$ (3.31) $\Rightarrow$ (3.32) from the end of Section 3.4 (with $\psi_1 := \overline{\mathrm{co}}\rho_f$, and $\psi_2 := \psi_1$), while (iii) $\Rightarrow$ (iv) is obvious, with $\psi_3 := \psi_2$.

(iv) $\Rightarrow$ (v) is obvious, with $\psi_4 := 2\psi_3$.

(v) $\Rightarrow$ (vi) is obvious with $\varphi(t) := t^{-1}\psi_3(t)$ for $t > 0$.

(vi) $\Rightarrow$ (vii) and (vii) $\Rightarrow$ (vi) are immediate with $\theta := \varphi^h$ and $\varphi := \theta^e$, respectively.

(vi) $\Rightarrow$ (iii) as in the proof of Theorem 3.4.3, with $\psi_2(t) := (1 - \gamma)\int_0^t \varphi(\gamma s)\, ds$, where $\gamma \in]0,1[$ is fixed.

(vii) $\Leftrightarrow$ (viii) is obvious; for $\Leftarrow$ take θ the gage of uniform continuity of $(\partial f)^{-1}$.

(i) $\Leftrightarrow$ (ix) follows from Theorem 3.5.5.

(iii) $\Leftrightarrow$ (x) as in the proof Theorem 3.4.3, with $\sigma_1 := \psi_2^{\#}$ and $\psi_2 := \sigma_1^{\#}$.

(x) $\Rightarrow$ (ix) First of all let us note that $\operatorname{dom} f^* = X^*$. Indeed, in our conditions there exists $t_0 > 0$ such that $\sigma_1^{\#}(t_0) < \infty$. From Eq. (3.53) we have that $\operatorname{Im} \partial f + t_0 B_X \subset \operatorname{dom} f^*$. On the other hand, from the Brøndsted–Rockafellar theorem we have that $\operatorname{dom} f^* \subset \overline{\operatorname{Im} \partial f}$ (the closure being taken in the norm topology of X^*). Therefore $\operatorname{dom} f^* = X^*$. Since f^* is w^*-lsc, f^* is $\|\cdot\|$-lsc, and so f^* is $\|\cdot\|$-continuous.

Let $x^*, y^* \in X^*$ and $\lambda \in]0,1[$ consider $x_\lambda^* := (1-\lambda)x^* + \lambda y^*$. Suppose that $\sigma_1(\|y^* - x^*\|) < \infty$. Since $\overline{\operatorname{Im} \partial f} = X^*$, there exists $((u_n, u_n^*)) \subset \operatorname{gr} \partial f$ such that $(u_n^*) \to x_\lambda^*$. From the hypothesis we have that

$$f^*(x^*) \le f^*(u_n^*) + \langle u_n, x^* - u_n^* \rangle + \sigma_1(\|u_n^* - x^*\|),$$
$$f^*(y^*) \le f^*(u_n^*) + \langle u_n, y^* - u_n^* \rangle + \sigma_1(\|u_n^* - y^*\|).$$

Multiplying the first relation with $1 - \lambda$ and the second one with λ, then adding them, we obtain that

$$(1-\lambda)f^*(x^*) + \lambda f^*(y^*)$$
$$\le f^*(u_n^*) + \langle u_n, x_\lambda^* - u_n^* \rangle + (1-\lambda)\sigma_1(\|u_n^* - x^*\|) + \lambda\sigma_1(\|u_n^* - y^*\|).$$

Since f^* is continuous, $\partial f^* : X^* \rightrightarrows X^{**}$ is locally bounded (see Theorem 2.4.13); as $\operatorname{gr}(\partial f)^{-1} \subset \operatorname{gr}\partial f^*$, $(\partial f)^{-1}$ is locally bounded, too. It follows that the sequence (u_n) is bounded. Furthermore we have that $\|u_n^* - x^*\| \to \lambda \|y^* - x^*\| < \|y^* - x^*\|$ and $\|u_n^* - y^*\| \to (1-\lambda)\|y^* - x^*\| < \|y^* - x^*\|$. Passing to the limit in the above relation and taking into account the convexity of σ_1, we obtain that

$$(1-\lambda)f^*(x^*) + \lambda f^*(y^*) \leq f^*(x_\lambda^*) + 2\lambda(1-\lambda)\sigma_1(\|y^* - x^*\|).$$

Therefore f^* is $2\sigma_1$-uniformly smooth, and so it is uniformly smooth.

(x) $\Rightarrow$ (xi) is obvious, with $\sigma_2 := \sigma_1$.

(xi) $\Rightarrow$ (x) As in the proof of the implication (x) $\Rightarrow$ (ix) we have that f^* is continuous and $\overline{\operatorname{dom}(\partial f)^{-1}} = X^*$. Let $(x, x^*) \in \operatorname{gr}\partial f$ and $y^* \in X^*$. There exists $((y_n, y_n^*)) \subset \operatorname{gr}\partial f$ with $(y_n^*) \to y^*$. It is clear that for $n \in \mathbb{N}$ we have

$$f^*(y_n^*) \leq f^*(x^*) + \langle y_n, y_n^* - x^*\rangle, \qquad 0 \leq \langle y_n - x, x^* - y_n^*\rangle + \sigma_2(\|y_n^* - x^*\|).$$

Adding these two inequalities side by side we obtain that

$$f^*(y_n^*) \leq f^*(x^*) + \langle x, y_n^* - x^*\rangle + \sigma_2(\|y_n^* - y^*\|) \quad \forall n \in \mathbb{N}.$$

Because σ_2 is continuous on $[0,\infty[\setminus\{t_0\}$, where $t_0 := \sup(\operatorname{dom}\sigma_2)$, we obtain Eq. (3.53) passing to the limit for $\|y^* - x^*\| \neq t_0$, with $\sigma_1 := \sigma_2$. Let $\|y^* - x^*\| = t_0$; then

$$f^*(\lambda y^* + (1-\lambda)x^*) \leq f^*(x^*) + \lambda\langle x, y^* - x^*\rangle + \sigma_2(\lambda\|y^* - x^*\|) \ \forall \lambda \in]0,1[.$$

Since $\sigma_2|_{[0,t_0]}$ is continuous, we obtain again Eq. (3.53) for $\lambda \to 1$.

(ix) $\Leftrightarrow$ (xii) This equivalence follows from the equivalence (i) $\Leftrightarrow$ (xi) of Theorem 3.5.6 because $f^* \in \Gamma(X^*)$ and X^* is a Banach space. □

To the characterizations of the uniform convexity of f in Theorem 3.5.10 we can add, besides (xii), the other characterizations of uniform smoothness of f^* obtained in Theorem 3.5.6.

Corollary 3.5.11 *Let X be a Banach space, $f \in \Gamma(X)$ and $p, q \in \mathbb{R}$, with $1 < p \leq 2 \leq q$ and $p^{-1} + q^{-1} = 1$. The following statements are equivalent:*

(i) *there exists $c_1 > 0$ such that f is ρ-convex with $\rho(t) := \frac{c_1}{q}t^q$ for $t \geq 0$;*

(ii) $\exists\, c_2 > 0, \forall x, y \in \operatorname{dom} f : f(y) \geq f(x) + f'(x; y - x) + \frac{c_2}{q}\|x - y\|^q$;

(iii) $\exists\, c_3 > 0,\ \forall\,(x, x^*) \in \operatorname{gr}\partial f,\ \forall\, y \in X$:

$$f(y) \geq f(x) + \langle y - x, x^* \rangle + \tfrac{c_3}{q} \|x - y\|^q ;$$

(iv) $\exists\, c_4 > 0,\ \forall\,(x, x^*), (y, y^*) \in \operatorname{gr}\partial f$:

$$f(y) \geq f(x) + \langle y - x, x^* \rangle + \tfrac{c_4}{q} \|x - y\|^q ;$$

(v) $\exists\, c_5 > 0,\ \forall\,(x, x^*), (y, y^*) \in \operatorname{gr}\partial f\ :\ \langle y - x, y^* - x^* \rangle \geq \frac{2c_5}{q} \|x - y\|^q$;

(vi) $\exists\, c_6 > 0,\ \forall\,(x, x^*), (y, y^*) \in \operatorname{gr}\partial f\ :\ \|x^* - y^*\| \geq \frac{2c_6}{q} \|x - y\|^{q-1}$;

(vii) *there exists* $c_7 > 0$ *such that* f^* *is* σ*-smooth, where* $\sigma(t) := \frac{c_7^{1-p}}{p} t^p$;

(viii) $\exists\, c_8 > 0,\ \forall\,(x, x^*) \in \operatorname{gr}\partial f,\ \forall\, y^* \in X^*$:

$$f^*(y^*) \leq f^*(x^*) + \langle x, y^* - x^* \rangle + \frac{c_8^{1-p}}{p} \|x^* - y^*\|^p ;$$

(ix) $\operatorname{dom} f^* = X^*$ *and* $\exists\, c_9 > 0,\ \forall\,(x, x^*), (y, y^*) \in \operatorname{gr}\partial f$:

$$\langle y - x, y^* - x^* \rangle \leq \frac{2c_9^{1-p}}{p} \|x^* - y^*\|^p ;$$

(x) $\operatorname{dom} f^* = X^*$, f^* *is Fréchet differentiable on* X^* *and*

$$\exists\, c_{10} > 0,\ \forall\, x^*, y^* \in X^*\ :\ \|\nabla f^*(x^*) - \nabla f^*(y^*)\| \leq c_{10}^{1-p} \|x^* - y^*\|^{p-1} .$$

Proof. The conditions (i)–(vi), (vii)–(x) of the corollary correspond to conditions (i)–(vi) and (ix)–(xii) of Theorem 3.5.10. One must only put in evidence the relations among the different constants. These follow from the relations among the gauges in (the proof of) Theorem 3.5.10 and among the constants in Corollary 3.5.7. So, the implications (i) $\Rightarrow$ (ii) $\Rightarrow$ (iii) $\Rightarrow$ (iv) $\Rightarrow$ (v) $\Rightarrow$ (vi) $\Rightarrow$ (x) and (viii) $\Rightarrow$ (ix) hold, with $c_2 := c_1$, $c_3 := c_2$, $c_4 := c_3$, $c_5 := c_4$, $c_6 := c_5$, $c_{10} := 2c_6/q$ and $c_9 := c_8$, respectively. For the equivalence (vii) $\Leftrightarrow$ (i) apply Corollary 3.5.4; one obtains the implications with $c_7 := c_1$ and $c_1 := c_7$.

Applying again Theorem 3.5.10, everyone of the conditions (vii)–(x) implies that f^* is finite, Fréchet differentiable on X^* and $\operatorname{gr}\partial f^* = (\operatorname{gr}\partial f)^{-1}$ (the former set being considered for the duality (X^*, X^{**})). So, the equivalence of conditions (vii)–(x) with the corresponding relations among the constants c_7–c_{10}, is obtained applying Corollary 3.5.7. □

Remark 3.5.3 For $p = 2$ we may take the same $c_k = c > 0$ for all $k \in \overline{1, 10}$ in the preceding corollary, as the proof shows.

In the next result we add some characterizations of uniformly smooth convex functions on Banach spaces which differ slightly from those of Theorem 3.5.6.

Theorem 3.5.12 *Let X be a Banach space and $f \in \Gamma(X)$. The following statements are equivalent:*

(i) *f is uniformly smooth;*

(ii) *f^* is uniformly convex;*

(iii) $\exists\, \psi_1 \in \Gamma_0,\ \forall\, y^* \in X^*,\ \forall\, (x, x^*) \in \operatorname{gr} \partial f$:

$$f^*(y^*) \geq f^*(x^*) + \langle x, y^* - x^* \rangle + \psi_1(\|y^* - x^*\|);$$

(iv) $\exists\, \sigma_1 \in \Sigma_1,\ \forall\, y \in X,\ \forall\, (x, x^*) \in \operatorname{gr} \partial f$:

$$f(y) \leq f(x) + \langle y - x, x^* \rangle + \sigma_1(\|y - x\|);$$

(v) $\exists\, \psi_2 \in \Gamma_0,\ \forall\, (x, x^*),\ (y, y^*) \in \operatorname{gr} \partial f$:

$$\langle y - x, y^* - x^* \rangle \geq \psi_2(\|y^* - x^*\|);$$

(vi) $\operatorname{dom} f = X$ *and* $\exists\, \sigma_2 \in \Sigma_1,\ \forall\, (x, x^*),\ (y, y^*) \in \operatorname{gr} \partial f$:

$$\langle y - x, y^* - x^* \rangle \leq \sigma_2(\|y - x\|);$$

(vii) $\exists\, \varphi \in \mathcal{A}_0 \cap \mathcal{N}_0,\ \forall\, (x, x^*),\ (y, y^*) \in \operatorname{gr} \partial f\ :\ \|y - x\| \geq \varphi(\|y^* - x^*\|)$;

(viii) $\exists\, \theta \in \mathcal{N}_0 \cap \Omega_0,\ \forall\, (x, x^*),\ (y, y^*) \in \operatorname{gr} \partial f\ :\ \|y^* - x^*\| \leq \theta(\|y - x\|)$.

In (iii) *and* (v) *one can take ψ_1 and ψ_2 in Γ_0^2 instead of Γ_0, while in* (iv) *and* (vi) *one can take σ_1 and σ_2 in Σ_1^2 instead of Σ_1, respectively.*

Proof. (i) $\Leftrightarrow$ (ii) is obtained in Theorem 3.5.5. Moreover f^* is $\sigma_f^\#$-convex.

Because X is a Banach space and f is lsc conditions (iv) and (vi) coincide with conditions (iv) and (v) of Theorem 3.5.6; so the equivalence of conditions (i), (iv) and (vi) follows using this theorem.

(ii) $\Rightarrow$ (iii) $\Rightarrow$ (v) $\Rightarrow$ (vii) are obvious, with $\psi_1 := \sigma_f^\#$, $\psi_2 := 2\psi_1$ and $\varphi(t) := t^{-1}\psi_2(t)$ for $t > 0$, respectively.

(vii) $\Leftrightarrow$ (viii) is obvious with $\theta := \varphi^h$ and $\varphi := \theta^e$, respectively.

(viii) $\Rightarrow$ (i) Let $(x, x^*), (y, y^*) \in \operatorname{gr} \partial f$; from the hypothesis we have that

$$\langle y - x, y^* - x^* \rangle \leq \|y - x\| \cdot \|y^* - x^*\| \leq \|y - x\|\, \theta(\|y - x\|) = \sigma_6(\|y - x\|),$$

where $\sigma_6(t) := t\theta(t)$, and so Eq. (3.42) holds. It is obvious that $\sigma_6 \in \Omega_1$. We must only prove that $A := \operatorname{dom} \partial f = X$. Because $\theta(0) = 0$ we have that ∂f is single-valued. Let $((x_n, x_n^*)) \subset \operatorname{gr} \partial f$ with $(x_n) \to x$, *i.e.* $x \in \overline{\operatorname{dom} \partial f}$. Since $\|x_n^* - x_m^*\| \le \theta(\|x_n - x_m\|)$ and $\lim_{t \downarrow 0} \theta(t) = 0$ it follows that (x_n^*) is a Cauchy sequence, and so it is norm-convergent to an element $x^* \in X^*$. Since $\operatorname{gr} \partial f$ is (norm-) closed we have that $(x, x^*) \in \operatorname{gr} \partial f$, and so $x \in A$. Therefore A is a closed convex set. Suppose that $A \neq X$. By the Bishop–Phelps theorem, A has support points. Let $\overline{x} \in \operatorname{Bd} A \subset A$ be a support point and $\overline{x}^*$ $(\neq 0)$ a support functional at $\overline{x}$. Taking $\overline{x}_0^* \in \partial f(\overline{x})$ $(\neq \emptyset)$, it follows immediately that $\overline{x}_0^* + \overline{x}^* \in \partial f(\overline{x})$, contradicting the fact that ∂f is single-valued on $\operatorname{dom} \partial f$. Hence $\operatorname{dom} \partial f = X$. □

The following two results show that the natural framework for uniformly convex and uniformly smooth functions is that of reflexive Banach spaces.

Theorem 3.5.13 *Let X be a Banach space. If there exists a uniformly convex function $f \in \Gamma(X)$ whose domain has nonempty interior then X is reflexive.*

Proof. By Theorem 2.2.20 f is continuous on $\operatorname{int}(\operatorname{dom} f)$. Taking $x_0 \in \operatorname{int}(\operatorname{dom} f)$, $x_0^* \in \partial f(x_0)$, and replacing f by $f - x_0^*$, we may assume that $f(x_0) \le f(x)$ for every $x \in X$. Because f is continuous at x_0, there exist $r, M > 0$ such that $f(x) \le f(x_0) + M$ for every $x \in D(x_0, r)$. Since $\rho_f(\|x - x_0\|) \le f(x) - f(x_0)$ for $x \in X$ (see the proof of Proposition 3.5.8), we have that $[f \le \gamma] \subset D(x_0, r/2)$, where $\gamma := f(x_0) + \rho_f(r/2)$. Taking $x_1 \in [f \le \gamma] \setminus \{x_0\}$, by Theorem 2.2.11, f is Lipschitz on $[f \le f(x_1)]$. Taking $r' > 0$ such that $f(x) < f(x_1)$ for every $x \in D(x_0, r')$ we have that $[f \le f(x_0)] + r'U_X = D(x_0, r') \subset [f \le f(x_1)]$. Therefore condition (iii) of Theorem 2.5.2 holds. By Proposition 3.5.8 we have that f is coercive and so also condition (i) holds. Since for any nonempty closed convex set $C \subset X$, $f + \iota_C$ is either identically $+\infty$ or uniformly convex, using again Proposition 3.5.8, f attains its infimum on C. Therefore all conditions of Theorem 2.5.2 hold. Hence X is reflexive. □

Corollary 3.5.14 *Let X be a Banach space. If there exists a uniformly smooth convex function $f \in \Gamma(X)$ such that* $\liminf_{\|x\| \to \infty} f(x)/\|x\| > 0$ *then X is reflexive.*

Proof. Taking $\rho := \liminf_{\|x\| \to \infty} f(x)/\|x\|$, one obtains easily (see also Exercise 2.45) that $\rho B_{X^*} \subset \operatorname{dom} f^*$. By Theorem 3.5.12 the function f^* is uniformly convex. Applying the preceding theorem we obtain that X^* is reflexive, and so X is reflexive, too. □

3.6 Uniformly Convex and Uniformly Smooth Convex Functions on Bounded Sets

In the sequel we study the uniform convexity or uniform smoothness on bounded subsets. Having $f \in \Lambda(X)$ and $r > 0$, we denote by $\rho_{f,r}$, $\sigma_{f,r}$ and $\sigma^0_{f,r}$ the gages of uniform convexity, uniform smoothness and midpoint uniform smoothness of f on rU_X, respectively. We say that f is **uniformly convex on bounded sets** if f is uniformly convex on rU_X for all $r > 0$. Similarly, we say that f is **uniformly smooth on bounded sets** if $\operatorname{dom} f = X$ and f is uniformly smooth on rU_X for all $r > 0$.

We begin with the following auxiliary results.

Lemma 3.6.1 *Let $f \in \Gamma(X)$. Then the following statements are equivalent:*

(i) *f is strongly coercive,*

(ii) $\forall r > 0,\ \exists \alpha \in \mathbb{R},\ \forall x \in X\ :\ f(x) \geq r\|x\| + \alpha$,

(iii) $\operatorname{dom} f^* = X^*$ *and f^* is bounded on bounded sets.*

The statements obtained from (i)–(iii) *interchanging f, f^* and X, X^*, respectively are also equivalent.*

Proof. Since f and f^* are proper, convex and lower semicontinuous, they are bounded from below by affine continuous functions, and so they are bounded below on bounded sets.

(i) ⇒ (ii) Take $r > 0$; since f is strongly coercive, there exists $\rho > 0$ such that $f(x) \geq r\|x\|$ for $\|x\| \geq \rho$. Since f is bounded from below by some $\alpha \leq 0$ on ρU_X, we have that $f(x) \geq r\|x\| + \alpha$ for all $x \in X$.

(ii) ⇒ (i) Assume that $f(x) \geq r\|x\| + \alpha$ for every $x \in X$; then, obviously, $\liminf_{\|x\|\to\infty} f(x)/\|x\| \geq r$. The conclusion is obvious.

(ii) ⇔ (iii) Let $g := r\|\cdot\| + \alpha$. We have that $f \geq g \Leftrightarrow f^* \leq g^* = \iota_{rU_{X^*}} - \alpha$ $\Leftrightarrow f^*(x^*) \leq -\alpha$ for every $x^* \in rU_{X^*}$. The conclusion follows.

The proof of the other statements are similar. □

Proposition 3.6.2 *Let $f \in \Gamma(X)$.*

(i) *Assume that f is strongly coercive. If f is uniformly convex on bounded sets then f^* is uniformly smooth on bounded sets, while if f is* *uniformly smooth on bounded sets then f^* is uniformly convex on bounded sets.*

(ii) *Assume that f^* is strongly coercive. Then the assertions in* (i) *remain valid interchanging f and f^*.*

Proof. We prove only (i), the proof of (ii) being similar.

Let f be strongly coercive. Let $r > 0$ be fixed. Of course, there exists $\alpha \in \mathbb{R}$ such that $f^*(x^*) \geq \alpha + 1$ for every $x^* \in (r+1)U_{X^*}$. Because f is strongly coercive, there exists $r' > 0$ such that $f(x) \geq (r+1)\|x\| - \alpha$ for every $x \in X$, $\|x\| \geq r'$. Since for $x^* \in (r+1)U_{X^*}$ and $\|x\| \geq r'$ we have that $\langle x, x^* \rangle - f(x) \leq (r+1)\|x\| - (r+1)\|x\| + \alpha < \alpha + 1$, it follows that

$$\forall x^* \in (r+1)U_{X^*} \ : \ f^*(x^*) = \sup\{\langle x, x^* \rangle - f(x) \mid x \in r'U_X\}. \qquad (3.54)$$

Assume first that f is uniformly convex on bounded sets and take $r > 0$. The above argument shows that there exists $r' > 0$ satisfying (3.54). Take t, x^*, y^*, λ, x_0^*, x_1^*, γ_0, γ_1, x_0 and x_1 like in the proof of Proposition 3.5.3(i), but with $t \in]0,1]$ and $x^* \in rU_{X^*}$. Then $x_0^*, x_1^* \in (r+1)U_{X^*}$, and so, by Eq. (3.54), we may take $x_0, x_1 \in r'U_X$. Continuing in the same way like in the proof of Proposition 3.5.3(i), but with $\rho_{f,r'}$ instead of ρ, we obtain that $\sigma_{f^*,r}(t) \leq (\rho_{f,r'})^{\#}(t)$ for every $t \in]0,1]$. It follows that f^* is uniformly smooth on rU_{X^*}.

Assume now that f is uniformly smooth on bounded sets and take $r > 0$. As above, there exists $r' > 0$ satisfying Eq. (3.54). Consider $x_0^*, x_1^* \in rU_{X^*}$, $x \in r'U_X$, $y \in X$, $\lambda \in]0,1[$ and denote $x_\lambda^* := (1-\lambda)x_0^* + \lambda x_1^*$. Proceeding like in the proof of Proposition 3.5.3(ii), but with $\sigma_{f,r'}$ instead of σ, and taking into account Eq. (3.54), one obtains that f^* is $(\sigma_{f,r'})^{\#}$-convex on rU_{X^*}. The conclusion follows from Lemma 3.3.1. □

In the next two results we point out the relationships among uniform convexity on bounded sets, uniform smoothness on bounded sets and uniform continuity on bounded sets of the gradient.

Proposition 3.6.3 *Let $f \in \Gamma(X)$ be convex. Consider the following statements:*

(i) *f is bounded and uniformly smooth on bounded sets,*

(ii) *f is Fréchet differentiable on $X = \operatorname{dom} f$ and ∇f is uniformly continuous on bounded sets,*

(iii) *f^* is strongly coercive and uniformly convex on bounded sets.*

Then (i) $\Leftrightarrow$ (ii) $\Leftarrow$ (iii). *Moreover, if f is strongly coercive then* (i) $\Rightarrow$ (iii); *in this case X^* is reflexive (also X is reflexive if X is a Banach space).*

Proof. (i) $\Rightarrow$ (ii) Since f is bounded on bounded sets, by Theorem 2.4.13, f is Lipschitz on bounded sets; since f is uniformly smooth on bounded sets, f is smooth at any $x \in X$, and so f is Fréchet differentiable on X (see Theorem 3.3.2). In order to show that ∇f is uniformly continuous on bounded sets, let $r > 0$ and $L > 0$ be a constant Lipschitz for f on $(r+1)U_X$. Consider $x, x' \in rU_X$, $t \in]0,1]$ and $y \in S_X$. Taking into account the inequalities

$$f(x+ty) + f(x-ty) - 2f(x) \geq f(x+ty) - f(x) - t\langle y, \nabla f(x)\rangle \geq 0,$$

and the similar one for x', we have that

$$\begin{aligned}
&|\nabla f(x)(ty) - \nabla f(x')(ty)| \\
&\quad \leq |f(x) - f(x')| + |f(x+ty) - f(x'+ty)| \\
&\qquad + |f(x+ty) - f(x) - \nabla f(x)(ty)| + |f(x'+ty) - f(x') - \nabla f(x')(ty)| \\
&\quad \leq 2\sigma^0_{f,r}(t) + 2L\|x - x'\|.
\end{aligned}$$

Hence

$$\begin{aligned}
\|\nabla f(x) - \nabla f(x')\| &= \sup\nolimits_{y\in S_X} t^{-1}|\nabla f(x)(ty) - \nabla f(x')(ty)| \\
&\leq 2t^{-1}\sigma^0_{f,r}(t) + 2Lt^{-1}\|x - x'\|
\end{aligned}$$

for all $x, x' \in rU_X$ and $t \in]0,1]$. Let $\varepsilon > 0$; since f is uniformly smooth on rU_X, there exists $t_0 \in]0,1]$ such that $2t_0^{-1}\sigma^0_{f,r}(t_0) < \varepsilon/2$. Taking $\delta := \varepsilon t_0/(4L)$ we have that $\|\nabla f(x) - \nabla f(x')\| < \varepsilon$ for all $x, x' \in rU_X$ with $\|x - x'\| < \delta$, *i.e.* ∇f is uniformly continuous on rU_X.

(ii) $\Rightarrow$ (i) It is known that a uniformly continuous function on a bounded convex set is bounded (Exercise!). Hence ∇f is bounded on bounded sets in our conditions. Using the mean value theorem (see Exercise 2.31 or [Vainberg (1964)]) one obtains immediately that f is Lipschitz on bounded sets, and so f is also bounded on bounded sets. In order to show that f is uniformly smooth on rU_X for (every) $r > 0$, consider the modulus $\omega : [0,2] \to \overline{\mathbb{R}}_+$ of uniform continuity of ∇f on $(r+1)U_X$ (restricted to $[0,2]$) defined by

$$\omega(t) := \sup\{\|\nabla f(x) - \nabla f(x')\| \mid x, x' \in (r+1)U_X,\ \|x - x'\| \leq t\}.$$

Then for $t \in]0,1]$, $x \in rU_X$ and $y \in S_X$ we have that

$$\begin{aligned}
&f(x+ty)+f(x-ty)-2f(x)\\
&\quad=(f(x+ty)-f(x))+(f(x-ty)-f(x))\\
&\quad=\nabla f(x+\theta y)(ty)-\nabla f(x-\theta' y)(ty)=t\left(\nabla f(x+\theta y)-\nabla f(x-\theta' y)\right)(y)\\
&\quad\le t\,\|\nabla f(x+\theta y)-\nabla f(x-\theta' y)\|\le t\omega(2t),
\end{aligned}$$

where $\theta, \theta' \in [0,t]$ are obtained by using the mean value theorem. It follows that $\sigma^0_{f,r}(t) \le t\omega(2t)$ for $t \in [0,1]$, and so, because ∇f is uniformly continuous on rU_X, f is uniformly smooth on rU_X.

(iii) ⇒ (i) Because f^* is strongly coercive, by Lemma 3.6.1 f is bounded on bounded sets; using Proposition 3.6.2(ii) we obtain also that f is uniformly smooth on bounded sets.

(i) ⇒ (iii) (When f is strongly coercive.) Again, by Proposition 3.6.2(i) f^* is uniformly smooth on bounded sets, while from Lemma 3.6.1 we obtain that f^* is strongly coercive. In this case $f^* + \iota_{rU_{X^*}}$ is uniformly convex and continuous at 0, and so, by Theorem 3.5.13, X^* is reflexive. Of course, if X is a Banach space, X is reflexive, too. □

Proposition 3.6.4 *Let $f \in \Gamma(X)$. Consider the following statements:*

(i) *f is strongly coercive and uniformly convex on bounded sets,*

(ii) *f^* is bounded and uniformly smooth on bounded sets,*

(iii) *f^* is Fréchet differentiable on $X^* = \operatorname{dom} f^*$ and ∇f^* is uniformly continuous on bounded sets.*

Then (i) ⇒ (ii) ⇔ (iii). *Moreover, if f is bounded on bounded sets then* (ii) ⇒ (i)*; in this case X^* is reflexive (also X is reflexive if X is a Banach space).*

Proof. (ii) ⇔ (iii) follows from the preceding proposition applied to f^*.

(i) ⇒ (ii) follows immediately applying Lemma 3.6.1 and Proposition 3.6.2.

(ii) ⇒ (i) (When f is bounded on bounded sets.) By Lemma 3.6.1 f^* is strongly coercive; being also uniformly smooth on bounded sets, using Proposition 3.6.2, we have that f is uniformly convex on bounded sets. In this case, since X^* is a Banach space, we obtain that X^* is reflexive applying the preceding proposition; so, if X is a Banach space then X is reflexive. □

In the next result we characterize uniform convexity on bounded sets without asking coercivity of the function.

Proposition 3.6.5 *Let X be a Banach space and $f \in \Gamma(X)$. Then f is uniformly convex on bounded sets if and only if $\vartheta_{f,B}(t) > 0$ for every $t > 0$ and every bounded set $B \subset X$ with $B \cap \operatorname{dom} f \neq \emptyset$, where*

$$\vartheta_{f,B}(t) := \inf\{f(y) - f(x) - f'(x, y - x) \mid x \in B \cap \operatorname{dom} f, \\ y \in \operatorname{dom} f,\ \|y - x\| = t\}.$$

Proof. Assume that f is uniformly convex on bounded sets and consider $B \subset X$ a bounded set with $B \cap \operatorname{dom} f \neq \emptyset$. Let $r > 1$ be such that $B \subset (r-1)U_X$. By hypothesis $g := f + \iota_{rU_X}$ is uniformly convex. Hence, from Theorem 3.5.10, there exists $\psi \in \Gamma_0$ such that $g(y) \geq g(x) + g'(x, y - x) + \psi(\|y - x\|)$ for all $x, y \in \operatorname{dom} g$, whence $f(y) \geq f(x) + g'(x, y - x) + \psi(\|y - x\|)$ for all $x \in B \cap \operatorname{dom} f$ and $y \in rU_X \cap \operatorname{dom} f$. But for such x and y we have that $g'(x, y - x) = f'(x, y - x)$. It follows that $f(y) \geq f(x) + f'(x, y - x) + \psi(t)$ for all $x \in B \cap \operatorname{dom} f$, $y \in \operatorname{dom} f$ with $\|y - x\| = t$ and $t \in (0, 1]$. Since $\vartheta_{f,B} = \inf_{x \in B \cap \operatorname{dom} f} \vartheta_{f,x}$ (see page 201 for the definition of $\vartheta_{f,x}$) and $\vartheta_{f,x} \in N_1$, $\vartheta_{f,B} \in N_1$, too; in particular $\vartheta_{f,B}$ is nondecreasing. But $\vartheta_{f,B}(t) \geq \psi(t)$ for $t \in [0, 1]$, and so $\vartheta_{f,B}(t) > 0$ for every $t > 0$.

Conversely, assume that $\vartheta_{f,B}(t) > 0$ for every $t > 0$ and every bounded set $B \subset X$ with $B \cap \operatorname{dom} f \neq \emptyset$. Consider $r > 0$ such that $rU_X \cap \operatorname{dom} f \neq \emptyset$. Then $f(y) \geq f(x) + f'(x, y - x) + \vartheta_{f,rU_X}(\|y - x\|)$ for all $x \in rU_X \cap \operatorname{dom} f$ and $y \in \operatorname{dom} f$. Taking $g = f + \iota_{rU_X}$, we obtain that $g(y) \geq g(x) + g'(x, y - x) + \vartheta_{f,rU_X}(\|y - x\|)$ for all $x, y \in \operatorname{dom} g$. As seen above, $\vartheta_{f,B} \in N_1$. By our hypothesis we have that $\vartheta_{f,B} \in \mathcal{A}_0$; using Exercise 3.12 we obtain that $\psi = \overline{\operatorname{co}}\vartheta_{f,B} \in \Gamma_0$. It follows that condition (ii) of Theorem 3.5.10 holds, and so g is uniformly convex. Therefore f is uniformly convex on bounded sets. $\square$

We end this section with characterizations of uniform convexity and uniform smoothness on bounded sets in the case $X = \mathbb{R}^n$.

Proposition 3.6.6 *Let $f \in \Gamma(\mathbb{R}^n)$.*

(i) *Assume that either $n = 1$ or $\operatorname{dom} f$ is closed and $f|_{\operatorname{dom} f}$ is continuous. Then f is uniformly convex on bounded sets if and only if f is strictly convex.*

(ii) *Assume that $\operatorname{dom} f = \mathbb{R}^n$. Then f is uniformly smooth on bounded sets if and only if f is differentiable (Gâteaux or Fréchet).*

Proof. (i) Suppose first that f is uniformly convex on bounded sets. Taking into account that $\rho_{f,x}$ is nondecreasing for $x \in \operatorname{dom} f$, it follows that f is uniformly convex at any $x \in \operatorname{dom} f$. Then, as observed at the beginning of Section 3.5, f is strictly convex.

Assume now that $\operatorname{dom} f$ is closed and $f|_{\operatorname{dom} f}$ is continuous but f is not uniformly convex on bounded sets. Then there exists $r > 0$ such that $rU_X \cap \operatorname{dom} f \neq \emptyset$ and $f + \iota_{rU_X}$ is not uniformly convex. Taking into account Exercise 3.3 we have that for some $\varepsilon > 0$ and any $n \in \mathbb{N}$, there exist $x_n, y_n \in rU_X \cap \operatorname{dom} f$ such that $|x_n - y_n| = 2\varepsilon$ and

$$\tfrac{1}{2}f(x_n) + \tfrac{1}{2}f(y_n) < f\left(\tfrac{1}{2}(x_n + y_n)\right) + \tfrac{1}{n}. \tag{3.55}$$

Since (x_n) and (y_n) are bounded we may suppose that $(x_n) \to x \in \operatorname{cl}(\operatorname{dom} f)$ and $(y_n) \to y \in \operatorname{cl}(\operatorname{dom} f)$; of course, $\|x - y\| = 2\varepsilon$. Because $\operatorname{dom} f$ is closed and $f|_{\operatorname{cl}(\operatorname{dom} f)}$ is continuous, $x, y, \frac{1}{2}(x + y) \in \operatorname{dom} f$ and $\left(f(x_n)\right) \to f(x)$, $\left(f(y_n)\right) \to f(y)$, $\left(f(\frac{1}{2}(x_n + y_n))\right) \to f\left(\frac{1}{2}(x + y)\right)$. Taking the limit in Eq. (3.55) we get $\frac{1}{2}f(x) + \frac{1}{2}f(y) \leq f\left(\frac{1}{2}(x+y)\right)$, and so f is not strictly convex.

In the case $n = 1$, without any supplementary condition, by Proposition 2.1.6 we have that $f|_{\operatorname{cl}(\operatorname{dom} f)}$ is continuous. As above, we get $(x_n), (y_n) \subset [-r, r] \cap \operatorname{dom} f$ and $\varepsilon > 0$ for which Eq. (3.55) holds. Taking $x, y \in \operatorname{cl}(\operatorname{dom} f)$ the limits of (x_n) and (y_n), we have that $\frac{1}{2}f(x) + \frac{1}{2}f(y) \leq f\left(\frac{1}{2}(x+y)\right)$. Since $\frac{1}{2}(x + y) \in \operatorname{int}(\operatorname{dom} f)$, the preceding inequality implies that $x, y \in \operatorname{dom} f$, and so f is not strictly convex in this case, too.

(ii) Since $\operatorname{dom} f = \mathbb{R}^n$, by Corollary 2.2.21, f is continuous, and so f is bounded on bounded sets. On the other hand, because the Gâteaux and Fréchet bornologies coincide on $\mathbb{R}^n$, Theorem 3.3.2 shows that Gâteaux and Fréchet differentiability of f coincide in our case. Moreover, by Corollary 3.3.3, ∇f is continuous when f is differentiable, and so ∇f is uniformly continuous on bounded sets in such a case. The conclusion follows applying Proposition 3.6.3. □

3.7 Applications to the Geometry of Normed Spaces

The subdifferential of the norm has an interesting geometric interpretation. We have already seen that a hyperplane is a set $H_{x^*,\alpha} := \{x \in X \mid \langle x, x^* \rangle = \alpha\}$, where $x^* \in X^* \setminus \{0\}$ and $\alpha \in \mathbb{R}$; if $\dim X \geq 1$ then $H_{x^*,\alpha} \neq \emptyset$. Taking into account that for every $\beta \neq 0$, $H_{\beta x^*,\beta\alpha} = H_{x^*,\alpha}$, we can always suppose that $x^* \in S_{X^*}$ and $\alpha \geq 0$. One can establish easily (Exercise!) that for

$x^*, y^* \in S_{X^*}$, $\alpha, \beta > 0$ we have $H_{x^*,\alpha} = H_{y^*,\beta}$ if and only if $x^* = y^*$ and $\alpha = \beta$.

Now we can give the geometric interpretation of the subdifferential of the norm:

Let $\overline{x} \in S_X$, $x^* \in S_{X^*}$ and $\alpha \geq 0$; then $H_{x^*,\alpha}$ is a supporting hyperplane of U_X at $\overline{x} \in S_X$ if and only if $\alpha = 1$ and $x^* \in \partial\|\cdot\|(\overline{x})$.

Therefore the set of supporting hyperplanes of U_X at $\overline{x} \in S_X$ is

$$\{H_{x^*,1} \mid x^* \in \partial\|\cdot\|(\overline{x})\} \tag{3.56}$$

(Exercise!). If U_X has a unique supporting hyperplane at every point of its boundary S_X we say that X is **smooth**; note that this is a geometric property (of the norm), but not a topological property. The dual property (as we shall see later on) is that of strictly convex space. We say that X is **strictly convex** if

$$\forall\, x, y \in X,\ x \neq y,\ \|x\| = \|y\| = 1\ :\ \|\tfrac{1}{2}(x+y)\| < 1.$$

Of course, the above condition is equivalent to $\frac{1}{2}(x+y) \in B_X$ for all $x, y \in U_X$, $x \neq y$.

Throughout this section $\varphi : \mathbb{R}_+ \to \mathbb{R}_+$ is a **weight function**, *i.e.* φ is increasing, continuous, $\varphi(0) = 0$ and $\lim_{t\to\infty} \varphi(t) = \infty$. It is obvious that a weight function is bijective and its inverse is a weight function, too. In the following results we establish some properties of the function

$$f_\varphi : X \to \mathbb{R}, \qquad f_\varphi(x) := \int_0^{\|x\|} \varphi(s)\, ds. \tag{3.57}$$

In the sequel we shall consider also

$$\psi : \mathbb{R}_+ \to \mathbb{R}_+, \qquad \psi(t) := \int_0^t \varphi(s)\, ds, \tag{3.58}$$

and so $f_\varphi(x) = \psi(\|x\|)$ for every $x \in X$. An important example of weight function is given by $\varphi(t) = t^{p-1}$ (and so $\psi(t) = \frac{1}{p}t^p$) for $p \in]1, \infty[$. In the next lemma we collect some properties of the function ψ defined above.

Lemma 3.7.1 *Let φ be a weight function and ψ be defined by Eq. (3.58). Then ψ is strictly convex, derivable with $\psi' = \varphi$ on $\mathbb{R}_+$, $\lim_{t\to\infty} t^{-1}\psi(t) = \infty$, $\psi^\#(t) := \int_0^t \varphi^{-1}(s)\, ds$ and $\psi(t) + \psi^\#(s) = ts$ if and only if $s = \varphi(t)$.*

Proof. It is obvious that ψ is derivable and $\psi'(t) = \varphi(t)$ for $t \geq 0$. Moreover, $\lim_{t\to\infty} \psi(t)/|t| = \lim_{t\to\infty} \psi'(t) = \infty$. Let $s > 0$; the function $\theta : \mathbb{R}_+ \to \mathbb{R}$ defined by $\theta(t) := st - \psi(t)$ is derivable and $\theta'(t) = s - \varphi(t)$.

It follows that θ is increasing on $[0, \varphi^{-1}(s)]$ and decreasing on $[\varphi^{-1}(s), \infty[$, and so $\psi^{\#}(s) = \sup_{t\geq 0} \theta(t) = s\varphi^{-1}(s) - \psi(\varphi^{-1}(s))$, $t = \varphi^{-1}(s)$ being the unique point where the supremum is attained; it is obvious that this assertion is also valid for $s = 0$. Therefore

$$\psi^{\#}(s) = s\varphi^{-1}(s) - \int_0^{\varphi^{-1}(s)} \varphi(u)\, du = \int_0^{\varphi^{-1}(s)} u\, d\varphi(u) = \int_0^{s} \varphi^{-1}(v)\, dv,$$

where the second equality is obtained by integrating by parts in a Riemann–Stieltjes integral, while the last equality is obtained by changing the variable: $u = \varphi^{-1}(v)$. □

Having a weight function φ, throughout this section the functions f_φ and ψ are defined by Eqs. (3.57) and (3.58), respectively.

Theorem 3.7.2 *Let φ be a weight function. Then:*

(i) *f_φ is convex and continuous.*

(ii) *$(f_\varphi)^* = \psi^{\#} \circ \|\cdot\|$, i.e. $(f_\varphi)^*(x^*) = \psi^{\#}(\|x^*\|) = \int_0^{\|x^*\|} \varphi^{-1}(t)\, dt$ for all $x^* \in X^*$.*

(iii) *For every $x \in X$ one has*

$$\begin{aligned} \partial f_\varphi(x) &= \{x^* \in X^* \mid \langle x, x^*\rangle = \|x\| \cdot \|x^*\|,\ \|x^*\| = \varphi(\|x\|)\} \\ &= \varphi(\|x\|) \cdot \partial\|\cdot\|(x), \\ \partial f_\varphi(-x) &= -\partial f_\varphi(x). \end{aligned}$$

(iv) *The following statements are equivalent:* (a) *X is smooth,* (b) *f_φ is Gâteaux differentiable on X,* (c) *$\|\cdot\|$ is Gâteaux differentiable on $X \setminus \{0\}$. Moreover, if X is smooth then ∇f_φ is norm-weak* continuous.*

(v) *The following statements are equivalent:* (a) *X is strictly convex,* (b) *f_φ is strictly convex,* (c) *∂f_φ is strictly monotone,* (d) *$\forall x, y \in X$, $x \neq y$: $\partial f_\varphi(x) \cap \partial f_\varphi(y) = \emptyset$.*

(vi) *X is reflexive $\Leftrightarrow$ X is a Banach space and $\operatorname{Im}(\partial f_\varphi) = X^*$.*

Proof. Let us denote f_φ by f.

(i) Let $x, y \in X$ and $\lambda \in]0,1[$. Then

$$\begin{aligned} f(\lambda x + (1-\lambda)y) &= \psi(\|\lambda x + (1-\lambda)y\|) \leq \psi(\lambda\|x\| + (1-\lambda)\|y\|) \\ &\leq \lambda\psi(\|x\|) + (1-\lambda)\psi(\|y\|) = \lambda f(x) + (1-\lambda) f(y), \end{aligned}$$

the last but one inequality being strict if $\|x\| \neq \|y\|$ (because ψ is strictly convex). Therefore f is convex. It is obvious that f is continuous.

(ii) The conclusion is immediate from Lemma 3.3.1 and Lemma 3.7.1.

(iii) We have that

$$x^* \in \partial f(x) \Leftrightarrow f(x) + f^*(x^*) = \langle x, x^* \rangle \Leftrightarrow \psi(\|x\|) + \psi^{\#}(\|x^*\|) = \langle x, x^* \rangle .$$

Because $\langle x, x^* \rangle \leq \|x\| \cdot \|x^*\|$, using Lemma 3.7.1, we obtain that

$$x^* \in \partial f(x) \Leftrightarrow \langle x, x^* \rangle = \|x\| \cdot \|x^*\| \text{ and } \|x^*\| = \varphi(\|x\|).$$

The other statements are immediate.

(iv) Since the functions f and $\|\cdot\|$ are continuous, using Corollary 2.4.10, these functions are Gâteaux differentiable at $x \in X$ if and only if they have a unique subgradient at x. Since $\partial f(0) = \{0\}$ and $\partial f(x) = \varphi(\|x\|) \cdot \partial \|\cdot\|(x)$ for $x \neq 0$, we have that (b) $\Leftrightarrow$ (c). The expression of the set of supporting hyperplanes of U_X at a point of S_X given by Eq. (3.56) shows that (c) $\Rightarrow$ (a). Conversely, if X is smooth, using the expression of the set of hyperplanes mentioned above, $\partial \|\cdot\|(x)$ has only one element for every $x \in S_X$, and so $\partial \|\cdot\|(x)$ is a singleton for every $x \in X \setminus \{0\}$. Using (iii) we obtain that $\partial f(x)$ is a singleton for every $x \in X$, whence (a) $\Rightarrow$ (b). The last part follows from Corollary 3.3.3.

(v) (a) $\Rightarrow$ (b) Suppose that X is strictly convex and let $x, y \in X$, $x \neq y$, and $\lambda \in]0,1[$. We have seen in (i) that

$$f(\lambda x + (1-\lambda)y) < \lambda f(x) + (1-\lambda) f(y)$$

if $\|x\| \neq \|y\|$. Let now $\|x\| = \|y\| =: \rho$; of course $\rho > 0$ (otherwise $x = y = 0$!). We want to prove that $\|\theta(\lambda)\| < \rho$, where $\theta(\mu) := \mu x + (1-\mu)y$. Let us consider $\varepsilon := \min\{\lambda, 1-\lambda\} > 0$. It is obvious that $\|\theta(\lambda - \varepsilon)\| \leq \rho$, $\|\theta(\lambda + \varepsilon)\| \leq \rho$ and $\theta(\lambda) = \frac{1}{2}\theta(\lambda - \varepsilon) + \frac{1}{2}\theta(\lambda + \varepsilon)$; moreover $\theta(\lambda - \varepsilon) \neq \theta(\lambda + \varepsilon)$. Thus we have $\|\theta(\lambda)\| < \rho$. Therefore

$$\begin{aligned} f(\lambda x + (1-\lambda)y) &= \psi(\|\lambda x + (1-\lambda)y\|) < \psi(\rho) = \lambda\psi(\|x\|) + (1-\lambda)\psi(\|y\|) \\ &= \lambda f(x) + (1-\lambda) f(y). \end{aligned}$$

The implication (b) $\Rightarrow$ (c) follows from Theorem 2.4.4(ii), while the implication (c) $\Rightarrow$ (d) is obvious.

(d) $\Rightarrow$ (a) Let $x, y \in S_X$, $x \neq y$. Suppose that $\frac{1}{2}(x+y) \in S_X$ and let $x^* \in \partial f(\frac{x+y}{2})$. Then, by (iii), $\varphi(1) = \varphi(\|x\|) = \varphi(\|y\|) = \varphi(\|\frac{x+y}{2}\|) = \|x^*\|$ and

$$\varphi(1) = \langle \tfrac{1}{2}(x+y), x^* \rangle \leq \tfrac{1}{2}\|x\| \cdot \|x^*\| + \tfrac{1}{2}\|y\| \cdot \|x^*\| = \varphi(1).$$

Therefore $\|x\| \cdot \|x^*\| = \langle x, x^* \rangle$ and $\|y\| \cdot \|x^*\| = \langle y, x^* \rangle$. Using again (iii) we obtain the contradiction $x^* \in \partial f(x) \cap \partial f(y)$. Therefore $\|\frac{1}{2}(x+y)\| < 1$, which shows that X is strictly convex.

(vi) Suppose that X is reflexive, and let $x^* \in X^*$. Let us consider the function $g : X \to \mathbb{R}$, $g(x) := f(x) - \langle x, x^* \rangle$. It is obvious that g is convex, continuous and $g(x) \geq \psi(\|x\|) - \|x\| \cdot \|x^*\|$; therefore $\lim_{\|x\|\to\infty} g(x) = \infty$. By Theorem 2.5.1(ii) there exists $\overline{x} \in X$ such that $g(\overline{x}) \leq g(x)$ for every $x \in X$, *i.e.* $\langle x - \overline{x}, x^* \rangle \leq f(x) - f(\overline{x})$ for every x. This relation proves that $x^* \in \partial f(\overline{x})$. Thus we have $\operatorname{Im}(\partial f) = X^*$.

Conversely, suppose that X is a Banach space and $\operatorname{Im}(\partial f) = X^*$. Let $x^* \in X^*$; taking into account James' theorem (see [Diestel (1975), Th. 1.6]), it is sufficient to show that x^* attains its supremum on U_X. Of course, we may suppose that $\|x^*\| = \varphi(1)$. By hypothesis, there exists $\overline{x} \in X$ such that $x^* \in \partial f(\overline{x})$. Therefore $\langle \overline{x}, x^* \rangle = \|\overline{x}\| \cdot \|x^*\|$ and $\|x^*\| = \varphi(\|\overline{x}\|) = \varphi(1)$. Hence $\|\overline{x}\| = 1$ and

$$\forall x \in U_X \ : \ \langle \overline{x}, x^* \rangle = \|\overline{x}\| \cdot \|x^*\| = \|x^*\| \geq \langle x, x^* \rangle.$$

Thus x^* attains its supremum on U_X at $\overline{x}$. □

The multifunction $\Phi_X := \partial(\frac{1}{2}\|\cdot\|^2) : X \rightrightarrows X^*$ is called the **duality mapping** of X. Therefore

$$\Phi_X(x) = \left\{ x^* \in X^* \mid \langle x, x^* \rangle = \|x\|^2 = \|x^*\|^2 \right\}. \tag{3.59}$$

From the preceding theorem we obtain that

$$\begin{aligned} \Phi_X \text{ is single-valued } &\Leftrightarrow X \text{ is smooth} \\ &\Leftrightarrow \|\cdot\| \text{ is Gâteaux differentiable on } X \setminus \{0\}. \end{aligned}$$

Moreover, if X is smooth then Φ_X is norm-weak* continuous, identifying $\Phi_X(x)$ with its unique element for every $x \in X$.

Related to the relationships between strict convexity and smoothness of a normed space and its dual one has the following result.

Theorem 3.7.3 *If X^* is smooth (resp. strictly convex), then X is strictly convex (resp. smooth). If X is a reflexive Banach space then X^* is smooth (resp. strictly convex) if and only if X is strictly convex (resp. smooth).*

Proof. To begin with, let X^* be smooth and $x, y \in S_X$, $x \neq y$; suppose that $\frac{1}{2}(x+y) \in S_X$. Taking $\overline{x}^* \in \Phi_X(\frac{x+y}{2})$, as in the proof of assertion (v) of the preceding theorem (taking $\varphi(t) = t$ for $t \geq 0$), we obtain that

$$1 = \|x\|^2 = \|y\|^2 = \|\overline{x}^*\|^2 = \langle x, \overline{x}^* \rangle = \langle y, \overline{x}^* \rangle.$$

From this relation we obtain that $\{x^* \in X^* \mid \langle x, x^* \rangle = 1\}$ and $\{x^* \in X^* \mid \langle y, x^* \rangle = 1\}$ are distinct supporting hyperplanes of U_{X^*} at $\overline{x}^* \in S_{X^*}$, which contradicts the fact that X^* is smooth. Therefore X is strictly convex.

Suppose now that X^* is strictly convex but X is not smooth at $\overline{x} \in S_X$. Then there exist x^*, $y^* \in S_{X^*}$ such that $\{x \in X \mid \langle x, x^* \rangle = 1\}$ and $\{x \in X \mid \langle x, y^* \rangle = 1\}$ are distinct supporting hyperplanes of U_X at $\overline{x}$. Therefore $x^* \neq y^*$ and

$$\langle \overline{x}, x^* \rangle = \langle \overline{x}, y^* \rangle = \langle \overline{x}, \tfrac{1}{2}(x^* + y^*) \rangle \leq \|\overline{x}\| \cdot \|\tfrac{1}{2}(x^* + y^*)\| = \|\tfrac{1}{2}(x^* + y^*)\| \leq 1.$$

Because $\langle \overline{x}, x^* \rangle = 1$, we obtain that $\|\frac{1}{2}(x^* + y^*)\| = 1$, which contradicts the fact that X^* is strictly convex. Therefore X is smooth.

If X is reflexive and if X is strictly convex (resp. smooth) then X^{**} is strictly convex (resp. smooth), whence, using the first part, we have that X^* is smooth (resp. strictly convex). □

Let us introduce other possible geometric properties of a normed space. We say that X is **uniformly smooth at x** if

$$\forall \varepsilon > 0,\ \exists \delta > 0,\ \forall y \in \delta U_X \ : \ \|x + y\| + \|x - y\| - 2\|x\| \leq \varepsilon \|y\|,$$

locally uniformly smooth if X is uniformly smooth at any $x \in S_X$ and **uniformly smooth** if

$$\forall \varepsilon > 0,\ \exists \delta > 0,\ \forall x \in S_X,\ \forall y \in \delta U_X \ : \ \|x + y\| + \|x - y\| - 2\|x\| \leq \varepsilon \|y\|;$$

it is obvious that X is not uniformly smooth at 0.

Consider the *gage of uniform smoothness of the norm at x* defined by

$$\sigma_x : \mathbb{R}_+ \to \mathbb{R}_+, \quad \sigma_x(t) := \sup\{\|x + ty\| + \|x - ty\| - 2\|x\| \mid y \in S_X\};$$

it is clear that $\sigma_x(t) \leq 2t\|x\|$ for $t \geq 0$ and $\sigma_x \in \Gamma$. It is obvious that X is uniformly smooth at x if and only if $\sigma_x \in \Sigma_1$. Considering also the *gage of uniform smoothness of the norm* of X defined by

$$\sigma_X : \mathbb{R}_+ \to \mathbb{R}_+, \quad \sigma_X := \sup\{\sigma_x \mid x \in S_X\},$$

we have that $\sigma_X(t) \leq 2t$ for $t \geq 0$ and $\sigma_X \in \Gamma$, too. So, X is uniformly smooth if and only if $\sigma_X \in \Sigma_1$.

In the next result we give some characterizations of the previous notions.

Theorem 3.7.4 *Let φ be a weight function.*

(i) *Let $x \in X \setminus \{0\}$; X is uniformly smooth at x $\Leftrightarrow$ $\|\cdot\|$ is Fréchet differentiable at x $\Leftrightarrow$ f_φ is Fréchet differentiable at x;*

(ii) *X is locally uniformly smooth $\Leftrightarrow$ $\|\cdot\|$ is Fréchet differentiable on $X \setminus \{0\}$ $\Leftrightarrow$ f_φ is Fréchet differentiable $\Leftrightarrow$ $f_\varphi \in C^1(X)$;*

(iii) *X is uniformly smooth $\Leftrightarrow$ f_φ is uniformly smooth on rU_X for some (every) $r > 0$.*

Proof. Denote f_φ by f.

(i) The first equivalence is immediate from Theorem 3.3.2, while the second one is obvious because $f = \psi \circ \|\cdot\|$, where ψ is defined by Eq. (3.58); ψ and ψ^{-1} are derivable on $\mathbb{P}$.

(ii) This is an immediate consequence of (i); notice that f is always Fréchet differentiable at 0 and $\nabla f(0) = 0$. The last equivalence follows from the final part of Theorem 3.3.2.

(iii) Assume first that f is uniformly smooth on rU_X (even on rS_X) for some $r > 0$. Then there exists $\sigma \in \Omega_1$ such that $f(x+ty) + f(x-ty) - 2f(x) \leq \sigma(t)$ for all $x \in rS_X$, $y \in S_X$ and $t \geq 0$, or equivalently

$$\psi(\|x+ty\|) + \psi(\|x-ty\|) - 2\psi(\|x\|) \leq \sigma(t)$$

for all $x \in rS_X$, $y \in S_X$ and $t \geq 0$. Because $\psi(t) \geq \psi(t_0) + \varphi(t_0) \cdot (t - t_0)$ for all $t, t_0 \geq 0$, we obtain that

$$r\varphi(r)\left(\left\|u + tr^{-1}y\right\| + \left\|u - tr^{-1}y\right\| - 2\|u\|\right) \leq \sigma(t)$$

for all $u, y \in S_X$ and $t \geq 0$. It follows that $r\varphi(r) \cdot \sigma_X(t) \leq \sigma(tr)$ for $t \geq 0$, and so $\sigma_X \in \Omega_1$. Hence X is uniformly smooth.

Assume now that X is uniformly smooth; then $\sigma_X \in \Gamma \cap \Omega_1$. Let $r > 0$. Because the function $\widetilde{\psi} : \mathbb{R} \to \mathbb{R}$, $\widetilde{\psi}(t) := \psi(|t|)$, is uniformly smooth on $[0, r]$ (being derivable on $\mathbb{R}$), there exists $\sigma \in \Omega_1$ such that $\psi(s) \leq \psi(t) + (s-t) \cdot \psi'(t) + \sigma(|s-t|)$ for all $t \in [0, r]$, $s \in \mathbb{R}$. It follows that for $x \in rU_X$, $y \in S_X$ and $t \geq 0$,

$$\psi(\|x+ty\|) \leq \psi(\|x\|) + \varphi(\|x\|)\left(\|x+ty\| - \|x\|\right) + \sigma\left(\left|\|x+ty\| - \|x\|\right|\right),$$
$$\psi(\|x-ty\|) \leq \psi(\|x\|) + \varphi(\|x\|)\left(\|x-ty\| - \|x\|\right) + \sigma\left(\left|\|x-ty\| - \|x\|\right|\right),$$

and so, because $\left|\|x \pm ty\| - \|x\|\right| \leq t$ and σ is nondecreasing,

$$f(x+ty) + f(x-ty) - 2f(x) \leq \varphi(\|x\|)\left(\|x+ty\| + \|x-ty\| - 2\|x\|\right) + 2\sigma(t)$$

for $x \in rU_X$, $y \in S_X$ and $t \geq 0$. Taking $\gamma \in]0, r[$, we obtain that

$$\varphi(\|x\|)\left(\|x+ty\|+\|x-ty\|-2\|x\|\right) \leq 2t\varphi(\gamma)$$

for $x \in \gamma U_X$ and

$$\varphi(\|x\|)\left(\|x+ty\|+\|x-ty\|-2\|x\|\right) \leq r\varphi(r)\sigma_X(t/\|x\|) \leq r\varphi(r)\sigma_X(t/\gamma)$$

for $x \in rU_X \setminus \gamma U_X$. Therefore

$$\sigma^0_{f,r}(t) \leq \max\left(2t\varphi(\gamma), r\varphi(r)\sigma_X(t/\gamma)\right) + 2\sigma(t)$$

for all $t \geq 0$ and $\gamma \in]0, r[$. In order to show that $\sigma^0_{f,r} \in \Omega_1$, let $\varepsilon > 0$. Because $\lim_{t\downarrow 0}\varphi(t) = \varphi(0) = 0$, there exists $\gamma \in]0, r[$ such that $\varphi(\gamma) < \varepsilon/4$; with this γ, since $\sigma_X \in \Omega_1$, there exists $\delta' > 0$ such that $(t/\gamma)^{-1}\sigma_X(t/\gamma) < \varepsilon/\left(2r\varphi(r)\right)$ for $t \in]0, \delta']$; finally, because $\sigma \in \Omega_1$, there exists $\delta'' > 0$ such that $t^{-1}\sigma(t) < \varepsilon/4$ for $t \in]0, \delta'']$. Taking $\delta := \min(\delta', \delta'')$, it follows that $t^{-1}\sigma^0_{f,r}(t) < \varepsilon$ for all $t \in]0, \delta]$. Hence $\sigma^0_{f,r} \in \Omega_1$, and so f is uniformly smooth on rU_X. □

The relations among the above smoothness notions are stated next. We recall first that X has the **Kadec–Klee property** if for all sequences $(x_n) \subset X$ and $x \in X$,

$$(x_n) \stackrel{w}{\to} x, \; (\|x_n\|) \to \|x\| \;\Rightarrow\; (x_n) \stackrel{\|\cdot\|}{\to} x.$$

Similarly, X^* (with its dual norm) has the **weak* Kadec–Klee property** if the above property holds in X^* but with w replaced by w^*.

Proposition 3.7.5 *The following implications hold:*

X is uniformly smooth $\Rightarrow$ X is locally uniformly smooth $\Rightarrow$ X is smooth.

Moreover, if X^* is smooth and has the weak Kadec–Klee property, then X is locally uniformly smooth.*

Proof. The above implications obviously follow from Theorems 3.7.2(iv) and 3.7.4.

Assume now that X^* is smooth and has the weak* Kadec–Klee property. Let φ be a weight function. By Theorem 3.7.2 (iv) we have that f_φ is Gâteaux differentiable and ∇f_φ is norm-weak* continuous. Let $(x_n) \subset X$ converge to $x \in X$; then $(\nabla f_\varphi(x_n)) \stackrel{w^*}{\to} \nabla f_\varphi(x)$. Because $\|\nabla f_\varphi(x')\| = \varphi(\|x'\|)$ for every $x' \in X$, we obtain that $(\|\nabla f_\varphi(x_n)\|) \to \|\nabla f_\varphi(x)\|$, and so $(\nabla f_\varphi(x_n)) \overset{\|\cdot\|}{\to} \nabla f_\varphi(x)$. Using now Theorem 3.3.2 we obtain that f_φ is Fréchet differentiable. The conclusion follows from Theorem 3.7.4. $\square$

Consider now two other notions. We say that X is **locally uniformly convex** if

$$\forall x \in S_X,\ \forall \varepsilon > 0,\ \exists \delta > 0,\ \forall y \in S_X\ :\ \|y - x\| \geq \varepsilon \Rightarrow \left\|\tfrac{1}{2}(x+y)\right\| \leq 1-\delta,$$

and X is **uniformly convex** if

$$\forall \varepsilon > 0,\ \exists \delta > 0,\ \forall x, y \in S_X\ :\ \|y - x\| \geq \varepsilon \Rightarrow \left\|\tfrac{1}{2}(x+y)\right\| \leq 1-\delta.$$

Of course, every uniformly convex space is locally uniformly convex, and any locally uniformly convex space is strictly convex. Moreover, note that X is locally uniformly convex if and only if for every $x \in S_X$ and every sequence $(x_n) \subset S_X$, $\left(\left\|\frac{1}{2}(x+x_n)\right\|\right) \to 1 \Rightarrow (\|x_n - x\|) \to 0$, while X is uniformly convex if and only if for all sequences $(x_n), (y_n) \subset S_X$, $\left(\left\|\frac{1}{2}(x_n+y_n)\right\|\right) \to 1 \Rightarrow (\|x_n - y_n\|) \to 0$.

The next result shows that any locally uniformly convex space has the Kadec–Klee property (even in a stronger form).

Proposition 3.7.6 *Assume that X is a locally uniformly convex space. If the net $(x_i)_{i\in I} \subset X$ converges weakly to $x \in X$ and $(\|x_i\|)_{i\in I} \to \|x\|$, then $(\|x_i - x\|)_{i\in I} \to 0$.*

Proof. Consider first the net $(x_i)_{i\in I} \subset S_X$ such that $(x_i)_{i\in I} \overset{w}{\to} x$ and $(\|x_i\|)_{i\in I} \to \|x\|$; hence $x \in S_X$. Assume that $(\|x_i - x\|)_{i\in I} \not\to 0$. Then there exist $\varepsilon_0 > 0$ and a cofinal set $J \subset I$ such that $\|x_i - x\| \geq \varepsilon_0$. Since X is locally uniformly convex, there exists $\delta > 0$ such that $\left\|\frac{1}{2}(x_i + x)\right\| \leq 1-\delta$ for every $i \in J$. As $\|\cdot\|$ is w-lower semicontinuous and $\left(\frac{1}{2}(x_i+x)\right)_{i\in J} \overset{w}{\to} x$, we get the contradiction $1 \leq \liminf_{i\in J} \left\|\frac{1}{2}(x_i + x)\right\| \leq 1 - \delta$. Therefore $(\|x_i - x\|)_{i\in I} \to 0$.

Let now $(x_i)_{i\in I} \subset X$ be arbitrary such that $(x_i)_{i\in I} \overset{w}{\to} x$ and $(\|x_i\|)_{i\in I} \to \|x\|$. If $x = 0$ the conclusion is obvious. Thus take $x \neq 0$. Then $x_i \neq 0$ for $i \succeq i_0$ for some $i_0 \in I$. Take $x_i' := x_i / \|x_i\| \in S_X$ (for $i \succeq i_0$) and $x' := x/\|x\|$. It is obvious that $(x_i')_{i\in I} \overset{w}{\to} x'$ and $(\|x_i'\|)_{i\in I} \to \|x'\|$. From the first part we get $(x_i')_{i\in I} \overset{\|\cdot\|}{\to} x'$, whence $(\|x_i\| \cdot x_i')_{i\in I} \overset{\|\cdot\|}{\to} \|x\| \cdot x'$, *i.e.* $(x_i)_{i\in I} \overset{\|\cdot\|}{\to} x$. $\square$

In the next result we give dual characterizations for uniformly convex and locally uniformly convex spaces.

Theorem 3.7.7 *Let φ be a weight function. Then*

(i) *X is locally uniformly convex $\Leftrightarrow$ f_φ is uniformly convex at any $x \in X$,*

(ii) *X is uniformly convex $\Leftrightarrow$ f_φ is uniformly convex on rU_X for any (some) $r > 0$.*

Proof. Denote again f_φ by f.

(ii) Assume first that X is uniformly convex, but f_φ is not uniformly convex on rU_X for some $r > 0$. Taking into account Exercise 3.3, there exists $\varepsilon_0 > 0$ such that for every $\delta > 0$, there exist $x, y \in rU_X$ with

$$\|x - y\| = \varepsilon_0 \quad \text{and} \quad f(\tfrac{1}{2}x + \tfrac{1}{2}y) > \tfrac{1}{2}f(x) + \tfrac{1}{2}f(y) - \delta.$$

Taking $(\delta_n) \downarrow 0$, there exists $(x_n), (y_n) \subset rU_X$ such that $\|x_n - y_n\| = \varepsilon_0$ and $\psi(\|\frac{1}{2}x_n + \frac{1}{2}y_n\|) \geq \frac{1}{2}\psi(\|x_n\|) + \frac{1}{2}\psi(\|y_n\|) - \delta_n$ for every $n \in \mathbb{N}$. Let $x_n = \eta_n u_n$ and $y_n = \mu_n v_n$ with $\eta_n, \mu_n \in [0, r]$ and $u_n, v_n \in S_X$ for every $n \in \mathbb{N}$. Then for every $n \in \mathbb{N}$ we have that

$$\psi\left(\tfrac{1}{2}\eta_n + \tfrac{1}{2}\mu_n\right) \geq \psi\left(\left\|\tfrac{1}{2}\eta_n u_n + \tfrac{1}{2}\mu_n v_n\right\|\right) \geq \tfrac{1}{2}\psi(\eta_n) + \tfrac{1}{2}\psi(\mu_n) - \delta_n. \quad (3.60)$$

Passing to a subsequence if necessary, we may suppose that $(\eta_n) \to \eta \in [0, r]$, $(\mu_n) \to \mu \in [0, r]$ and $\left(\left\|\frac{1}{2}u_n + \frac{1}{2}v_n\right\|\right) \to \gamma \in [0, 1]$. From Eq. (3.60) we obtain that $\psi\left(\frac{1}{2}\eta + \frac{1}{2}\mu\right) \geq \frac{1}{2}\psi(\eta) + \frac{1}{2}\psi(\mu)$. Because ψ is strictly convex we obtain that $\eta = \mu$. Since

$$\varepsilon_0 = \|\eta_n u_n - \mu_n v_n\| \leq \eta_n \|u_n - v_n\| + |\eta_n - \mu_n| \leq 3\eta_n + \mu_n \quad (3.61)$$

for every $n \in \mathbb{N}$, it follows that $\eta > 0$. Since

$$\left\|\tfrac{1}{2}\eta_n u_n + \tfrac{1}{2}\mu_n v_n\right\| \leq \eta_n \left\|\tfrac{1}{2}u_n + \tfrac{1}{2}v_n\right\| + \tfrac{1}{2}|\mu_n - \eta_n| \quad \forall n \in \mathbb{N},$$

from Eq. (3.60) we obtain that $\psi(\eta\gamma) \geq \psi(\eta)$, and so $\gamma \geq 1$. It follows that $\left(\left\|\frac{1}{2}u_n + \frac{1}{2}v_n\right\|\right) \to 1$, and so $(\|u_n - v_n\|) \to 0$. Passing to the limit in Eq. (3.61), we get the contradiction $\varepsilon_0 \leq 0$. Therefore f is uniformly convex.

Assume now that f is uniformly convex on rU_X for some $r > 0$. Then for $\varepsilon > 0$ and $x, y \in S_X$ with $\|x - y\| \geq \varepsilon$ we have that $f\left(\frac{1}{2}rx + \frac{1}{2}ry\right) \leq \frac{1}{2}f(rx) + \frac{1}{2}f(ry) - \frac{1}{4}\rho_{f,r}(r\|x - y\|)$, and so $\psi\left(r\left\|\frac{1}{2}x + \frac{1}{2}y\right\|\right) \leq \psi(r) - \frac{1}{4}\rho_{f,r}(r\varepsilon) < \psi(r)$. Taking $r' \in]0, r[$ such that $\psi(r') = \psi(r) - \frac{1}{4}\rho_{f,r}(r\varepsilon)$ and $\delta := 1 - r'/r > 0$, we have that $\left\|\frac{1}{2}x + \frac{1}{2}y\right\| \leq 1 - \delta$. Therefore X is uniformly convex.

For the proof of (i) fix $x \in S_X$ for the sufficiency (and take $\rho_{f,x}$ instead of $\rho_{f,R}$) and fix $x \in X$ for the necessity. In this case the sequence (η_n) is constant while $(\mu_n) \subset [0, \infty[$ is bounded. Then we can proceed as in the proof of (ii). □

We end this section with two results which prove the complete duality between uniform smoothness and uniform convexity.

Theorem 3.7.8 *Let φ be a weight function. The following statements are equivalent:*

(i) *X is uniformly smooth,*

(ii) *f_φ is uniformly smooth on rU_X for some (any) $r > 0$,*

(iii) *f_φ is Fréchet differentiable and ∇f_φ is uniformly continuous on bounded sets,*

(iv) *X^* is uniformly convex,*

(v) *$(f_\varphi)^*$ is uniformly convex on rU_{X^*} for some (any) $r > 0$.*

If one of the above conditions holds and X is a Banach space then X is reflexive.

Proof. The equivalence of (i) and (ii) is given by Proposition 3.7.4, the equivalence of (ii), (iii) and (v) follows from Proposition 3.6.3 because f_φ and $(f_\varphi)^*$ are strongly coercive, and the equivalence of (iv) and (v) follows from Proposition 3.7.7. The last statement follows by Proposition 3.6.3. □

A dual result holds, too.

Theorem 3.7.9 *Let φ be a weight function. The following statements are equivalent:*

(i) *X is uniformly convex,*

(ii) *f_φ is uniformly convex on rU_X for some (any) $r > 0$,*

(iii) *X^* is uniformly smooth,*

(iv) *$(f_\varphi)^*$ is uniformly smooth on rU_{X^*} for some (any) $r > 0$,*

(v) *$(f_\varphi)^*$ is Fréchet differentiable and ∇f_φ^* is uniformly continuous on bounded sets.*

If one of the above conditions holds and X is a Banach space then X is reflexive.

Proof. The proof is similar to that of the preceding theorem (and follows from the preceding theorem when X is reflexive). □

We end this section with another sufficient condition for reflexivity of a Banach space.

Proposition 3.7.10 *Let X be a Banach space. If the dual norm is Fréchet differentiable on $X^* \setminus \{0\}$ then X is reflexive.*

Proof. Let φ be a weight function. Then, by Theorem 3.7.2(ii), $(f_\varphi)^* = f_{\varphi^{-1}}$ (on X^*). Because the dual norm is Fréchet differentiable on $X^* \setminus \{0\}$, using Theorem 3.7.4(ii) we have that $(f_\varphi)^*$ is Fréchet differentiable on X^*. By Corollary 3.3.4 we have that $\nabla(f_\varphi)^*(x^*) \in X$ for every $x^* \in X^*$, and so $\operatorname{Im}(\partial f_\varphi) = X^*$. By Theorem 3.7.2(vi) we obtain that X is reflexive. □

3.8 Applications to the Best Approximation Problem

Let $C \subset (X, \|\cdot\|)$ be a nonempty set and consider the **distance function**

$$d_C : X \to \mathbb{R}, \quad d_C(x) := d(x, C) := \inf\{\|x - c\| \mid c \in C\}. \tag{3.62}$$

It is well known that d_C is Lipschitz with Lipschitz constant 1 (Exercise!).

Having $x \in X$, an important problem consists in determining the set

$$P_C(x) = \{\overline{x} \in C \mid \|\overline{x} - x\| = d_C(x)\};$$

$\overline{x} \in P_C(x)$ is called a **best approximation** of x by elements of C. It follows that the multifunction P_C introduced above is bounded on bounded sets. Indeed, if $x \in rU_X$, $\overline{x} \in P_C(x)$ and $\overline{c}$ is a fixed element of C, then $\|\overline{x}\| \le \|x - \overline{x}\| + \|x\| \le \|x - \overline{c}\| + \|x\| \le 2r + \|\overline{c}\|$.

Remark 3.8.1 If $\overline{x} \in P_C(x)$ then $\overline{x} \in P_C(\lambda x + (1 - \lambda)\overline{x})$ for every $\lambda \in [0, 1]$.

Indeed, taking $x_\lambda := \lambda x + (1 - \lambda)\overline{x}$, this follows immediately from the inclusion $B(x_\lambda, \|x_\lambda - \overline{x}\|) \subset B(x, \|x - \overline{x}\|)$.

The following notion is important in the best approximation theory. We say that the nonempty set $C \subset X$ is a **Chebyshev set** if $P_C(x)$ is a singleton for every $x \in X$.

Note that for $x_0 \in X$, $P_C(x_0)$ is the set of optimal solutions for everyone of the following minimization problems:

(P_1) $\quad \min \|x - x_0\|, \ x \in C,$

(P_2) $\quad \min \frac{1}{2}\|x - x_0\|^2, \ x \in C.$

Moreover $v(P_1) = d_C(x_0)$, $v(P_2) = \frac{1}{2}d_C^2(x_0)$.

In the (best) approximation theory, the fundamental problems are: the existence, the uniqueness and the characterization of best approximations. Because $P_C(x) = \{x\}$ if $x \in C$, $P_C(x) = \emptyset$ if $x \in (\operatorname{cl} C) \setminus C$, and $P_C(x) = \emptyset$ for every $x \in X \setminus C$ if C is open (Exercise!), it is quite natural to consider, generally, elements $x_0 \in X \setminus C$ and to assume that C is a *closed* set (taking into account also that $d_C(x_0) = d_{\operatorname{cl} C}(x_0)$). Because we propose ourselves to apply the results on convex functions obtained till now (but this is not the sole reason), we shall assume in this section that C is *convex*, too.

The first result gives sufficient conditions for the existence and the uniqueness of best approximations, respectively.

Theorem 3.8.1 *Let $C \subset X$ be a nonempty closed convex set and $x_0 \in X$.*

(i) *If X is a reflexive Banach space then $P_C(x_0) \neq \emptyset$.*

(ii) *If $X_0 := \operatorname{lin} C$ is a space of finite dimension then $P_C(x_0) \neq \emptyset$.*

(iii) *If X is a strictly convex normed space then $P_C(x_0)$ has at most one element.*

Proof. (i) Let us consider the function $f := \|\cdot - x_0\| + \iota_C$. By hypothesis f is convex and lsc; moreover $\lim_{\|x\|\to\infty} f(x) = \infty$. By Theorem 2.5.1(ii), there exists $\overline{x} \in X$ such that $f(\overline{x}) \leq f(x)$ for every $x \in X$, *i.e.* $\overline{x} \in P_C(x_0)$.

(ii) There exists $(x_n)_{n\geq 1} \subset C \subset X_0$ such that $(\|x_n - x_0\|) \to d_C(x_0)$. It follows that (x_n) is bounded. The subspace X_0 being of finite dimension, X_0 is isomorphic to $\mathbb{R}^p$ ($p = \dim X_0$) whence (x_n) contains a subsequence (x_{n_k}) convergent to $\overline{x} \in X_0 \subset X$; hence $(\|x_{n_k} - x_0\|) \to \|x - x_0\| = d_C(x_0)$. Since the set C is closed, $\overline{x} \in C$. Therefore $\overline{x} \in P_C(x_0)$.

(iii) We have already seen that $P_C(x_0) = \{x_0\}$ for $x_0 \in C$. Let $x_0 \notin C$. Suppose that there are two distinct elements x_1, x_2 in $P_C(x_0)$. Then

$$\tfrac{1}{2}(x_1 + x_2) \in C, \quad \text{and} \quad \|x_1 - x_0\| = \|x_2 - x_0\| = d_C(x_0) > 0.$$

Since X is strictly convex, we obtain

$$\begin{aligned}\|\tfrac{1}{2}(x_1 + x_2) - x_0\| &= \|\tfrac{1}{2}(x_1 - x_0) + \tfrac{1}{2}(x_2 - x_0)\| \\ &< \tfrac{1}{2}\|x_1 - x_0\| + \tfrac{1}{2}\|x_2 - x_0\| = d_C(x_0).\end{aligned}$$

This contradiction proves that $P_C(x_0)$ has at most one element. □

From the preceding theorem we obtain that any closed convex subset of a strictly convex and reflexive Banach space is Chebyshev. In the next section we shall show that any weakly closed Chebyshev subset of a Hilbert space is convex.

The following duality result reveals itself to be useful sometimes.

Theorem 3.8.2 *Let $C \subset X$ be a nonempty closed convex set and $x_0 \in X \setminus C$. Then*

$$d_C(x_0) = \max_{x^* \in U_{X^*}} \inf_{x \in C} \langle x_0 - x, x^* \rangle,$$
$$\tfrac{1}{2} d_C^2(x_0) = \max_{x^* \in X^*} \inf_{x \in C} \left(\langle x_0 - x, x^* \rangle - \tfrac{1}{2} \|x^*\|^2 \right).$$

If C is a cone, then

$$d_C(x_0) = \max_{x^* \in U_{X^*} \cap -C^+} \langle x_0, x^* \rangle, \quad \tfrac{1}{2} d_C^2(x_0) = \max_{x^* \in -C^+} \left(\langle x_0, x^* \rangle - \tfrac{1}{2} \|x^*\|^2 \right).$$

Proof. Let

$$f_1, f_2 : X \to \mathbb{R}, \ f_1(x) := \|x - x_0\|, \ f_2(x) := \tfrac{1}{2} \|x - x_0\|^2. \tag{3.63}$$

Using Fenchel–Rockafellar duality formula (Corollary 2.8.5) and the formulas for the conjugates of the norm and its square, we get

$$\begin{aligned}
d_C(x_0) &= \inf_{x \in X} (f_1(x) + \iota_C(x)) = \max_{x^* \in X^*} (-f_1^*(-x^*) - \iota_C^*(x^*)) \\
&= \max_{x^* \in U_{X^*}} \left(\langle x_0, x^* \rangle - \sup_{x \in C} \langle x, x^* \rangle \right) = \max_{x^* \in U_{X^*}} \inf_{x \in C} \langle x_0 - x, x^* \rangle, \\
\tfrac{1}{2} d_C^2(x_0) &= \inf_{x \in X} (f_2(x) + \iota_C(x)) = \max_{x^* \in X^*} (-f_2^*(-x^*) - \iota_C^*(x^*)) \\
&= \max_{x^* \in X^*} \left(\langle x_0, x^* \rangle - \tfrac{1}{2} \|x^*\|^2 - \sup_{x \in C} \langle x, x^* \rangle \right) \\
&= \max_{x^* \in X^*} \inf_{x \in C} \left(\langle x_0 - x, , x^* \rangle - \tfrac{1}{2} \|x^*\|^2 \right).
\end{aligned}$$

When C is a cone, $\iota_C^*(x^*) = 0$ if $x^* \in -C^+$, $= \infty$ otherwise; the conclusion follows immediately. □

In the next result we summarize several properties of d_C and $\frac{1}{2} d_C^2$.

Proposition 3.8.3 *Let $C \subset X$ be a nonempty convex set. Then*

(i) $(d_C)^* = \iota_{U_{X^*}} + \iota_C^* = \iota_{U_{X^*}} + s_C$ *and* $(\frac{1}{2} d_C^2)^* = \frac{1}{2} \|\cdot\|^2 + s_C$.

(ii) *If $x \in X$ and $\overline{x} \in P_C(x)$ then*

$$\partial d_C(x) = \partial \|\cdot\| (x - \overline{x}) \cap N(C, \overline{x}), \ \partial (\tfrac{1}{2} d_C^2)(x) = \Phi_X(x - \overline{x}) \cap N(C, \overline{x}); \tag{3.64}$$

in particular, for every $x \in C$ we have that

$$\partial d_C(x) = U_{X^*} \cap N(C, x), \quad \partial (\tfrac{1}{2} d_C^2)(x) = \{0\}. \tag{3.65}$$

(iii) *For any* $x \in C$ *and any* $u \in X$ *we have that*

$$(d_C)'(x,u) = d\big(u, \mathcal{C}(C,x)\big), \quad (\tfrac{1}{2}d_C^2)'(x,u) = \tfrac{1}{2}\left[d\big(u,\mathcal{C}(C,x)\big)\right]^2. \quad (3.66)$$

Proof. (i) Since $d_C = \|\cdot\| \,\Box\, \iota_C$ and $\frac{1}{2}d_C^2 = (\frac{1}{2}\|\cdot\|^2)\Box\iota_C$, the given formulas follow immediately from Theorem 2.3.1(ix), Corollary 2.4.16 and Theorem 3.7.2(ii) for $\varphi(t) = t$.

(ii) If $x \in X$ and $\overline{x} \in P_C(x)$ then $d_C(x) = \|x - \overline{x}\| + \iota_C(\overline{x})$ and $\left(\frac{1}{2}d_C^2\right)(x) = \frac{1}{2}\|x - \overline{x}\|^2 + \iota_C(\overline{x})$. Using Corollary 2.4.7, Corollary 2.4.16 and Theorem 3.7.2(iii) for $\varphi(t) = t$ we get Eq. (3.64). When $x \in C$, $\overline{x} = x$, and so $\partial\|\cdot\|(x - \overline{x}) = U_{X^*}$; hence Eq. (3.65) holds.

(iii) Let $x \in C$ and $u \in X$. Because d_C is a continuous convex function, by Theorem 2.4.9 and (ii) we have that

$$\begin{aligned}(d_C)'(x,u) &= \max\{\langle u, x^*\rangle \mid x^* \in \partial d_C(x)\}\\ &= \max\{\langle u, x^*\rangle \mid x^* \in U_{X^*} \cap N(C,\overline{x})\} = d\big(u, \mathcal{C}(C,x)\big),\end{aligned}$$

the last equality being obtained applying the last part of the preceding theorem for C replaced by $\mathcal{C}(C,x)$. □

In the following theorem we give several characterizations of best approximations in general normed spaces.

Theorem 3.8.4 *Let* $C \subset X$ *be a nonempty convex set,* $x_0 \in X \setminus C$ *and* $\overline{x} \in C$. *The following statements are equivalent:*

(i) $\overline{x} \in P_C(x_0)$;

(ii) $\exists\, x^* \in S_{X^*}$, $\forall\, x \in C$: $\langle \overline{x} - x_0, x^*\rangle = \|\overline{x} - x_0\|$ *and* $\langle x - \overline{x}, x^*\rangle \geq 0$;

(iii) $\exists\, x^* \in S_{X^*}$, $\forall\, x \in C$: $\langle x - x_0, x^*\rangle \geq \|\overline{x} - x_0\|$;

(iv) $\exists\, x^* \in X^*$, $\forall\, x \in C$: $\langle \overline{x} - x_0, x^*\rangle = \|x^*\|^2 = \|\overline{x} - x_0\|^2$ *and* $\langle x - \overline{x}, x^*\rangle \geq 0$, *or, equivalently,* $\Phi_X(x_0 - \overline{x}) \cap N(C;\overline{x}) \neq \emptyset$;

(v) $\exists\, x^* \in X^*$, $\forall\, x \in C$: $\|x^*\| = \|\overline{x} - x_0\|$ *and* $\langle x - x_0, x^*\rangle \geq \|\overline{x} - x_0\|^2$;

(vi) *(if* X *is smooth)* $\forall\, x \in C$: $\langle x - \overline{x}, \Phi_X(\overline{x} - x_0)\rangle \geq 0$, *or, equivalently,* $\Phi_X(x_0 - \overline{x}) \in N(C;\overline{x})$ *(identifying* $\Phi_X(x)$ *with its unique element);*

(vii) *(if* C *is a cone)* $\exists\, x^* \in S_{X^*} \cap -C^+$: $\langle x_0, x^*\rangle = \|\overline{x} - x_0\|$.

Proof. (i) ⇒ (ii) If $\overline{x} \in P_C(x_0)$ then $0 \in \partial f_1(\overline{x}) + \partial\iota_C(\overline{x})$, where f_1 is defined in Eq. (3.63). The conclusion is immediate if we take into account the expression of the subdifferential of the norm.

(ii) $\Rightarrow$ (iii) With x^* from (ii) we have that

$$\langle x - x_0, -x^* \rangle = \langle x - \overline{x}, -x^* \rangle - \langle \overline{x} - x_0, x^* \rangle = \langle x - \overline{x}, -x^* \rangle - \|\overline{x} - x_0\| \geq 0$$

for every $x \in C$. The element $-x^*$ satisfies the desired condition.

(iii) $\Rightarrow$ (i) With x^* from (iii) we have that

$$\forall x \in C \ : \ \|x - x_0\| \geq \langle x - x_0, x^* \rangle \geq \|\overline{x} - x_0\|,$$

i.e. $\overline{x} \in P_C(x_0)$.

The implications (i) $\Rightarrow$ (iv) $\Rightarrow$ (v)$\Rightarrow$(i) are proved exactly as above, but using this time the function f_2 defined in Eq. (3.63).

Condition (vi) is identical to condition (iv) when X is smooth ($\partial f_2(x) = \Phi_X(x)$; if X is smooth $\Phi_X(x)$ is a singleton, whose element is denoted by $\Phi_X(x)$, too).

Suppose now that C is a cone.

(iii) $\Rightarrow$ (vii) With x^* from (iii) we have that

$$\forall t \geq 0, \ \forall x \in C \ : \ \langle tx, x^* \rangle \geq \langle x_0, x^* \rangle + \|\overline{x} - x_0\|.$$

Therefore $\langle x, x^* \rangle \geq 0$ for every $x \in C$, *i.e.* $x^* \in C^+$, and

$$\langle x_0, -x^* \rangle \geq \|\overline{x} - x_0\| \geq \langle x_0 - \overline{x}, -x^* \rangle \geq \langle x_0, -x^* \rangle,$$

whence $-x^*$ is the desired element.

(vii) $\Rightarrow$ (iii) With x^* from (vii) we have that

$$\forall x \in C \ : \ \langle x - x_0, -x^* \rangle \geq \langle x_0, x^* \rangle = \|\overline{x} - x_0\|,$$

which shows that $-x^*$ satisfies the condition of (iii). □

The following result is sometimes useful.

Corollary 3.8.5 *Let $C \subset X$ be a nonempty convex set and $\overline{x} \in C$. Then $\overline{x} \in P_C(\overline{x} + tu)$ for all $u \in \Phi_X^{-1}(N(C;\overline{x}))$ and $t \in \mathbb{R}_+$.*

Proof. The conclusion is obvious for $u = 0$ or $t = 0$. Assume that $u \neq 0$ and $t > 0$. Then $\overline{x} + tu \notin C$; otherwise, taking $u^* \in \Phi_X(u) \cap N(C, \overline{x})$, we obtain the contradiction $t\|u\|^2 = \langle \overline{x} + tu - \overline{x}, u^* \rangle \leq 0$. Since $\Phi_X((\overline{x} + tu) - \overline{x}) \cap N(C, \overline{x}) = t(\Phi_X(u) \cap N(C, \overline{x})) \neq \emptyset$, the conclusion follows from assertion (iv) of the preceding theorem. □

We end this section with the following useful result.

Proposition 3.8.6 *Assume that X is reflexive and strictly convex and let $C \subset X$ be a nonempty closed convex set. Then $P_C(x)$ is a singleton for every $x \in X$ and we identify it with its element. Moreover:*

(i) *P_C is norm-weak continuous.*

(ii) *If X has the Kadec–Klee property then P_C is norm-norm continuous.*

(iii) *If X is a Hilbert space then P_C has Lipschitz constant 1.*

Proof. We already observed that P_C is single-valued in our case.

(i) Let $(x_n)_{n\in\mathbb{N}} \subset X$ converge to $x \in X$. Because P_C is bounded on bounded sets, the sequence $(P_C(x_n))$ is bounded. The space X being reflexive, there exists a subsequence $(P_C(x_{n_k}))_{k\in\mathbb{N}}$ converging weakly to $\overline{x} \in C$ (C being closed). So, $\|x_{n_k} - P_C(x_{n_k})\| \leq \|x_{n_k} - c\|$ for every $c \in C$. Taking the liminf inferior for $k \to \infty$, we obtain that $\|x - \overline{x}\| \leq \|x - c\|$ for every $c \in C$, and so $\overline{x} = P_C(x)$. $P_C(x)$ being the unique weak limit point of $(P_C(x_n))$, it follows that $(P_C(x_n)) \to P_C(x)$ weakly. Therefore P_C is norm-weak continuous.

(ii) The conclusion is obvious from (i) and the definition of the Kadec–Klee property.

(iii) Assume that X is a Hilbert space with scalar product $(\cdot \mid \cdot)$; we identify X^* with X by the Riesz theorem, and so $\Phi_X(x) = x$ for every $x \in X$. From assertion (vi) of the preceding theorem we have that $(x - P_C(x) \mid P_C(x) - c) \geq 0$ for all $c \in C$ and $x \in X$. Taking $x, y \in X$ we have $(x - P_C(x) \mid P_C(x) - P_C(y)) \geq 0$ and $(y - P_C(y) \mid P_C(y) - P_C(x)) \geq 0$. Adding these two relations side by side we get

$$\big(x - y - P_C(x) + P_C(y) \mid P_C(x) - P_C(y)\big) \geq 0,$$

whence

$$\|P_C(x) - P_C(y)\|^2 \leq (x - y \mid P_C(x) - P_C(y)) \leq \|x - y\| \cdot \|P_C(x) - P_C(y)\|.$$

Hence $\|P_C(x) - P_C(y)\| \leq \|x - y\|$ for all $x, y \in X$. □

Remark 3.8.2 The proof of assertion (i) of Proposition 3.8.6 shows that P_C is norm-weak continuous for every weakly closed Chebyshev subset of a reflexive Banach space.

3.9 Characterizations of Convexity in Terms of Smoothness

We begin this section with the following interesting result.

Theorem 3.9.1 *Let X be a normed space, $f : X \to \overline{\mathbb{R}}$ be a proper function, $\overline{x} \in X$ and $\overline{x}^* \in X^*$. Consider the function $g : X \to \overline{\mathbb{R}}$, $g(x) := f(x) - \langle x, \overline{x}^* \rangle$. Consider the following conditions:*

(i) *f is lsc (resp. weakly lsc) at $\overline{x}$, $\overline{x}^* \in \operatorname{dom} f^*$, f^* is Fréchet (resp. Gâteaux) differentiable at $\overline{x}^*$ and $\nabla f^*(\overline{x}^*) = \overline{x}$.*

(ii) *$g(\overline{x}) = \inf g(X)$ and $(x_n) \to \overline{x}$ (resp. $(x_n) \xrightarrow{w} \overline{x}$) whenever $g(x_n) \to g(\overline{x})$, i.e. (g, X) is Tikhonov (resp. weakly Tikhonov) well posed.*

(iii) *$\overline{x} \in \operatorname{dom} f$ and $f^{**}(\overline{x}) = f(\overline{x})$.*

Then (i) $\Rightarrow$ (ii) $\Rightarrow$ (iii). *Furthermore, if f is convex then* (i) $\Leftrightarrow$ (ii).

Proof. (ii) $\Rightarrow$ (iii) Since g is proper and $g(\overline{x}) = \inf g(X)$, we have that $\overline{x} \in \operatorname{dom} g = \operatorname{dom} f$. Moreover

$$\begin{aligned}
[\forall x \in X \ : \ g(x) \geq g(\overline{x})] &\Leftrightarrow \overline{x}^* \in \partial f(\overline{x}) \\
&\Leftrightarrow \overline{x}^* \in \operatorname{dom} f^* \text{ and } f(\overline{x}) = \langle \overline{x}, \overline{x}^* \rangle - f^*(\overline{x}^*) \\
&\Rightarrow f(\overline{x}) \leq f^{**}(\overline{x}).
\end{aligned}$$

Since the other inequality is always true we obtain that $f^{**}(\overline{x}) = f(\overline{x})$.

Note that in cases (i) and (ii) $\overline{x}^* \in \operatorname{dom} f^*$ and $\overline{x} \in \partial f^*(\overline{x}^*)$. Replacing eventually the function f by

$$h : X \to \overline{\mathbb{R}}, \quad h(x) := f(\overline{x} + x) + f^*(\overline{x}^*) - \langle \overline{x} + x, \overline{x}^* \rangle$$

(hence $h^*(x^*) = f^*(\overline{x}^* + x^*) - f^*(\overline{x}^*) - \langle \overline{x}, x^* \rangle$), we may assume that $\overline{x} = 0$, $\overline{x}^* = 0$, $f^*(0) = 0$ and $0 \in \partial f^*(0)$ (what we do in the sequel). In this situation $g = f$. We get immediately that

$$\forall x \in X \ : \ f(x) \geq f^{**}(x) \geq f^{**}(0) = 0.$$

(i) $\Rightarrow$ (ii) To begin with, consider that f is lsc at $\overline{x} = 0$ and f^* is Fréchet differentiable at $\overline{x}^* = 0$. Since $\nabla f^*(0) = 0$,

$$\forall \varepsilon > 0, \ \exists \delta(\varepsilon) > 0, \forall x^* \in X^*, \ \|x^*\| \leq \delta(\varepsilon) \ : \ f^*(x^*) \leq \varepsilon \|x^*\|.$$

Consider the functions

$$\gamma, \varphi : \mathbb{R}_+ \to \overline{\mathbb{R}}_+, \quad \gamma(t) := \inf\{\varepsilon \mid \varepsilon > 0, \ \delta(\varepsilon) \geq t\}, \quad \varphi(t) := t\gamma(t).$$

It is obvious that $\gamma \in N_0 \cap \Omega_0$, and so $\varphi \in N_1 \cap \Omega_1$; moreover,

$$\forall x^* \in X^* \ : \ f^*(x^*) \leq \varphi(\|x^*\|).$$

From this relation and Lemma 3.3.1(v) we obtain that

$$\forall x \in X \ : \ f^{**}(x) \geq (\varphi \circ \|\cdot\|)^*(x) = \varphi^{\#}(\|x\|).$$

By Lemma 3.3.1(iii) we obtain that $\varphi^{\#} \in \mathcal{A}_0$. Let $(x_n) \subset X$ be such that $(f(x_n)) \to \inf f = -f^*(0) = 0$ (such a sequence exists!). Suppose that $(x_n) \not\to 0$. Then there exists $\alpha > 0$ such that $P := \{n \in \mathbb{N} \mid \|x_n\| \geq \alpha\}$ is infinite. Therefore

$$\forall n \in P \ : \ f(x_n) \geq f^{**}(x_n) \geq \varphi^{\#}(\|x_n\|) \geq \varphi^{\#}(\alpha) > 0,$$

which contradicts the fact that $(f(x_n)) \to 0$. Therefore $(x_n) \to 0$. Since f is lsc at $\overline{x} = 0$ we have that $0 \leq f(0) \leq \liminf f(x_n) = 0$, *i.e.* $f(0) = 0$. Thus we have obtained that $f(x) \geq f(0)$ for every $x \in X$ and $(f(x_n)) \to f(0) \Rightarrow (x_n) \to 0$.

Suppose now that f is w−lsc at $\overline{x} = 0$, f^* is Gâteaux differentiable at $\overline{x}^* = 0$ and $\nabla f^*(0) = 0$. Let $(x_n) \subset X$ be such that $(f(x_n)) \to \inf f = -f^*(0) = 0$ (such a sequence exists!). Let us prove that $(x_n) \xrightarrow{w} 0$. In the contrary case there exists $x^* \in X^*$ such that $P := \{n \in \mathbb{N} \mid \langle x_n, x^*\rangle \geq 1\}$ is infinite. Then

$$\forall t > 0, \ \forall n \in P \ : \ f^*(tx^*) \geq \langle x_n, tx^*\rangle - f(x_n) \geq t - f(x_n).$$

Taking the limit for $n \to \infty$ $(n \in P)$, we obtain that $f^*(tx^*) \geq t$ for every $t > 0$, and so $0 = \nabla f^*(0)(x^*) \geq 1$. This contradiction proves that $(x_n) \xrightarrow{w} 0$. Since f is w−lsc at 0 we obtain that $0 \leq f(0) \leq \liminf f(x_n) = 0$. Therefore $f(x) \geq f(0)$ for every $x \in X$ and $(f(x_n)) \to f(0) \Rightarrow (x_n) \xrightarrow{w} 0$.

(ii) $\Rightarrow$ (i) if f is convex. The lower semi-continuity (weak and strong) of f at $\overline{x} = 0$ is obvious since $f(0) = \inf f$ $(= -f^*(0) = 0)$.

To begin with, suppose that for every sequence $(x_n) \subset X$, $(f(x_n)) \to 0 \Rightarrow (x_n) \to 0$. Consider the function

$$\varphi : \mathbb{R}_+ \to \overline{\mathbb{R}}_+, \quad \varphi(t) := \inf\{f(x) \mid x \in X, \ \|x\| = t\}.$$

It is obvious that $\varphi(0) = 0$; moreover $\varphi(t) > 0$ for $t > 0$ (indeed, if $\varphi(t) = 0$, then there exists $(x_n) \subset X$ such that $(f(x_n)) \to 0$ and $\|x_n\| = t$ for every n; by hypothesis we have that $(x_n) \to 0$, and so $t = 0$). Furthermore $\varphi \in N_1$ and so $\varphi \in N_0$. Indeed, let $0 < \tau < t$ and $x \in X$ be such that $\|x\| = t$.

Then $\varphi(\tau) \le f(\frac{\tau}{t}x) \le \frac{\tau}{t}f(x)$, which proves that $\varphi(\tau)/\tau \le \varphi(t)/t$. Hence $\varphi \in \mathcal{A}_0 \cap N_1$. By Lemma 3.3.1(iii) we have that $\varphi^{\#\#} = \overline{\text{co}}\varphi \in \mathcal{A}_0$ and $\varphi^{\#} \in \Omega_1$.

The construction of φ shows that $f(x) \ge \varphi(\|x\|)$ for every $x \in X$, whence, using again Lemma 3.3.1(v), we have that $0 = f^*(0) \le f^*(x^*) \le \varphi^{\#}(\|x^*\|)$ for every $x^* \in X^*$. Hence $\lim_{x^* \to 0}(f^*(x^*) - f^*(0))/\|x^*\| = \lim_{t \downarrow 0} \varphi^{\#}(t)/t = 0$, which proves that f^* is Fréchet differentiable at 0 and $\nabla f^*(0) = 0$.

Suppose now that $(f(x_n)) \to 0 \Rightarrow (x_n) \xrightarrow{w} 0$ for every sequence $(x_n) \subset X$. Let $x^* \in X^*$ be fixed. Suppose that $f^{*\prime}_+(0; x^*) > 0$, *i.e.* there exists $\mu > 0$ such that $f^*(tx^*) > t\mu$ for every $t > 0$. It follows that $f^*(\frac{1}{n}x^*) > \frac{1}{n}\mu$ for every $n \in \mathbb{N}$. Therefore there exists $(x_n) \subset X$ such that

$$\forall n \in \mathbb{N} : \frac{\|x_n\| \cdot \|x^*\|}{n} \ge \frac{1}{n}\langle x_n, x^*\rangle > \frac{1}{n}\langle x_n, x^*\rangle - \frac{\mu}{n} > f(x_n) \ge 0. \quad (3.67)$$

It follows that $\langle x_n, x^*\rangle > \mu$ for every $n \in \mathbb{N}$, which shows that (x_n) has no subsequences weakly converging to 0. Suppose that (x_n) contains a bounded subsequence (x_{n_k}). From relation (3.67) we have that $0 \le f(x_{n_k}) \le \|x_{n_k}\| \cdot \|x^*\|/n_k \to 0$, whence, from our hypothesis we get the contradiction $x_{n_k} \xrightarrow{w} 0$. Therefore $\alpha_n := \|x_n\| \to \infty$. Suppose that $\limsup_{n\to\infty} \alpha_n/n < \infty$. Let $t_n := \sqrt{\alpha_n}$ $(\to \infty)$ and $y_n := \frac{1}{t_n}x_n$; of course, $(\|y_n\|) \to \infty$. Then $0 \le f(y_n) \le \frac{1}{t_n}f(x_n) \le \frac{\alpha_n}{nt_n}\|x^*\| \to 0$. Therefore $(f(y_n)) \to 0$, and so $(y_n) \xrightarrow{w} 0$; this contradicts the fact that $(\|y_n\|) \to \infty$. Therefore $\limsup_{n\to\infty} \alpha_n/n = \infty$. Let $(n_k) \subset \mathbb{N}$ be an increasing sequence such that $(\alpha_{n_k}/n_k) \to \infty$. Let us take $t_k := \alpha_{n_k}/\sqrt{n_k}$ $(\to \infty)$ and $y_k := \frac{1}{t_k}x_{n_k}$; of course $(\|y_k\|) \to \infty$. Then

$$0 \le f(y_k) \le \frac{1}{t_k}f(x_{n_k}) \le \frac{\alpha_{n_k}}{t_k n_k}\|x^*\| = \frac{1}{\sqrt{n_k}}\|x^*\| \to 0.$$

Therefore $(f(y_k)) \to 0$, whence $(y_k) \xrightarrow{w} 0$, contradicting the fact that (y_k) is not bounded.

What we showed above proves that $f^{*\prime}_+(0; x^*) \le 0$ for every $x^* \in X^*$. Therefore, $0 \in (\text{dom}\, f^*)^i$. Since $f^{*\prime}_+(0, \cdot)$ is sublinear (and finite in this case), we have $f^{*\prime}_+(0; x^*) = 0$ for every $x^* \in X^*$. It follows that f^* is Gâteaux differentiable at 0 and $\nabla f^*(0) = 0$. □

Theorem 3.9.2 *Assume that X is a Banach space and $f : X \to \overline{\mathbb{R}}$ is a proper function with* $\text{dom}\,\partial f^*$ *nonempty and open, the subdifferential* *being taken for the duality* (X^*, X^{**}). *Then* f *is convex provided one of the following conditions holds:*

(i) f *is lsc and* f^* *is Fréchet differentiable at any* $x^* \in \operatorname{dom} \partial f^*$,

(ii) X *is weakly sequentially complete,* f *is weakly lsc and* f^* *is Gâteaux differentiable at any* $x^* \in \operatorname{dom} \partial f^*$.

Proof. Let $g := f^{**}$ (for the duality (X, X^*)). Because $g^* = f^*$ is proper we have that $g \in \Gamma(X)$ and $g \leq f$. Using Corollaries 3.3.4 and 3.3.5 for g, in both cases we have that $\nabla f^*(x^*) \in X$ for any $x^* \in \operatorname{dom} \partial f^*$. Let $D := \{\nabla f^*(x^*) \mid x^* \in \operatorname{dom} \partial f^*\} \subset X$. Using the preceding theorem, in both cases we have that $D \subset \operatorname{dom} f$ and $f(x) = g(x)$ for every $x \in D$. Let $x \in \operatorname{dom} g$. Using the Brøndsted–Rockafellar theorem, there exists $(x_n) \subset \operatorname{dom} \partial g$ such that $(x_n) \to x$ and $(g(x_n)) \to g(x)$. Since $\operatorname{Im} \partial g \subset \operatorname{dom} \partial f^*$, it follows that $(x_n) \subset D$. Thus $f(x) \leq \liminf f(x_n) = \liminf g(x_n) = g(x)$, which, together with $g \leq f$, implies that $f(x) = g(x)$. As for $x \notin \operatorname{dom} g$ we have that $f(x) \geq g(x) = \infty$, we obtain that $f(x) = g(x)$ for all $x \in X$, and so f is convex. □

Remark 3.9.1 (a) The condition "$\operatorname{dom} \partial f^*$ is nonempty and open" in the preceding theorem can be replaced by "$\operatorname{dom} \partial f^* = (\operatorname{dom} \partial f^*)^i \neq \emptyset$"; (b) under the conditions of the preceding theorem f is strictly convex on $\operatorname{int}(\operatorname{dom} f)$.

Indeed, when $\operatorname{dom} \partial f^* = (\operatorname{dom} \partial f^*)^i \neq \emptyset$, we have $\emptyset \neq (\operatorname{dom} \partial f^*)^i \subset (\operatorname{dom} f^*)^i = \operatorname{int}(\operatorname{dom} f^*)$ and f^* is continuous on $\operatorname{int}(\operatorname{dom} f^*)$ (see Theorem 2.2.20). Hence, by Theorem 2.4.9, $\operatorname{int}(\operatorname{dom} f^*) \subset \operatorname{dom} \partial f^*$ (for the duality (X^*, X^{**})). It follows that $\operatorname{dom} \partial f^*$ is open. The second statement follows from Exercise 2.32.

In the next result the convexity of a closed subset of a Hilbert space is characterized by the differentiability of the square of the distance function.

Theorem 3.9.3 *Let* $(X, (\cdot \mid \cdot))$ *be a Hilbert space and* $\emptyset \neq C \subset X$ *be a closed set. Then* C *is convex if, and only if,* d_C^2 *is Fréchet differentiable.*

Proof. We identify X^* with X by the Riesz theorem. Consider $f = \frac{1}{2} \|\cdot\|^2 + \iota_C$. Then

$$\begin{aligned} f^*(y) &= \sup \left\{ (x \mid y) - \tfrac{1}{2} \|x\|^2 \mid x \in C \right\} = \tfrac{1}{2} \|y\|^2 - \tfrac{1}{2} \inf \left\{ \|y - x\|^2 \mid x \in C \right\} \\ &= \tfrac{1}{2} \|y\|^2 - \tfrac{1}{2} d_C^2(y) \end{aligned}$$

for every $y \in X$. Therefore

$$d_C^2 = \|\cdot\|^2 - 2f^*, \tag{3.68}$$

and so $\operatorname{dom} f^* = X$ and f^* is continuous on X. In particular $\operatorname{dom} \partial f^* = X$ is open. From Eq. (3.68) we obtain that f^* is Fréchet differentiable if d_C^2 is so.

Assume that d_C^2 is Fréchet differentiable; then f^* is also Fréchet differentiable. Because f is lower semicontinuous (the set C being closed), assertion (i) of the preceding theorem shows that f is convex, and so $C = \operatorname{dom} f$ is a convex set.

Assume now that C is a convex set. By Theorems 2.8.7 (iii) and 3.7.2 (i) we have that $f^* = \frac{1}{2}\|\cdot\|^2 \,\square\, \iota_C^*$ and the convolution is exact. Since $\frac{1}{2}\|\cdot\|^2$ is Fréchet differentiable and $\operatorname{dom} \partial f^* = X$, by Corollary 2.4.8 we have that f^* is Fréchet differentiable. □

Remark 3.9.2 Formula (3.68) shows that d_C^2 is the difference of two (finite and continuous) convex functions for every nonempty subset C of a Hilbert space; in particular d_C^2 (and also d_C) has directional derivatives at any point.

The next result refers to Chebyshev subsets of Hilbert spaces.

Theorem 3.9.4 *Let $(X, (\cdot \mid \cdot))$ be a Hilbert space and $\emptyset \neq C \subset X$ be a weakly closed set. Then the following four conditions are equivalent:*

(i) *C is convex;*

(ii) *C is a Chebyshev set;*

(iii) *d_C^2 is Gâteaux differentiable;*

(iv) *d_C^2 is Fréchet differentiable.*

Proof. The implication (i) ⇒ (ii) follows from Theorem 3.8.1, the equivalence of (i) and (iv) follows from the preceding theorem, while (iv) ⇒ (iii) is immediate.

As in the proof of the preceding theorem, consider $f = \frac{1}{2}\|\cdot\|^2 + \iota_C$ and identify X^* with X. Because C is weakly lower semicontinuous, f is weakly lsc, too.

Assuming that (iii) holds, from Eq. (3.68) we have that f^* is Gâteaux differentiable on $X^* = X$. Taking into account that X is weakly sequentially complete, by Theorem 3.9.2 (ii) we obtain that C is convex. Hence (iii) ⇒ (i).

(ii) $\Rightarrow$ (iii) Assume that C is a Chebyshev set and denote by $P_C(x)$ the best approximation of $x \in X$ by elements of C. We shall show that

$$\forall x \in X \; : \; \partial f^*(x) = \{P_C(x)\}, \tag{3.69}$$

and so, by the continuity and convexity of f^*, we obtain that f^* is Gâteaux differentiable, or, equivalently, d_C^2 is Gâteaux differentiable.

Indeed, let $x \in X$. Then, by Eq. (3.68),

$$\begin{aligned} f^*(x) &= \tfrac{1}{2}\|x\|^2 - \tfrac{1}{2}\|x - P_C(x)\|^2 = (x \mid P_C(x)) - \tfrac{1}{2}\|P_C(x)\|^2 \\ &= (x \mid P_C(x)) - f(P_C(x)), \end{aligned}$$

and so, by Theorem 2.4.2 (iii), $P_C(x) \in \partial f^*(x)$ for every $x \in X$.

Let now $x \in X$ be fixed and $y \in \partial f^*(x)$. Take $x_n := x + n^{-1}(y - P_C(x))$; of course, $(x_n) \to x$. As observed in Remark 3.8.2, P_C is norm-weak continuous, and so $(P_C(x_n)) \xrightarrow{w} P_C(x)$. Because ∂f^* is monotone, we have that

$$0 \le (x_n - x \mid P_C(x_n) - y) = n^{-1}(y - P_C(x) \mid P_C(x_n) - y),$$

whence $(y - P_C(x) \mid P_C(x_n) - y) \ge 0$ for every $n \in \mathbb{N}$. Taking the limit for $n \to \infty$, we get $-\|y - P_C(x)\| \ge 0$, *i.e.* $y = P_C(x)$, and so Eq. (3.69) holds for every $x \in X$. □

3.10 Weak Sharp Minima, Well-behaved Functions and Global Error Bounds for Convex Inequalities

As in the other sections of this chapter, $(X, \|\cdot\|)$ is a normed vector space. Let $f \in \Lambda(X)$ and $S \subset X$ be a nonempty set. One says that S is a **set of weak sharp minima** if there exists $\alpha > 0$ such that

$$f(x) \ge f(\overline{x}) + \alpha \cdot d_S(x), \quad \forall x \in X, \; \forall \overline{x} \in S; \tag{3.70}$$

this notion generalizes that of *sharp minimum point*, obtained when $S = \{\overline{x}\}$. It is clear that $S \subset \operatorname{argmin} f \subset \operatorname{cl} S$ if S is a set of weak sharp minima for f. Hence $S = \operatorname{argmin} f$ when S is closed. This is a reason for taking in the sequel $S = \operatorname{argmin} f$. In this case Eq. (3.70) becomes

$$f(x) \ge \inf f + \alpha \cdot d_S(x) \quad \forall x \in X,$$

i.e. f is ψ_α-conditioned, with $\psi_\alpha(t) = \alpha t$ for $t \geq 0$. Applying Theorem 3.4.3 we obtain a part of the next result, where, as usual, $d(x, \emptyset) := \infty$.

Theorem 3.10.1 *Let $f \in \Gamma(X)$ be such that $S := \operatorname{argmin} f \neq \emptyset$ and $\alpha > 0$. Consider the following statements:*

(i) $\forall x \in X \ : \ f(x) \geq \inf f + \alpha \cdot d_S(x)$;

(ii) $\forall x^* \in \alpha U_{X^*} \ : \ f^*(x^*) = f^*(0) + \iota_S^*(x^*)$;

(iii) $\forall \overline{x} \in S,\ \forall (x, x^*) \in \operatorname{gr} \partial f \ : \ \langle x - \overline{x}, x^* \rangle \geq \alpha \cdot d_S(x)$;

(iv) $d(0, \partial f(X \setminus S)) \geq \alpha$;

(v) $\forall x^* \in \alpha B_{X^*} \ : \ \partial f^*(x^*) \subset S$ *[$\partial f^*(x^*)$ being taken for the duality (X, X^*)];*

(vi) $\forall x \in X,\ \forall \overline{x} \in P_S(x) \ : \ f'(\overline{x}, x - \overline{x}) \geq \alpha \cdot d_S(x)$;

(vii) $\forall \overline{x} \in S,\ \forall u \in \Phi_X^{-1}(N(S, \overline{x})) \ : \ f'(\overline{x}, u) \geq \alpha \|u\|$;

(viii) $\forall \overline{x} \in S,\ \forall u \in \Phi_X^{-1}(N(S, \overline{x})) \cap \operatorname{cone}(\operatorname{dom} f - \overline{x}) \ : \ f'(\overline{x}, u) \geq \alpha \|u\|$;

(ix) $\forall \overline{x} \in S,\ \forall u \in X \ : \ f'(\overline{x}, u) \geq \alpha \cdot d(u, \mathcal{C}(S, \overline{x}))$;

(x) $\forall \overline{x} \in S,\ \alpha U_{X^*} \cap N(S, \overline{x}) \subset \partial f(\overline{x})$.

Then (i) $\Leftrightarrow$ (ii) $\Rightarrow$ (iii) $\Rightarrow$ (iv) $\Leftrightarrow$ (v), (i) $\Rightarrow$ (vi) $\Rightarrow$ (vii) $\Leftrightarrow$ (viii) *and* (vi) $\Rightarrow$ (ix) $\Rightarrow$ (x). *If X is a Banach space then* (v) $\Rightarrow$ (i). *Moreover, if X is a reflexive Banach space then* (vii) $\Rightarrow$ (i) *and* (x) $\Rightarrow$ (i)*; hence all ten conditions are equivalent in this case.*

Proof. The implications (i) $\Rightarrow$ (iii) $\Rightarrow$ (iv) follow immediately from the implications (i) $\Rightarrow$ (v) $\Rightarrow$ (vi) of Theorem 3.4.3 (taking $\psi := \psi_\alpha$ defined above), while (iv) $\Leftrightarrow$ (v) is obvious. The equivalence (vii) $\Leftrightarrow$ (viii) is also obvious because $\operatorname{dom} f'(\overline{x}, \cdot) = \operatorname{cone}(\operatorname{dom} f - \overline{x})$.

(i) $\Rightarrow$ (ii) Taking $\psi = \psi_\alpha$, condition (iii) of Theorem 3.4.3 is satisfied; hence $f^*(0) \in \mathbb{R}$ and

$$f^*(x^*) \leq f^*(0) + \iota_S^*(x^*) + \psi^\#(\|x^*\|) = f^*(0) + \iota_S^*(x^*) + \iota_{\alpha U_{X^*}}(x^*) \ \forall x^* \in X^*.$$

Because for $a \in S$ we have that $\langle a, x^* \rangle \leq f(a) + f^*(x^*) = -f^*(0) + f^*(x^*)$, we obtain that $f^*(0) + \iota_S^*(x^*) \leq f^*(x^*)$ for every $x^* \in X^*$. Therefore the conclusion holds.

(ii) $\Rightarrow$ (i) From the hypothesis, taking again $\psi = \psi_\alpha$, we have that

$$\forall x^* \in X^* \ : \ f^*(x^*) \leq f^*(0) + \iota_S^*(x^*) + \psi^\#(\|x^*\|).$$

Using the implication (iv) $\Rightarrow$ (iii) of Theorem 3.4.3, the conclusion holds.

(i) $\Rightarrow$ (vi) Take $x \in X$ and $\overline{x} \in P_S(x)$. Then for every $t \in]0,1[$ we have that $\overline{x} \in P_S\big((1-t)\overline{x} + tx\big)$ (see Corollary 3.8.5), and so

$$f\left(\overline{x} + t\left(x - \overline{x}\right)\right) - f(\overline{x}) \geq \alpha d_S\left((1-t)\overline{x} + tx\right) = \alpha t \left\|x - \overline{x}\right\| = \alpha t d_S(x).$$

Dividing by t and letting $t \to 0$ we get the conclusion.

(vi) $\Rightarrow$ (vii) Let $\overline{x} \in S$ and $u \in \Phi_X^{-1}\big(N(S,\overline{x})\big)$. By Corollary 3.8.5 we have that $\overline{x} \in P_S(\overline{x} + u)$. From (vi) we get the conclusion.

(vi) $\Rightarrow$ (ix) Let $\overline{x} \in S$ and $u \in X$. From (vi) and Proposition 3.8.3(iii) we have that

$$f'(\overline{x}, u) \geq \alpha \cdot d_S(\overline{x} + u) \geq \alpha \left(d_S(\overline{x}) + (d_S)'(\overline{x}, u)\right) \geq \alpha \cdot d\big(u, \mathcal{C}(S, \overline{x})\big).$$

(ix) $\Rightarrow$ (x) Let $\overline{x} \in S$. From (ix) and Proposition 3.8.3 (iii) we have that $f'(\overline{x}, u) \geq \alpha (d_S)'(\overline{x}, u)$ for every $u \in X$, and so

$$\alpha U_{X^*} \cap N(S, \overline{x}) = \alpha \cdot \partial d_S(\overline{x}) = \alpha \cdot (d_S)'(\overline{x}, \cdot)(0) \subset f'(\overline{x}, \cdot)(0) = \partial f(\overline{x}).$$

(v) $\Rightarrow$ (i) when X is a Banach space. Suppose that (i) does not hold. Then there exists $\overline{x} \in X$ such that $f(\overline{x}) < \inf f + \alpha \cdot d_S(\overline{x})$. It follows that $\overline{x} \notin S$. Therefore there exists $\alpha' \in]0, \alpha[$ such that $f(\overline{x}) < \inf f + \alpha' \cdot d_S(\overline{x})$. We fix $\varepsilon \in]\alpha', \alpha[$. Using Ekeland's variational principle, we obtain the existence of $u \in X$ such that

$$f(u) + \varepsilon \cdot \|u - \overline{x}\| \leq f(\overline{x}) \text{ and } f(u) \leq f(x) + \varepsilon \cdot \|u - x\| \quad \forall x \in X. \tag{3.71}$$

The last relation is equivalent to $\partial f(u) \cap \varepsilon U_{X^*} \neq \emptyset$. Hence there exists $u^* \in \partial f(u)$ such that $\|u^*\| \leq \varepsilon < \alpha$. From our hypothesis it follows that $\partial f^*(u^*) \subset S$, and so $u \in S$. By the choice of α' and the first relation of (3.71) we have that

$$\varepsilon \cdot d_S(\overline{x}) \leq \varepsilon \cdot \|u - \overline{x}\| < \alpha' \cdot d_S(\overline{x}).$$

As $d_S(\overline{x}) > 0$, we obtain the contradiction $\varepsilon < \alpha'$.

Assume now that X is a reflexive Banach space.

(vii) $\Rightarrow$ (i) Let $x \in \operatorname{dom} f$ and take $\overline{x} \in P_S(x)$ (such an element exists by Theorem 3.8.1). By Theorem 3.8.4(iv) $\Phi_X(x - \overline{x}) \cap N(S, \overline{x}) \neq \emptyset$. It follows that $x - \overline{x} \in \Phi_X^{-1}\left(N(S, \overline{x})\right)$, and so $f'(\overline{x}, x - \overline{x}) \geq \alpha \left\|x - \overline{x}\right\| = \alpha d_S(x)$. The conclusion follows from the inequality $f'(\overline{x}, x - \overline{x}) \leq f(x) - f(\overline{x})$.

(x) $\Rightarrow$ (i) Let $x \in X \setminus S$ and take $\overline{x} \in P_S(x)$. By Theorem 3.8.4 (iv), there exists $x^* \in \Phi_X(x - \overline{x}) \cap N(S, \overline{x})$. Then $u^* := \alpha\|x - \overline{x}\|^{-1}(x - \overline{x}) \in$

$\alpha U_{X^*} \cap N(S, \overline{x})$, whence $u^* \in \partial f(\overline{x})$. It follows that

$$f(x) = f(x) - f(\overline{x}) \geq \langle x - \overline{x}, u^* \rangle = \alpha \|x - \overline{x}\| = \alpha \cdot d_S(x).$$

The proof is complete. □

Using the preceding theorem we get the following estimate for the best coefficient in Eq. (3.70).

Corollary 3.10.2 *Let X be a Banach space and $f \in \Gamma(X)$ such that $S := \operatorname{argmin} f \neq \emptyset$. Then*

$$\inf_{x \in X \setminus S} \frac{f(x) - \inf f}{d(x, S)} = d\big(0, \partial f(X \setminus S)\big).$$

Moreover, if X is reflexive, then

$$\begin{aligned}
\inf_{x \in X \setminus S} \frac{f(x) - \inf f}{d(x, S)} &= \inf \left\{ f'\left(\overline{x}, \frac{x - \overline{x}}{\|x - \overline{x}\|}\right) \,\middle|\, x \in \operatorname{dom} f \setminus S,\ \overline{x} \in P_S(x) \right\} \\
&= \inf \left\{ f'(\overline{x}, u) \mid \overline{x} \in S,\ u \in S_X \cap \Phi_X^{-1}\big(N(S, \overline{x})\big) \right\} \\
&= \inf \left\{ \frac{f'(\overline{x}, u)}{d\big(u, C(S, \overline{x})\big)} \,\middle|\, \overline{x} \in S,\ u \in X \setminus C(S, \overline{x}) \right\}.
\end{aligned}$$

Proof. It is obvious that anyone of the numbers appearing in the statement of the corollary is nonnegative. By the equivalences (i) ⇔ (vi) ⇔ (vii) ⇔ (ix) ⇔ (iv) of the preceding theorem, if one of them is positive then all of them are equal. The conclusion follows. □

Remark 3.10.1 For every nonempty closed convex set $A \subset X$ we have

$$d\big(u, C(A, x)\big) \geq \|u\|, \quad \forall x \in A,\ \forall u \in \Phi_X^{-1}\big(N(A, x)\big). \tag{3.72}$$

Indeed, we have that $A = \operatorname{argmin} d_A$ and $d_A(x) \geq \inf d_A + 1 \cdot d_A(x)$ for every $x \in A$. From the implication (i) ⇒ (vii) of the preceding theorem and Proposition 3.8.3(iii) one gets Eq. (3.72).

As we shall see later the notion of weak sharp minima is closely related to that of global error bound for (convex) inequality systems.

We introduce now another notion which is related to well-conditioning and weak sharp minima. As seen in Theorem 3.4.3, when $\operatorname{argmin} f$ is nonempty, $f \in \Gamma(X)$ is well-conditioned if and only if $\big(d\left(0, \partial f(x_n)\right)\big) \to 0 \Rightarrow$

$\big(d(x_n, S)\big) \to 0$. So, it is natural to consider a class of functions with a similar property:

$$\mathcal{F} := \left\{ f \in \Gamma(X) \mid \big(d\left(0, \partial f(x_n)\right)\big) \to 0 \Rightarrow \big(f(x_n)\big) \to \inf f \right\}. \tag{3.73}$$

In the next result we give a characterization of the elements of $\mathcal{F}$.

Proposition 3.10.3 *Let $f \in \Gamma(X)$. Then*

$$f \in \mathcal{F} \;\Leftrightarrow\; \overline{r}_f(t) > 0, \ \forall\, t > \inf f,$$

where, for $t \in \mathbb{R}$,

$$\overline{r}_f(t) := \inf \left\{ d\left(0, \partial f(x)\right) \mid x \in X, \ f(x) \geq t \right\} = d\big(0, \partial f(X \setminus [f < t])\big). \tag{3.74}$$

Proof. It is obvious that

$$\begin{aligned} f \notin \mathcal{F} &\Leftrightarrow \exists\,(x_n) \subset X \ : \ \big(d(0, \partial f(x_n))\big) \to 0, \ \big(f(x_n)\big) \not\to \inf f \\ &\Leftrightarrow \exists\, t > \inf f, \ \exists\,(x_n) \subset X \setminus [f < t] \ : \ \big(d(0, \partial f(x_n))\big) \to 0 \\ &\Leftrightarrow \exists\, t > \inf f \ : \ \overline{r}_f(t) = 0. \end{aligned}$$

Hence the conclusion holds. □

It is obvious that $\overline{r}_f$ is nondecreasing. Moreover, for $t \leq \inf f$ we have that $\overline{r}_f(t) = d\left(0, \operatorname{Im} \partial f\right)$; in particular, by Brøndsted–Rockafellar's theorem, $\overline{r}_f(\inf f) = d\left(0, \operatorname{dom} f^*\right)$ when X is a Banach space. For a more detailed discussion of the position of 0 and $\operatorname{dom} f^*$ see Exercise 2.45.

In order to give other expressions for $\overline{r}_f(t)$ we introduce several related quantities. First, we mention a simple property which will be used repetitively in this section.

Lemma 3.10.4 *Let $f \in \Gamma(X)$, $t \in \mathbb{R}$, and $x, y \in \operatorname{dom} f$ such that $f(y) < t < f(x)$. Then there exists $z \in]x, y[$ such that $f(z) = t$. In particular $\|y - z\| < \|y - x\|$ and $\|x - z\| < \|y - x\|$.*

Proof. Of course, the mapping $\varphi : \mathbb{R} \to \overline{\mathbb{R}}$, $\varphi(\lambda) := f(\lambda x + (1 - \lambda) y)$ is a lsc convex function and $[0, 1] \subset \operatorname{dom} \varphi$. By Proposition 2.1.6 φ is continuous on $[0, 1]$. As $\varphi(0) < t < \varphi(1)$, there exists $\overline{\lambda} \in]0, 1[$ such that $\varphi(\overline{\lambda}) = t$. Taking $z := \overline{\lambda} x + (1 - \overline{\lambda}) y$ the conclusion follows. □

Proposition 3.10.5 *Let $f \in \Gamma(X)$ and $x \in \operatorname{dom} f$ with $f(x) > \inf f$. Consider the following numbers:*

$$\gamma_1(x) := \sup\left\{ \frac{f(x)-f(y)}{\|x-y\|} \;\middle|\; y \in [f < f(x)] \right\},$$
$$\gamma_2(x) := \sup\left\{ \frac{f(x)-t}{d(x,[f \le t])} \;\middle|\; t < f(x) \right\},$$
$$\gamma_3(x) := \sup\left\{ -f'\left(x, \frac{y-x}{\|y-x\|}\right) \;\middle|\; y \in [f < f(x)] \right\},$$
$$\gamma_4(x) := \sup\left\{ -f'(x,u) \mid u \in S_X \right\},$$
$$\gamma_5(x) := d(0, \partial f(x)),$$
$$\gamma_6(x) := \sup\left\{ -f'(x,-u) \mid u \in S_X \cap N([f \le f(x)], x) \right\}.$$

Then $\gamma_1(x) = \gamma_2(x) = \gamma_3(x) = \gamma_4(x) = \gamma_5(x) \in]0,\infty]$. Moreover, if f is continuous at x and X is a Hilbert space, identified with X^* by Riesz theorem, then $\gamma_1(x) = \gamma_6(x)$.

Proof. Let us denote, during the proof, the set $[f < f(x)]$ by S. It is obvious that $\gamma_1(x) > 0$, $\gamma_2(x) > 0$ and $\gamma_5(x) > 0$ (because $S \neq \emptyset$). Since $f'(x, y-x) \le f(y) - f(x) < 0$ for $y \in S$, we also have that $\gamma_3(x) > 0$, and so $\gamma_4(x) > 0$ as, obviously, $\gamma_4(x) \ge \gamma_3(x)$.

Let first $t < f(x)$ be such that $[f \le t] \neq \emptyset$, and show that

$$\sup_{y \in [f \le t]} \frac{f(x)-f(y)}{\|x-y\|} = \frac{f(x)-t}{d(x,[f \le t])}. \tag{3.75}$$

Consider $(y_n) \subset [f \le t]$ such that $(\|x-y_n\|) \to d(x,[f \le t])$. Then

$$\sup_{y \in [f \le t]} \frac{f(x)-f(y)}{\|x-y\|} \ge \frac{f(x)-f(y_n)}{\|x-y_n\|} \ge \frac{f(x)-t}{\|x-y_n\|} \to \frac{f(x)-t}{d(x,[f \le t])},$$

which proves the inequality $\ge$ in Eq. (3.75). Conversely, let $y \in [f \le t]$. Using Lemma 3.10.4 we get $\lambda \in [0,1[$ such that $f(y') = t$, where $y' := \lambda x + (1-\lambda)y$. Of course, $\|x-y'\| = (1-\lambda)\|x-y\|$ and $t \le \lambda f(x) + (1-\lambda) f(y)$, whence $(1-\lambda)(f(x)-f(y)) \le f(x) - t$. It follows that

$$\frac{f(x)-f(y)}{\|x-y\|} = \frac{(1-\lambda)(f(x)-f(y))}{(1-\lambda)\|x-y\|} \le \frac{f(x)-t}{\|x-y'\|} \le \frac{f(x)-t}{d(x,[f \le t])},$$

which proves the inequality $\le$ in Eq. (3.75). Hence Eq. (3.75) holds.

Now, by Eq. (3.75), we have

$$\gamma_1(x) = \sup_{t<f(x)} \sup_{y\in[f\le t]} \frac{f(x)-f(y)}{\|x-y\|} = \sup_{t<f(x)} \frac{f(x)-t}{d(x,[f\le t])} = \gamma_2(x).$$

(Note that the values of t with $[f \le t] = \emptyset$ influence neither $\gamma_1(x)$ nor $\gamma_2(x)$.)

Let $y \in S$ and take $u := (y-x)/\|y-x\|$. Since $f'_+(x, y-x) \le f(y) - f(x)$, we have that $(f(x)-f(y))/\|x-y\| \le -f'(x,u)$. It follows that $\gamma_1(x) \le \gamma_3(x)$. On the other hand, we have that $x + tu \in S$ for $t \in]0, \|y-x\|^{-1}]$, and so

$$-f'(x,u) = \sup\left\{ \frac{f(x)-f(x+tu)}{\|x-(x+tu)\|} \;\middle|\; t \in]0, \|y-x\|^{-1}] \right\} \le \gamma_1(x),$$

which implies that $\gamma_3(x) \le \gamma_1(x)$. Hence $\gamma_1(x) = \gamma_3(x)$.

It was observed above that $0 < \gamma_3(x) \le \gamma_4(x)$. Let $u \in S_X$. Suppose first that there exists $\bar{t} > 0$ such that $y := x + \bar{t}u \in S$; then $-f'(x,u) = -f'(x, \frac{y-x}{\|y-x\|}) \le \gamma_3(x)$. In the contrary case $f(x+tu) \ge f(x)$ for every $t > 0$, and so $f'(x,u) \ge 0$, whence $-f'(x,u) \le 0 < \gamma_3(x)$. Hence $\gamma_3(x) = \gamma_4(x)$.

Assume that $\gamma \in \mathbb{R}$ is such that $\gamma_4(x) \le \gamma$ (thus $\gamma > 0$). It follows that $-f'(x,u) \le \gamma\|u\|$ for all $u \in X$, and so $0 \le \gamma\|u\| + f'(x,u)$ for $u \in X$. Hence 0 is a minimum point for $\gamma\|\cdot\| + f'(x,\cdot)$, and so $0 \in \partial(\gamma\|\cdot\| + f'(x,\cdot))(0)$. But

$$\partial(\gamma\|\cdot\| + f'(x,\cdot))(0) = \gamma U_{X^*} + \partial f'(x,\cdot)(0) = \gamma U_{X^*} + \partial f(x). \quad (3.76)$$

Therefore $\gamma U_{X^*} \cap \partial f(x) \ne \emptyset$, which implies that $d(0, \partial f(x)) \le \gamma$. Conversely, if $d(0,\partial f(x)) \le \gamma$ (hence $\gamma > 0$ because $0 \notin \partial f(x)$), then $\gamma U_{X^*} \cap \partial f(x) \ne \emptyset$ (since the norm of X^* is w^*-lsc and $\partial f(x)$ is w^*-closed), and so $0 \in \gamma U_{X^*} + \partial f(x)$. By Eq. (3.76) we obtain that 0 is a minimum point of $\gamma\|\cdot\| + f'(x,\cdot)$, which shows that $\gamma_4(x) \le \gamma$. Hence $\gamma_4(x) = \gamma_5(x)$.

Assume now that X is a Hilbert space (identified with its topological dual) and f is continuous at x; then, by Theorem 2.4.9

$$f'(x,u) = \max\{(u \mid x^*) \mid x^* \in \partial f(x)\} \quad \forall u \in X, \quad (3.77)$$

and, by Corollary 2.9.5, $K := N([f \le f(x)], x) = \mathbb{R}_+ \partial f(x)$. It is clear that $\gamma_6(x) \le \gamma_5(x)$. Let $y \in S$ and take $u := (x-y)/\|x-y\|$; it was observed above that $\gamma := -f'(x,-u) > 0$. Then for $x^* \in \partial f(x)$ we have that

$(u \mid x^*) = -(-u \mid x^*) \geq -f'(x, -u) \geq \gamma$. Take $\overline{u} = P_K(u)$. By Theorem 3.8.4(vi) we have that $(u - \overline{u} \mid v - \overline{u}) \leq 0$ for every $v \in K$. Replacing v by tv with $t \geq 0$ we get $(u - \overline{u} \mid v) \leq 0 = (u - \overline{u} \mid \overline{u})$ for all $v \in K$. In particular $\gamma \leq (u \mid x^*) \leq (\overline{u} \mid x^*)$ for all $x^* \in \partial f(x)$, and so $\overline{u} \neq 0$. Since P_K is Lipschitz with Lipschitz constant 1 [see Proposition 3.8.6(iii)] and $0 = P_K(0)$, we have that $\|\overline{u}\| \leq 1$. Taking $u_0 := \overline{u}/\|\overline{u}\|$, we have that $u_0 \in S_X \cap N([f \leq f(x)], x)$ and, by (3.77), $f'(x, -u) \leq -\gamma$. Hence $\gamma_6(x) \geq \gamma$, whence $\gamma_6(x) \geq \gamma_3(x)$. □

With the aid of the numbers $\gamma_1(x)$–$\gamma_6(x)$ we introduce the functions

$$r_f^i :]\inf f, \infty[\to [0, \infty], \quad r_f^i(t) = \inf \{\gamma_i(x) \mid x \in [f = t]\}, \quad i \in \overline{1,6}.$$

Of course, if $[f = t] = \emptyset$ then $r_f^i(t) = +\infty$.

From the preceding result we obtain immediately

Corollary 3.10.6 *Let $f \in \Gamma(X)$. Then $r_f^1(t) = r_f^2(t) = r_f^3(t) = r_f^4(t) = r_f^5(t)$ for every $t > \inf f$. Moreover, if X is a Hilbert space identified with its dual and f is continuous at every point of $[f = t] \neq \emptyset$ then $r_f^1(t) = r_f^6(t)$.*

In the sequel we shall denote by r_f any of the functions r_f^1–r_f^5 defined above. From the preceding proposition we obtain that

$$\overline{r}_f(t) = \inf\{r_f(s) \mid s \geq t\}, \quad \forall t > \inf f, \tag{3.78}$$

where $\overline{r}_f$ is defined by Eq. (3.74).

Example 3.10.1 1) Any coercive function $f \in \Gamma(X)$ with X a reflexive Banach space is in $\mathcal{F}$; see the solution of Exercise 3.15 for details.

2) Let $f : \mathbb{R}^2 \to \mathbb{R}$ be defined by $f(x, y) := \sqrt{x^2 + y^2} - x$. It is obvious that f is a sublinear functional with $\inf f = 0$ and $\operatorname{argmin} f = \mathbb{R}_+ \times \{0\}$. One has $\overline{r}_f(t) = r_f(t) = 0$ for $t \geq 0$ (see the solution of Exercise 3.18 for details).

Consider now the function $l_f : [\inf f, \infty[\to [0, \infty]$ defined by

$$l_f(t) := \inf \left\{ \frac{f(x) - t}{d(x, [f \leq t])} \;\middle|\; x \in \operatorname{dom} f \setminus [f \leq t] \right\}. \tag{3.79}$$

As usual, we take $\inf \emptyset = +\infty$. Hence $l_f(\inf f) = 0$ if $\inf f$ is not attained; if $\inf f$ is attained then $\operatorname{argmin} f$ is a set of weak sharp minima if and only if $l_f(\inf f) > 0$. Even for $t > \inf f$ the number l_f is related to weak sharp minima. Indeed, for $t > \inf f$, taking $h : X \to \overline{\mathbb{R}}$ with $h(x) := \max\{f(x) -$

$t, 0\}$ we have that $\inf h = 0$ and $\operatorname{argmin} h = [f \leq t]$. Then $[f \leq t]$ is a set of weak sharp minima for h if and only if $l_f(t) > 0$. This remark suggests having formulas for $l_f(t)$ similar to those in Corollary 3.10.2.

Proposition 3.10.7 *Let X be a Banach space and $f \in \Gamma(X)$. Consider $t \in [\inf f, \infty[$ with $[f \leq t] \neq \emptyset$. Then*

$$l_f(t) = d\big(0, \partial f(X \setminus [f \leq t])\big). \tag{3.80}$$

Assume now that X is reflexive. Then

$$l_f(t) = \inf\left\{ f'\left(y, \frac{x-y}{\|x-y\|}\right) \,\middle|\, x \in \operatorname{dom} f \setminus [f \leq t],\ y \in P_{[f\leq t]}(x) \right\} \tag{3.81}$$

$$= \inf \left\{ f'(y,u) \mid y \in [f = t],\ u \in S_X \cap \Phi_X^{-1}\left(N([f \leq t], y)\right) \right\} \tag{3.82}$$

$$= \inf \left\{ \frac{f'(y,u)}{d\big(u, C([f \leq t], y)\big)} \,\middle|\, y \in [f = t],\ u \in X \setminus C([f \leq t], y) \right\}. \tag{3.83}$$

Moreover, if either (i) $t > \inf f$ *or* (ii) $N([f \leq t], x) = \mathbb{R}_+ \partial f(x)$ *for every $x \in [f = t]$, then*

$$l_f(t) = k_f(t) := \inf \left\{ f'\left(y, \frac{u}{\|u\|}\right) \,\middle|\, y \in [f = t],\ u \in \Phi_X^{-1}\big(\partial f(y)\big) \right\}. \tag{3.84}$$

Proof. If $t = \inf f$ the conclusion is given by Corollary 3.10.2. Assume that $t > \inf f$ and denote by $l_1(t)$, $l_2(t)$, $l_3(t)$, and $l_4(t)$ the quantities on the right-hand side of Eqs. (3.80)–(3.83), respectively. Consider $h : X \to \overline{\mathbb{R}}$ with $h(x) := \max\{f(x) - t, 0\}$; we have that $\inf h = 0$ and $\operatorname{argmin} h = [f \leq t]$. The equality of $l_f(t)$ and $l_1(t)$ follows directly from Corollary 3.10.2 because $U := X \setminus [f \leq t]$ is open and $f|_U = h|_U + t$, and so $\partial f(x) = \partial h(x)$ for every $x \in U$. Since, obviously, $l_f(t) = l_h(0)$, we apply again Corollary 3.10.2 for h. Hence, in order to obtain Eqs. (3.81)–(3.83), it is sufficient to observe the following:

a) Let $x \in \operatorname{dom} f \setminus [f \leq t]$ and $y \in P_{[f\leq t]}(x)$. Then, by Lemma 3.10.4, we have that $f(y) = t$. Since $y + s(x-y) \in \operatorname{dom} f \setminus [f \leq t]$ for every $s \in]0,1]$, we obtain that $f'(y, x-y) = h'(y, x-y)$.

b) Let $y \in [f = t]$ and $u \in S_X \cap \Phi_X^{-1}\left(N([f \leq t], y)\right)$. Then $y \in P_{[f\leq t]}(y + su)$ for every $s > 0$. It follows that either $y + su \notin \operatorname{dom} f$ for every $s > 0$, and so $f'(y,u) = h'(y,u) = \infty$, or $y + s_0 u \in \operatorname{dom} f$ for some $s_0 > 0$; in this case $f(y) = t$ and again $f'(y, x-y) = h'(y, x-y)$.

c) Let $y \in [f = t]$ and $u \in X \setminus C([f \le t], y)$. Because $u \notin C([f \le t], y)$, we have that $y + su \notin [f \le t]$ for every $s > 0$. Hence $f(y+su) = h(y+su)+t$ for all $s > 0$. If $y + su \notin \operatorname{dom} f$ for $s > 0$ then $f'(y,u) = h'(y,u) = \infty$. Assume that $y + s_0 u \in \operatorname{dom} f$ for some $s_0 > 0$; as above we obtain that $f(y) = t$ and so, once again, $f'(y, x-y) = h'(y, x-y)$.

Let us prove now relation (3.84). The case (ii) is an immediate consequence of Eq. (3.82). So, let $t > \inf f$. Because $\partial f(y) \subset N([f \le t], y)$ for every $y \in [f = t]$, we have that $k_f(t) \ge l_3(t) = l_f(t)$. Let now $x \in \operatorname{dom} f \setminus [f \le t]$ and take $y \in P_{[f \le t]}(x)$; of course, $y \ne x$. By Lemma 3.10.4 we have that $y \in [f = t]$. Since y is a solution of the problem

$$\min \ \tfrac{1}{2}\|z - x\|^2 \text{ subject to } f(z) - t \le 0,$$

and the Slater condition holds, we have that $0 \in \partial\left(\frac{1}{2}\|\cdot - x\|^2 + \lambda(f - t)\right)$ for some $\lambda \ge 0$, *i.e.* y is a minimum point of $\frac{1}{2}\|\cdot - x\|^2 + \lambda(f-t)$. Hence $\Phi_X(x-y) \cap \partial(\lambda f)(y) \ne \emptyset$. Assuming that $\lambda = 0$, we obtain that $\|y - x\| \le \|z - x\|$ for every $z \in \operatorname{dom} f$. Taking $z := x$ we obtain the contradiction $\|y - x\| \le 0$, *i.e.* $y = x$. Therefore $\lambda > 0$. Then $u := \lambda^{-1}(x-y) \in \Phi_X^{-1}\left(\partial f(y)\right)$. It follows that $l_f(t) = l_2(t) \ge k_f(t)$. The proof is complete. □

Note that the equality $N([f \le t], x) = \mathbb{R}_+ \partial f(x)$ for $x \in [f = t]$ is not guaranteed for arbitrary $f \in \Gamma(X)$ when $t > \inf f$ (compare with Corollary 2.9.5). However it holds if f is continuous at $x \in [f = t]$.

Relation (3.80) shows that l_f is nondecreasing on $[\inf f, \infty[$ when X is a Banach space. In fact l_f has this property in any normed vector space.

Proposition 3.10.8 *Let $f \in \Gamma(X)$ and $\inf f \le t_1 < t_2 < \infty$. Then*

$$l_f(t_1) \le l_f(t_2) \quad \text{and} \quad l_f(t_1) \le r_f(t_2).$$

In particular l_f is nondecreasing on $[\inf f, \infty[$. Moreover, if X is a reflexive Banach space then $r_f(t) \le l_f(t)$ for every $t \in]\inf f, \infty[$, and so r_f is nondecreasing on $]\inf f, \infty[$; hence $r_f = \overline{r}_f$ in this case.

Proof. Let first $\overline{x} \in \operatorname{dom} f$ and take $t_1, t_2 \in \mathbb{R}$ such that $t_1 < t_2 < f(\overline{x})$; then

$$\frac{f(\overline{x}) - t_1}{d(\overline{x}, [f \le t_1])} \le \frac{f(\overline{x}) - t_2}{d(\overline{x}, [f \le t_2])}. \tag{3.85}$$

It is obvious that Eq. (3.85) holds if $[f \le t_1] = \emptyset$. Assume that $[f \le t_1] \ne \emptyset$ and take $x \in [f \le t_1]$. Let $\overline{\lambda} := (f(\overline{x}) - t_2)/(f(\overline{x}) - t_1) \in]0,1[$. Then

$$f\left(\overline{\lambda}x + (1-\overline{\lambda})\overline{x}\right) \le \overline{\lambda}f(x) + (1-\overline{\lambda})f(\overline{x}) \le \overline{\lambda}t_1 + (1-\overline{\lambda})f(\overline{x}) = t_2.$$

Hence $\overline{\lambda}x + (1-\overline{\lambda})\overline{x} \in [f \le t_2]$, and so $d(\overline{x}, [f \le t_2]) \le \overline{\lambda}\,\|\overline{x} - x\|$. Taking the infimum for $x \in [f \le t_1]$ we get Eq. (3.85). The inequality Eq. (3.85) is obvious if $\overline{x} \notin \operatorname{dom} f$ and $[f \le t_1] \ne \emptyset$.

Using Eq. (3.85) we obtain that

$$\begin{aligned}\inf_{x\in\operatorname{dom} f\setminus[f\le t_1]} \frac{f(x)-t_1}{d(x,[f\le t_1])} &\le \inf_{x\in\operatorname{dom} f\setminus[f\le t_2]} \frac{f(x)-t_1}{d(x,[f\le t_1])} \\ &\le \inf_{x\in\operatorname{dom} f\setminus[f\le t_2]} \frac{f(x)-t_2}{d(x,[f\le t_2])},\end{aligned}$$

and

$$\begin{aligned}\inf_{x\in\operatorname{dom} f\setminus[f\le t_1]} \frac{f(x)-t_1}{d(x,[f\le t_1])} &\le \inf_{x\in[f=t_2]} \frac{f(x)-t_1}{d(x,[f\le t_1])} \\ &\le \inf_{x\in[f=t_2]} \sup_{t<t_2} \frac{f(x)-t}{d(x,[f\le t])}.\end{aligned}$$

The first inequality shows that $l_f(t_1) \le l_f(t_2)$, while the second one shows that $l_f(t_1) \le r_f(t_2)$.

Assume now that X is reflexive and take $t \in]\inf f, \infty[$. If there is no $x \in \operatorname{dom} f$ with $f(x) > t$ then $l_f(t) = \infty \ge r_f(t)$. Let $x \in \operatorname{dom} f$ with $f(x) > t$. Because X is reflexive and $[f \le t]$ is closed and convex, there exists $y \in P_{[f\le t]}(x)$. Using Lemma 3.10.4 we obtain that $f(y) = t$. Let $z \in [f < t]$. Taking $\mu := \|x-y\|/(\|x-y\| + \|y-z\|) \in]0,1[$, we have that $f((1-\mu)x + \mu z) \ge t$. In the contrary case $u := (1-\mu)x + \mu z \in [f < t]$. Then, using again Lemma 3.10.4, there exists $\lambda \in]0,1[$ with $f(\lambda u + (1-\lambda)x) = t$. It follows that

$$\|x-y\| \le \lambda\,\|u-x\| < \|u-x\| = \mu\,\|x-z\| = \frac{\|x-y\|\cdot\|x-z\|}{\|x-y\| + \|y-z\|},$$

whence the contradiction $\|x-y\| + \|y-z\| < \|x-z\|$. Hence $t \le f\big((1-\mu)x + \mu z\big) \le (1-\mu)f(x) + \mu f(z)$; it follows that $\|x-y\|\,(t - f(z)) \le \|y-z\|\,(f(x) - t)$. Therefore

$$\frac{t-f(z)}{\|y-z\|} \le \frac{f(x)-t}{\|x-y\|} = \frac{f(x)-t}{d(x,[f\le t])}.$$

Since $z \in [f < t]$ is arbitrary, we get $r_f(t) = r_f^1(t) \leq l_f(t)$. $\square$

Corollary 3.10.9 *Let X be a reflexive Banach space, $f \in \Gamma(X)$ and $t > \inf f$. If f is Gâteaux differentiable on $[f = t] \subset \operatorname{int}(\operatorname{dom} f)$ then*

$$l_f(t) = r_f(t) = \inf \{\|\nabla f(x)\| \mid x \in [f = t]\}. \tag{3.86}$$

Proof. If $[f = t] = \emptyset$ then, obviously, $r_f(t) = l_f(t) = \infty$, and so Eq. (3.86) holds. Let $[f = t]$ be nonempty and take $x_0 \in [f < t]$ and $x \in [f = t]$. Since $x \in \operatorname{int}(\operatorname{dom} f)$ we obtain that $]x_0, x] \subset \operatorname{int}(\operatorname{dom} f)$ by using Theorem 1.1.2(ii). As f is continuous on $\operatorname{int}(\operatorname{dom} f)$, it follows that $]x_0, x[\subset \operatorname{int}[f \leq t]$. Because $x \in [f = t]$, $x \notin \operatorname{int}[f \leq t]$; hence, by Theorem 1.1.3 applied for x and $[f \leq t]$, we get $u^* \in S_{X^*} \cap N([f \leq t], x)$. Let $u \in \Phi_X^{-1}(u^*)$. Taking into account relation (3.82) we get $\|\nabla f(x)\| \geq \langle u, \nabla f(x) \rangle = f'(x, u)$. It follows that

$$\begin{aligned} r_f(t) &= r_f^5(t) = \inf \{\|\nabla f(x)\| \mid x \in [f = t]\} \\ &\geq \inf \left\{ f'(x,u) \mid x \in [f = t],\ u \in S_X \cap \Phi_X^{-1} \left(N([f \leq t], x) \right) \right\} = l_f(t). \end{aligned}$$

Using Proposition 3.10.8 we obtain that Eq. (3.86) holds. $\square$

One says that $f \in \Gamma(X)$ is **well behaved (asymptotically)** or *has good asymptotic behavior* if $l_f(t) > 0$ for every $t > \inf f$.

The next result shows that the class $\mathcal{F}$ of functions defined in Eq. (3.73) coincides with the class of functions with good asymptotic behavior when the space is reflexive.

Theorem 3.10.10 *Let X be a reflexive Banach space and $f \in \Gamma(X)$. Then*

$$f \in \mathcal{F} \ \Leftrightarrow \ r_f(t) > 0 \ \forall t \in]\inf f, \infty[\ \Leftrightarrow \ l_f(t) > 0 \ \forall t \in]\inf f, \infty[.$$

Proof. The conclusion is an immediate consequence of Proposition 3.10.8 and relation (3.78). $\square$

As mentioned before, the set $\operatorname{argmin} f$ is a set of weak sharp minima of f (*i.e.* f is well-conditioned with linear rate) exactly when $\inf f$ is attained and $l_f(\inf f) > 0$. One can ask if there are other relations between well-conditioning and well-behaving of convex functions.

Proposition 3.10.11 *Let $f \in \Gamma(X)$. If f is well-conditioned then f has good asymptotic behavior. Moreover, $\lim_{t \downarrow \inf f} (t - \inf f)^{-1} l_f(t) = \infty$.*

Proof. Let $S := \operatorname{argmin} f = [f \leq \inf f] \neq \emptyset$. Of course, if $\operatorname{dom} f = S$ it is nothing to prove ($l_f(t) = \infty$ for all $t > \inf f$). We may suppose that $\inf f = 0$ and $\operatorname{dom} f \neq S$. Since f is well conditioned, by Proposition 3.4.1, the function $\psi := \psi_f \in \mathcal{A}_0 \cap \mathcal{N}_1$; moreover, because $\operatorname{dom} f \neq S$, ψ is finite at some $t_0 > 0$, and so $\psi \in \Omega_0$. Therefore, by Lemma 3.3.1, $\psi^e \in \Omega_0 \cap \mathcal{A}_0$. The choice of ψ shows that $f(x) \geq \psi(d(x,S))$ for every $x \in X$. Let $t > 0 = \inf f$ and $x \in \operatorname{dom} f \setminus [f \leq t]$. Taking $\alpha > 0$, we have either $d(x,S) \geq \alpha$, and so $\frac{f(x)}{d(x,S)} \geq \frac{\psi(d(x,S))}{d(x,S)} \geq \frac{\psi(\alpha)}{\alpha}$ (because $\psi \in \mathcal{N}_1$), or $d(x,S) \leq \alpha$, and so $\frac{f(x)}{d(x,S)} \geq \frac{t}{\alpha}$. Hence

$$\frac{f(x)}{d(x,S)} \geq \sup\left\{\min\left(\frac{t}{\alpha}, \frac{\psi(\alpha)}{\alpha}\right) \,\Big|\, \alpha > 0\right\} \geq \sup\left\{\frac{t}{\alpha} \,\Big|\, \alpha > 0,\ \psi(\alpha) \geq t\right\}$$
$$= \frac{t}{\inf\{\alpha > 0 \mid \psi(\alpha) \geq t\}} = \frac{t}{\psi^e(t)}.$$

Taking $s_1 = 0$ and $s_2 = t$ in Eq. (3.85), we have that

$$\frac{f(x) - t}{d(x,[f \leq t])} \geq \frac{f(x)}{d(x,S)} \geq \frac{t}{\psi^e(t)}.$$

It follows that $l_f(t) \geq t/\psi^e(t) > 0$ for every $t > 0$, and so f is well-behaved. Furthermore, $t^{-1}l_f(t) \geq 1/\psi^e(t)$, and so $\lim_{t\downarrow 0} t^{-1}l_f(t) = \infty$ because $\psi^e \in \Omega_0$. □

It is obvious that any function $f \in \Gamma(\mathbb{R})$ has good asymptotic behavior, but, for example, $f := \exp$ is not well conditioned because $\operatorname{argmin} f = \emptyset$. The next result furnishes a sufficient condition on the function l_f in order that f be well conditioned.

Theorem 3.10.12 *Let X be a Banach space and $f \in \Gamma(X)$ be such that $\inf f > -\infty$. Let l_f be defined by Eq. (3.79). If $l_f(t) > 0$ for every $t > \inf f$, i.e. f has good asymptotic behavior, and there exists $\alpha > 0$ such that $\theta(\alpha) < \infty$, where*

$$\theta(s) := \int_{\inf f}^{s+\inf f} \frac{dt}{l_f(t)} \quad \forall\, s > 0,$$

then f is well-conditioned and $\psi_f \geq \theta^e$.

Proof. We may assume that $\inf f = 0$. If f is constant on $\operatorname{dom} f$ there is nothing to prove. Because l_f is nondecreasing and positive on $]0,\infty[$ (eventually taking the value ∞) and $\theta(\alpha) < \infty$, we have that $\theta(s) > 0$

for every $s > 0$; we set $\theta(0) := 0$. It follows that θ is continuous and nondecreasing on $\mathbb{R}_+$; θ is even increasing on $f(\operatorname{dom} f)$. It follows that $\theta \in \mathcal{A}_0 \cap \mathcal{N}_0 \cap \Omega_0$, and so $\theta^e \in \mathcal{A}_0$ (by Lemma 3.3.1). Let $x_0 \in \operatorname{dom} f$ with $f(x_0) > 0$ and $p \in]1, \infty[$. Consider the sequence $(s_n) \subset \mathbb{P}$, $s_n := p^{-n} f(x_0)$ for every $n \in \mathbb{N} \cup \{0\}$. Then the series $\sum_{n\geq 1} \frac{s_{n-1}-s_n}{l_f(s_n)}$ is convergent. Indeed, l_f being nondecreasing (see Proposition 3.10.8), $l_f(s_n) \leq l_f(t) \leq l_f(s_{n-1})$, and so $1/l_f(s_{n-1}) \leq 1/l_f(t) \leq 1/l_f(s_n)$ for $s_n \leq t \leq s_{n-1}$ and $n \in \mathbb{N}$. Hence

$$\frac{s_{n-1}-s_n}{l_f(s_{n-1})} \leq \int_{s_n}^{s_{n-1}} \frac{dt}{l_f(t)} \leq \frac{s_{n-1}-s_n}{l_f(s_n)} = p\frac{s_n - s_{n+1}}{l_f(s_n)} \quad \forall n \in \mathbb{N}.$$

Therefore the series $\sum_{n\geq 1} \frac{s_{n-1}-s_n}{l_f(s_{n-1})}$ is convergent and its sum is smaller than $\int_0^{s_0} \frac{dt}{l_f(t)}$. The relations above show that

$$\begin{aligned}\sum\nolimits_{n\geq 1} \frac{s_{n-1}-s_n}{l_f(s_n)} &= p\left(\sum\nolimits_{n\geq 1} \frac{s_{n-1}-s_n}{l_f(s_{n-1})} - \frac{s_0-s_1}{l_f(s_0)}\right) \\ &\leq p\left(\int_0^{s_0} \frac{dt}{l_f(t)} - \frac{s_0-s_1}{l_f(s_0)}\right) < \infty. \qquad (3.87)\end{aligned}$$

Let $\varepsilon > 0$ be fixed. Since $s_0 = f(x_0) > s_1 > \inf f$, from the definition of l_f, there exists $x_1' \in [f \leq s_1]$ such that $\|x_0 - x_1'\| \leq (s_0 - s_1)/l_f(s_1) + 2^{-1}\varepsilon$. By Lemma 3.10.4 there exists $x_1 \in [x_0, x_1']$ with $f(x_1) = s_1$; of course $\|x_0 - x_1\| \leq (s_0 - s_1)/l_f(s_1) + 2^{-1}\varepsilon$. Continuing in this way we find a sequence $(x_n)_{n\geq 0} \subset \operatorname{dom} f$ such that

$$f(x_n) = s_n, \quad \|x_n - x_{n+1}\| \leq \frac{s_n - s_{n+1}}{l_f(s_{n+1})} + \frac{\varepsilon}{2^{n+1}} \quad \forall n \geq 0. \qquad (3.88)$$

From Eq. (3.87) we obtain that the series $\sum_{n\geq 0} \|x_n - x_{n+1}\|$ is convergent, and so (x_n) is a Cauchy sequence. As X is a Banach space, the sequence (x_n) converges to some $x \in X$. It follows that $0 = \inf f \leq f(x) \leq \liminf f(x_n) \leq \lim s_n = 0$. Hence $x \in S := \operatorname{argmin} f$, and so $S \neq \emptyset$. Using Eq. (3.88) we obtain that

$$\|x_0 - x_n\| \leq \sum\nolimits_{i=1}^{n} \|x_{i-1} - x_i\| \leq \sum\nolimits_{i=1}^{n} \left(\frac{s_{i-1}-s_i}{l_f(s_i)} + \frac{\varepsilon}{2^i}\right),$$

whence, by Eq. (3.87),

$$\|x_0 - x\| \leq p\left(\int_0^{s_0} \frac{dt}{l_f(t)} - \frac{s_0 - s_1}{l_f(s_0)}\right) + \varepsilon \leq p\int_0^{s_0} \frac{dt}{l_f(t)} + \varepsilon.$$

As $\varepsilon > 0$ and $p > 1$ are arbitrary, we obtain that $d(x_0, S) \leq \|x_0 - x\| \leq \theta(s_0) = \theta(f(x_0))$. It follows that $f(x_0) \geq \theta^e(d(x_0, S))$. As $x_0 \in \operatorname{dom} f \setminus S$ was arbitrary and $\theta^e \in \mathcal{A}_0$, f is well conditioned. □

Consider now the system of inequalities

$$g_i(x) \leq 0, \quad \text{for } i = 1, \ldots m, \tag{3.89}$$

where $g_1, \ldots, g_m \in \Lambda(X)$. We assume that the solution set

$$C := \{x \in X \mid g_i(x) \leq 0, \ \forall i \in \overline{1, m}\}$$

of the system (3.89) is nonempty. We are interested in the following *global error bound* for (the solutions set) C:

$$d_C(x) \leq \alpha \cdot \max_{1 \leq i \leq m} [g_i(x)]_+ \quad \forall x \in X, \tag{3.90}$$

for some $\alpha > 0$ (which, of course, depends on the functions g_i), where $\gamma_+ := \gamma \vee 0$ for $\gamma \in \mathbb{R}$.

Of course, $g := \max\{g_i \mid i \in \overline{1, m}\} \in \Lambda(X)$, the system (3.89) is equivalent to

$$g(x) \leq 0, \quad x \in X, \tag{3.91}$$

and $C = [g \leq 0]$; Eq. (3.90) becomes

$$d_C(x) \leq \alpha \cdot [g(x)]_+ \quad \forall x \in X. \tag{3.92}$$

Taking $h := g_+ := g \vee 0$, we have that $\inf h = 0$ (under our assumption that $C \neq \emptyset$) and $C = \operatorname{argmin} h$. Hence the global error bound for C holds if and only if C is a set of weak sharp minima for h. On the other hand Eq. (3.92) is equivalent to $l_g(0) \geq \alpha^{-1}$, and so the global error bound for C holds if and only if $[g \leq 0] \neq \emptyset$ and $l_g(0) > 0$.

In the next result we collect several sufficient conditions for the existence of $\alpha > 0$ verifying Eq. (3.92).

Theorem 3.10.13 *Let $g \in \Gamma(X)$ be such that $C := [g \leq 0] \neq \emptyset$. Consider the following conditions:*

(i) *there exists $u \in X$ such that $g_\infty(u) < 0$;*

(ii) *there exists $\eta > 0$ such that $\sup_{x \in [g \leq 0]} d(x, [g \leq -\eta]) < \infty$;*

(iii) *there exists $\eta > 0$ such that $\sup_{x \in [g = 0]} d(x, [g \leq -\eta]) < \infty$;*

(iv)

$$\inf_{\eta>0} \sup_{x\in[g=0]} \inf_{y\in[g\leq-\eta]} \frac{\|x-y\|}{-g(y)} = \inf_{\eta>0} \sup_{x\in[g=0]} \frac{d(x,[g\leq-\eta])}{\eta} < \infty; \tag{3.93}$$

(v) $0 > \inf g$ *and* $r_g(0) > 0$;

(vi) $l_g(0) > 0$.

Then (i) $\Rightarrow$ (ii) $\Rightarrow$ (iii) $\Leftrightarrow$ (iv) $\Rightarrow$ (v) *and* (ii) $\Rightarrow$ (vi). *If* g *is not constant on* $\operatorname{dom} g$ *and* $[g \leq 0] \subset \operatorname{int}(\operatorname{dom} g)$ *then* (iii) $\Rightarrow$ (ii). *If* X *is a reflexive Banach space, then* (v) $\Rightarrow$ (vi)*; moreover, if* g *is not constant,* $[g = 0] \subset \operatorname{int}(\operatorname{dom} g)$ *and* g *is Gâteaux differentiable on* $[g = 0]$ *then* (vi) $\Rightarrow$ (v) *and* $l_g(0) = r_g(0)$.

Proof. First note that the equality in Eq. (3.93) follows from Eq. (3.75) when $[g = 0] \neq \emptyset$. When $[g = 0] = \emptyset$, both sides of the equality are $-\infty$.

The implications (ii) $\Rightarrow$ (iii) $\Leftrightarrow$ (iv) are obvious.

(i) $\Rightarrow$ (ii) Let us take $u \in S_X$ such that $g_\infty(u) < 0$ and $\eta > 0$. Consider $x \in [g \leq 0]$. Taking $\gamma := -\eta/g_\infty(u) > 0$, we have that

$$g(x + \gamma u) \leq g(x) + \gamma g_\infty(u) \leq \gamma g_\infty(u) = -\eta,$$

and so $d(x, [g \leq -\eta]) \leq \|x + \gamma u - x\| = \gamma < \infty$.

(iii) $\Rightarrow$ (ii) Assume that g is not constant on $\operatorname{dom} g$ and $[g \leq 0] \subset \operatorname{int}(\operatorname{dom} g)$. We shall show that

$$\sup\{d(x,[g\leq -\eta]) \mid x \in [g=0]\} = \sup\{d(x,[g\leq -\eta]) \mid x\in[g\leq 0]\} > 0 \tag{3.94}$$

for any $\eta > 0$.

If $[g < 0] = \emptyset$ then $[g \leq 0] = [g = 0]$ and Eq. (3.94) is obvious. Assume that $g(x_0) < 0$ and take $x \in \operatorname{dom} g$ with $g(x) \neq g(x_0)$; we may assume that $g(x_0) < g(x)$. If $g(x) \geq 0$, by Lemma 3.10.4, there exists $y \in]x_0, x]$ such that $g(y) = 0$. Assume that $g(x) < 0$. Since $0 < g(x) - g(x_0) \leq t^{-1}\big(g(x_0 + t(x-x_0)) - g(x_0)\big)$ for $t \geq 1$, there exists $t > 1$ such that $g(x_0 + t(x - x_0)) > 0$. The set $\Lambda := \{t \geq 1 \mid g(x_0 + t(x - x_0)) \leq 0\}$ is a closed bounded interval; take $\bar{t} := \max \Lambda$. Since $x_0 + \bar{t}(x - x_0) \in [g \leq 0] \subset \operatorname{int}(\operatorname{dom} g)$, there exists $t > \bar{t}$ such that $x_0 + t(x - x_0) \in \operatorname{dom} g$. It follows that $g(x_0 + t(x - x_0)) > 0$, and so necessarily $x_0 + \bar{t}(x - x_0) \in [g = 0]$. Hence there exists $y \in [g = 0]$ such that $x \in]x_0, y[$.

Let us prove Eq. (3.94) for $\eta > 0$. This relation is obvious if $[g \leq -\eta] = \emptyset$. Assume that $[g \leq -\eta] \neq \emptyset$ and take $x \in [g < 0] \setminus [g \leq -\eta]$. Consider

$x_0 \in [g \leq -\eta]$. As above, there exist $y \in [g = 0]$ and $\lambda \in]0,1[$ such that $x = \lambda y + (1-\lambda)x_0$. Take now $z \in [g \leq -\eta]$; then $\lambda z + (1-\lambda)x_0 \in [g \leq -\eta]$, and so

$$d(x, [g \leq -\eta]) \leq \|x - \lambda z - (1-\lambda)x_0\| = \lambda \|y - z\| \leq \|y - z\|,$$

whence $d(x, [g \leq -\eta]) \leq d(y, [g \leq -\eta])$. It follows that $\sup_{x\in[g=0]} d(x, [g \leq -\eta]) \geq \sup_{x\in[g\leq 0]} d(x, [g \leq -\eta])$. As the converse inequality is obvious, Eq. (3.94) holds. Hence (iii) $\Rightarrow$ (ii) in our case.

(iv) $\Rightarrow$ (v) If $0 = \inf g$ then $C = [g = 0] \neq \emptyset$ and $[g \leq -\eta] = \emptyset$ for any $\eta > 0$. Hence $d(x, [g \leq -\eta]) = \infty$ for $x \in [g = 0]$ and $\eta > 0$, whence the contradiction $\inf_{\eta>0} \sup_{x\in[g=0]} \eta^{-1} d(x, [g \leq -\eta]) = \infty$. Hence $0 > \inf g$. The conclusion then follows from the relation $r_g(0) = r_g^2(0)$ in Corollary 3.10.6 and the obvious inequality

$$\inf_{\eta>0} \sup_{x\in[g=0]} \frac{d(x, [g \leq -\eta])}{\eta} \geq \sup_{x\in[g=0]} \inf_{\eta>0} \frac{d(x, [g \leq -\eta])}{\eta} = \frac{1}{r_g^2(0)}.$$

(ii) $\Rightarrow$ (vi) Because $\sup_{x\in[g\leq 0]} d(x, [g \leq -\eta]) < \infty$, we have that $[g \leq -\eta] \neq \emptyset$. Consider $0 < \mu < \eta$. Since $[g \leq -\mu] \subset [g \leq -\eta]$, we obtain that $\sup_{x\in[g\leq -\mu]} d(x, [g \leq -\eta]) < \infty$. From the implication (ii) $\Rightarrow$ (v) with 0 replaced by $-\mu$ we obtain that $r_g(-\mu) > 0$. Using Proposition 3.10.8 we obtain that $l_g(0) \geq r_g(-\mu) > 0$.

Assume now that X is a reflexive Banach space. Then the implication (v) $\Rightarrow$ (vi) is true because, by Proposition 3.10.8, $l_g(0) \geq r_g(0)$.

(vi) $\Rightarrow$ (v) Assume that g is not constant, $l_g(0) > 0$ and g is Gâteaux differentiable on $[g = 0] \subset \operatorname{int}(\operatorname{dom} g)$. It is sufficient to show that $0 > \inf g$; then the conclusion follows from Corollary 3.10.9. If $[g = 0] = \emptyset$ then $0 > \inf g$ (because $C \neq \emptyset$). Assume that $[g = 0] \neq \emptyset$; if $[g < 0] = \emptyset$ then every $\overline{x} \in C$ is a minimum point of g, and so $0 \in \partial g(\overline{x}) = \{\nabla g(\overline{x})\}$. Assuming that g is 0 on $\operatorname{dom} g$ then $[g = 0] = \operatorname{dom} g = X$ ($\operatorname{dom} g$ being closed and open), a contradiction. It follows that $\operatorname{dom} g \setminus [g \leq 0] \neq \emptyset$, and so, by relation (3.81) in Proposition 3.10.7, $l_g(0) = 0$. Hence $[g < 0] \neq \emptyset$.□

Coming back to the system (3.89), several conditions in the preceding theorem can be formulated in terms of the functions g_i, $i \in \overline{1,m}$, when $g := \max_{i\in\overline{1,m}} g_i$. Let us denote the set $\{j \in \overline{1,m} \mid g_j(x) = g(x)\}$ by $I(x)$ for $x \in \operatorname{dom} g$. Assume that $g_i \in \Lambda(X)$ is continuous on $\operatorname{int}(\operatorname{dom} g_i)$ and $\emptyset \neq [g \leq 0] \subset \operatorname{int}(\operatorname{dom} g_i)$ for every $i \in \overline{1,m}$. Then, by Corollary 2.8.13, we have that

$$\partial g(x) = \operatorname{co}\left(\bigcup\nolimits_{i \in I(x)} \partial g_i(x)\right) \quad \forall x \in [g \le 0],$$

and so, by Theorem 2.4.9,

$$g'(x,u) = \max_{i \in I(x)} \max\{\langle u, x^* \rangle \mid x^* \in \partial g_i(x)\} \quad \forall x \in [g \le 0],\ \forall u \in X.$$

Moreover, since $\operatorname{epi} g = \bigcap_{i \in \overline{1,m}} \operatorname{epi} g_i \ne \emptyset$, the relation $g_\infty = \max_{i \in \overline{1,m}} (g_i)_\infty$ holds. For example, condition (i) of the preceding theorem reduces to the existence of $u \in X$ such that $(g_i)_\infty(u) < 0$ for every $i \in \overline{1,m}$. The condition $[g < 0] \ne \emptyset$ in (iv) says that the Slater condition holds for the system of inequalities (3.74), while the condition $r_g(0) > 0$ may be described in several ways, taking into account that $r_g = r_g^j$ for $1 \le j \le 5$. One of them (corresponding to r_g^5) is

$$0 \notin \operatorname{cl}\left(\bigcup\nolimits_{x \in [g=0]} \operatorname{co}\left(\bigcup\nolimits_{i \in I(x)} \partial g_i(x)\right)\right),$$

another one (corresponding to r_g^4) being

$$\inf_{x \in [g=0]} \sup_{u \in S_X} \min_{i \in I(x)} \min_{x^* \in \partial g_i(x)} \langle u, x^* \rangle > 0.$$

We can consider (3.91) as being an unconstrained inequality, in contrast with

$$g(x) \le 0, \quad x \in A, \tag{3.95}$$

where $g \in \Gamma(X)$ and $A \subset X$ is a nonempty closed convex set. This inequality is equivalent with the system

$$g(x) \le 0, \quad d_A(x) \le 0, \quad x \in X. \tag{3.96}$$

Of course, the Slater condition does not hold for the system (3.96). We assume that the solution set $C := \{x \in A \mid g(x) \le 0\}$ of the system (3.95) is nonempty. We say that the *global error bound holds for* (3.95) if there exists $\alpha > 0$ such that

$$d_C(x) \le \alpha \cdot \max\{g(x), d_A(x)\} \quad \forall x \in X. \tag{3.97}$$

Theorem 3.10.14 *Let X be a reflexive Banach space, $g \in \Gamma(X)$ and $A \subset X$ be a closed convex set such that $C := \{x \in A \mid g(x) \le 0\}$ is* *nonempty. Assume that one of the conditions below holds:*

$$P_A(x) \cap \operatorname{dom} g \neq \emptyset \quad \forall x \in [g<0], \tag{3.98}$$

$$[g<0] \cap A \subset \operatorname{int}(\operatorname{dom} g). \tag{3.99}$$

Then the following statements are equivalent:

(i) *there exists $\alpha > 0$ such that Eq. (3.97) holds,*

(ii) $\exists \gamma > 0,\ \forall x \in A \cap [g=0],\ \forall u \in \Phi_X^{-1}(N(C,x))$:

$$\max\{g'(x,u), d(u, C(A,x))\} \geq \gamma \|u\|.$$

Proof. Consider $h := \max\{g, d_A\}$. By Proposition 3.10.7 we have that

$$l_h(0) = \inf\{h'(x,u) \mid x \in C,\ u \in S_X \cap \Phi_X^{-1}(N(C,x))\}.$$

Let $x \in C$; we wish to evaluate $h'(x,u)$ for $u \in X$. Because

$$\operatorname{dom} h'(x,\cdot) = \operatorname{dom} g'(x,\cdot) = \operatorname{cone}(\operatorname{dom} g - x),$$

we take $u \in \operatorname{cone}(\operatorname{dom} g - x)$. Let first $x \in [g=0] \cap A$. Taking into account Proposition 3.8.3(iii), we have that

$$\begin{aligned} h'(x,u) &= \max\left\{\lim_{t\downarrow 0} \frac{g(x+tu)-g(x)}{t}, \lim_{t\downarrow 0} \frac{d_A(x+tu)-d_A(x)}{t}\right\} \\ &= \max\{g'(x,u), d(u, C(A,x))\}. \end{aligned}$$

Let now $x \in [g<0] \cap A$. Since $u \in \operatorname{cone}(\operatorname{dom} g - x)$, there exists $t' > 0$ such that $x' := x + t'u \in \operatorname{dom} g$. As $g|_{[x,x']}$ is continuous, there exists $\bar{t} > 0$ such that $x + tu \in [g<0]$ for $t \in [0, \bar{t}]$. It follows that

$$\begin{aligned} h'(x,u) &= \lim_{t\downarrow 0} \frac{\max\{g(x+tu), d_A(x+tu)\}}{t} = \lim_{t\downarrow 0} \frac{d_A(x+tu)}{t} \\ &= d(u, C(A,x)). \end{aligned} \tag{3.100}$$

Let $x \in [g<0] \cap A$ and $u \in S_X \cap \Phi_X^{-1}(N(C,x)) \cap \operatorname{cone}(\operatorname{dom} g - x)$. Let us show in our conditions that $h'(x,u) \geq 1$. As seen above, there exists $\bar{t} > 0$ such that $x + tu \in [g<0]$ for $t \in [0, \bar{t}]$. By Corollary 3.8.5 we have that $x + tu \notin C$ and $x \in P_C(x+tu)$ for $t > 0$. It follows that $x + \bar{t}u \notin A$. Assume that $x \notin P_A(x + \bar{t}u)$. Suppose first that Eq. (3.98) holds. Then there exists $\overline{x} \in P_A(x+\bar{t}u) \cap \operatorname{dom} g$. Of course, $\|x + \bar{t}u - \overline{x}\| < \|x + \bar{t}u - x\| = \bar{t}$. Since $g|_{[\overline{x},x]}$ is continuous and $g(x) < 0$, there exists $\lambda \in]0,1[$ such that

$g(\lambda\overline{x}+(1-\lambda)x)<0$. Of course, $x':=\lambda\overline{x}+(1-\lambda)x\in A$, and so $x'\in C$. The contradiction follows:

$$\overline{t}=d(x+\overline{t}u,C)\leq\|x+\overline{t}u-x'\|\leq\lambda\|x+\overline{t}u-\overline{x}\|+(1-\lambda)\|x+\overline{t}u-x\|<\overline{t}.$$

Suppose now that Eq. (3.98) holds. Because $x\notin P_A(x+\overline{t}u)$, there exists $\overline{x}\in A$ such that $\left\|x+\overline{t}u-\overline{x}\right\|<\left\|x+\overline{t}u-x\right\|=\overline{t}$. From Eq. (3.98) we have that $x\in\operatorname{int}(\operatorname{dom}g)$, and so there exists $\lambda\in]0,1[$ such that $\lambda\overline{x}+(1-\lambda)x\in\operatorname{dom}g$. Decreasing λ if necessary, we may assume that $g(\lambda\overline{x}+(1-\lambda)x)<0$. Proceeding as above we get the same contradiction. Hence $x\in P_A(x+\overline{t}u)$; using Eqs. (3.100) and (3.72) we obtain that $h'_+(x,u)\geq 1$. Concluding the above discussion we obtain that $\mu\geq l_h(0)\geq\min\{1,\mu\}$, where

$$\mu:=\inf\left\{\max\left(g'(x,u),d\left(u,\mathcal{C}(a,x)\right)\right)\mid x\in[g=0]\cap A,\right.$$
$$\left.u\in S_X\cap\Phi_X^{-1}\left(N(C,x)\right)\right\}.$$

The conclusion follows. □

In the next result we present two sufficient conditions in order that the global error bound holds for the constrained system (3.95).

Proposition 3.10.15 *Let X be a reflexive Banach space, $g\in\Gamma(X)$ and A be a closed convex subset of X such that $C:=[g\leq 0]\cap A\neq\emptyset$. Assume that either $C\subset\operatorname{int}(\operatorname{dom}g)$, or (3.98) and $A\cap[g=0]\subset\operatorname{int}(\operatorname{dom}g)$ holds. Consider the following statements:*

(i) *there exists $\alpha>0$ such that Eq. (3.97) holds,*

(ii) $\exists\,\kappa>0,\ \forall\,x\in A\cap[g=0],\ \forall\,x^*\in N(A,x),\ \forall\,y^*\in\partial g(x)$:

$$\|x^*\|+1\leq\kappa\,\|x^*+y^*\|\,. \tag{3.101}$$

(iii) *the set $\partial g\,(A\cap[g=0])$ is bounded and*

$$0\notin\operatorname{cl}\left(\bigcup\nolimits_{x\in A\cap[g=0]}\left(\partial g(x)+N(A,x)\right)\right).$$

Then (iii) $\Rightarrow$ (ii) $\Rightarrow$ (i).

Proof. It is obvious that Eq. (3.98) or Eq. (3.99) holds, and so conditions (i) and (ii) of the preceding theorem are equivalent. It follows that everyone of the statements (i), (ii) and (iii) above is true if $A\cap[g=0]=\emptyset$, and so the conclusion holds in this case.

Assume that $A\cap[g=0]\neq\emptyset$; in our conditions $A\cap[g=0]\subset\operatorname{int}(\operatorname{dom}g)$.

(iii) ⇒ (ii) By hypothesis, there exist $\gamma, \eta > 0$ such that $\|x^* + y^*\| \geq \gamma$ and $\|y^*\| \leq \eta$ for all $x \in A \cap [g = 0]$, $x^* \in N(A, x)$ and $y^* \in \partial g(x)$. Take $\kappa = (1 + \eta + \gamma)/\gamma$. Consider x, x^* and y^* as before. If $\|x^*\| \leq \kappa\gamma - 1$ then, obviously, $\|x^*\| + 1 \leq \kappa \|x^* + y^*\|$. Assume that $\|x^*\| > \kappa\gamma - 1$. Then $\kappa \|x^* + y^*\| \geq \kappa (\|x^*\| - \eta) \geq \|x^*\| + 1 + (\kappa - 1)(\kappa\gamma - 1) - \kappa\eta - 1 = \|x^*\| + 1$. Therefore Eq. (3.101) holds.

(ii) ⇒ (i) As observed at the beginning of the proof, it is sufficient to show that condition (ii) of the preceding theorem holds. First of all note that in our conditions there exists $x_0 \in A \cap [g < 0]$; otherwise, taking $\overline{x} \in C = A \cap [g = 0]$, $\overline{x}$ is a minimum point of $g + \iota_A$; since $\overline{x} \in \operatorname{int}(\operatorname{dom} g)$, we have that g is continuous at $\overline{x} \in \operatorname{dom} \iota_A$, and so $0 \in \partial(g + \iota_A)(\overline{x}) = \partial g(\overline{x}) + N(A, \overline{x})$ which contradicts Eq. (3.101). Taking $\overline{x} \in A \cap [g = 0]$, we have that $\overline{x} \in \operatorname{int}(\operatorname{dom} g)$, and so $]x_0, \overline{x}[\subset A \cap [g < 0] \cap \operatorname{int}(\operatorname{dom} g)$. It follows that g is continuous at any point of $]x_0, \overline{x}[$, whence $A \cap \operatorname{int}[g \leq 0] \neq \emptyset$. Using Corollaries 2.8.4(i) and 2.9.5 we obtain that

$$N(C, x) = N(A, x) + N([g \leq 0], x) = N(A, x) + \mathbb{R}_+ \partial g(x) \qquad (3.102)$$

for every $x \in A \cap [g = 0]$.

Let $\overline{x} \in A \cap [g = 0]$ and $u \in S_X \cap \Phi_X^{-1}(N(C, \overline{x}))$. As observed in the proof of the preceding theorem, $h'(\overline{x}, u) = \max\{g'(\overline{x}, u), d(u, \mathcal{C}(A, \overline{x}))\}$, where $h := \max\{g, d_A\}$. Since g and d_A are continuous at $\overline{x}$, h is so. It follows that $h'(\overline{x}, u) = \max\{\langle u, u^* \rangle \mid u^* \in \partial h(\overline{x})\}$. From Corollary 2.8.13 and Proposition 3.8.3(ii) we have that

$$\partial h(\overline{x}) = \operatorname{co}(\partial g(\overline{x}) \cup \partial_A(\overline{x})) = \operatorname{co}(\partial g(\overline{x}) \cup (U_{X^*} \cap N(A, \overline{x}))).$$

Take $u^* \in \Phi_X(u) \cap N(C, \overline{x})$. From Eq. (3.102), there exist $x^* \in N(A, \overline{x})$, $y^* \in \partial g(\overline{x})$ and $\lambda \geq 0$ such that $u^* = x^* + \lambda y^*$. Since $u^* \neq 0$, $\mu := \|x^*\| + \lambda > 0$. It follows that $\mu^{-1} u^* \in \partial h(\overline{x})$. On the other hand, by Eq. (3.101) we have that $1 = \|u^*\| = \|x^* + \lambda y^*\| \geq \kappa^{-1} \mu$. Hence

$$\max\{g'(\overline{x}, u), d(u, \mathcal{C}(A, \overline{x}))\} = h'(\overline{x}, u) \geq \langle u, \mu^{-1} u^* \rangle = \mu^{-1} \geq \kappa^{-1}.$$

The proof is complete. □

During the proof of the preceding result we obtained formula (3.102) under the hypothesis that $A \cap \operatorname{int}[g \leq 0] \neq \emptyset$ and g is continuous at $x \in A \cap [g = 0]$. Another sufficient condition for it, which is a kind of local error bound for $A \cap [g \leq 0]$, can be given.

Proposition 3.10.16 *Let $g \in \Lambda(X)$ and $C := \{x \in A \mid g(x) \leq 0\}$. Assume that the system (3.95) is* metrically regular at $x \in A \cap [g = 0]$, *i.e. there exist $\delta, \gamma > 0$ such that*

$$\gamma d_C(y) \leq \max\{g(y), d_A(y)\} \quad \forall y \in B(x, \delta). \tag{3.103}$$

If $x \in (\operatorname{dom} g)^i$, then Eq. (3.102) holds.

Proof. The inclusions $\supset$ are easy (and true always).

Conversely, let $x^* \in N(C, x) \cap S_{X^*}$. By Proposition 3.8.3(ii) we have that $x^* \in \partial d_C(x)$, and so $\gamma x^* \in \partial(h + \iota_{B(x,\delta)})(x) = \partial h(x)$, where $h = \max\{g, d_A\}$ (see Remark 2.4.1). By Corollary 2.8.11,

$$\partial h(x) = \bigcup\nolimits_{\lambda \in [0,1]} \left((1 - \lambda)\partial d_A(x) + \lambda \partial g(x)\right);$$

the equality $\partial(\lambda g)(x) = \lambda \partial g(x)$ holds also for $\lambda = 0$ because $x \in (\operatorname{dom} g)^i$. Using again Proposition 3.8.3(ii) we obtain that $x^* \in N(A, x) + \mathbb{R}_+ \partial g(x)$. Hence (3.102) holds. □

A particular case of Proposition 3.10.16 is when $g = d_B$ with $B \subset X$ a nonempty closed convex set. Then Eq. (3.103) implies that

$$N(A \cap B, x) = N(A, x) + N(B, x).$$

(Compare with Corollary 2.8.4 for $X = Y$ and $A = \mathrm{Id}_X$.)

3.11 Monotone Multifunctions

Consider the subset

$$\varsigma_E := \left\{\mu : E \to [0, \infty[\;\middle|\; \operatorname{supp} \mu \text{ is finite}, \sum\nolimits_{x \in E} \mu(x) = 1\right\}$$

of $\mathbb{R}^E$, where E is a nonempty set and $\operatorname{supp} \mu := \{x \in E \mid \mu(x) \neq 0\}$; if $A \subset E$ is a nonempty set, then $\varsigma_A := \{\mu \in \varsigma_E \mid \operatorname{supp} \mu \subset A\}$. It is obvious that ς_A is a convex subset of $\mathbb{R}^E$; if $\emptyset \neq A \subset B \subset E$ then $\varsigma_A \subset \varsigma_B$. If $u \in E$ is fixed, we denote by δ_u the element $\mu \in \varsigma_E$ defined by $\mu(x) = 1$ for $x = u$ and $\mu(x) = 0$ for $x \neq u$.

In the sequel we take $E = X \times X^*$. Consider also the functions

$$p : \varsigma_{X\times X^*} \to X, \quad p(\mu) := \sum\nolimits_{(x,x^*)\in X\times X^*} \mu(x,x^*)\cdot x,$$
$$q : \varsigma_{X\times X^*} \to X^*, \quad q(\mu) := \sum\nolimits_{(x,x^*)\in X\times X^*} \mu(x,x^*)\cdot x^*,$$
$$r : \varsigma_{X\times X^*} \to \mathbb{R}, \quad r(\mu) := \sum\nolimits_{(x,x^*)\in X\times X^*} \mu(x,x^*)\cdot \langle x,x^*\rangle .$$

Note that p, q and r are affine functions, *i.e.* $p(\lambda\mu_1 + (1-\lambda)\mu_2) = \lambda p(\mu_1) + (1-\lambda)p(\mu_2)$ for every $\mu_1, \mu_2 \in \varsigma_{X\times X^*}$, $\lambda \in [0,1]$, and similarly for q, r.

The following characterization of monotone subsets of $X \times X^*$ holds.

Lemma 3.11.1 *Let $M \subset X \times X^*$ be a nonempty set; M is monotone if and only if*

$$\forall \mu \in \varsigma_M \ : \ r(\mu) \geq \langle p(\mu), q(\mu)\rangle .$$

Proof. Suppose that M is monotone and let $\mu \in \varsigma_M$; then $\operatorname{supp}\mu = \{(x_1,x_1^*),\ldots,(x_n,x_n^*)\}$ for some $n \in \mathbb{N}$. Take $\alpha_i := \mu(x_i,x_i^*)$ for $1 \leq i \leq n$. Then:

$$\begin{aligned}
r(\mu) - \langle p(\mu), q(\mu)\rangle &= \sum\nolimits_{i=1}^n \alpha_i \langle x_i, x_i^*\rangle - \left\langle \sum\nolimits_{i=1}^n \alpha_i x_i, \sum\nolimits_{j=1}^n \alpha_j x_j^*\right\rangle \\
&= \sum\nolimits_{i=1}^n \sum\nolimits_{j=1}^n \alpha_i\alpha_j \langle x_i, x_i^*\rangle - \sum\nolimits_{i=1}^n \sum\nolimits_{j=1}^n \alpha_i\alpha_j \langle x_i, x_j^*\rangle \\
&= \sum\nolimits_{i=1}^n \sum\nolimits_{j=1}^n \alpha_i\alpha_j \langle x_i, x_i^* - x_j^*\rangle \\
&= \sum\nolimits_{i=1}^n \sum\nolimits_{j=1}^n \alpha_i\alpha_j \langle x_j, x_j^* - x_i^*\rangle \\
&= \frac{1}{2}\sum\nolimits_{i=1}^n \sum\nolimits_{j=1}^n \alpha_i\alpha_j \langle x_i - x_j, x_i^* - x_j^*\rangle \geq 0.
\end{aligned}$$

Conversely, taking $(x_1,x_1^*), (x_2,x_2^*) \in M$ and $\mu = \frac{1}{2}\delta_{(x_1,x_1^*)} + \frac{1}{2}\delta_{(x_2,x_2^*)} \in \varsigma_M$, we have that

$$\begin{aligned}
r(\mu) - \langle p(\mu), q(\mu)\rangle &= \tfrac{1}{2}\langle x_1, x_1^*\rangle + \tfrac{1}{2}\langle x_2, x_2^*\rangle - \langle \tfrac{1}{2}x_1 + \tfrac{1}{2}x_2, \tfrac{1}{2}x_1^* + \tfrac{1}{2}x_2^*\rangle \\
&= \tfrac{1}{4}\langle x_1 - x_2, x_1^* - x_2^*\rangle \geq 0,
\end{aligned}$$

and so M is monotone. □

To $M \subset X \times X^*$ we associate the function

$$\chi_M : X \to \overline{\mathbb{R}}, \qquad \chi_M(x) := \sup_{\mu\in\varsigma_M} \frac{\langle x, q(\mu)\rangle - r(\mu)}{1 + \|p(\mu)\|}.$$

It is obvious that χ_M is a lsc convex function as supremum of a family of continuous affine functions. In the sequel $\operatorname{dom} M$ denotes $\operatorname{Pr}_X(M)$.

Lemma 3.11.2 *Let $\emptyset \neq M \subset X \times X^*$ be a monotone set and $\widetilde{M} = M - (w, w^*)$, where $(w, w^*) \in X \times X^*$ is fixed. Then:*

(i) $\operatorname{dom} M \subset \operatorname{co}(\operatorname{dom} M) \subset \operatorname{dom} \chi_M$.

(ii) $\operatorname{dom} \widetilde{M} = \operatorname{dom} M - w$ *and* $\operatorname{dom} \chi_{\widetilde{M}} = \operatorname{dom} \chi_M - w$.

(iii) *Let $X_0 \subset X$ be a closed linear subspace such that* $\operatorname{dom} M \subset X_0$ *and* $M + \{0\} \times X_0^\perp = M$. *Consider* $M_0 := \{(x, x^*|_{X_0}) \mid (x, x^*) \in M\} \subset X_0 \times X_0^*$. *Then*

$$\chi_{M_0} = \chi_M|_{X_0}, \quad \operatorname{dom} \chi_{M_0} = \operatorname{dom} \chi_M \subset X_0$$

and

$$M \text{ is maximal monotone} \Leftrightarrow M_0 \text{ is maximal monotone (in } X_0 \times X_0^*).$$

(iv) *Suppose that M is maximal monotone and* $\operatorname{dom} M \subset x_0 + X_0$, *where $X_0 \subset X$ is a closed linear subspace and $x_0 \in X$. Then $M + \{0\} \times X_0^\perp = M$ and* $\operatorname{dom} \chi_M \subset \overline{\operatorname{aff}(\operatorname{dom} M)}$.

Proof. (i) Let $(x, x^*) \in M$ and $\mu \in \varsigma_M$ be fixed. For $(y, y^*) \in M$, from the monotonicity of M, we have that

$$\langle y - x, y^* - x^* \rangle = \langle y, y^* \rangle - \langle y, x^* \rangle - \langle x, y^* \rangle + \langle x, x^* \rangle \geq 0.$$

Multiplying with $\mu(y, y^*)$ for $(y, y^*) \in \operatorname{supp} \mu$, then adding the corresponding relations, we obtain that

$$r(\mu) - \langle p(\mu), x^* \rangle - \langle x, q(\mu) \rangle + \langle x, x^* \rangle \geq 0,$$

and so

$$|\langle x, x^* \rangle| + \|p(\mu)\| \cdot \|x^*\| \geq \langle x, x^* \rangle - \langle p(\mu), x^* \rangle \geq \langle x, q(\mu) \rangle - r(\mu).$$

It follows that

$$\chi_M(x) \leq \max\{|\langle x, x^* \rangle|, \|x^*\|\} < \infty.$$

Therefore $\operatorname{dom} M \subset \operatorname{dom} \chi_M$; since $\operatorname{dom} \chi_M$ is a convex set, we have $\operatorname{co}(\operatorname{dom} M) \subset \operatorname{dom} \chi_M$, too.

(ii) It is obvious that $\operatorname{dom}\widetilde{M} = \operatorname{dom} M - w$. We note that if $\widetilde{\mu} \in \varsigma_{\widetilde{M}}$, taking $\mu(x,x^*) := \widetilde{\mu}((x,x^*) - (w,w^*))$ $[\Leftrightarrow \widetilde{\mu}(x,x^*) = \mu((x,x^*) + (w,w^*))]$, then $\mu \in \varsigma_M$. So, for $\widetilde{\mu} \in \varsigma_{\widetilde{M}}$ and the corresponding $\mu \in \varsigma_M$, we have that

$$\begin{aligned} p(\widetilde{\mu}) &= \sum\nolimits_{(x,x^*)\in X\times X^*} \widetilde{\mu}(x,x^*)\cdot x = \sum\nolimits_{(x,x^*)\in X\times X^*} \mu(x,x^*)\cdot(x-w) \\ &= p(\mu) - w, \end{aligned}$$

and similarly

$$q(\widetilde{\mu}) = q(\mu) - w^*, \quad r(\widetilde{\mu}) = r(\mu) - \langle w, q(\mu)\rangle - \langle p(\mu), w^*\rangle + \langle w, w^*\rangle.$$

Let $x \in X$ and $\widetilde{\mu} \in \varsigma_{\widetilde{M}}$. From the above relations we have that

$$\begin{aligned} \langle x - w, q(\widetilde{\mu})\rangle - r(\widetilde{\mu}) &= \langle x - w, q(\mu) - w^*\rangle - r(\mu) + \langle w, q(\mu)\rangle \\ &\qquad + \langle p(\mu), w^*\rangle - \langle w, w^*\rangle \\ &= \langle x, q(\mu)\rangle - r(\mu) - \langle x, w^*\rangle + \langle p(\mu), w^*\rangle \\ &\le \chi_M(x)\cdot(1 + \|p(\mu)\|) + \beta(1 + \|p(\mu)\|), \end{aligned}$$

where $\beta := \max\{|\langle x, w^*\rangle|, \|w^*\|\}$, and so

$$\frac{\langle x - w, q(\widetilde{\mu})\rangle - r(\widetilde{\mu})}{1 + \|p(\widetilde{\mu})\|} \le (\chi_M(x) + \beta)\cdot\frac{1 + \|p(\mu)\|}{1 + \|p(\widetilde{\mu})\|} \le (1 + \|w\|)\cdot(\chi_M(x) + \beta).$$

Therefore $\operatorname{dom}\chi_M - w \subset \operatorname{dom}\chi_{\widetilde{M}}$. Since $M = \widetilde{M} + (w, w^*)$, it follows also the converse inclusion.

(iii) It is obvious that M_0 is a monotone subset of $X_0 \times X_0^*$. (It is well-known that X_0^* is identified with $\{x^*|_{X_0} \mid x^* \in X^*\}$ endowed with the norm $\|u^*\| := \inf\{\|x^*\| \mid x^*|_{X_0} = u^*\}$.)

Let $x \in X \setminus X_0$ and fix $(x_0, x_0^*) \in M$. Since $x - x_0 \notin X_0$, there exists $u^* \in X_0^\perp$ such that $\langle x - x_0, u^*\rangle > 0$. Therefore, taking $\mu = \delta_{(x_0, x_0^* + tu^*)}$,

$$\chi_M(x) \ge \frac{\langle x, x_0^* + tu^*\rangle - \langle x_0, x_0^* + tu^*\rangle}{1 + \|x_0\|} = \frac{\langle x - x_0, x_0^*\rangle + t\langle x - x_0, u^*\rangle}{1 + \|x_0\|}$$

for every $t > 0$, and so $\chi_M(x) = \infty$. It follows that $\operatorname{dom}\chi_M \subset X_0$.

Let $\mu \in \varsigma_M$; consider

$$\mu_0 : X_0 \times X_0^* \to \mathbb{R}, \qquad \mu_0(x, u^*) := \sum\nolimits_{x^*|_{X_0} = u^*} \mu(x, x^*).$$

It is clear that $\mu_0 \in \varsigma_{M_0}$. Furthermore, we have that

$$p(\mu_0) = p(\mu), \quad q(\mu_0) = q(\mu)|_{X_0} \text{ and } r(\mu_0) = r(\mu).$$

It follows that for $x \in X_0$,

$$\frac{\langle x, q(\mu)\rangle - r(\mu)}{1+\|p(\mu)\|} = \frac{\langle x, q(\mu_0)\rangle - r(\mu_0)}{1+\|p(\mu_0)\|}.$$

Conversely, let $\mu_0 \in \varsigma_{M_0}$ with $\operatorname{supp}\mu_0 = \{(x_1, u_1^*), ..., (x_n, u_n^*)\}$, $n \in \mathbb{N}$. For every $i \in \overline{1,n}$ we take (only) one $x_i^* \in X^*$ such that $x_i^*|_{X_0} = u_i^*$, and define $\mu \in \varsigma_M$ by $\mu(x, x^*) = \mu_0(x_i, u_i^*)$ if $(x, x^*) = (x_i, x_i^*)$, $\mu(x, x^*) = 0$ otherwise. We note that to this μ corresponds, by the procedure described above, μ_0. Therefore, $\chi_{M_0}(x) = \chi_M(x)$ for every $x \in X_0$, *i.e.* $\chi_{M_0} = \chi_M|_{X_0}$.

Suppose that M is maximal monotone and let $(x, u^*) \in X_0 \times X_0^*$ be such that $\langle y - x, v^* - u^*\rangle \geq 0$ for every $(y, v^*) \in M_0$. Then there exists $x^* \in X^*$ such that $x^*|_{X_0} = u^*$. It follows that $\langle y - x, y^* - x^*\rangle = \langle y - x, y^*|_{X_0} - u^*\rangle \geq 0$ for every $(y, y^*) \in M$, and so $(x, x^*) \in M$; hence $(x, u^*) \in M_0$.

Suppose now that M_0 is maximal monotone and let $(x, x^*) \in X \times X^*$ be such that $\langle y - x, y^* - x^*\rangle \geq 0$ for every $(y, y^*) \in M$. Fixing $(y_0, y_0^*) \in M$, then $(y_0, y_0^* + z^*) \in M$ for every $z^* \in X_0^\perp$. Therefore $\langle y_0 - x, y_0^* + z^* - x^*\rangle \geq 0$ for every $z^* \in X_0^\perp$, which implies that $x \in (X_0^\perp)^\perp = X_0$. We obtain so that $\langle y - x, y^*|_{X_0} - x^*|_{X_0}\rangle \geq 0$ for every $(y, y^*) \in M$, and so $(x, x^*|_{X_0}) \in M_0$. From the definition of M_0 there exists $z^* \in X^*$ such that $(x, z^*) \in M$ and $z^*|_{X_0} = x^*|_{X_0}$ ($\Leftrightarrow z^* - x^* \in X_0^\perp$). From our hypothesis it follows that $(x, x^*) \in M$, and so M is maximal monotone.

(iv) Suppose now that M is maximal monotone and $\operatorname{dom} M \subset x_0 + X_0$, where $X_0 \subset X$ is a closed linear subspace and $x_0 \in X$. Consider $(x, x^*) \in M$ and $u^* \in X_0^\perp$. Then

$$\forall\,(y, y^*) \in M \ :\ \langle y - x, y^* - (x^* + u^*)\rangle = \langle y - x, y^* - x^*\rangle \geq 0,$$

because $y - x \in X_0$, and so $(x, x^* + u^*) \in M$. Therefore $M + \{0\} \times X_0^\perp = M$.

Fix $x_1 \in \operatorname{dom} M$; consider $M_1 := M - (x_1, 0)$ and $X_1 := \overline{\operatorname{lin}(\operatorname{dom} M_1)} = \overline{\operatorname{aff}(\operatorname{dom} M)} - x_1$. Since M_1 is maximal monotone, from what precedes it follows that $M_1 + \{0\} \times X_1^\perp = M_1$, while from (iii) it follows that $\operatorname{dom}\chi_{M_1} \subset X_1$. Using (ii) we obtain that $\operatorname{dom}\chi_M \subset x_1 + X_1 = \overline{\operatorname{aff}(\operatorname{dom} M)}$. □

Consider the function

$$h : \varsigma_{X\times X^*} \times X^* \to \mathbb{R}, \qquad h(\mu, x^*) := r(\mu) - \langle p(\mu), x^*\rangle. \tag{3.104}$$

The function h is affine with respect to μ and x^*, respectively, and w^*-continuous with respect to x^*.

Lemma 3.11.3 *Let X be a Banach space and M_1, M_2 be nonempty monotone subsets of $X \times X^*$. Suppose that*

$$0 \in (\operatorname{dom} \chi_{M_1} - \operatorname{dom} \chi_{M_2})^i . \tag{3.105}$$

Then there exists $\gamma > 0$ such that

$$\forall (\mu_1, \mu_2) \in \varsigma_{M_1} \times \varsigma_{M_2}, \exists \lambda \in [0, 2\gamma], (x_1^*, x_2^*) \in X^* \times X^* : \|x_1^*\|, \|x_2^*\| \le \gamma,$$
$$h(\mu_1, x_1^*) + h(\mu_2, x_2^*) + \lambda \|q(\mu_1) + q(\mu_2)\| \ge \tfrac{1}{2}\left(\lambda^2 + \|x_1^* + x_2^*\|^2\right). \tag{3.106}$$

Proof. By Proposition 2.8.9, there exists $\eta \in (0, 1]$ and $\nu \in [1, \infty)$ such that

$$\eta U \subset \{x \in X \mid \chi_{M_1}(x) \le \nu, \|x\| \le \nu\} - \{x \in X \mid \chi_{M_2}(x) \le \nu, \|x\| \le \nu\}. \tag{3.107}$$

Let $\gamma := 5\nu^2/\eta$ and fix $(\mu_1, \mu_2) \in \varsigma_{M_1} \times \varsigma_{M_2}$. Consider first the case in which $\max\{\|q(\mu_1)\|, \|q(\mu_2)\|\} \le \gamma$. In this situation we take $(x_1^*, x_2^*) = (q(\mu_1), q(\mu_2))$ and $\lambda = \|q(\mu_1) + q(\mu_2)\|$. From Lemma 3.11.1 we have that $h(\mu_i, x_i^*) = r(\mu_i) - \langle p(\mu_i), q(\mu_i)\rangle \ge 0$ for $i = 1, 2$, and so Eq. (3.106) holds.

Consider now the case $\max\{\|q(\mu_1)\|, \|q(\mu_2)\|\} > \gamma$. We may assume that $\gamma < \|q(\mu_1)\|$. There exists $v \in X$ such that

$$\|v\| = \eta \quad \text{and} \quad \langle v, q(\mu_1)\rangle > \gamma\eta = 5\nu^2. \tag{3.108}$$

From Eq. (3.107) there exist $x, y \in X$ such that

$$\chi_{M_1}(x) \le \nu, \quad \chi_{M_2}(y) \le \nu, \quad \|y\| \le \nu \quad \text{and} \quad x - y = v.$$

There exist also $x_1^*, x_2^* \in X^*$ such that

$$\|x_i^*\| = \nu \quad \text{and} \quad \langle p(\mu_i), x_i^*\rangle = -\nu \|p(\mu_i)\|, \ i = 1, 2.$$

Since $\gamma \ge \nu$ we have that $\|x_i^*\| \le \gamma$ for $i = 1, 2$; since $\chi_{M_1}(x) \le \nu$ and $x = v + y$, we have that

$$\begin{aligned} h(\mu_1, x_1^*) &= r(\mu_1) - \langle p(\mu_1), x_1^*\rangle = r(\mu_1) + \nu \|p(\mu_1)\| \ge \langle x, q(\mu_1)\rangle - \nu \\ &= \langle v + y, q(\mu_1)\rangle - \nu = \langle v, q(\mu_1)\rangle + \langle y, q(\mu_1)\rangle - \nu. \end{aligned}$$

From Eq. (3.108) we obtain that

$$h(\mu_1, x_1^*) \ge 5\nu^2 + \langle y, q(\mu_1)\rangle - \nu \ge 4\nu^2 + \langle y, q(\mu_1)\rangle . \tag{3.109}$$

Since $\chi_{M_2}(y) \le \nu$,

$$h(\mu_2, x_2^*) = r(\mu_2) - \langle p(\mu_2), x_2^* \rangle = r(\mu_2) + \nu \|p(\mu_2)\| \ge \langle y, q(\mu_2) \rangle - \nu. \quad (3.110)$$

From Eqs. (3.109) and (3.110) we obtain

$$h(\mu_1, x_1^*) + h(\mu_2, x_2^*) \ge 4\nu^2 + \langle y, q(\mu_1) + q(\mu_2) \rangle - \nu \ge 3\nu^2 - \nu \|q(\mu_1) + q(\mu_2)\|,$$

and so

$$2h(\mu_1, x_1^*) + 2h(\mu_2, x_2^*) + 2\nu \cdot \|q(\mu_1) + q(\mu_2)\| - \nu^2 - \|x_1^* + x_2^*\|^2$$
$$\ge 6\nu^2 - \nu^2 - \|x_1^* + x_2^*\|^2 \ge 5\nu^2 - 4\nu^2 > 0.$$

Hence Eq. (3.106) holds with $\lambda = \nu$ ($\le 2\gamma$) in this case. □

Theorem 3.11.4 *Let X be a reflexive Banach space and M_1, M_2 be nonempty monotone subsets of $X \times X^*$. Suppose that condition (3.105) is satisfied.*

(i) *Then there exist $(x_1^*, x_2^*) \in X^* \times X^*$ and $z \in X$ such that for every $(y_1, y_1^*) \in M_1$, $(y_2, y_2^*) \in M_2$ we have*

$$2\langle y_1 - z, y_1^* - x_1^* \rangle + 2\langle y_2 - z, y_2^* - x_2^* \rangle \ge \|z\|^2 + \|x_1^* + x_2^*\|^2 + 2\langle z, x_1^* + x_2^* \rangle. \quad (3.111)$$

(ii) *If moreover M_1 and M_2 are maximal monotone then there exist $z \in X$ and $x_1^*, x_2^* \in X^*$ such that $(z, x_1^*) \in M_1$, $(z, x_2^*) \in M_2$ and*

$$\|z\|^2 + \|x_1^* + x_2^*\|^2 + 2\langle z, x_1^* + x_2^* \rangle = 0. \quad (3.112)$$

In particular $\operatorname{dom} M_1 \cap \operatorname{dom} M_2 \ne \emptyset$.

Proof. (i) Consider $\gamma > 0$ given by the preceding lemma, $A := \varsigma_{M_1} \times \varsigma_{M_2}$, $B := \{(x_1^*, x_2^*, z) \in X^* \times X^* \times X \mid \|x_1^*\|, \|x_2^*\| \le \gamma,\ \|z\| \le 2\gamma\}$ and the function $k : A \times B \to \mathbb{R}$ defined by

$$k((\mu_1, \mu_2), (x_1^*, x_2^*, z)) := 2(h(\mu_1, x_1^*) + h(\mu_2, x_2^*) - \langle z, q(\mu_1) + q(\mu_2) \rangle)$$
$$- \|z\|^2 - \|x_1^* + x_2^*\|^2.$$

It is obvious that the sets A, B are convex, and k is convex (even affine) with respect to $(\mu_1, \mu_2) \in A$ and concave with respect to $(x_1^*, x_2^*, z) \in B$. Furthermore, endowing $X^* \times X^* \times X$ with the product topology $w^* \times w^* \times w$,

B is compact and k is upper semicontinuous with respect to (x_1^*, x_2^*, z). Using the minimax theorem (Theorem 2.10.2) we have that

$$\inf_A \max_B k = \max_B \inf_A k.$$

Let $(\mu_1, \mu_2) \in A$. Using the preceding lemma, there exist $\lambda \in [0, 2\gamma]$ and $x_1^*, x_2^* \in X^*$ with $\|x_1^*\|, \|x_2^*\| \leq \gamma$ such that

$$2h(\mu_1, x_1^*) + 2h(\mu_2, x_2^*) + 2\lambda \cdot \|q(\mu_1) + q(\mu_2)\| - \lambda^2 - \|x_1^* + x_2^*\|^2 \geq 0.$$

Since X is reflexive, there exists $z \in X$ such that

$$\|z\| = \lambda, \ \langle z, q(\mu_1) + q(\mu_2)\rangle = -\lambda \cdot \|q(\mu_1) + q(\mu_2)\|.$$

It follows that $k((\mu_1, \mu_2), (x_1^*, x_2^*, z)) \geq 0$. Therefore $\inf_A \max_B k \geq 0$. Hence $\max_B \inf_A k \geq 0$, *i.e.* there exists $(x_1^*, x_2^*, z) \in B$ such that

$$2\left(h(\mu_1, x_1^*) + h(\mu_2, x_2^*) - \langle z, q(\mu_1) + q(\mu_2)\rangle\right) \geq \|z\|^2 + \|x_1^* + x_2^*\|^2$$

for all $(\mu_1, \mu_2) \in A$. Taking $\mu_i = \delta_{(y_i, y_i^*)}$ with $(y_i, y_i^*) \in M_i$, $i = 1, 2$, we obtain that

$$2\left(\langle y_1, y_1^*\rangle - \langle y_1, x_1^*\rangle + \langle y_2, y_2^*\rangle - \langle y_2, x_2^*\rangle - \langle z, y_1^* + y_2^*\rangle\right) \geq \|z\|^2 + \|x_1^* + x_2^*\|^2$$

i.e. Eq. (3.111) holds.

(ii) For (x_1^*, x_2^*, z) found at (i), from Eq. (3.111) we obtain that

$$\inf_{(y_1, y_1^*) \in M_1} \langle y_1 - z, y_1^* - x_1^*\rangle + \inf_{(y_2, y_2^*) \in M_2} \langle y_2 - z, y_2^* - x_2^*\rangle \geq \tfrac{1}{2}\left(\|z\|^2 + \|x_1^* + x_2^*\|^2 + 2\langle z, x_1^* + x_2^*\rangle\right).$$

Since for $x \in X$ and $x^* \in X^*$ the inequality

$$\|x\|^2 + \|x^*\|^2 + 2\langle x, x^*\rangle \geq \|x\|^2 + \|x^*\|^2 - 2\|x\| \cdot \|x^*\| = (\|x\| - \|x^*\|)^2 \quad (3.113)$$

holds, we obtain that

$$\inf_{(y_1, y_1^*) \in M_1} \langle y_1 - z, y_1^* - x_1^*\rangle + \inf_{(y_2, y_2^*) \in M_2} \langle y_2 - z, y_2^* - x_2^*\rangle \geq 0.$$

Because M_1 is monotone, it is obvious that

$$\inf_{(y_1, y_1^*) \in M_1} \langle y_1 - z, y_1^* - x_1^*\rangle = 0$$

for $(z, x_1^*) \in M_1$, and, since M_1 is maximal monotone,

$$\inf_{(y_1,y_1^*)\in M_1} \langle y_1 - z, y_1^* - x_1^* \rangle < 0$$

for $(z, x_1^*) \notin M_1$. The same argument being true also for M_2, we obtain that $(z, x_1^*) \in M_1$ and $(z, x_2^*) \in M_2$. Furthermore Eq. (3.112) holds. □

It is obvious that $X \times \{0\} \subset X \times X^*$ is a maximal monotone set whose domain is X. Taking in the preceding theorem one of the monotone sets to be $X \times \{0\}$ we obtain the following result.

Theorem 3.11.5 *Let X be a reflexive Banach space and $M \subset X \times X^*$ be a nonempty monotone set. Then*

(i) *there exists $(x, x^*) \in X \times X^*$ such that*

$$\forall (y, y^*) \in M \ : \ 2 \langle y - x, y^* - x^* \rangle \geq \|x\|^2 + \|x^*\|^2 + 2 \langle x, x^* \rangle ;$$

(ii) *M is maximal monotone if and only if for every $(w, w^*) \in X \times X^*$ there exists $(x, x^*) \in M$ such that*

$$\|x - w\|^2 + \|x^* - w^*\|^2 + 2 \langle x - w, x^* - w^* \rangle = 0. \tag{3.114}$$

Proof. (i) Taking $M_1 := M$ in the preceding theorem and $M_2 := X \times \{0\}$, condition (3.105) is satisfied, and so there exists $(x_1^*, x_2^*, z) \in X^* \times X^* \times X$ verifying Eq. (3.111). Fixing $(y_1, y_1^*) \in M_1 = M$ we obtain that

$$\forall y \in X \ : \ \langle y_1 - z, y_1^* - x_1^* \rangle + \langle z, x_2^* \rangle \geq \langle y, x_2^* \rangle ;$$

therefore $x_2^* = 0$. Taking $(x, x^*) = (z, x_1^*)$, the conclusion is immediate.

(ii) Suppose that M is maximal monotone and consider $(w, w^*) \in X \times X^*$. It is obvious that $M_1 := M - (w, w^*)$ and $M_2 := X \times \{0\}$ are maximal monotone. Using Theorem 3.11.4(ii) we obtain the existence of $(u, u^*) \in M_1$ such that $\|u\|^2 + \|u^*\|^2 + 2 \langle u, u^* \rangle = 0$, and so there exists $(x, x^*) \in M$ satisfying Eq. (3.114).

We prove now the converse implication. Let $(w, w^*) \in X \times X^*$ be such that $\langle y - w, y^* - w^* \rangle \geq 0$ for every $(y, y^*) \in M$. For this $(w, w^*) \in X \times X^*$, using our hypothesis, there exists $(x, x^*) \in M$ satisfying Eq. (3.114), and so $\|x - w\|^2 + \|x^* - w^*\|^2 \leq 0$, *i.e.* $(w, w^*) = (x, x^*) \in M$. □

Taking into account the expression of the duality mapping Φ_X given in Eq. (3.59), from Eq. (3.113) we obtain that

$$\|x\|^2 + \|x^*\|^2 + 2 \langle x, x^* \rangle = 0 \Leftrightarrow x^* \in -\Phi_X(x) \Leftrightarrow x^* \in \Phi_X(-x).$$

This remark gives us the possibility to obtain other characterizations of maximal monotone subsets (multifunctions).

Theorem 3.11.6 *Let X be a reflexive Banach space and M be a non-empty monotone subset of $X \times X^*$. Then M is maximal monotone if and only if $M + \operatorname{gr}(-\Phi_X) = X \times X^*$.*

Proof. Assertion (ii) of the preceding theorem shows that the maximal monotonicity of M is equivalent to each of the following relations:

$$\forall (w, w^*) \in X \times X^*, \ \exists (x, x^*) \in M \ : \ w^* - x^* \in -\Phi_X(w - x),$$
$$\forall (w, w^*) \in X \times X^* \ : \ (w, w^*) \in M + \operatorname{gr}(-\Phi_X).$$

Therefore the conclusion of the theorem holds. □

Theorem 3.11.7 *Let X be a reflexive Banach space and $T : X \rightrightarrows X^*$ be a monotone multifunction. If T is maximal monotone then $\operatorname{Im}(\Phi_X + T) = X^*$. Conversely, if X is smooth and strictly convex, and $\operatorname{Im}(\Phi_X + T) = X^*$, then T is maximal monotone.*

Proof. To begin with, suppose that T is maximal monotone. Let $x^* \in X^*$; because $(0, x^*) \in X \times X^*$, from the preceding theorem we get the existence of the elements $(y, y^*) \in \operatorname{gr} T$ and $(u, u^*) \in \operatorname{gr} \Phi_X$ such that $(y, y^*) + (-u, u^*) = (0, x^*)$. Therefore $u = y$ and $x^* = y^* + u^*$, *i.e.* $x^* \in T(y) + \Phi_X(y)$.

Suppose now that X is strictly convex and smooth. From Theorem 3.7.2 we have that Φ_X is single-valued and strictly monotone. Let $(x, x^*) \in X \times X^*$ be such that $\langle y - x, y^* - x^* \rangle \geq 0$ for every $(y, y^*) \in \operatorname{gr} T$. Since $x^* + \Phi_X(x) \in X^* = \operatorname{Im}(\Phi_X + T)$, there exists $u \in X$ such that $x^* + \Phi_X(x) \in \Phi_X(u) + T(u)$. It follows that

$$0 \leq \langle u - x, x^* + \Phi_X(x) - \Phi_X(u) - x^* \rangle = - \langle x - u, \Phi_X(x) - \Phi_X(u) \rangle \leq 0,$$

and so $\langle x - u, \Phi_X(x) - \Phi_X(u) \rangle = 0$. Since Φ_X is strictly monotone, we have that $x = u$, and therefore $(x, x^*) \in \operatorname{gr} T$. Hence T is maximal monotone. □

Another version of the preceding result is the following.

Theorem 3.11.8 *Let X be a reflexive Banach space and $T : X \rightrightarrows X^*$ be a monotone multifunction. If T is maximal monotone then*

$$\forall y \in X, \ \exists x \in X \ : \ 0 \in T(x) + \Phi_X(x - y). \tag{3.115}$$

Conversely, if X is smooth and strictly convex, and Eq. (3.115) holds, then T is maximal monotone.

Proof. It is sufficient to note (identifying X^{**} with X) that $\Phi_{X^*} = \Phi_X^{-1}$, T is maximal monotone if and only if $T^{-1} : X^* \rightrightarrows X$ is maximal monotone and

$$\begin{aligned} \operatorname{Im}(\Phi_{X^*} + T^{-1}) = X &\Leftrightarrow \forall y \in X, \exists x^* \in X^* \ : \ y \in \Phi_{X^*}(x^*) + T^{-1}(x^*) \\ &\Leftrightarrow \forall y \in X, \exists x \in X \ : \ \Phi_X(y - x) \cap T(x) \neq \emptyset \\ &\Leftrightarrow \forall y \in X, \exists x \in X \ : \ 0 \in T(x) + \Phi_X(x - y). \end{aligned}$$

The proof is complete. □

Remark 3.11.1 Since $T \subset X \times X^*$ is maximal monotone if and only if λT is maximal monotone for some/any $\lambda > 0$, it follows that, when X is strictly convex and smooth, we have that T is maximal monotone $\Leftrightarrow$ $\operatorname{Im}(\lambda\Phi_X + T) = X^*$ for some/any $\lambda > 0 \Leftrightarrow \forall y \in X, \exists x \in X \ : \ 0 \in \lambda T(x) + \Phi_X(x - y)$ for some/any $\lambda > 0$.

Another very important result is the following one.

Theorem 3.11.9 *Let X be a reflexive Banach space and $T_1, T_2 : X \rightrightarrows X^*$ be maximal monotone multifunctions. Suppose that*

$$0 \in (\operatorname{dom} \chi_{T_1} - \operatorname{dom} \chi_{T_2})^i .$$

Then $T_1 + T_2$ is maximal monotone. In particular $\operatorname{dom} T_1 \cap \operatorname{dom} T_2 \neq \emptyset$.

Proof. Let $M_1 = \operatorname{gr} T_1$, $M_2 = \operatorname{gr} T_2$ and $M = \operatorname{gr}(T_1 + T_2)$. The hypothesis shows that M_1 and M_2 are maximal monotone and condition (3.105) is satisfied. From Theorem 3.11.4 it follows that $\operatorname{dom} M_1 \cap \operatorname{dom} M_2 = \operatorname{dom} T_1 \cap \operatorname{dom} T_2 \neq \emptyset$, and so M is a nonempty monotone set. To prove the maximal monotonicity of M we use Theorem 3.11.5. Let $(w, w^*) \in X \times X^*$ and $\widetilde{T}_1, \widetilde{T}_2 : X \rightrightarrows X^*$, $\widetilde{T}_i(x) := T_i(x + w) - \frac{1}{2}w^*$ $(i = 1, 2)$. It is obvious that $\widetilde{T}_i$ $(i = 1, 2)$ is maximal monotone, while from Lemma 3.11.2 we get that $\operatorname{gr} \widetilde{T}_1$, $\operatorname{gr} \widetilde{T}_2$ satisfy condition (3.105). We can apply Theorem 3.11.4; so there exists $(z_1^*, z_2^*, z) \in X^* \times X^* \times X$ such that $(z, z_i^*) \in \operatorname{gr} \widetilde{T}_i$ $(i = 1, 2)$ and

$$\|z\|^2 + \|z_1^* + z_2^*\|^2 + 2\langle z, z_1^* + z_2^* \rangle = 0.$$

Denoting $z + w$ by x and $z_i^* + \frac{1}{2}w^*$ by x_i^*, we obtain that $(x, x_i^*) \in \operatorname{gr} T_i$ $(i = 1, 2)$ and

$$\|x - w\|^2 + \|x_1^* + x_2^* - w^*\|^2 + 2\langle x - w, x_1^* + x_2^* - w^*\rangle = 0.$$

Taking $x^* = x_1^* + x_2^*$, we obtain that $(x, x^*) \in M$ and

$$\|x - w\|^2 + \|x^* - w^*\|^2 + 2\langle x - w, x^* - w^*\rangle = 0.$$

Using Theorem 3.11.5 we have that M is maximal monotone, and so $T_1 + T_2$ is maximal monotone. □

The sufficient condition from the preceding theorem is formulated using the auxiliary function χ, which is not easy to calculate. In the following theorem we establish other sufficient conditions (in fact equivalent with the mentioned one) for the maximal monotonicity of the sum.

Theorem 3.11.10 *Let X be a reflexive Banach space and $T_1, T_2 : X \rightrightarrows X^*$ be maximal monotone multifunctions. Then*

$$\operatorname{int}(\operatorname{dom} T_1 - \operatorname{dom} T_2) = \operatorname{int}(\operatorname{dom} \chi_{T_1} - \operatorname{dom} \chi_{T_2}) = (\operatorname{dom} \chi_{T_1} - \operatorname{dom} \chi_{T_2})^i. \tag{3.116}$$

Therefore $\operatorname{int}(\operatorname{dom} T_1 - \operatorname{dom} T_2)$ *is a convex set; moreover, the following statements are equivalent:*

$$0 \in (\operatorname{dom} \chi_{T_1} - \operatorname{dom} \chi_{T_2})^i, \tag{3.117}$$

$$0 \in \left(\operatorname{co}(\operatorname{dom} T_1) - \operatorname{co}(\operatorname{dom} T_2)\right)^i, \tag{3.118}$$

$$0 \in (\operatorname{dom} T_1 - \operatorname{dom} T_2)^i, \tag{3.119}$$

$$0 \in \operatorname{int}(\operatorname{dom} T_1 - \operatorname{dom} T_2), \tag{3.120}$$

each of these conditions ensuring that $T_1 + T_2$ is maximal monotone.

Furthermore, if $\operatorname{int}(\operatorname{dom} T_1 - \operatorname{dom} T_2) \neq \emptyset$ *(equivalently, if* $(\operatorname{dom} \chi_{T_1} - \operatorname{dom} \chi_{T_2})^i$ *is nonempty), then*

$$\overline{\operatorname{int}(\operatorname{dom} T_1 - \operatorname{dom} T_2)} = \overline{\operatorname{dom} T_1 - \operatorname{dom} T_2} = \overline{\operatorname{dom} \chi_{T_1} - \operatorname{dom} \chi_{T_2}}, \tag{3.121}$$

and so $\overline{\operatorname{dom} T_1 - \operatorname{dom} T_2}$ *is a convex set.*

Proof. Taking into account Lemma 3.11.2, we have that

$$\begin{aligned}\operatorname{int}(\operatorname{dom} T_1 - \operatorname{dom} T_2) &\subset (\operatorname{dom} T_1 - \operatorname{dom} T_2)^i \\ &\subset \left(\operatorname{co}(\operatorname{dom} T_1) - \operatorname{co}(\operatorname{dom} T_2)\right)^i \\ &\subset (\operatorname{dom} \chi_{T_1} - \operatorname{dom} \chi_{T_2})^i,\end{aligned}$$

which shows that the inclusions $\subset$ from Eq. (3.116) hold and (3.120) $\Rightarrow$ (3.119) $\Rightarrow$ (3.118) $\Rightarrow$ (3.117). Furthermore, by Proposition 2.8.9 we have that

$$\operatorname{int}(\operatorname{dom} \chi_{T_1} - \operatorname{dom} \chi_{T_2}) = (\operatorname{dom} \chi_{T_1} - \operatorname{dom} \chi_{T_2})^i. \tag{3.122}$$

Let us show that

$$(\operatorname{dom} \chi_{T_1} - \operatorname{dom} \chi_{T_2})^i \subset \operatorname{dom} T_1 - \operatorname{dom} T_2,$$

which together with Eq. (3.122) will prove Eq. (3.116). Let w be an element of $(\operatorname{dom} \chi_{T_1} - \operatorname{dom} \chi_{T_2})^i$. Consider the multifunction $\widetilde{T}_1$ defined by $\widetilde{T}_1(x) = T(x+w)$. We have that $\operatorname{dom} \widetilde{T}_1 = \operatorname{dom} T_1 - w$ and $\operatorname{dom} \chi_{\widetilde{T}_1} = \operatorname{dom} \chi_{T_1} - w$. Therefore $\widetilde{T}_1$ and T_2 verify condition (3.105). From Theorem 3.11.4 it follows that $\operatorname{dom} \widetilde{T}_1 \cap \operatorname{dom} T_1 \neq \emptyset$, *i.e.* $w \in \operatorname{dom} T_1 - \operatorname{dom} T_2$. Therefore Eq. (3.116) holds. From this relation we obtain that (3.117) $\Rightarrow$ (3.120). From what was proved above and Theorem 3.11.9 we have that each of the conditions (3.117)–(3.120) ensures that $T_1 + T_2$ is a maximal monotone multifunction. Furthermore, if $\operatorname{int}(\operatorname{dom} T_1 - \operatorname{dom} T_2) \neq \emptyset$, from Eq. (3.116) and

$$\operatorname{int}(\operatorname{dom} T_1 - \operatorname{dom} T_2) \subset \operatorname{dom} T_1 - \operatorname{dom} T_2 \subset \operatorname{dom} \chi_{T_1} - \operatorname{dom} \chi_{T_2}$$

we get Eq. (3.121). The convexity of the set $\operatorname{int}(\operatorname{dom} T_1 - \operatorname{dom} T_2)$ and, when this is nonempty, of the set $\overline{\operatorname{dom} T_1 - \operatorname{dom} T_2}$ follows from Eqs. (3.116) and (3.121). □

Conditions (3.117)–(3.120) from preceding theorem can be "relativized."

Theorem 3.11.11 *Let X be a reflexive Banach space and $T_1, T_2 : X \rightrightarrows X^*$ be maximal monotone multifunctions. Then*

$$^{ic}(\operatorname{dom} T_1 - \operatorname{dom} T_2) = {}^{ic}(\operatorname{dom} \chi_{T_1} - \operatorname{dom} \chi_{T_2}). \tag{3.123}$$

Therefore ${}^{ic}(\operatorname{dom} T_1 - \operatorname{dom} T_2)$ *is a convex set and the following statements are equivalent:*

$$0 \in {}^{ic}(\operatorname{dom} \chi_{T_1} - \operatorname{dom} \chi_{T_2}), \tag{3.124}$$

$$0 \in {}^{ic}\big(\operatorname{co}(\operatorname{dom} T_1) - \operatorname{co}(\operatorname{dom} T_2)\big), \tag{3.125}$$

$$0 \in {}^{ic}(\operatorname{dom} T_1 - \operatorname{dom} T_2)\,, \tag{3.126}$$

$$\operatorname{dom} T_1 - \operatorname{dom} T_2 \textit{ is neighborhood of the origin in } \overline{\operatorname{lin}(\operatorname{dom} T_1 - \operatorname{dom} T_2)}, \tag{3.127}$$

$$\bigcup\nolimits_{\lambda \geq 0} \lambda\,(\operatorname{dom} T_1 - \operatorname{dom} T_2) \textit{ is a closed linear subspace,} \tag{3.128}$$

each of these conditions ensuring that $T_1 + T_2$ *is maximal monotone.*

Furthermore, if ${}^{ic}(\operatorname{dom} T_1 - \operatorname{dom} T_2) \neq \emptyset$ *(equivalently, if* ${}^{ic}(\operatorname{dom} \chi_{T_1} - \operatorname{dom} \chi_{T_2})$ *is nonempty), then*

$$\overline{{}^{ic}(\operatorname{dom} T_1 - \operatorname{dom} T_2)} = \overline{\operatorname{dom} T_1 - \operatorname{dom} T_2} = \overline{\operatorname{dom} \chi_{T_1} - \operatorname{dom} \chi_{T_2}}, \tag{3.129}$$

and so $\overline{\operatorname{dom} T_1 - \operatorname{dom} T_2}$ *is a convex set.*

Proof. To begin with, suppose that condition (3.124) holds; we wish to prove that $T_1 + T_2$ is maximal monotone; if we succeed we shall have that $\operatorname{dom} T_1 \cap \operatorname{dom} T_2 \neq \emptyset$, *i.e.* $0 \in \operatorname{dom} T_1 - \operatorname{dom} T_2$.

In our hypothesis $X_0 := \operatorname{aff}(\operatorname{dom} \chi_{T_1} - \operatorname{dom} \chi_{T_2})$ is a closed linear subspace. Replacing eventually T_i by $\widetilde{T}_i$ where $\widetilde{T}_i(x) = T_i(x+w)$ $(i = 1, 2)$ with $w \in \operatorname{dom} \chi_{T_1} \cap \operatorname{dom} \chi_{T_2}$, we may assume that $0 \in \operatorname{dom} \chi_{T_1} \cap \operatorname{dom} \chi_{T_2}$, and so $\operatorname{dom} T_1 \cup \operatorname{dom} T_2 \subset X_0$. Let $i \in \{1, 2\}$. Since T_i is maximal monotone, it follows that $\operatorname{gr} T_i + \{0\} \times X_0^{\perp} = \operatorname{gr} T_i$. Consider

$$T_i^0 : X_0 \rightrightarrows X_0^{\perp}, \quad T_i^0(x) = \{x^*|_{X_0} \mid x^* \in T_i(x)\}.$$

From Lemma 3.11.2 (iii) it follows that T_i^0 is maximal monotone and $\operatorname{dom} \chi_{T_i} = \operatorname{dom} \chi_{T_i^0}$. X_0 being a reflexive Banach space and $0 \in (\operatorname{dom} \chi_{T_1^0} - \operatorname{dom} \chi_{T_2^0})^i$, from the preceding theorem we have that $T_1^0 + T_2^0$ is maximal monotone and

$$(\operatorname{dom} \chi_{T_1^0} - \operatorname{dom} \chi_{T_2^0})^i = \operatorname{int}(\operatorname{dom} T_1^0 - \operatorname{dom} T_2^0) \quad (\text{with respect to } X_0),$$

i.e.

$${}^{ic}(\operatorname{dom} \chi_{T_1} - \operatorname{dom} \chi_{T_2}) = {}^{ic}(\operatorname{dom} T_1 - \operatorname{dom} T_2) = \operatorname{int}_{X_0}(\operatorname{dom} T_1 - \operatorname{dom} T_2). \tag{3.130}$$

It is obvious that $(T_1+T_2)^0 = T_1^0 + T_2^0$. From Lemma 3.11.2 (iii) we have that T_1+T_2 is maximal monotone.

Let us prove now the other statements. Taking into account Lemma 3.11.2(iii), (iv), we have that

$$\begin{aligned}\operatorname{dom} T_1 - \operatorname{dom} T_2 &\subset \operatorname{dom} \chi_{T_1} - \operatorname{dom} \chi_{T_2} \subset \overline{\operatorname{aff}(\operatorname{dom} T_1)} - \overline{\operatorname{aff}(\operatorname{dom} T_2)} \\ &\subset \overline{\operatorname{aff}(\operatorname{dom} T_1 - \operatorname{dom} T_2)},\end{aligned}$$

which imply that

$$\operatorname{aff}(\operatorname{dom} T_1 - \operatorname{dom} T_2) \subset \operatorname{aff}(\operatorname{dom} \chi_{T_1} - \operatorname{dom} \chi_{T_2}) \subset \overline{\operatorname{aff}(\operatorname{dom} T_1 - \operatorname{dom} T_2)}.$$

Therefore $\operatorname{aff}(\operatorname{dom} T_1 - \operatorname{dom} T_2)$ is closed if and only if $\operatorname{aff}(\operatorname{dom} \chi_{T_1} - \operatorname{dom} \chi_{T_2})$ is closed; in this situation these sets are equal. Taking into account this remark it follows that the inclusion $\subset$ holds in Eq. (3.123).

Let us prove the converse inclusion. Let $w \in {}^{ic}(\operatorname{dom} \chi_{T_1} - \operatorname{dom} \chi_{T_2})$; this means that $X_0 := \operatorname{aff}(\operatorname{dom} \chi_{T_1} - \operatorname{dom} \chi_{T_2})$ is a closed set and $w \in {}^{i}(\operatorname{dom} \chi_{T_1} - \operatorname{dom} \chi_{T_2})$. Taking $\widetilde{T}_1$ as in the proof of the preceding theorem, we have that $0 \in {}^{ic}(\operatorname{dom} \chi_{\widetilde{T}_1} - \operatorname{dom} \chi_{T_2})$. From the first part of the proof it follows that $\operatorname{dom} \widetilde{T}_1 \cap \operatorname{dom} T_2 \neq \emptyset$, whence we have that $w \in \operatorname{dom} T_1 - \operatorname{dom} T_2$. We have obtained so that

$${}^{i}(\operatorname{dom} \chi_{T_1} - \operatorname{dom} \chi_{T_2}) \subset \operatorname{dom} T_1 - \operatorname{dom} T_2 \ \ (\subset \operatorname{dom} \chi_{T_1} - \operatorname{dom} \chi_{T_2} \subset X_0),$$

and so the inclusion $\supset$ holds in Eq. (3.123), too.

Note that relation Eq. (3.130), proved above when Eq. (3.124) holds, shows that (3.124) $\Rightarrow$ (3.127). Condition (3.127) ensures that $0 \in {}^{i}(\operatorname{dom} T_1 - \operatorname{dom} T_2)$ and $\operatorname{aff}(\operatorname{dom} T_1 - \operatorname{dom} T_2) = \operatorname{lin}(\operatorname{dom} T_1 - \operatorname{dom} T_2)$ is closed. Hence (3.127) $\Rightarrow$ (3.126). The implications (3.126) $\Rightarrow$ (3.125) $\Rightarrow$ (3.124) are obvious. The implication (3.127) $\Rightarrow$ (3.128) is obvious, too. Suppose that (3.128) holds. Denoting by X_0 the closed linear space $\bigcup_{\lambda \geq 0}(\operatorname{dom} T_1 - \operatorname{dom} T_2)$, we obtain that $X_0 = \bigcup_{\lambda \geq 0} \left(\operatorname{co}(\operatorname{dom} T_1) - \operatorname{co}(\operatorname{dom} T_2)\right)$, *i.e.* condition (3.125) holds. Therefore conditions (3.124)–(3.128) are equivalent.

If ${}^{ic}(\operatorname{dom} T_1 - \operatorname{dom} T_2) \neq \emptyset$ (or equivalently ${}^{ic}(\operatorname{dom} \chi_{T_1} - \operatorname{dom} \chi_{T_2}) \neq \emptyset$), then relation (3.129) follows similarly as in the proof of the preceding theorem (or one reduces to this theorem). □

When X is reflexive, if $T : X \rightrightarrows X^*$ is maximal monotone, taking also the maximal monotone set $\{0\} \times X^*$, we obtain that

$$(\operatorname{dom} \chi_T)^i = \operatorname{int}(\operatorname{dom} \chi_T) = \operatorname{int}(\operatorname{co}(\operatorname{dom} T)) = \operatorname{int}(\operatorname{dom} T) = (\operatorname{dom} T)^i, \tag{3.131}$$

and so $\operatorname{int}(\operatorname{dom} T)$ is a convex set; if this set is nonempty, then

$$\overline{(\operatorname{dom} T)^i} = \overline{\operatorname{dom} T} = \overline{\operatorname{co}(\operatorname{dom} T)} = \overline{\operatorname{dom} \chi_T}, \tag{3.132}$$

and so $\overline{\operatorname{dom} T}$ is a convex set, too. In the above relations we may replace ^{i} with ^{ic}.

The convexity of the set $\overline{\operatorname{dom} T}$ also holds without interiority conditions.

Theorem 3.11.12 *Let X be a reflexive Banach space and $T : X \rightrightarrows X^*$ be a maximal monotone multifunction. Then $\overline{\operatorname{dom} T}$ and $\overline{\operatorname{Im} T}$ are convex sets.*

Proof. It is obvious that for every $x \in X$ and $\lambda > 0$ the multifunction $T_{x,\lambda}$ defined by $T_{x,\lambda}(u) = \lambda T(u+x)$ is maximal monotone. Using Theorem 3.11.7 we have that $0 \in \operatorname{Im}(\Phi_X + T_{x,\lambda})$, *i.e.* there exists $y \in X$ such that $0 \in \Phi_X(y - x) + \lambda T(y)$. We have obtained so that

$$\forall x \in X,\ \lambda > 0,\ \exists\, (x_\lambda, x_\lambda^*) \in X \times X^* \ :\ \lambda^{-1} x_\lambda^* \in T(x_\lambda),\ -x_\lambda^* \in \Phi_X(x_\lambda - x). \tag{3.133}$$

Let $x \in X$ be fixed; for every $\lambda > 0$ consider (x_λ, x_λ^*) given by Eq. (3.133). Consider also $u \in \operatorname{dom} T$ and $u^* \in T(u)$ (fixed for the moment). From $\langle x_\lambda - u, \lambda^{-1} x_\lambda^* - u^* \rangle \geq 0$ we obtain that

$$\|x_\lambda - x\|^2 = \langle x_\lambda - x, -x_\lambda^* \rangle \leq \langle x - u, x_\lambda^* \rangle + \lambda \langle x - x_\lambda, u^* \rangle + \lambda \langle u - x, u^* \rangle, \tag{3.134}$$

and so, for $\lambda \in (0, 1]$ we have that

$$\begin{aligned} \|x_\lambda - x\|^2 &\leq \|x - u\| \cdot \|x_\lambda^*\| + \lambda \|x_\lambda - x\| \cdot \|u^*\| + \lambda \|x - u\| \cdot \|u^*\| \\ &\leq \alpha \|x_\lambda - x\| + \beta, \end{aligned} \tag{3.135}$$

where $\alpha := \|x - u\| + \lambda \|u^*\|$ and $\beta := \lambda \|x - u\| \cdot \|u^*\|$ (we have taken into account that $\|x_\lambda^*\| = \|x_\lambda - x\|$). From Eq. (3.135) it follows that there exists $M > 0$ such that $\|x_\lambda^*\| = \|x_\lambda - x\| \leq M$ for every $\lambda \in (0, 1]$.

Let $\gamma := \limsup_{\lambda \to 0_+} \|x_\lambda - x\| < \infty$ and $(\lambda_n) \subset (0, 1]$ be such that $(\lambda_n) \to 0$ and $(\|x_{\lambda_n} - x\|) \to \gamma$. Since $(x_{\lambda_n}^*)$ is bounded and X is reflexive, we may suppose that $x_{\lambda_n}^* \xrightarrow{w^*} x^* \in X^*$. Considering $\lambda = \lambda_n$ in Eq. (3.134) and then taking the limit, we obtain that $\gamma^2 \leq \langle x - u, x^* \rangle$ for every $u \in \operatorname{dom} T$; from this we get immediately that $\gamma^2 \leq \langle x - u, x^* \rangle$ for every $u \in \overline{\operatorname{co}}(\operatorname{dom} T)$. Therefore, $\lim_{\lambda \to 0} x_\lambda = x$ (with respect to the norm) for every $x \in \overline{\operatorname{co}}(\operatorname{dom} T)$. Since $(x_\lambda) \subset \operatorname{dom} T$, we obtain that $\overline{\operatorname{co}}(\operatorname{dom} T) \subset \overline{\operatorname{dom} T}$. Hence $\overline{\operatorname{dom} T}$ is a convex set. Since $\operatorname{Im} T = \operatorname{dom} T^{-1}$ and T^{-1} is maximal monotone, we have that $\overline{\operatorname{Im} T}$ is a convex set, too. □

The hypothesis that X be reflexive for having Eqs. (3.131) and (3.132) is not necessary (under interiority conditions).

Theorem 3.11.13 *Let X be a Banach space and $T : X \rightrightarrows X^*$ be a maximal monotone multifunction. Then condition (3.131) holds. In particular* $\operatorname{int}(\operatorname{dom} T)$ *is a convex set. Furthermore, if* $\operatorname{int}(\operatorname{dom} T) \neq \emptyset$ *then condition (3.132) holds, and so* $\overline{\operatorname{dom} T}$ *is a convex set.*

Proof. To prove condition (3.131) it is sufficient to show that $\operatorname{dom} T \supset (\operatorname{dom} \chi_T)^i$. Taking into account Lemma 3.11.2(ii) it is even sufficient to prove that $0 \in \operatorname{dom} T$ if $0 \in (\operatorname{dom} \chi_T)^i$. Let $0 \in (\operatorname{dom} \chi_T)^i$. From Theorem 2.2.20 we have that χ_T is continuous at 0, and so there exists $\eta, \nu > 0$, $\eta \leq 1 \leq \nu$, such that $\chi_T(x) \leq \nu$ for $\|x\| \leq \eta$. Therefore

$$\forall \mu \in \varsigma_T,\ \forall x,\ \|x\| \leq \eta\ :\ \langle x, q(\mu) \rangle \leq r(\mu) + \nu(1 + \|p(\mu)\|\,,$$

i.e.

$$\forall \mu \in \varsigma_T\ :\ \eta\,\|q(\mu)\| \leq r(\mu) + \nu(1 + \|p(\mu)\|\,. \tag{3.136}$$

Let $\kappa := \nu/\eta \geq 1$. Let us prove that

$$\forall \mu \in \varsigma_T\ :\ r(\mu) + \kappa\,\|p(\mu)\| \geq 0. \tag{3.137}$$

Let $\mu \in \varsigma_T$. If $\|q(\mu)\| \geq \kappa$, then we obtain immediately Eq. (3.137) from Eq. (3.136). So, suppose that $\|q(\mu)\| \leq \kappa$. Then, from Lemma 3.11.1, we have that

$$r(\mu) + \kappa\,\|p(\mu)\| \geq \langle p(\mu), q(\mu) \rangle + \kappa\,\|p(\mu)\| \geq \kappa\,\|p(\mu)\| - \|p(\mu)\| \cdot \|q(\mu)\| \geq 0.$$

Consider the function h defined by Eq. (3.104). Since ς_T is convex and κU_{X^*} is w^*-compact and convex, from the minimax theorem (Theorem 2.10.2) we have that

$$\begin{aligned}\max_{x^* \in \kappa U_{X^*}} \inf_{\mu \in \varsigma_T} h(\mu, x^*) &= \inf_{\mu \in \varsigma_T} \max_{x^* \in \kappa U_{X^*}} \left(r(\mu) - \langle p(\mu), x^* \rangle \right) \\ &= \inf_{\mu \in \varsigma_T} \left(r(\mu) + \kappa\,\|p(\mu)\| \right) \geq 0.\end{aligned}$$

Hence there exists $u^* \in \kappa U_{X^*}$ such that $h(\mu, u^*) \geq 0$ for every $\mu \in \varsigma_T$. Taking $\mu = \delta_{(x,x^*)}$ with $(x, x^*) \in \operatorname{gr} T$, we obtain that $\langle x, x^* \rangle - \langle x, u^* \rangle = \langle x - 0, x^* - u^* \rangle \geq 0$ for every $(x, x^*) \in \operatorname{gr} T$. Since T is maximal monotone, it follows that $(0, u^*) \in \operatorname{gr} T$, and so $0 \in \operatorname{dom} T$. Therefore Eq. (3.131) holds.

Suppose now that $\operatorname{int}(\operatorname{dom} T) \neq \emptyset$ (equivalently, $(\operatorname{dom} \chi_T)^i \neq \emptyset$). Since $\operatorname{int}(\operatorname{dom} T) \subset \operatorname{dom} T \subset \operatorname{dom} \chi_T$, from Eq. (3.131) and Theorem 1.1.2 we obtain that Eq. (3.132) holds. □

The next result refers to the local boundedness of monotone multifunctions. The multifunction $T : X \rightrightarrows X^*$ is called **locally bounded at** $\overline{x} \in X$ if there exists $\eta > 0$ such that $T(B(\overline{x}, \eta))$ is a bounded set; of course, T is **locally bounded** if T is locally bounded at any $x \in X$.

Theorem 3.11.14 *Let X be a Banach space and $T : X \rightrightarrows X^*$ be a monotone multifunction. Suppose that $x_0 \in (\operatorname{co}(\operatorname{dom} T))^i$. Then T is locally bounded at x_0.*

Proof. From Lemma 3.11.2 we have that $\operatorname{co}(\operatorname{dom} T) \subset \operatorname{dom} \chi_T$. Therefore $x_0 \in (\operatorname{dom} \chi_T)^i$, and so χ_T is continuous at x_0. Then there exists $\eta, \gamma > 0$ such that $\chi_T(x) \leq \gamma$ for every $x \in D(x_0, 2\eta)$. In particular,

$$\forall\, (x, x^*) \in \operatorname{gr} T,\ \|x - x_0\| \leq \eta,\ \forall\, u \in \eta U_X \ :\ \frac{\langle x + u, x^* \rangle - \langle x, x^* \rangle}{1 + \|x\|} \leq \gamma,$$

and so $\|x^*\| \leq \gamma\,(1 + \eta + \|x\|)\,/\eta =: \nu$. □

The following partial converse of the preceding theorem holds.

Theorem 3.11.15 *Let X be a Banach space and $T : X \rightrightarrows X^*$ be a maximal monotone multifunction for which $\overline{\operatorname{dom} T}$ is a convex set. If T is locally bounded at $\overline{x} \in \overline{\operatorname{dom} T}$, then $\overline{x} \in \operatorname{int}(\operatorname{dom} T)$.*

Proof. Let $C := \overline{\operatorname{dom} T}$; therefore C is a nonempty closed convex set. From our hypothesis there exists $\eta > 0$ such that $T(B(\overline{x}, \eta))$ is a bounded set. Since $\overline{x} \in \overline{\operatorname{dom} T}$, there exists $(x_n) \subset \operatorname{dom} T$ such that $(x_n) \to \overline{x}$; we can suppose that $(x_n) \subset B(\overline{x}, \eta)$. Let $(x_n^*) \subset X^*$ be such that $x_n^* \in T(x_n)$ for every $n \in \mathbb{N}$. Of course (x_n^*) is bounded, and so it has a subnet $(x_i^*)_{i \in I}$ which w^*-converges to $x^* \in X^*$. Let $(x_i)_{i \in I}$ be the corresponding subnet of (x_n). We have that $\langle x_i - y, x_i^* - y^* \rangle \geq 0$ for every $(y, y^*) \in \operatorname{gr} T$ and $i \in I$. Passing to the limit we obtain that $\langle \overline{x} - y, x^* - y^* \rangle \geq 0$ for every $(y, y^*) \in \operatorname{gr} T$. Since T is maximal monotone we have that $(\overline{x}, x^*) \in \operatorname{gr} T$, and so $\overline{x} \in \operatorname{dom} T$. Because $T(B(x,\eta/2)) \subset T(B(\overline{x},\eta))$ for every $x \in \overline{\operatorname{dom} T} \cap B(\overline{x},\eta/2)$, with similar arguments as above we obtain that

$$C \cap B(\overline{x},\eta/2) = \overline{\operatorname{dom} T} \cap B(\overline{x},\eta/2) \subset \operatorname{dom} T. \tag{3.138}$$

Suppose that $\overline{x} \notin \operatorname{int} C$, and so $\overline{x} \in \operatorname{Bd} C$. From Bishop–Phelps theorem, it follows that there exists $\widetilde{x} \in \operatorname{Bd} C \cap B(\overline{x},\eta/2)$ and $\widetilde{x}^* \in X^* \setminus \{0\}$ such that

$$\langle \widetilde{x} - x, \widetilde{x}^* \rangle \geq 0 \quad \forall x \in C \supset \operatorname{dom} T.$$

From Eq. (3.138) we have that $\widetilde{x} \in \operatorname{dom} T$; let $\widetilde{x}_0^* \in T(\widetilde{x})$. Taking into account the preceding relation, for $t > 0$ we have that

$$\langle \widetilde{x} - y, \widetilde{x}_0^* + t\widetilde{x}^* - y^* \rangle = \langle \widetilde{x} - y, \widetilde{x}_0^* - y^* \rangle + t \langle \widetilde{x} - x, \widetilde{x}^* \rangle \geq 0$$

for all $(y, y^*) \in \operatorname{gr} T$, and so $\{\widetilde{x}_0^* + t\widetilde{x}^* \mid t \geq 0\} \subset T(\widetilde{x})$; this contradicts the boundedness of $T(B(\overline{x},\eta))$. Therefore $\overline{x} \in \operatorname{int} C$; from Eq. (3.138) it follows immediately that $\overline{x} \in \operatorname{int}(\operatorname{dom} T)$. □

The following result shows that the converse of Theorem 2.4.13 holds in Banach spaces for lsc convex functions.

Corollary 3.11.16 *Let X be a Banach space and $f \in \Gamma(X)$.*

(i) *Let $\overline{x} \in \overline{\operatorname{dom} f}$. Then ∂f is locally bounded at $\overline{x}$ if and only if $\overline{x} \in \operatorname{int}(\operatorname{dom} f)$.*

(ii) *The following assertions are equivalent:* (a) *∂f is locally bounded at any $x \in \operatorname{dom} f$,* (b) *$\operatorname{dom} f$ is open,* (c) *f is continuous.*

(iii) *∂f is locally bounded at any $x \in X$ if and only if $\operatorname{dom} f = X$ and f is continuous.*

Proof. (i) The sufficiency is proved in Theorem 2.4.13. Assume that ∂f is locally bounded at $\overline{x}$. By the Brøndsted–Rockafellar theorem we have that $\overline{x} \in \overline{\operatorname{dom} \partial f}$, while from the Rockafellar theorem we have that ∂f is maximal monotone. Using the preceding theorem we obtain that $\overline{x} \in \operatorname{int}(\operatorname{dom} \partial f) \subset \operatorname{int}(\operatorname{dom} f)$.

(ii) (a) $\Leftrightarrow$ (b) is obvious from (i); (c) $\Rightarrow$ (b) because $\operatorname{dom} f = \{x \in X \mid f(x) < \infty\}$.

(b) $\Rightarrow$ (c) Note first that f is continuous at any $x \in X \setminus \operatorname{dom} f$ because f is lsc. The space X being complete and f lsc, by Theorem 2.2.20 we have that f is continuous at any $x \in (\operatorname{dom} f)^i = \operatorname{int}(\operatorname{dom} f)$. Therefore f is continuous. □

We close this section with the following important theorem.

Theorem 3.11.17 (Rockafellar) *Let X be a Banach space and $T : X \rightrightarrows X^*$ be a maximal monotone multifunction. Suppose that either X is reflexive and T is locally bounded at some point of* $\operatorname{dom} T$ *or* $\operatorname{int}(\operatorname{co}(\operatorname{dom} T)) \neq \emptyset$. *Then* $\operatorname{int}(\operatorname{dom} T) \neq \emptyset$ *and its closure is equal to* $\overline{\operatorname{dom} T}$. *Furthermore T is locally bounded at every point of* $\operatorname{int}(\operatorname{dom} T)$ *and is not locally bounded at any point of* $\operatorname{Bd}(\operatorname{dom} T)$.

Proof. If X is Banach space and $\operatorname{int}(\operatorname{co}(\operatorname{dom} T)) \neq \emptyset$, from Theorem 3.11.13 we obtain that $\operatorname{int}(\operatorname{dom} T) \neq \emptyset$ and $\overline{\operatorname{dom} T}$ is a convex set. The rest of the conclusion follows from Theorems 3.11.14 and 3.11.15.

If X is a reflexive Banach space, using Theorem 3.11.12, we have that $\overline{\operatorname{dom} T}$ is convex, and so, from Theorem 3.11.15, we have that $\operatorname{int}(\operatorname{dom} T) \neq \emptyset$. The rest of the conclusion follows from the first part. □

3.12 Exercises

Exercise 3.1 Consider the function

$$f : \ell^1 \to \mathbb{R}, \quad f(x) := \sum\nolimits_{k \geq 1} |x_k|^{1+1/k},$$

where $x = (x_k)_{k \geq 1}$. Prove that f is well-defined, continuous and strictly convex. Moreover show that f is Gâteaux differentiable on ℓ^1 with

$$\nabla f(x) = \left(\left(1 + k^{-1}\right) |x_k|^{1/k} \operatorname{sgn}(x_k) \right)_{k \geq 1} \qquad \forall x = (x_k)_{k \geq 1} \in \ell^1,$$

where $\operatorname{sgn}(\gamma) := 1$ for $\gamma \geq 0$ and $\operatorname{sgn}(\gamma) := -1$ for $\gamma < 0$, but f is nowhere Fréchet differentiable.

Exercise 3.2 Let $f : \ell^\infty \to \overline{\mathbb{R}}$ be defined by $f(x) := \limsup |x_n|$. Prove that $\operatorname{dom} f = \ell^\infty$, f is convex and continuous but nowhere Gâteaux differentiable.

Exercise 3.3 Let $(X, \|\cdot\|)$ be a normed space and $f \in \Lambda(X)$. Prove that the statements (i), (ii) and (iii) below are equivalent:

(i) f is uniformly convex;

(ii) $\forall \varepsilon > 0,\ \exists \delta > 0,\ \forall x, y \in \operatorname{dom} f,\ \|x - y\| \geq (>, =)\delta,\ \forall \lambda \in [0,1]$: $f(\lambda x + (1-\lambda)y) \leq \lambda f(x) + (1-\lambda) f(y) - \lambda(1-\lambda)\delta$;

(iii) $\forall \varepsilon > 0,\ \exists \delta > 0,\ \forall x, y \in \operatorname{dom} f,\ \|x - y\| \geq (>, =)\delta$: $f(\frac{1}{2}x + \frac{1}{2}y) \leq \frac{1}{2}f(x) + \frac{1}{2}f(y) - \delta$.

(iv) Prove that the statements obtained from (ii) and (iii) by fixing $x \in \operatorname{dom} f$ are equivalent to the fact that f is uniformly convex at x.

Exercise 3.4 Let $f : \ell^1 \to \mathbb{R}$ be defined by $f(x) := \sum_{n\geq 1} |x_k|^{1+1/k}$. By Exercise 3.1 f is strictly convex, continuous and Gâteaux differentiable but nowhere Fréchet differentiable. Prove that $\rho_{f,x}(t) = 0$ for every $t \geq 0$, $\vartheta_{f,x}(t) > 0$ for all $t > 0$ and $x \in A := \{x \in \ell^1 \mid \limsup_{k\to\infty} |x_k|^{1/k} < 1\}$, and $\vartheta_{f,x}(t) = 0$ for all $t > 0$ and $x \in \ell^1 \setminus A$.

Exercise 3.5 Let $(X, \|\cdot\|)$ be a normed space.

(i) Let $A := \{x_n^* \mid n \in \mathbb{N}\} \subset S_{X^*}$ and consider the function

$$f : X \to \mathbb{R}, \quad f(x) := \sum\nolimits_{n\geq 1} 2^{-n} \langle x, x_n^* \rangle^2 .$$

Prove that f is convex and uniformly Fréchet differentiable, and $\sqrt{f}$ is a continuous semi-norm. Moreover, f is strictly convex $\Leftrightarrow$ $\sqrt{f}$ is a norm $\Leftrightarrow$ $\operatorname{lin} A$ is w^*-dense in X^*. In particular $\sqrt{f}$ is a norm if A is norm-dense in S_{X^*}.

(ii) If X^* is $\|\cdot\|$-separable (or even w^*-separable), prove that there exists an equivalent norm $\|\cdot\|_1$ on X such that $(X, \|\cdot\|_1)$ is strictly convex. If X is reflexive and $\|\cdot\|$-separable, prove that there exists an equivalent norm $\|\cdot\|_2$ on X such that $(X, \|\cdot\|_2)$ is strictly convex and smooth.

Exercise 3.6 Let $\varphi \in \mathcal{A}$ (see page 195) and $p \in \mathbb{P}$. Show that the mappings $\mathbb{P} \ni t \mapsto t^{-1/p}\varphi^e(t)$ and $\mathbb{P} \ni t \mapsto t^{-1/p}\varphi^h(t)$ are nonincreasing (nondecreasing) if the mapping $\mathbb{P} \ni t \mapsto t^{-p}\varphi(t)$ is nondecreasing (nonincreasing).

Exercise 3.7 Let $C \subset (X, \|\cdot\|)$ be a nonempty closed convex set, $g := d_C$, and $\psi \in \Gamma$ (see page 195). Prove that

$$g^* = \iota_{U_{X^*}} + \iota_C^*, \quad (\psi \circ d_C)^* = \iota_C^* + \psi^\# \circ \|\cdot\|, \quad \left(\iota_C^* + \psi^\# \circ \|\cdot\|\right)^* = \psi \circ d_C.$$

Moreover, if $x_0 \in X \setminus C$ and $\overline{x} \in P_C(x_0)$ then $\partial g(x_0) = \{x^* \in S_{X^*} \cap N(C, \overline{x}) \mid \langle x_0 - \overline{x}, x^* \rangle = \|x_0 - \overline{x}\|\}$.

Exercise 3.8 Assume that X is a reflexive Banach space and $f \in \Gamma(X)$ is such that $S := \operatorname{argmin} f$ is nonempty and bounded. Prove that $\partial f(S)$ is a neighborhood of 0 (for the norm topology) if S is a set of weak sharp minima of F.

Exercise 3.9 Let X be a normed vector space, $g \in \Lambda(X)$ be such that $S := \{x \in X \mid g(x) \leq 0\}$ is nonempty and $h := \max(g, 0)$. We say that g is metrically regular at the nonempty set $\widehat{S} \subset S$ if there exist $\delta, \alpha > 0$ such that $d_S(x) \leq \alpha \cdot h(x)$ for all $x \in X$ with $d_{\widehat{S}}(x) \leq \delta$; when $\widehat{S} = \{z\}$, one says that g is metrically regular at z. Consider the following conditions:

(i) S is a set of weak sharp minima for h;

(ii) g is metrically regular at any nonempty set $\widehat{S} \subset S$;

(iii) g is metrically regular at S;

(iv) g is metrically regular at any $z \in S$.

Prove the following implications: (i) $\Leftrightarrow$ (ii) $\Leftrightarrow$ (iii) $\Rightarrow$ (iv). Moreover, if S is compact prove that (iv) $\Rightarrow$ (i).

Exercise 3.10 Let X be a normed space, $f \in \Gamma(X)$ and $\mu > 0$. Let us consider the *Baire regularization* of f defined by

$$f_B(\cdot,\mu) : X \to \overline{\mathbb{R}}, \qquad f_B(x,\mu) := \inf\{f(y) + \mu\|x-y\| \mid y \in X\}.$$

1) Prove that for every $\mu > 0$ the function $f_B(\cdot,\mu)$ is either identical $-\infty$, or finite, convex and Lipschitz with Lipschitz constant μ. Moreover, $f_B(\cdot,\mu_1) \le f_B(\cdot,\mu_2) \le f$ for $0 < \mu_1 < \mu_2$ and $\lim_{\mu\to\infty} f_B(x,\mu) = f(x)$ for every $x \in X$.

2) Let $x_0 \in X$ be fixed. Prove the equivalence of the following statements: (a) $\partial f(x_0) \cap \mu U_{X^*} \ne \emptyset$, (b) $f(x_0) = f_B(x_0,\mu)$, (c) $f_B(x_0,\mu) \in \mathbb{R}$ and $\partial f_B(x_0,\mu) = \partial f(x_0) \cap \mu U_{X^*}$.

Exercise 3.11 Let X be a normed space, $f \in \Gamma(X)$ and $\mu > 0$. Let us consider the *Moreau regularization* of f defined by

$$f_M(\cdot,\mu) : X \to \overline{\mathbb{R}}, \qquad f_M(x,\mu) := \inf\left\{ f(y) + \tfrac{\mu}{2}\|x-y\|^2 \mid y \in X\right\}.$$

1) Prove that for every $\mu > 0$ the function $f_M(\cdot,\mu)$ is finite, convex and continuous, $f_M(\cdot,\mu_1) \le f_M(\cdot,\mu_2) \le f$ if $0 < \mu_1 < \mu_2$ and $\lim_{\mu\to\infty} f_M(x,\mu) = f(x)$ for every $x \in X$. Moreover, if $x^* \in \partial f(x)$ then $f(x) - \frac{1}{2\mu}\|x^*\|^2 \le f_M(x,\mu)$.

In the sequel we assume that X is a reflexive Banach space.

2) Prove that there exists $J_\mu : X \to X$ (unique if X is also strictly convex) such that

$$\forall x \in X \ : \ f_M(x,\mu) = f\left(J_\mu(x)\right) + \tfrac{\mu}{2}\|x - J_\mu(x)\|^2;$$

furthermore, $\lim_{\mu\to\infty} J_\mu(x) = x$ and $\lim_{\mu\to\infty} f\left(J_\mu(x)\right) = f(x)$ for every $x \in \operatorname{dom} f$.

3) Assume that $\partial f(x) \ne \emptyset$. Prove that $\|x_\mu^*\| \le d_{\partial f(x)}(0)$ for all $\mu > 0$ and $x_\mu^* \in \partial f_M(x,\mu)$. Moreover, if $x_\mu^* \in \partial f_M(x,\mu)$ for all $\mu > 0$ and $x_{\mu_i}^* \xrightarrow{w^*} x^*$ for some net $(\mu_i)_{i\in I} \to \infty$ then $x^* \in P_{\partial f(x)}(0)$; conclude that $\lim_{\mu\to\infty}\|x_\mu^*\| = d_{\partial f(x)}(0)$.

4) Assume that X is smooth. Prove that $f_M(\cdot,\mu)$ is Gâteaux differentiable on X and $\nabla f_M(x,\mu) = \mu\Phi_X(x - J_\mu(x))$ for every $x \in X$, and $w^* - \lim_{\mu\to\infty}\nabla f_M(x,\mu) = x_0^*$ if $\partial f(x) \ne \emptyset$, where x_0^* is the unique element of minimal norm of $\partial f(x)$. Moreover, if X^* has the Kadec–Klee property (see p. 233) then $x_0^* = \lim_{\mu\to\infty}\nabla f_M(x,\mu)$.

5) Prove that $f_M(\cdot,\mu)$ is Fréchet differentiable on X and $\nabla f_M(\cdot,\mu)$ is continuous when X is locally uniformly smooth.

Exercise 3.12 Let $\varphi \in \mathcal{A}_0 \cap N_0$ and $\psi = \overline{\operatorname{co}}\varphi$. Prove that the following statements are equivalent: 1) there exists $x > 0$ such that $\psi(x) > 0$, 2) $\psi(x) > 0$ for every $x > 0$, 3) $\liminf_{x\to\infty}\varphi(x)/x > 0$.

Exercise 3.13 Let $(X, (\cdot \mid \cdot))$ be a Hilbert space and $K \subset X$ be a closed convex cone. For every $x \in X$ we denote by $P_K(x)$ the projection of x on K; $P_K(x)$ exists and is unique. Let us consider $K^- := -K^+ = \{y \in X \mid \forall x \in K \ : \ (x \mid y) \leq 0\}$. Prove that

$$\forall x \in X \ : \ x = P_K(x) + P_{K^-}(x) \ \text{ and } \ (P_K(x) \mid P_{K^-}(x)) = 0.$$

Conversely, prove that $x_1 = P_K(x)$ and $x_2 = P_{K^-}(x)$ if $x = x_1 + x_2$ with $x_1 \in K$, $x_2 \in K^-$ and $(x_1 \mid x_2) = 0$.

Exercise 3.14 Let X, Y be Hilbert spaces, $A \in \mathcal{L}(X, Y)$ be a surjective operator, $C \subset Y$ be a nonempty closed convex set and $y_0 \in Y$ be fixed.

(a) Prove that the operator $T := A \circ A^*$ is bijective (X^* and Y^* being identified with X and Y, respectively).

(b) Prove that the problem (P) min $\|x\|^2$, $Ax + y_0 \in C$, has a unique solution. Moreover, $\overline{x} \in X$ is the solution of (P) if and only if $\overline{x} = (A^* \circ T^{-1})(\overline{y} - y_0)$, where $\overline{y}$ is the solution of the equation $\overline{y} = P_C(\overline{y} - \rho T^{-1}\overline{y} + \rho T^{-1} y_0)$ for some (any) $\rho > 0$.

Exercise 3.15 Let X be a reflexive Banach space and $g \in \Gamma(X)$ be a coercive function, *i.e.* $\lim_{\|x\| \to \infty} g(x) = \infty$. Prove that $g \in \mathcal{F}$ (see Eq. (3.73) on page 252).

Exercise 3.16 Let X be a reflexive Banach space and $g \in \Gamma(X)$. Prove that $g \in \mathcal{F}$ whenever (a) $0 \in (\operatorname{dom} g^*)^i$, or more generally, (b) $0 \in {}^i(\operatorname{dom} g^*)$ and $\operatorname{lin}(\operatorname{dom} g^*)$ is w^*-closed.

Exercise 3.17 Let $(X, \|\cdot\|)$ be a normed space, $g \in \Gamma(X)$ and $A \subset X$ be a closed convex set such that $C := A \cap [g \leq 0]$ is nonempty. Assume that $g_\infty(\overline{u}) < 0$ for some $\overline{u} \in A_\infty \cap S_X$. Prove that for $\alpha := (-g_\infty(\overline{u}))^{-1}$ one has

$$d_C(x) \leq \alpha \cdot g_+(x) \quad \forall x \in A.$$

Exercise 3.18 Let $g_1, g_2, g_3 : \mathbb{R}^2 \to \mathbb{R}$ be defined by:

$$g_1(x, y) := \sqrt{x^2 + y^2} - x - 1, \quad g_2(x, y) := y^2 - 2x - 1, \quad g_3 := \max\{g_1, g_2\}.$$

The functions g_1, g_2, g_3 are convex, g_1 and g_2 being differentiable on $\mathbb{R}^2 \setminus \{(0, 0)\}$ and $\mathbb{R}^2$, respectively. Prove that

$$r_{g_1}(t) = k_{g_1}(t) = 0 \ \forall t > -1 = \inf g_1, \quad r_{g_2}(t) = k_{g_2}(t) = 2 \ \forall t > -\infty = \inf g_2,$$
$$r_{g_3}(0) = r_{g_1}(0) = 0, \ k_{g_3}(0) = k_{g_2}(0) = 2,$$

where r_g and k_g are defined on page 255.

3.13 Bibliographical Notes

Section 3.1: We used the book [Phelps (1989)] in the presentation of the theorems of Borwein, Brøndsted–Rockafellar and Bishop–Phelps, as well as Proposition 3.1.3; in the statement of Borwein's theorem we added an estimation given in [Penot (1996b), Prop. 1]. The statement (i) of Theorem 3.1.4 can be found in [Phelps (1993)], while (ii) is established by Attouch and Beer (1993). For Simons' theorem and the proof of the Rockafellar theorem we used [Simons (1994b); Simons (1994a)]. Note that in many books one prefers to give the proof of Rockafellar's theorem in reflexive Banach spaces using Theorem 3.11.7. Proposition 3.1.5 is stated by Zălinescu (1999); for $X = Y$, $A = \mathrm{Id}_X$ and $F(x,y) = f(x)+g(y)$, Attouch and Théra (1996) (for X reflexive) and Simons (1998b) (in an equivalent formulation) obtained the weaker formula $\mathrm{int}(\mathrm{dom}\, f - \mathrm{dom}\, g) = \mathrm{int}(\mathrm{dom}\, \partial f - \mathrm{dom}\, \partial g)$. Theorems 3.1.6 and 3.1.7 are obtained by Penot (1996b). A result similar to Theorems 3.1.7 is obtained by Thibault (1995a).

Section 3.2: The results on the Clarke tangent cone, the Clarke–Rockafellar directional derivative and the Clarke subdifferential can be found in the books [Clarke (1983); Rockafellar (1981)]. Theorem 3.2.5 (with the conclusions (i), (iii) and (iv)) was originally stated by Zagrodny (1988) for the Clarke–Rockafellar subdifferential and $r = f(b)$. The present statement subsumes those in [Thibault (1995b)] and [Aussel *et al.* (1995), Th. 4.2]; the proof follows the lines of the corresponding result in [Thibault (1995b)]. Theorem 3.2.7 was stated for the first time by Poliquin (1990) in $\mathbb{R}^n$ for ∂_C; the present statement subsumes the corresponding results in [Luc (1993), Th. 3.1], [Correa *et al.* (1994)] and [Aussel *et al.* (1995), Th. 5.5] (see also Remark 3.2.3). The proof of Theorem 3.2.8 belongs to Luc (1993). The statements and proofs of Theorem 3.2.9 and Corollaries 3.2.10, 3.2.11 are from [Thibault and Zagrodny (1995)]; there it is assumed that the subdifferential $\overline{\partial}$ satisfies several conditions (among them conditions (P4) and (P5)) which imply our condition (P1). Note that Corollary 3.2.11 is a famous result of Rockafellar (1966).

Section 3.3: The lowest and greatest quasi-inverses were introduced by Penot and Volle (1990). The equivalence of (i) and (ii) of Theorem 3.3.2 for the Fréchet bornology can be found in the book [Phelps (1989)], while that for the Gâteaux bornology in [Giles (1982)]; the equivalence of conditions (i), (iii), and (v) for the Gâteaux and Fréchet bornologies can be found in [Giles (1982)]; the equivalence of (i), (vi), (vii) and (viii) are obtained by Borwein and Vanderwerff (2000). Corollaries 3.3.4 and 3.3.5 are stated by Asplund and Rockafellar (1969).

Section 3.4: In writing this section we followed [Cornejo *et al.* (1997)]. Note that Proposition 3.4.1 was obtained by Penot (1996a) under more general conditions. Corollary 3.4.4 is established, mainly, by Zălinescu (1983b). Theorem 3.4.3 generalizes Corollary 3.4.4 to the case when $\mathrm{argmin}\, f$ is not a singleton; the equivalence of conditions (i), (ii), (v) and (viii) is established by Lemaire (1992). For a detailed study of the problems considered in this section for the nonconvex case see [Penot (1998b)]. Proposition 3.4.5 is established by Butnariu and Iusem (2000) for Fréchet differentiable functions.

Section 3.5: The results of this section are mainly from [Zălinescu (1983b)] and [Azé and Penot (1995)]. The strongly convex functions were introduced by Polyak (1966), the uniformly convex functions by Levitin and Polyak (1966) and the uniformly smooth convex functions by Azé and Penot (1995); the notion of uniformly smooth function introduced by Shioji (1995) coincides, for convex functions, with that of Azé and Penot (1995). Proposition 3.5.1 was established by Vladimirov *et al.* (1978), while Propositions 3.5.2, 3.5.3, Corollary 3.5.4 and Theorem 3.5.5 were established by Azé and Penot (1995). Theorem 3.5.6 is new for X a general normed space. The equivalences of the conditions of Theorem 3.5.12 to which one adds the conditions (iv) and (v) of Theorem 3.5.6 are established by Azé and Penot (1995), too. Theorem 3.5.10 and the implication (i) $\Rightarrow$ (iii) of Proposition 3.6.4 were established, in reflexive Banach spaces, by Zălinescu (1983b). Theorem 3.5.13 and its corollary shows that the existence of uniformly convex or uniformly smooth convex functions on a Banach space is very close to the reflexivity of the space; Theorem 3.5.13 was established by Vladimirov *et al.* (1978) under the supplementary hypothesis that the function f is finite valued and bounded on bounded sets. The equivalence of conditions (i) and (vi) in Corollary 3.5.7 is mentioned in [Borwein and Preiss (1987)], while the equivalence of conditions (iii), (v) and (vi) (see also Remark 3.5.2) can be found in [Hiriart-Urruty (1998)] for $X = \mathbb{R}^n$, $p = 2$ and f Fréchet differentiable. The equivalence of conditions (i), (iii) and (v) in Corollary 3.5.11 was established by Rockafellar (1976) in Hilbert spaces for $p = 2$ with the same constant c (see also Remark 3.5.3). Applications to the study of uniformly convex and locally uniformly convex Banach spaces can be found in [Vladimirov *et al.* (1978); Zălinescu (1983b); Azé and Penot (1995)].

Sections 3.6, 3.7: The results of Section 3.6 are new. The characterizations of the geometrical properties of Banach spaces presented in Section 3.7 are classical for $\varphi(t) = t$ and can be found in the books [Diestel (1975)] and/or [Cioranescu (1990)]; for general weight functions one can find Theorem 3.7.2 and Theorem 3.7.8 ((i) $\Leftrightarrow$ (iii) $\Leftrightarrow$ (iv)) in [Cioranescu (1990)]. Note that Proposition 3.7.10 is stated in [Cioranescu (1990)] under the supplementary assumption that the norm of X is Fréchet differentiable. Theorem 3.7.7 was established in [Zălinescu (1983b)] for slightly more general functions φ; when $\varphi(t) = \frac{1}{p}t^p$ ($p > 1$), (ii) was established in [Vladimirov *et al.* (1978)] and [Azé and Penot (1995)].

Section 3.8: The results on best approximation are classical. Most of them can be found in the books [Holmes (1972); Precupanu (1992)]; the other results are folklore. However we mention that the formula (3.66) for the directional derivative of the distance function was obtained by Lewis and Pang (1998)) in Euclidean spaces.

Section 3.9: The implication (a) $\Rightarrow$ (c) of Theorem 3.9.1 is the main result of [Soloviov (1993)] (see also [Soloviov (1997)]); the convex case is established (even in a more general framework) in [Asplund and Rockafellar (1969)] (see also

[Dontchev and Zolezzi (1993)]). The other results in this section can be found in the survey paper [Soloviov (1997)]. Apparently Theorem 3.9.3 was stated by Vlasov (1972), while Theorem 3.9.4 was stated by Efimov and Stechkin (1959).

Section 3.10: The presentation of this section follows, mainly, [Zălinescu (2001)]. The notion of set of weak sharp minima, in the present form, was introduced by Burke and Ferris (1993), generalizing the notion of sharp minimum point introduced by Polyak (1980); in the same paper Polyak introduces the notion of set of weak sharp minima in a slightly different form. The equivalence of conditions (i), (vi), (ix) and (x) is stated by Burke and Ferris (1993) for $\dim X < \infty$; (i) $\Leftrightarrow$ (x) is mentioned in [Jourani (2000)] for X a Hilbert space for $f \in \Lambda(X)$ with S closed (an inspection of the proof shows that this is true also in our case). The equivalence of (i) and (ii) of Theorem 3.10.1 is established by Lemaire (1989). The proof for the equivalence of (i), (ii) and (v) is taken from [Cornejo *et al.* (1997)]. The class $\mathcal{F}$ of functions defined by Eq. (3.73) was introduced and studied by Auslender and Crouzeix (1989) in finite dimensional spaces. The numbers $l_f(t)$, $k_f(t)$ and $r_f^5(t)$ were introduced in [Auslender and Crouzeix (1989)] (as in Eqs. (3.79) and (3.82)); in this paper Propositions 3.10.3, 3.10.8, Eq. (3.84) of Proposition 3.10.7 and Theorem 3.10.10 were stated, too. Formula (3.84) Auslender *et al.* (1993) simplify several proofs from [Auslender and Crouzeix (1989)] and mention that those results hold in reflexive Banach spaces; one can find the reflexive case of Proposition 3.10.8 and Theorem 3.10.10 in [Cominetti (1994)]. Proposition 3.10.11 and Theorem 3.10.12 are from [Dolecki and Angleraud (1996)]. The problem of global error bounds for convex inequalities is studied in many articles. The results stated in this section are related to those in the articles [Auslender and Crouzeix (1988); Deng (1997); Deng (1998); Lewis and Pang (1998); Klatte and Li (1999)]. Excepting [Deng (1997); Deng (1998)], the results in the above mentioned papers are stated in finite dimensional spaces. Klatte and Li (1999) introduced the numbers $r_f^i(0)$ $(1 \le i \le 6)$ for f finite valued with $0 > \inf f$ and established Corollary 3.10.6. The implications (i) $\Rightarrow$ (vi) and (ii) $\Rightarrow$ (vi) of Theorem 3.10.13 can be found in [Deng (1997)] and [Deng (1998)], respectively, while the other ones can be found in [Klatte and Li (1999)] (in the mentioned framework of finite valued functions on $\mathbb{R}^n$). Theorem 3.10.14 and Propositions 3.10.15, 3.10.16 were established by Lewis and Pang (1998).

Section 3.11: Coodey and Simons (1996) associated an adequate convex function to a multifunction. Simons (1998b) gave an interesting approach of monotone multifunctions theory by using these associated convex functions. Excepting Theorem 3.11.15, which is an amelioration of a result in [Azé and Penot (1995)], its corollary, and Theorem 3.11.17, which is established in [Rockafellar (1969)] (with a different proof), all the results can be found in the book [Simons (1998a)] and partially in [Coodey and Simons (1996); Simons (1998b)].

Exercises: Exercises 3.1 and 3.4 are from [Butnariu and Iusem (2000)], Exercise 3.2 is from [Phelps (1989)], Exercises 3.3 and 3.12 are from [Zălinescu (1983b)], for more general renorming theorems than those in Exercise 3.5 see

[Deville *et al.* (1993)], the second formula from Exercise 3.7 is proved in [Cornejo *et al.* (1997)], Exercise 3.8 is from [Burke and Ferris (1993)] (there $X = \mathbb{R}^n$), Exercise 3.9 is from [Deng (1998)] (with g finite and continuous), Exercise 3.10 is from [Borwein and Vanderwerff (1995)], Exercise 3.11 (without (v)) is from [Barbu and Precupanu (1986)], Exercise 3.13 is a classical result from [Moreau (1965)], Exercise 3.14 is from [Kassara (2000)], Exercise 3.15 (for $X = \mathbb{R}^n$) is from [Auslender and Crouzeix (1989)], Exercise 3.17 for X a Banach space can be found in [Deng (1997)] for g finite and continuous and in [Jourani (2000)] for g satisfying the condition $\partial(g+\iota_A)(x) = \partial g(x)+N(A,x)$ for all $x \in (A \cap \operatorname{dom} g)\setminus C$, the functions considered in Exercise 3.18 are from [Klatte and Li (1999)].

Exercises – Solutions

Exercise 1.1 Let $x \in \operatorname{co} A$ and consider $P := \{m \in \mathbb{N} \mid \exists\, \lambda_1, \ldots, \lambda_m \in \mathbb{P}, x_1, \ldots, x_m \in A : \sum_{k=1}^{m} \lambda_k = 1, \sum_{k=1}^{m} \lambda_k x_k = x\}$. Of course, taking into account the formula for $\operatorname{co} A$ on page 2, $P \neq \emptyset$. Let $p := \min P \in \mathbb{N}$. Assume that $p > n+1$. Since $p \in P$, there exist $\lambda_1, \ldots, \lambda_p \in \mathbb{P}$ and $x_1, \ldots, x_p \in A$ such that $\sum_{k=1}^{p} \lambda_k = 1$ and $\sum_{k=1}^{p} \lambda_k x_k = x$. Since $x_1 - x_p, \ldots, x_{p-1} - x_p$ are linearly dependent, there exists $(\mu_1, \ldots, \mu_{p-1}) \in \mathbb{R}^{p-1} \setminus \{0\}$ such that $\sum_{k=1}^{p-1} \mu_k (x_k - x_p) = 0$, or equivalently, there exists $(\mu_1, \ldots, \mu_p) \in \mathbb{R}^p \setminus \{0\}$ such that $\sum_{k=1}^{p} \mu_k = 0$ and $\sum_{k=1}^{p} \mu_k x_k = 0$. It follows that $x = \sum_{k=1}^{p} (\lambda_k + t\mu_k) x_k$ for every $t \in \mathbb{R}$. Taking $t = \bar{t} := \min\{-\lambda_k/\mu_k \mid \mu_k < 0\} = -\lambda_{\bar{k}}/\mu_{\bar{k}} > 0$, we obtain that $x = \sum_{k \neq \bar{k}} (\lambda_k + \bar{t}\mu_k) x_k$, and so $\min P \leq p-1$, a contradiction.

Exercise 1.2 Without loss of generality we suppose that $0 \in A$. Let $X_0 := \operatorname{lin} A$. If $X_0 = \{0\}$, then $A = \{0\}$ and the properties of the statement are obvious. Suppose that $\dim X_0 \geq 1$. Taking into account that $\dim X_0 \leq \dim X < \infty$, X_0 is a closed linear subspace, and so we may replace X by X_0, which is the same as saying that $\operatorname{lin} A = X$; in this situation ${}^iA = A^i$. Since X is finite dimensional and A is convex, to prove the relations of the statement it is sufficient, taking into account Theorem 1.1.2, to verify that $A^i \neq \emptyset$.

Since $\operatorname{lin} A = X$, there exists a base $\{e_1, \ldots, e_k\}$, $k \geq 1$, of X with elements of A. The element $\bar{x} := \frac{1}{2k}(e_1 + \cdots + e_k) \in A^i$. Indeed, let $x = \lambda_1 e_1 + \cdots + \lambda_k e_k \in X$ and $\delta := \min\{\frac{1}{2k|\lambda_1|}, \ldots, \frac{1}{2k|\lambda_k|}\}$; for every $t \in]-\delta, \delta[$ we have that $\bar{x} + tx \in A$. Therefore $\bar{x} \in A^i$.

Exercise 1.3 Doing a translation, we may suppose that $0 \in H$, and so $H = \ker x^*$ for some $x^* \in X^* \setminus \{0\}$. Let $x_0 \in \operatorname{int}_H(A \cap H)$ and $x_1 \in A \setminus H$; without loss of generality we suppose that $\langle x_1, x^* \rangle = 1$. From the hypothesis, there exists $U \in \mathcal{N}_X^c$ such that $(x_0 + U) \cap H = x_0 + U \cap H \subset A$. Let $U_0 := \{x \in U \mid |\langle x, x^* \rangle| \leq 1/4\} \in \mathcal{N}_X^c$. Consider $\mu > 0$ such that $\mu(x_0 - x_1) \in U_0$ and take $0 < \gamma \leq \frac{\mu}{2(\mu+1)}$.

Let $u \in U_0$ be fixed. Then

$$\tfrac{1}{2}x_0 + \tfrac{1}{2}x_1 + \gamma u = \left(\tfrac{1}{2} - \gamma \langle u, x^* \rangle\right)(x_0 + u') + \left(\tfrac{1}{2} + \gamma \langle u, x^* \rangle\right) x_1, \qquad (^*)$$

where $u' := \frac{\gamma}{1/2-\gamma\langle u,x^*\rangle}u + \frac{\gamma\langle u,x^*\rangle}{1/2-\gamma\langle u,x^*\rangle}(x_0 - x_1)$; it follows immediately that $u' \in \ker x^* = H$. But

$$\frac{\gamma}{1/2 - \gamma \langle u, x^* \rangle} + \frac{\gamma\,|\langle u, x^* \rangle|}{\mu(1/2 - \gamma \langle u, x^* \rangle)} \le 1 \Leftrightarrow \gamma\,(\mu + |\langle u, x^* \rangle| + \langle u, x^* \rangle) \le \frac{\mu}{2},$$

the last inequality being valid for our choice of γ. Since 0, u and $\mu(x_0 - x_1)$ are in the convex set U_0, we have that $u' \in U$, and so $x_0 + u' \in A$. From $(^*)$ we obtain that $\frac{1}{2}x_0 + \frac{1}{2}x_1 + \gamma u \in A$ by the convexity of A. Therefore $\frac{1}{2}x_0 + \frac{1}{2}x_1 + \gamma U_0 \subset A$, which proves that $\operatorname{int} A \neq \emptyset$.

Let now $M \subset X$ be as in the statement of the exercise. If $A \subset M$ it is nothing to prove. Otherwise let $x_1 \in A \setminus M$ and consider $M_1 := \operatorname{aff}(M \cup \{x_1\})$. Doing a translation, we assume that $x_1 = 0$. Therefore $M_1 = \operatorname{lin} M = (M - x_0) + \mathbb{R}x_0$ for some $x_0 \in M$. By the Dieudonné theorem (Theorem 1.1.8) M_1 is a closed linear subspace of X. Since M is a closed hyperplane of M_1 and $(A \cap M_1) \cap M = A \cap M$, by what precedes we have that $\operatorname{int}_{M_1}(A \cap M_1) \neq \emptyset$. If $M_1 = \operatorname{aff} A$ the proof is complete. Otherwise, since M_1 is a closed affine subset of X in a finite number of steps ($\le \operatorname{codim} M$) we obtain that $\operatorname{rint} A = \operatorname{int}_{\operatorname{aff} A} A \neq \emptyset$.

Exercise 1.4 Since $A + \operatorname{int} C = \bigcup_{a \in A}(a + \operatorname{int} C)$, the set $A + \operatorname{int} C$ is open. Hence $A + \operatorname{int} C \subset \operatorname{int}(A + C)$. Let $\overline{x} \notin A + \operatorname{int} C$. Since the set $A + \operatorname{int} C$ is a convex set with nonempty interior, there exists $x^* \in X^* \setminus \{0\}$ such that

$$\forall a \in A,\ \forall c \in \operatorname{int} C\ :\ \langle \overline{x}, x^* \rangle \le \langle a, x^* \rangle + \langle c, x^* \rangle.$$

Since $C \subset \operatorname{cl}(\operatorname{int} C)$ we have that $\langle \overline{x}, x^* \rangle \le \langle x, x^* \rangle$ for every $x \in A + C$. Therefore $\overline{x} \notin \operatorname{int}(A + C)$, which proves that $\operatorname{int}(A + C) = A + \operatorname{int} C$.

Suppose now that $A \cap \operatorname{int} C \neq \emptyset$. The inclusion $\operatorname{cl}(A \cap C) \subset \operatorname{cl} A \cap \operatorname{cl} C$ is obvious (and true without any hypothesis on A and C); so it is sufficient to prove the converse inclusion. Without loss of generality, we suppose that $0 \in A \cap \operatorname{int} C$ (doing, if needed, a translation). Let p_C be the Minkowski gauge associated to C. By Proposition 1.1.1 p_C is continuous and

$$\operatorname{int} C = \{x \in X \mid p_C(x) < 1\}, \quad \operatorname{cl} C = \{x \in X \mid p_C(x) \le 1\}.$$

Let $x \in \operatorname{cl} A \cap \operatorname{cl} C$; there exists $(x_n) \subset A$ such that $(x_n) \to x$. Since p_C is continuous, we have $p_C(x_n) \to p_C(x) \le 1$. If the set $P := \{n \in \mathbb{N} \mid p_C(x_n) < 1\}$ is infinite, then $x_n \in A \cap C$ for every $n \in P$, whence $x = \lim_{n \in P} x_n \in \operatorname{cl}(A \cap C)$. If P is finite, there exists $n_0 \in \mathbb{N}$ such that $p_C(x_n) \ge 1$ for every $n \ge n_0$. Since $x \in \operatorname{cl} C$, we have that $p_C(x_n) \to 1$. For every $n \ge n_0$, consider $\overline{x}_n := \frac{n}{np_C(x_n)+1}x_n$. Since A is convex and $0, x_n \in A$, we have that $\overline{x}_n \in A$. Since

$p_C(\overline{x}_n) = np_C(x_n)/(np_C(x_n)+1) < 1$, we have that $\overline{x}_n \in A \cap C$ for every $n \geq n_0$. But $(x_n) \to x$, and so $x \in \mathrm{cl}(A \cap C)$ in this case, too. Therefore $\mathrm{cl}(A \cap C) = \mathrm{cl}\, A \cap \mathrm{cl}\, C$.

Suppose now that C is a convex cone with nonempty interior; we have $\mathrm{int}\, C = \mathrm{int}(\mathrm{cl}\, C)$. Using what we proved above, taking $A = \mathrm{cl}\, C$ we have that

$$\mathrm{cl}\, C + \mathrm{int}\, C = \mathrm{cl}\, C + \mathrm{int}(\mathrm{cl}\, C) = \mathrm{int}(\mathrm{cl}\, C + \mathrm{cl}\, C) = \mathrm{int}(\mathrm{cl}\, C) = \mathrm{int}\, C.$$

Exercise 1.5 (a) It is obvious that $X_0 + C$ is a convex cone. Let us show by mathematical induction on p that $X_0 + C$ is closed.

Let $p = 1$. If $a_1 \in X_0$ we have that $X_0 + C = X_0$, and so $X_0 + C$ is closed. Suppose that $a_1 \notin X_0$. Let us consider $(x_n) \subset X_0 + C$, $(x_n) \to x \in X$. Then there exist $(y_n) \subset X_0$, $(t_n) \subset \mathbb{R}_+$ such that $x_n = y_n + t_n a_1$ for every $n \in \mathbb{N}$. If $(t_n) \to \infty$ then $\left(-\frac{1}{t_n} y_n\right) \to a_1$, whence we get the contradiction $a_1 \in \mathrm{cl}\, X_0 = X_0$. Therefore (t_n) contains a bounded subsequence, and so it contains a subsequence $(t_{n_k})_{k \in N}$ convergent to $t \in \mathbb{R}_+$. Therefore $(y_{n_k}) \to x - ta_1 =: y \in X_0$, which proves that $x = y + ta_1 \in X_0 + C$. Therefore the statement is proved for $p = 1$.

Suppose now that the statement is true for $p \geq 1$; we are going to prove it for $p+1$. As for $p = 1$, consider two situations. To begin with, suppose that $-a_1, \ldots, -a_{p+1} \in X_0 + C$; then, as one can easily verify, $X_0 + C = X_0 + \mathrm{lin}\{a_1, \ldots, a_{p+1}\}$. Since the dimension of $\mathrm{lin}\{a_1, \ldots, a_{p+1}\}$ is less than or equal to $p + 1 < \infty$, using Theorem 1.1.8, we obtain that $X_0 + C$ is a closed linear subspace. Suppose now that there exists i such that $-a_i \notin X_0 + C$; without loss of generality, we assume that $i = p + 1$. Let

$$\widetilde{C} := \left\{ \sum\nolimits_{i=1}^{p} \lambda_i a_i \;\middle|\; \forall i,\ 1 \leq i \leq p \ :\ \lambda_i \geq 0 \right\}.$$

By the hypothesis of mathematical induction we have that $X_0 + \widetilde{C}$ closed. Let $(x_n) \subset X_0 + C$, $(x_n) \to x \in X$. There exist $(y_n) \subset X_0 + \widetilde{C}$, $(t_n) \subset \mathbb{R}_+$ such that $x_n = y_n + t_n a_{p+1}$ for every $n \in \mathbb{N}$. If $(t_n) \to \infty$ then $X_0 + \widetilde{C} \ni \frac{1}{t_n} y_n \to -a_{p+1}$, whence $-a_{p+1} \in X_0 + \widetilde{C} \subset X_0 + C$, which is a contradiction. Therefore (t_n) contains a subsequence $(t_{n_k})_{k \in N}$ convergent to $t \in \mathbb{R}_+$. Therefore $(y_{n_k}) \to x - ta_{p+1} =: y \in X_0 + \widetilde{C}$ which proves that $x = y + ta_{p+1} \in X_0 + C$. Therefore the statement is true for $p + 1$. Thus we obtained that $X_0 + C$ is a closed set for every $p \geq 1$.

(c) Let $K := \{x \in X \mid \forall i \in \overline{1,k}\ :\ \langle x, \varphi_i \rangle \leq \alpha_i\}$. If the set K is empty, the conclusion is obvious. Suppose that $K \neq \emptyset$. Consider the functions

$$T, A : X \to \mathbb{R}^k, \quad T(x) := (\langle x, \varphi_1 \rangle, \ldots, \langle x, \varphi_k \rangle), \quad A(x) := \alpha - T(x),$$

where $\alpha := (\alpha_1, \ldots, \alpha_k) \in \mathbb{R}^k$. It is obvious that T is a continuous linear operator. Therefore A is continuous and $T(X_0)$ is a linear subspace of $\mathbb{R}^k$. It follows that $T(X_0)$ is closed. Let us consider $P := \{y \in \mathbb{R}^k \mid \forall i \in \overline{1,k}\ :\ y_i \geq 0\}$; then

$K = A^{-1}(P)$ and $A(K) \subset P$. Let $(x_n) \subset X_0 + K$, $(x_n) \to x \in X$. There exist $(y_n) \subset X_0$ and $(z_n) \subset K$ such that $x_n = y_n + z_n$ for every $n \in \mathbb{N}$. Therefore $T(X_0) + P \ni A(z_n) - T(y_n) = A(x_n) \to A(x)$. By (b) we have that $T(X_0) + P$ is closed, whence $A(x) \in T(X_0) + P$. Therefore there exists $x_0 \in X_0$ such that $A(x + x_0) \in P$, *i.e.* $x \in X_0 + K$. Hence $X_0 + K$ is a closed set.

Exercise 1.6 Without loss of generality, we assume that $\|x_0\| = 1$. It is clear that $\lambda x \in P(\alpha)$ if $\lambda \in \mathbb{R}$ and $x \in P(\alpha)$. Let $x, y \in P(\alpha)$; we have that

$$(x \mid x_0) \geq \|x\| \cdot \cos\alpha, \quad (y \mid x_0) \geq \|y\| \cdot \cos\alpha,$$

and so

$$(x + y \mid x_0) \geq (\|x\| + \|y\|) \cdot \cos\alpha \geq \|x + y\| \cdot \cos\alpha.$$

Therefore $x + y \in P(\alpha)$. It is evident that $P(\alpha)$ is closed.

Note that

$$\begin{aligned} P(\alpha) &= \{x = \lambda x_0 + y \in X \mid \lambda = (x \mid x_0),\ y = x - \lambda x_0,\ (x \mid x_0) \geq \|x\| \cdot \cos\alpha\} \\ &= \left\{\lambda x_0 + y \,\middle|\, \lambda \geq 0,\ (y \mid x_0) = 0,\ \lambda \geq \sqrt{\lambda^2 + \|y\|^2} \cdot \cos\alpha \right\} \\ &= \{\lambda x_0 + y \mid \lambda \geq 0,\ (y \mid x_0) = 0,\ \lambda \sin\alpha \geq \|y\| \cdot \cos\alpha\}. \end{aligned}$$

Therefore

$$P(0) = \{\lambda x_0 \mid \lambda \geq 0\}, \quad P(\pi/2) = \{\lambda x_0 + y / \lambda \geq 0,\ (y \mid x_0) = 0\}.$$

Since $P(\alpha)$ is a cone, we have that $P(\alpha)^\circ = P(\alpha)^+$. To begin with, let us show that $P(0)^+ = P(\pi/2)$. Let $x = \lambda x_0$ and $y = \mu x_0 + v$, with $\lambda, \mu \geq 0$, $(v \mid x_0) = 0$. Then $(x \mid y) = \lambda\mu \geq 0$, an so $P(\pi/2) \subset P(0)^+$. Let now $y \in P(0)^+$. There exists $\mu \in \mathbb{R}$ and $v \in X$ such that $(v \mid x_0) = 0$, $y = \mu x_0 + v$. Then $(x_0 \mid \mu x_0 + v) = \mu \geq 0$, which proves that $y \in P(\pi/2)$. Therefore $P(0)^+ = P(\pi/2)$.

Let us consider now $\alpha \in]0, \frac{\pi}{2}[$. Let $x \in P(\alpha)$ and $y \in P(\pi/2 - \alpha)$. By what was proved above, there exist $\lambda, \mu \geq 0$, $u, v \in X$ such that

$$\begin{aligned} x &= \lambda x_0 + u, \quad (u \mid x_0) = 0,\ \lambda \sin\alpha \geq \|u\| \cdot \cos\alpha, \\ y &= \mu x_0 + v, \quad (v \mid x_0) = 0,\ \mu \cos\alpha \geq \|v\| \cdot \sin\alpha. \end{aligned}$$

Therefore $\lambda\mu \geq \|u\| \cdot \|v\|$. We obtain that

$$(x \mid y) = \lambda\mu + (u \mid v) \geq \lambda\mu - \|u\| \cdot \|v\| \geq 0.$$

It follows that $y \in P(\alpha)^+$, whence $P(\pi/2 - \alpha) \subset P(\alpha)^+$.

Let now $y \in P(\alpha)^+$; there exists $\mu \in \mathbb{R}$ and $v \in X$ such that $(v \mid x_0) = 0$, $y = \mu x_0 + v$. If $v = 0$, since $x_0 \in P(\alpha)$, we have that $(x_0 \mid y) = (x_0 \mid \mu x_0) = \mu \geq 0$, whence $y \in P(\pi/2-\alpha)$. Suppose that $v \neq 0$ and consider $x := x_0 - \frac{\sin\alpha}{\|v\|\cdot\cos\alpha}\cdot v$. Then $x \in P(\alpha)$, and so

$$(x \mid y) = \mu - \frac{\sin\alpha}{\|v\|\cdot\cos\alpha}\cdot(v \mid v) = \mu - \frac{\sin\alpha}{\cos\alpha}\cdot\|v\| \geq 0.$$

We get so that $\mu\sin(\pi/2-\alpha) \geq \|v\|\cdot\cos(\pi/2-\alpha)$, *i.e.* $y \in P(\pi/2-\alpha)$. Therefore $P(\alpha)^+ = P(\pi/2-\alpha)$ in this case, too.

Furthermore, we remark that $P(\alpha)$ is pointed if and only if $\alpha \in [0, \pi/2[$, and $\operatorname{int} P(\alpha) = \{x \in X \mid \angle(x, x_0) < \alpha\}$.

Exercise 1.7 Let $(u_1, \ldots, u_n, u_{n+1}) \in K_{\rho'}$ and $(x_1, \ldots, x_n, x_{n+1}) \in K_\rho$. Then, by the generalized means inequality (see Exercise 2.11),

$$\begin{aligned}
x_1u_1 + \ldots + x_nu_n &= \alpha_1\left(\alpha_1^{-1}x_1u_1\right) + \ldots + \alpha_n\left(\alpha_n^{-1}x_nu_n\right)\\
&\geq \left(\alpha_1^{-1}x_1u_1\right)^{\alpha_1}\cdots\left(\alpha_n^{-1}x_nu_n\right)^{\alpha_n}\\
&= \left(\alpha_1^{\alpha_1}\cdots\alpha_n^{\alpha_n}\right)^{-1}\left(x_1^{\alpha_1}\cdots x_n^{\alpha_n}\right)\left(u_1^{\alpha_1}\cdots u_n^{\alpha_n}\right)\\
&\geq \left(\alpha_1^{\alpha_1}\cdots\alpha_n^{\alpha_n}\right)^{-1}\rho^{-1}\rho'^{-1}\left|x_{n+1}\right|\left|u_{n+1}\right|\\
&= \left|x_{n+1}u_{n+1}\right| \geq -x_{n+1}u_{n+1}.
\end{aligned}$$

Therefore $x_1u_1 + \ldots + x_nu_n + x_{n+1}u_{n+1} \geq 0$, and so $K_{\rho'} \subset (K_\rho)^+$.

Let now $(u_1, \ldots, u_n, u_{n+1}) \in (K_\rho)^+$. Then $x_1u_1 + \ldots + x_nu_n + x_{n+1}u_{n+1} \geq 0$ for every $(x_1, \ldots, x_n, x_{n+1}) \in K_\rho$, and so

$$x_1u_1 + \ldots + x_nu_n \geq \rho x_1^{\alpha_1}\cdots x_n^{\alpha_n}\left|u_{n+1}\right| \quad \forall x_1, \ldots, x_n \geq 0.$$

Taking $x_i = 1$ and $x_j = 0$ for $j \neq i$, we obtain that $u_i \geq 0$ for every $i \in \overline{1,n}$. Moreover, if $u_1 = 0$ then, taking $x_2 = \ldots = x_n = 1$, we obtain $u_2 + \ldots + u_n \geq \rho^{-1}x^{\alpha_1}\left|u_{n+1}\right|$ for every $x > 0$. Leting $x \to \infty$ we get $u_{n+1} = 0$. The same conclusion is true if $u_i = 0$ for some $i \in \overline{2,n}$. Assume that $u_i > 0$ for every $i \in \overline{1,n}$ and take $x_i := \alpha_i/u_i$. The above inequality implies that $1 \geq \rho\alpha_1^{\alpha_1}\cdots\alpha_n^{\alpha_n}\left(u_1^{\alpha_1}\cdots u_n^{\alpha_n}\right)^{-1}\left|u_{n+1}\right|$, whence $\left|u_{n+1}\right| \leq \rho' u_1^{\alpha_1}\cdots u_n^{\alpha_n}$. Hence $(u_1, \ldots, u_n, u_{n+1}) \in K_{\rho'}$. It follows that $K_{\rho'} = (K_\rho)^+$. Hence K_ρ is a closed convex cone for every $\rho > 0$.

Exercise 1.8 Taking $x_0 := (1, 0, 1)$ in Exercise 1.6, we get easily $P = P(\pi/4)$. Also, taking $n = 2$ and $\alpha_1 = \alpha_2 = 1/2$, it is obvious that $P = K_{\sqrt{2}}$.

Exercise 1.9 Let $f : X \to \mathbb{R}$, $f(x) = \alpha\|x\| - \varphi(x)$. It is obvious that f is a sublinear continuous functional. Therefore C is a closed convex cone. If x, $-x \in C$ then

$$\varphi(x) \geq \alpha\|x\|, \; -\varphi(x) = \varphi(-x) \geq \alpha\|-x\| = \alpha\|x\|,$$

and so, adding these relations, we have that $0 \geq 2\alpha\|x\|$, *i.e.* $x = 0$. Therefore C is pointed. Since $\alpha < \|\varphi\|$, there exists $x \in X$ such that $\alpha\|x\| < \varphi(x)$, and so $0 > \inf f$. Therefore, using Exercise 2.4, we have that

$$\operatorname{int} C = \{x \in X \mid f(x) < 0 = f(0)\} \neq \emptyset.$$

Furthermore, using Corollary 2.9.5, we have that

$$N(C;0) = \mathbb{R}_+ \cdot \partial f(0) = \mathbb{R}_+ \cdot (\alpha U_{X^*} - \varphi),$$

and so $C^+ = -N(C;0) = \mathbb{R}_+ \cdot (\varphi + \alpha U_{X^*})$.

Exercise 1.10 Let $\overline{y} \in \operatorname{int} \mathcal{R}(\overline{x})$ and $V \in \mathcal{N}_X$ be such that $\overline{y}+V \subset \mathcal{R}(\overline{x})$. Because $y_0 \in (\operatorname{Im} \mathcal{R})^i$, there exists $\lambda > 0$ such that $y_1 := (1+\lambda)y_0 - \lambda\overline{y} \in \operatorname{Im} \mathcal{R}$. It follows that $y_0 = (1-\overline{\lambda})\overline{y} + \overline{\lambda} y_1$, where $\overline{\lambda} := (1+\lambda)^{-1} \in]0,1[$. Let $x_1 \in \mathcal{R}^{-1}(y_1)$ and take $u := (1-\overline{\lambda})\overline{x} + \overline{\lambda} x_1$. Consider $\mu \in]0,1]$ and $(x_\mu, y_\mu) := (1-\mu)(x_0, y_0) + \mu(x_1, y_1) \in \operatorname{gr} \mathcal{R}$. Then $y_0 = \gamma_\mu \overline{y} + (1-\gamma_\mu) y_\mu$, where $\gamma_\mu := \frac{\mu - \mu\overline{\lambda}}{\mu + \overline{\lambda} - \mu\overline{\lambda}} \in]0,1]$. It follows that for every $v \in V$ we have

$$(x'_\mu, y_0 + \gamma_\mu v) = \gamma_\mu(\overline{x}, \overline{y} + v) + (1-\gamma_\mu)(x_\mu, y_\mu) \in \operatorname{gr} \mathcal{R} \quad \forall \mu \in]0,1],$$

where

$$x'_\mu := \gamma_\mu \overline{x} + (1-\gamma_\mu) x_\mu = \frac{\overline{\lambda} - \mu\overline{\lambda}}{\mu + \overline{\lambda} - \mu\overline{\lambda}} x_0 + \frac{\mu}{\mu + \overline{\lambda} - \mu\overline{\lambda}} u.$$

Hence $y_0 + \gamma_\mu V \subset \mathcal{R}(x'_\mu)$, and so $y_0 \in \operatorname{int} \mathcal{R}(x'_\mu)$. Since every element of the interval $]0,1]$ can be written as $\frac{\mu}{\mu + \overline{\lambda} - \mu\overline{\lambda}}$ with $\mu \in]0,1]$, we obtain that $y_0 \in \operatorname{int} \mathcal{R}(x)$ for every $x \in]x_0, u]$.

Exercise 1.11 Suppose that $\operatorname{dom} \mathcal{C}$ is dense and $x_1^*, x_2^* \in \mathcal{C}^*(y^*)$; this means (see page 26) that $\langle x, x_1^* \rangle = \langle y, y^* \rangle = \langle x, x_2^* \rangle$ for every $(x,y) \in \operatorname{gr} \mathcal{C}$ (because $\operatorname{gr} \mathcal{C}$ is a linear subspace), and so $\langle x, x_1^* - x_2^* \rangle = 0$ for every $x \in \operatorname{dom} \mathcal{C}$. As $\operatorname{dom} \mathcal{C}$ is dense, it follows that $x_1^* = x_2^*$. Suppose now that $\operatorname{dom} \mathcal{C}$ is not dense and take $\overline{x} \in X \setminus \operatorname{cl}(\operatorname{dom} \mathcal{C})$. By Theorem 1.1.5, there exists $\overline{x}^* \in X^*$ such that $\langle \overline{x}, \overline{x}^* \rangle < 0 = \langle x, \overline{x}^* \rangle$ for every $x \in \operatorname{dom} \mathcal{C}$ (because $\operatorname{dom} \mathcal{C}$ is a linear subspace). So, $(\overline{x}^*, 0) \in (\operatorname{gr} \mathcal{C})^+$, whence $0, \overline{x}^* \in \mathcal{C}^*(0)$. Because $\overline{x}^* \neq 0$, the statement is proved.

Applying the preceding statement for $\mathcal{C}$ and $\mathcal{C}^*$ we get the last assertion.

Exercise 1.12 Let $(x_n) \subset X$ be a Cauchy sequence. Since $|d(x_n, x) - d(x_m, x)| \leq d(x_n, x_m)$ for all $n, m \in \mathbb{N}$ and $x \in X$, we have that the sequence $(d(x_n, x)) \subset \mathbb{R}$ is Cauchy, and so it is convergent. Let $f : X \to \mathbb{R}_+$ be defined by $f(x) := 2 \lim d(x_n, x)$. From the inequality $d(x_n, x) \leq d(x_n, x') + d(x', x)$ for $n \in \mathbb{N}$,

$x, x' \in X$, we obtain that f is Lipschitz with Lipschitz constant 2. Moreover, we have that $(f(x_n)) \to 0$. By hypothesis, there exists $\overline{x} \in X$ such that

$$\forall x \in X \ : \ f(\overline{x}) \leq f(x) + d(x, \overline{x}).$$

Taking $x = x_n$, we get $f(\overline{x}) \leq f(x_n) + d(x_n, \overline{x})$ for every $n \in \mathbb{N}$. Taking the limit, we get $f(\overline{x}) \leq \frac{1}{2} f(\overline{x})$, and so $f(\overline{x}) = 0$. This shows that $(x_n) \to \overline{x}$.

Exercise 1.13 Applying Ekeland's variational principle for $x_0 \in \operatorname{dom} f$ and $\varepsilon = 1$, there exists $x_1 \in X$ such that $f(x_1) + d(x_0, x_1) \leq f(x_0)$ and $f(x_1) < f(x) + d(x, x_1)$ for each $x \in X \setminus \{x_1\}$. Suppose that $x_1 \notin \operatorname{argmin} f$. Then there exists $\overline{x} \in X \setminus \{x_1\}$ such that $f(\overline{x}) + d(x_1, \overline{x}) \leq f(x_1) < f(\overline{x}) + d(\overline{x}, x_1)$, whence the contradiction $0 < 0$. Therefore $x_1 \in \operatorname{argmin} f$, and so $d(x_0, \operatorname{argmin} f) \leq d(x_0, x_1) \leq f(x_0) - f(x_1) = f(x_0) - \inf f$. As this inequality is obvious for $x_0 \notin \operatorname{dom} f$, the conclusion holds.

Exercise 1.14 As in the proof of Exercise 1.13, there exists $x_1 \in X$ such that $f(x_1) < f(x) + d(x, x_1)$ for each $x \in X \setminus \{x_1\}$. From the hypothesis there exists $x_2 \in \mathcal{R}(x_1)$ such that $d(x_2, x_1) + f(x_2) \leq f(x_1)$. If $x_1 \neq x_2$ we get the contradiction $d(x_2, x_1) + f(x_2) < f(x_2) + d(x_1, x_2)$. Therefore $x_2 = x_1 \in \mathcal{R}(x_1)$.

Exercise 1.15 (i) Indeed,

$$\lim_{\|x\| \to \infty} f(x) = \infty \Leftrightarrow \forall \lambda \in \mathbb{R},\ \exists \rho > 0,\ \forall x \in X,\ \|x\| > \rho \ : \ f(x) > \lambda$$
$$\Leftrightarrow \forall \lambda \in \mathbb{R},\ \exists \rho > 0 \ : \ [f \leq \lambda] \subset D(0, \rho).$$

(ii) Let $\lambda > \inf_{x \in \mathbb{R}^n} f(x)$. By hypothesis $[f \leq \lambda]$ is a closed. By (i) $[f \leq \lambda]$ is also bounded, and so it is compact. Applying the well-known Weierstrass' theorem for $f|_{[f \leq \lambda]}$ we get the desired conclusion.

Exercise 2.1 Let $t_1, t_2 \in]0, \infty[$ and $\lambda \in]0, 1[$. Consider $t := \lambda t_1 + (1 - \lambda) t_2 > 0$. Because $\frac{\lambda t_1}{t}(x + t_1^{-1} u) + \frac{(1-\lambda) t_2}{t}(x + t_2^{-1} u) = x + \frac{1}{t} u$, the convexity of f gives

$$f\left(x + \tfrac{1}{t} u\right) \leq \tfrac{\lambda t_1}{t} f(x + t_1^{-1} u) + \tfrac{(1-\lambda) t_2}{t} f(x + t_2^{-1} u),$$

whence $\psi(\lambda t_1 + (1 - \lambda) t_2) \leq \lambda \psi(t_1) + (1 - \lambda) \psi(t_2)$. Hence ψ is convex.

Exercise 2.2 (a) Assume that the conclusion does not hold. Then there exist $x_1, x_2 \in I$ such that $x_1 < x_2$ and $f(x_1) > f(x_2)$. Let $A := \{x \in [x_1, x_2] \mid f(x_1) \leq f(x)\}$ and take $\overline{x} := \sup A$. Of course, $\overline{x} \leq x_2$ and, from the hypothesis, $\overline{x} > x_1$. If $\overline{x} \notin A$ then $f(\overline{x}) < f(x_1)$. But there exists $\overline{\varepsilon} > 0$ such that $f|_{[\overline{x} - \overline{\varepsilon}, \overline{x}]}$ is nondecreasing, and so $f(x) \leq f(\overline{x}) < f(x_1)$ for all $x \in [\overline{x} - \overline{\varepsilon}, \overline{x}]$, contradicting the choice of $\overline{x}$. Therefore $\overline{x} \in A$, and so $\overline{x} < x_2$. Using again the hypothesis, there exists $0 < \overline{\varepsilon} < x_2 - \overline{x}$ such that $f|_{[\overline{x}, \overline{x} + \overline{\varepsilon}]}$ is nondecreasing. Because $f(x_1) \leq f(\overline{x})$

we obtain that $f(x_1) \le f(\overline{x}+\overline{\varepsilon})$, and so $\overline{x}+\overline{\varepsilon} \in A$, contradicting again the choice of $\overline{x}$. Hence f is nondecreasing on I.

(b) 1) Consider $\mu := \big(g(b)-g(a)\big)/(b-a)$ and $h(x) := g(x)-\mu x$. Then h is continuous and there exists $h'_+ = g'_+ - \mu$. There exists $c \in [a,b]$ such that $h(c) \le h(x)$ for every $x \in [a,b]$. Because $h(a) = h(b)$ we can take $c \in [a,b[$. It follows that $f'_+(c) = h'_+(c) - \mu \ge 0$, whence the conclusion. For 2) take $c \in]a,b]$.

(c) It is sufficient to consider only the case $X = \mathbb{R}$. So $A \subset \mathbb{R}$ is an open interval and for every $a \in A$ there exists $\varepsilon > 0$ such that the restriction of f to $]a-\varepsilon, a+\varepsilon[\subset A$ is convex. It follows that there exists $f'_+(b) = \lim_{x\downarrow b}\big(f(x) - f(b)\big)/(x-b) \in \mathbb{R}$ and $f'_-(b) = \lim_{x\uparrow b}\big(f(x)-f(b)\big)/(x-b) \le f'_+(b)$ for every $b \in]a-\varepsilon, a+\varepsilon[$. So the functions $f'_-, f'_+ : A \to \mathbb{R}$ are locally nondecreasing. From (a) we obtain that f'_- and f'_+ are nondecreasing on A.

Because f is locally convex, it is continuous. Let now $a, b \in A$, $a < b$, and $\lambda \in]0,1[$. Consider $c := (1-\lambda)a + \lambda b$. Using (b) we obtain $x' \in [a,c[$ and $x'' \in]c,b]$ such that $f(a) - f(c) \ge f'_+(x')(a-c) \ge f'_-(c)(a-c)$ and $f(b) - f(c) \ge f'_-(x'')(b-c) \ge f'_-(c)(b-c)$. Multiplying the first relation with λ and the second one with $1-\lambda$, then adding them we obtain that $f(c) \le (1-\lambda)f(a) + \lambda f(b)$. Therefore f is convex.

Exercise 2.3 It is obvious that the implications "$\Rightarrow$" are true. Suppose that f_x is usc at 0 for every $x \in \mathbb{R}^n$ and let $\mathbb{R} \ni \lambda > f(\overline{x}) = f_x(0)$. Since f_x is usc at 0, it follows immediately that $\overline{x} \in [f \le \lambda]^i$. The function f being quasi-convex, $[f \le \lambda]$ is convex; since $\dim \mathbb{R}^n < \infty$, we obtain that $\overline{x} \in \operatorname{int}([f \le \lambda])$. Therefore f is usc at $\overline{x}$.

Suppose now that f_x is lsc at 0 for every $x \in \mathbb{R}^n$. Suppose, by way of contradiction, that f is not lsc at $\overline{x}$. Then there exists $\mathbb{R} \ni \lambda < f(\overline{x})$ such that $\overline{x} \in \operatorname{cl}[f \le \lambda]$. Therefore $[f \le \lambda] \ne \emptyset$. Let $a \in {}^i[f \le \lambda]$ ($\ne \emptyset$ by Exercise 1.2). Using Exercise 1.2 we obtain that $]a, \overline{x}[\subset {}^i[f \le \lambda] \subset [f \le \lambda]$, contradicting the fact that the function $f_{a-\overline{x}}$ is lsc at 0. Therefore f is lsc at $\overline{x}$.

If X is an infinite dimensional normed space, there exists a linear functional $\varphi : X \to \mathbb{R}$ which is not continuous. It is obvious that φ is quasi-convex (in fact is convex!), but φ_x is continuous (being affine) on $\mathbb{R}$ for every $x \in X$.

The existence of the linear function mentioned above can be proved as follows: let $(e_i)_{i\in I}$ be an algebraic base of X; changing eventually e_i by $e_i/\|e_i\|$, we may consider that $\|e_i\| = 1$ for every $i \in I$. Since I is infinite, we can suppose that $\mathbb{N} \subset I$ and replacing X by $\operatorname{lin}\{e_n \mid n \in \mathbb{N}\}$, we may consider that $I = \mathbb{N}$. So, for every $x \in X$ there exists $n \in \mathbb{N}$ and $\lambda_0, \ldots, \lambda_n \in \mathbb{R}$ such that $x = \lambda_0 e_0 + \ldots + \lambda_n e_n$. Consider

$$\varphi : X \to \mathbb{R}, \quad \varphi(x) = \sum\nolimits_{k=0}^{n} k\lambda_k.$$

It is clear that φ is well defined and linear. Moreover

$$\sup\{|\varphi(x)| \mid \|x\| \le 1\} \ge \sup\{|\varphi(e_n)| \mid n \in \mathbb{N}\} = \sup\{n \mid n \in \mathbb{N}\} = \infty,$$

which shows that φ is not continuous.

Exercise 2.4 Let $\overline{x} \in X$ be such that $f(\overline{x}) < \lambda$ and take $x \in X$. If $f(\overline{x}+x) \leq \lambda$ we have that $\overline{x} + tx \in [f \leq \lambda]$ for $t \in [0,1[$, while if $f(\overline{x}+x) > \lambda$ we have that $\overline{x} + tx \in [f \leq \lambda]$ for $t \in [0,\overline{t}\,[$, where $\overline{t} := \frac{\lambda - f(\overline{x})}{f(\overline{x}+x)-f(\overline{x})}$. Therefore $\overline{x} \in [f \leq \lambda]^i$.

Let us prove now the converse inclusion. Since $\lambda > \inf f$, there exists $x_0 \in X$ such that $f(x_0) < \lambda$. Let $\overline{x} \in [f \leq \lambda]^i$. There exists $\overline{t} > 0$ such that $x := \overline{x} + \overline{t}(\overline{x} - x_0) \in [f \leq \lambda]$. So we obtain that $\overline{x} = \frac{1}{1+\overline{t}}u + \frac{\overline{t}}{1+\overline{t}}x_0$ and

$$f(\overline{x}) \leq \tfrac{1}{1+\overline{t}} f(x) + \tfrac{\overline{t}}{1+\overline{t}} f(x_0) < \tfrac{1}{1+\overline{t}}\lambda + \tfrac{\overline{t}}{1+\overline{t}}\lambda = \lambda,$$

which proves the converse inclusion. If f is continuous then

$$\operatorname{int}[f \leq \lambda] \supset \{x \in X \mid f(x) < \lambda\} \neq \emptyset,$$

and so, since $[f \leq \lambda]$ is a convex set, $\operatorname{int}[f \leq \lambda] = [f \leq \lambda]^i$.

Exercise 2.5 (a) Assume that f is lsc and $t \in \,]\inf f, \infty[$. Since $[f \leq t]$ is closed, the inclusion $[f \leq t] \supset \operatorname{cl}[f < t]$ is obvious. Let $x \in [f \leq t]$. Take $x_0 \in [f < t]$. Then $x_\lambda := (1-\lambda)x_0 + \lambda x \in [f < t]$ for $\lambda \in [0,1[$, whence $x = \lim_{\lambda \uparrow 1} x_\lambda \in \operatorname{cl}[f < t]$.

Assume now that $[f \leq t] = \operatorname{cl}[f < t]$ for $t \in \,]\inf f, \infty[$. If $t_0 := \inf f \in \mathbb{R}$, we have that $[f \leq t_0] = \bigcap_{t > t_0} [f \leq t]$. It follows that $[f \leq t]$ is closed for every $t \in \mathbb{R}$. Hence f is lower semicontinuous.

(b) Assume that f is continuous on $\operatorname{dom} f$ and $t \in \,]\inf f, \infty[$. Let $x \in [f < t]$; since f is continuous at x, $[f < t]$ is a neighborhood of x, which shows that $[f \leq t]$ is a neighborhood of x, too. Hence $x \in \operatorname{int}[f \leq t]$. It follows that $[f < t] \subset \operatorname{int}[f \leq t]$. Let now $x \in \operatorname{int}[f \leq t]$ and take $x_0 \in [f < t]$. As the mapping $\lambda \mapsto x_\lambda := (1-\lambda)x_0 + \lambda x$ is continuous at $1 \in \mathbb{R}$, there exists $\lambda > 1$ such that $x_\lambda \in [f \leq t]$. It follows that $x = \frac{1}{\lambda}x_\lambda + \frac{\lambda-1}{\lambda}x_0 \in [f < t]$. Hence $[f < t] = \operatorname{int}[f \leq t]$.

Assume now that $[f < t] = \operatorname{int}[f \leq t]$ for every $t \in \,]\inf f, \infty[$. Consider $x \in \operatorname{dom} f$ and take $t > f(x)$. Hence $x \in \operatorname{int}[f \leq t]$, and so f is bounded above on a neighborhood of x. Therefore, by Theorem 2.2.9, f is continuous at x.

(c) Assume that f is continuous on X and $t \in \,]\inf f, \infty[$. Since $[f \leq t]$ is closed in our case, from (b) we obtain that $\operatorname{bd}[f \leq t] = (\operatorname{cl}[f \leq t]) \setminus (\operatorname{int}[f \leq t]) = [f \leq t] \setminus [f < t] = [f = t]$.

Assume now that $[f = t] = \operatorname{bd}[f \leq t]$ for every $t \in \,]\inf f, \infty[$. Let $x \in X$; assume that f is not lsc at x. Then $\liminf_{y \to x} f(y) := \sup_{V \in \mathcal{N}(x)} \inf_{y \in V} f(y) < f(x)$. Let t be such that $\liminf_{y \to x} f(y) < t < f(x)$. Then, obviously, $x \in \operatorname{cl}[f \leq t]$. Since $x \notin [f \leq t]$, $x \notin \operatorname{int}[f \leq t]$, and so we get the contradiction $x \in \operatorname{bd}[f \leq t] = [f = t]$. Since f is usc at any $x \in X \setminus \operatorname{dom} f$, let $x \in \operatorname{dom} f$ and $t > f(x)$. Then $x \in [f < t] \subset \operatorname{cl}[f \leq t]$, but $x \notin [f = t] = \operatorname{bd}[f \leq t]$. It follows that $x \in \operatorname{int}[f \leq t]$. By Theorem 2.2.9, f is continuous at x, and so f is usc at x. Hence f is continuous on X.

Exercise 2.6 The hypothesis shows that $f'(a, -e_i) = -f'(a, e_i) \in \mathbb{R}$ for any $i \in \overline{1,n}$, where $e_i = (0, \ldots, 0, 1, 0, \ldots, 0)$ (1 being on the ith place). So $f'(a, te_i) =$

$tf'(a,e_i)$ for any $t \in \mathbb{R}$. Let $x \in \mathbb{R}^n$; then $x = x_1e_1 + \ldots + x_ne_n$ with $x_1, \ldots, x_n \in \mathbb{R}$. By the sublinearity of $f'(a,\cdot)$ it follows that $f'(a,x) \leq f'(a,x_1e_1) + \cdots + f'(a,x_ne_n) = x_1f'(a,e_1) + \cdots + x_nf'(a,e_n)$, and $f'(a,-x) \leq f'(a,-x_1e_1) + \cdots + f'(a,-x_ne_n) = -x_1f'(a,e_1) - \cdots - x_nf'(a,e_n)$. Therefore $0 \leq f'(a,x) + f'(a,-x) \leq 0$, and so $f'(a,tx) = tf'(a,x)$ for every $t \in \mathbb{R}$. It follows that $f'(a,\cdot)$ is linear, and so f is Gâteaux differentiable.

Exercise 2.7 It is obvious that $\varphi_2(t) = 0$ for $t \in \mathbb{R}_+$, while $\varphi_1(t) = |1-t| - (1+t) + 2t = \max\{0, 2t-2\}$. Hence φ_1 is convex and nondecreasing.

Let $p \in]1,2[\cup]2,\infty[$. Then

$$\varphi_p'(t) = \begin{cases} p\left(2 - (1-t)^{p-1} - (1+t)^{p-1}\right) & \text{if} \quad t \in [0,1[, \\ p\left(2 + (t-1)^{p-1} - (t+1)^{p-1}\right) & \text{if} \quad t \in]1,\infty[, \end{cases}$$

$$\varphi_p''(t) = \begin{cases} p(p-1)\left((1-t)^{p-2} - (1+t)^{p-2}\right) & \text{if} \quad t \in [0,1[, \\ p(p-1)\left((t-1)^{p-2} - (t+1)^{p-2}\right) & \text{if} \quad t \in]1,\infty[. \end{cases}$$

Because $\lim_{t\uparrow 1} \varphi_p'(t) = p(2-2^{p-1}) = \lim_{t\downarrow 1} \varphi_p'(t)$, φ is derivable on $[0,\infty[$ and φ_p' is continuous.

Let first $p \in]1,2[$; then $\varphi_p''(t) > 0$ for every $t \in \mathbb{R}_+ \setminus \{1\}$. It follows that φ_p' is increasing on $\mathbb{R}_+$, and so φ_p is strictly convex and $\varphi_p'(t) > \varphi_p'(0) = 0$ for $t > 0$. Hence φ_p is increasing on $\mathbb{R}_+$.

Let now $p \in]2,\infty[$; then $\varphi_p''(t) < 0$ for every $t \in \mathbb{R}_+ \setminus \{1\}$. It follows that φ_p' is decreasing on $\mathbb{R}_+$, and so φ_p is strictly concave and $\varphi_p'(t) < \varphi_p'(0) = 0$ for $t > 0$. Hence φ_p is decreasing on $\mathbb{R}_+$.

Exercise 2.8 Let $D :=]0,\infty[^2$; D is an open convex set and f is twice differentiable on D. Consider $\varphi(x,y) := \arctan \frac{x}{y}$ for $(x,y) \in D$. Then, for $(x,y) \in D$, we have

$$\frac{\partial f}{\partial x}(x,y) = \beta\,(\varphi(x,y))^{\beta-1}\,\frac{y}{\sqrt{x^2+y^2}} + (\varphi(x,y))^{\beta}\,\frac{x}{\sqrt{x^2+y^2}},$$
$$\frac{\partial f}{\partial y}(x,y) = \beta\,(\varphi(x,y))^{\beta-1}\,\frac{-x}{\sqrt{x^2+y^2}} + (\varphi(x,y))^{\beta}\,\frac{y}{\sqrt{x^2+y^2}},$$

and so

$$\frac{\partial^2 f}{\partial x^2}(x,y) = (\varphi(x,y))^{\beta-2}\,\frac{y^2}{(x^2+y^2)\sqrt{x^2+y^2}}\left(\beta(\beta-1) + (\varphi(x,y))^2\right),$$
$$\frac{\partial^2 f}{\partial x \partial y}(x,y) = (\varphi(x,y))^{\beta-2}\,\frac{-xy}{(x^2+y^2)\sqrt{x^2+y^2}}\left(\beta(\beta-1) + (\varphi(x,y))^2\right),$$
$$\frac{\partial^2 f}{\partial y^2}(x,y) = (\varphi(x,y))^{\beta-2}\,\frac{x^2}{(x^2+y^2)\sqrt{x^2+y^2}}\left(\beta(\beta-1) + (\varphi(x,y))^2\right).$$

Hence

$$d^2f(x,y)(h,k) = (\varphi(x,y))^{\beta-2}\left(\beta(\beta-1)+(\varphi(x,y))^2\right)\frac{(yh-xk)^2}{(x^2+y^2)\sqrt{x^2+y^2}} \geq 0$$

for all $(x,y) \in D$ and $(h,k) \in \mathbb{R}^2$. Using Theorem 2.1.11 we obtain that $f|_D$ is convex, and so the function $g : \mathbb{R}^2 \to \overline{\mathbb{R}}$ defined by $g(x,y) := f(x,y)$ for $(x,y) \in D$, $g(x,y) := \infty$ otherwise, is convex. It is obvious that for any $a > 0$, $\liminf_{(x,y)\to(a,0)} g(x,y) = \left(\frac{\pi}{2}\right)^\beta a = f(a,0)$ and $\liminf_{(x,y)\to(0,0)} g(x,y) = 0 = f(a,0)$. It follows that $f = \overline{g}$, and so f is convex and lsc. Because $f(t,t) = (\pi/4)^\beta\sqrt{2}\,t$ for $t \geq 0$, it is clear that f is not strictly convex.

Exercise 2.9 Consider $X := C[0,1]$; X is a Banach space for the Chebyshev norm: $\|x\| := \max_{t\in[0,1]}|x(t)|$ for $x \in X$.

Consider also the function $\varphi : \mathbb{R} \to \mathbb{R}$, $\varphi(t) = \sqrt{1+t^2}$. Then

$$\forall t \in \mathbb{R} \ : \ \varphi'(t) = t(1+t^2)^{-\frac{1}{2}}, \quad \varphi''(t) = (1+t^2)^{-\frac{3}{2}}, \quad \varphi'''(t) = -3t(1+t^2)^{-\frac{5}{2}}.$$

It follows that φ is (strictly) convex since $\varphi''(t) > 0$ for every $t \in \mathbb{R}$. By this remark the convexity of f is immediate. Using Taylor's formula for functions of a real variable, we have that for all $t, \tau \in \mathbb{R}$, there exists $\theta \in]0,1[$ such that

$$\varphi(t+\tau) - \varphi(t) - \varphi'(t)\tau = \tfrac{1}{2}\varphi''(t+\theta\tau)\tau^2,$$

and so

$$\forall t,\tau \in \mathbb{R} \ : \ |\varphi(t+\tau) - \varphi(t) - \varphi'(t)\tau| \leq \tfrac{1}{2}\tau^2.$$

Let $x, u \in X$ be fixed elements. Replacing t by $x(t)$ and τ by $u(t)$, $t \in [0,1]$ in the above inequality, then integrating on $[0,1]$, we obtain that

$$\left| f(x+u) - f(x) - \int_0^1 \frac{xu}{\sqrt{1+x^2}}\,dt \right| \leq \frac{1}{2}\int_0^1 u^2\,dt \leq \frac{1}{2}\|u\|^2.$$

This inequality proves that f is Fréchet differentiable at x and that $\nabla f(x)$ has the announced expression.

Let us prove that ∇f is Fréchet differentiable. For all $t, \tau \in \mathbb{R}$ there exists $\theta \in]0,1[$ such that

$$\varphi'(t+\tau) - \varphi'(t) - \varphi''(t)\tau = \frac{1}{2}\varphi'''(t+\theta\tau)\tau^2.$$

From the expression of φ''' given above we have that $|\varphi'''(t)| \leq 3$ for every $t \in \mathbb{R}$. Therefore

$$\forall t,\tau \in \mathbb{R} \ : \ |\varphi'(t+\tau) - \varphi'(t) - \varphi''(t)\tau| \leq \tfrac{3}{2}\tau^2.$$

Denoting by $B(x)$ the bilinear functional defined by

$$B(x) : X \times X \to \mathbb{R}, \quad B(x)(u, v) := \int_0^1 \frac{uv}{(1 + x^2)\sqrt{1 + x^2}}\, dt,$$

and using the preceding relation we obtain, as above, that

$$\forall\, x, u, v \in X \;:\; |(\nabla f(x+u) - \nabla f(x) - B(x)(u, \cdot))(v)| \le \tfrac{3}{2} \int_0^1 u^2 |v|\, dt \le \tfrac{3}{2} \|u\|^2 \|v\|,$$

and so

$$\forall\, x, u \in X \;:\; \|\nabla f(x + u) - \nabla f(x) - B(x)(u, \cdot)\| \le \tfrac{3}{2} \|u\|^2.$$

Therefore ∇f is Fréchet differentiable on X and $\nabla^2 f(x) = B(x)$ for every $x \in X$. From the expression of $\nabla^2 f(x)$ and the expression of φ''' we obtain easily that $\nabla^2 f$ is Lipschitz with Lipschitz constant 3. Therefore $\nabla^2 f$ is continuous on X.

Exercise 2.10 Consider $X := L^1(0, 1)$; X is a Banach space for the norm: $\|u\| := \int_0^1 |u(t)|\, dt$ for $u \in X$.

Using the function φ defined in the solution of the preceding exercise, taking also into account that φ is convex, f is convex. Moreover

$$\forall\, t, \tau \in \mathbb{R} \;:\; -|\tau| \le \varphi'(t)\tau \le \varphi(t + \tau) - \varphi(t) \le |\tau|;$$

the inequality from the left is obvious, while that from the right follows easily. Let us fix $x, u \in X$ and a sequence $\mathbb{R}\setminus\{0\} \supset (\tau_n) \to 0$, then consider the functions g_n, g defined (a.e.) by the formulas

$$g_n := \frac{\sqrt{1 + (x + \tau_n u)^2} - \sqrt{1 + x^2}}{\tau_n}, \quad g := \frac{xu}{\sqrt{1 + x^2}}.$$

It is obvious that $\lim_{n\to\infty} g_n = g$ a.e.; from the above inequality we obtain that $|g_n| \le |u|$ a.e. for every $n \in \mathbb{N}$. Using Lebesgue's theorem (see [Precupanu (1976)]) we obtain that

$$\lim_{n\to\infty} \int_0^1 g_n\, dt = \int_0^1 g\, dt.$$

This proves that f is Gâteaux differentiable at x, and the expression of $\nabla f(x)$ is that of the statement.

Let $x \in X$ be arbitrary; we show that f it is not Fréchet differentiable at x. We may assume that x is finite a.e. Of course,

$$[0, 1] = \bigcup\nolimits_{m\in\mathbb{N}} A_m, \text{ where } A_m := \{t \in [0, 1] \mid |x(t)| \le m\}.$$

Since $\lim_{m\to\infty} \text{meas}(A_m) = 1$, there exists $m_0 \in \mathbb{N}$ such that $\text{meas}(A_{m_0}) > \frac{1}{2}$.

But

$$\sqrt{1+(t+\tau)^2}-\sqrt{1+t^2}-\frac{t\tau}{\sqrt{1+t^2}}$$
$$=\frac{\tau^2}{\sqrt{1+(t+\tau)^2}+\sqrt{1+t^2}}\left(1-\frac{2t^2+t\tau}{\sqrt{1+t^2}\left(\sqrt{1+(t+\tau)^2}+\sqrt{1+t^2}\right)^2}\right);$$

for $\tau=n$ and $|t|\le m_0$ we have that

$$\frac{\tau}{\sqrt{1+(t+\tau)^2}+\sqrt{1+t^2}}\ge\frac{n}{2\sqrt{1+(n+m_0)^2}}\to\frac{1}{2},$$
$$\frac{2t^2+\tau t}{\sqrt{1+t^2}\left(\sqrt{1+(t+\tau)^2}+\sqrt{1+t^2}\right)^2}\le\frac{m_0}{\sqrt{1+m_0^2}}\cdot\frac{n+2m_0}{\sqrt{1+(n-m_0)^2}}\to\frac{m_0}{\sqrt{1+m_0^2}},$$

the limits being taken for $n\to\infty$. Therefore there exists $n_0\ge m_0$ and $\varepsilon_0>0$ such that

$$\forall t\in A_{m_0},\ \forall n\ge n_0\ :\ \sqrt{1+(t+n)^2}-\sqrt{1+t^2}-\frac{tn}{\sqrt{1+t^2}}\ge\varepsilon_0 n.$$

For every $n\ge n_0$ there exists a measurable set B_n such that $B_n\subset A_{m_0}$ and $\operatorname{meas}(B_n)=n^{-2}$. Let us consider the functions

$$u_n:[0,1]\to\mathbb{R},\quad u_n:=\begin{cases} n & \text{if } t\in B_n,\\ 0 & \text{if } t\in[0,1]\setminus B_n.\end{cases}$$

Then $\|u_n\|=\frac{1}{n}\to 0$. From the above inequality we have that

$$\int_0^1\left(\sqrt{1+(x+u_n)^2}-\sqrt{1+x^2}-\frac{xu_n}{\sqrt{1+x^2}}\right)dt\ge\varepsilon_0\int_0^1 u_n\,dt=\varepsilon_0\|u_n\|,$$

which shows that f it is not Fréchet differentiable at x.

Exercise 2.11 The function $f:\mathbb{R}\to\mathbb{R}$, $f(t):=\exp t$ is strictly convex. Consequently, for $a,b\in\mathbb{P}$, $a^p\ne b^q$ and $p,q\in]1,\infty[$, $\frac{1}{p}+\frac{1}{q}=1$ we have that

$$ab=f(\tfrac{1}{p}\ln a^p+\tfrac{1}{q}\ln b^q)<\tfrac{1}{p}f(\ln a^p)+\tfrac{1}{q}f(\ln b^q)=\tfrac{1}{p}a^p+\tfrac{1}{q}b^q.$$

Therefore the statement is true. It is obvious for $a=0$ or $b=0$.

The function $g:\mathbb{P}\to\mathbb{R}$, $g(t):=-\ln t$ is strictly convex. Let $n\in\mathbb{N}\setminus\{1\}$, $\alpha_1,\dots,\alpha_n\in]0,1[$ with $\alpha_1+\dots+\alpha_n=1$ and $x_1,\dots,x_n\in\mathbb{P}$, not all equal. Then

$$-\ln(\alpha_1x_1+\cdots+\alpha_nx_n)=g(\alpha_1x_1+\cdots+\alpha_nx_n)$$
$$<\alpha_1g(x_1)+\cdots+\alpha_ng(x_n)=-\ln\left(x_1^{\alpha_1}\cdots x_n^{\alpha_n}\right),$$

and so

$$\alpha_1 x_1 + \cdots + \alpha_n x_n > x_1^{\alpha_1} \cdots x_n^{\alpha_n}.$$

It is obvious that the above inequality is still valid when $x_1, \ldots, x_n \in [0, \infty[$ are not all equal. The final statement is now obvious.

Exercise 2.12 The implication $\Rightarrow$ is obvious. Assume that $f + x^*$ is quasi-convex for every $x^* \in X^*$ and consider $x, y \in \operatorname{dom} f$ with $x \neq y$ and $\lambda \in]0,1[$; set $z := (1-\lambda)x + \lambda y$. There exists $x^* \in X^*$ such that $\langle x - y, x^* \rangle = f(y) - f(x)$. Since $f + x^*$ is quasi-convex, we have that

$$\begin{aligned} f(z) + \langle z, x^* \rangle &\leq \max\{f(x) + \langle x, x^* \rangle, f(y) + \langle y, x^* \rangle\} \\ &= (1-\lambda)\left(f(x) + \langle x, x^* \rangle\right) + \lambda\left(f(y) + \langle y, x^* \rangle\right), \end{aligned}$$

whence $f((1-\lambda)x + \lambda y) \leq (1-\lambda)f(x) + \lambda f(y)$.

Exercise 2.13 Note that

$$\frac{\partial f}{\partial x_i}(x) = \frac{\alpha_i}{x_i} f(x), \quad \frac{\partial^2 f}{\partial {x_i}^2}(x) = \frac{\alpha_i(\alpha_i - 1)}{x_i^2} f(x),$$

while for $i \neq j$,

$$\frac{\partial^2 f}{\partial x_i \partial x_j}(x) = \frac{\alpha_i \alpha_j}{x_i x_j} f(x).$$

The function f is concave (*i.e.* $-f$ is convex) if and only if $\nabla^2 f(x)$ is nonnegatively defined for every $x \in \mathbb{P}^n$, *i.e.*

$$\forall\, u = (u_1, \ldots, u_n) \in \mathbb{R}^n \ : \ \sum\nolimits_{i=1}^n \frac{\alpha_i(\alpha_i - 1)}{x_i^2} u_i^2 + 2 \sum\nolimits_{1 \leq i < j \leq n} \frac{\alpha_i \alpha_j}{x_i x_j} u_i u_j \leq 0,$$

or, equivalently,

$$\forall\, u \in \mathbb{R}^n \ : \ \left(\sum\nolimits_{i=1}^n \alpha_i u_i\right)^2 \leq \sum\nolimits_{i=1}^n \alpha_i u_i^2.$$

Taking $\alpha_{n+1} = 1 - \sum_{i=1}^n \alpha_i$ and $u_{n+1} = 0$, the preceding inequality follows from the (strict) convexity of the function $t \to t^2$ on $\mathbb{R}$; the inequality is strict if $\sum_{i=1}^n \alpha_i < 1$ and $u \neq 0$, and so f is strictly concave in this case.

Exercise 2.14 We have that $\partial f / \partial x_i = (\exp x_i) \left(\sum_{k=1}^n \exp x_k\right)^{-1}$, and so

$$\begin{aligned} \frac{\partial^2 f}{\partial x_i^2} &= \left((\exp x_i) \sum\nolimits_{k=1}^n \exp x_k - (\exp x_i)^2\right) \left(\sum\nolimits_{k=1}^n \exp x_k\right)^{-2}, \\ \frac{\partial^2 f}{\partial x_i x_j} &= -(\exp x_i)(\exp x_j) \left(\sum\nolimits_{k=1}^n \exp x_k\right)^{-2} \quad \forall\, i, j \in \overline{1,n},\ i \neq j. \end{aligned}$$

So, for all $x, u \in \mathbb{R}^n$ we have that $\nabla^2 f(x)(u,u)$ equals

$$\left(\sum\nolimits_{k=1}^n y_k\right)^{-2} \left(\left(\sum\nolimits_{k=1}^n y_k\right) \left(\sum\nolimits_{k=1}^n y_k (u_k)^2\right) - \left(\sum\nolimits_{k=1}^n y_k u_k\right)^2 \right),$$

where $y_k = \exp x_k$ for $k \in \overline{1,n}$. Applying the well-known Cauchy–Schwarz inequality $\left(\sum_{k=1}^n a_k b_k\right)^2 \le \left(\sum_{k=1}^n a_k^2\right)\left(\sum_{k=1}^n b_k^2\right)$ for $a_k = \sqrt{y_k}$ and $b_k = \sqrt{y_k} u_k$, we obtain that $\nabla^2 f(x)(u,u) \ge 0$. Hence f is convex.

Exercise 2.15 One obtains easily that

$$\frac{\partial f}{\partial t}(t,x) = \frac{p}{t} f(t,x), \quad \frac{\partial f}{\partial x_i}(t,x) = -\frac{1}{x_i} f(t,x),$$
$$\frac{\partial^2 f}{\partial t^2}(t,x) = \frac{p(p-1)}{t^2} f(t,x), \quad \frac{\partial^2 f}{\partial {x_i}^2}(t,x) = \frac{2}{x_i^2} f(t,x), \quad \frac{\partial^2 f}{\partial t\, \partial x_i}(t,x) = -\frac{p}{x_i t} f(t,x),$$

while for $i \ne j$,

$$\frac{\partial^2 f}{\partial x_i\, \partial x_j}(t,x) = \frac{1}{x_i x_j} f(t,x).$$

Since $f(t,x) > 0$ for every $(t,x) \in \mathbb{P} \times \mathbb{P}^n$, the fact that f is strictly convex (resp. convex) is equivalent to the fact that the quadratic form

$$\frac{p(p-1)}{t^2} s^2 + \sum\nolimits_{i=1}^n \frac{2}{x_i^2} u_i^2 - 2\sum\nolimits_{i=1}^n \frac{p}{x_i t} s u_i + 2\sum\nolimits_{1 \le i < j \le n} \frac{1}{x_i x_j} u_i u_j$$

is positively (resp. nonnegatively) defined, or, equivalently, that the quadratic form

$$\frac{p-1}{p} s^2 + \sum\nolimits_{i=1}^n 2u_i^2 - 2\sum\nolimits_{i=1}^n s u_i + 2\sum\nolimits_{1 \le i < j \le n} u_i u_j$$

is positively (resp. nonnegatively) defined. The matrix associated to the last quadratic form is

$$A = \begin{pmatrix} \frac{p-1}{p} & -1 & -1 & \cdots & -1 \\ -1 & 2 & 1 & \cdots & 1 \\ -1 & 1 & 2 & \cdots & 1 \\ \vdots & \vdots & \vdots & \ddots & \vdots \\ -1 & 1 & 1 & \cdots & 2 \end{pmatrix}.$$

To study the positiveness of the matrix associated to this quadratic form, we have to determine its eigenvalues. Let $\delta_n := \det(\lambda \iota_{n-1} - A)$. After some computations we obtain that

$$\delta_n = \begin{vmatrix} \lambda - \frac{p-1}{p} & \lambda + \frac{1}{p} & \lambda + \frac{1}{p} & \cdots & \lambda + \frac{1}{p} \\ 1 & \lambda - 1 & 0 & \cdots & 0 \\ 1 & 0 & \lambda - 1 & \cdots & 0 \\ \vdots & \vdots & \vdots & \ddots & \vdots \\ 1 & 0 & 0 & \cdots & \lambda - 1 \end{vmatrix}.$$

Developing with respect to the last row we obtain that

$$\delta_n = (\lambda - 1)\delta_{n-1} - (\lambda - 1)^{n-1} (\lambda + 1/p),$$

and so

$$\begin{aligned} \delta_n &= (\lambda - 1)^{n-1}\delta_1 - (n-1)(\lambda - 1)^{n-1} (\lambda + 1/p) \\ &= (\lambda - 1)^{n-1} \left[\lambda^2 - (n + 2 - 1/p)\,\lambda + 1 - (n+1)/p\right]. \end{aligned}$$

Let λ_1, λ_2 be the roots of the polynomial between the square brackets. Because the other roots of the characteristic polynomial are all 1, the quadratic form under consideration is positively (resp. nonnegatively) defined if and only if λ_1 and λ_2 are positive (resp. nonnegative). For $p < 0$ we have that

$$\lambda_1 + \lambda_2 = n + 2 - 1/p > 0, \quad \lambda_1\lambda_2 = 1 - (n+1)/p > 0,$$

and so $\lambda_1, \lambda_2 > 0$. When $p > 0$ the roots λ_1, λ_2 are positive (resp. nonnegative) if and only if $p > n+1$ (resp. $p \geq n+1$). Therefore f is strictly convex if and only if $p \in]-\infty, 0[\cup]n+1, \infty[$ and is convex if and only if $p \in]-\infty, 0[\cup [n+1, \infty[$. Since $g(x) = f((x \mid c), x)$ and the function $x \mapsto (x \mid c)$ is linear, it follows immediately that g is (strictly) convex for $(p > n+1)$ $p \geq n+1$.

Since $h(x) = f(1, x)$ (for $p = n+2$), we obtain immediately that h is strictly convex.

Exercise 2.16 We give only the proof for ψ'_+, the proof for ψ'_- being similar. The function ψ is increasing, while ψ'_+ is nondecreasing and $0 < \psi'(t) < \infty$ for $t > 0$ (if $\psi'(t_0) = 0$ for some $t_0 > 0$ then $\psi'(t) = 0$ for $t \in [0, t_0]$, and so $\psi(t) = \psi(0) = 0$ for t$\in [0, t_0]$). It follows that the mapping $\mathbb{P} \ni t \mapsto 1/\left(\psi'_+ \circ \psi^{-1}\right)$ is positive and nonincreasing on $\mathbb{P}$.

Let us fix $\alpha > 0$ and take $\beta := \psi(\alpha) > 0$. From what was said above, the integral $\int_\gamma^\beta \frac{dt}{\psi'(\psi^{-1}(t))}$ exists in the sense of Riemann for all $\gamma \in]0, \beta[$. Considering it as a Riemann–Stieltjes integral, we have that

$$\int_\gamma^\beta \frac{dt}{\psi'\left(\psi^{-1}(t)\right)} = \int_{\psi^{-1}(\gamma)}^{\psi^{-1}(\beta)} \frac{1}{(\psi'_+ \circ \psi^{-1} \circ \psi)(s)} d\,\psi(s) = \int_\delta^\alpha \frac{1}{\psi'(s)} d\,\psi(s), \quad (^*)$$

where $\delta := \psi^{-1}(\gamma)$. Let $\delta = s_0 < s_1 < \ldots < s_n = \alpha$ be a division Δ of the interval $[\delta, \alpha]$ and take $\sigma_i := s_{i-1}$ for $i \in \overline{1,n}$. Then $\psi'(\sigma_i) \in \partial\psi(s_{i-1})$, and so

$$\Sigma(\Delta, (\sigma_i)) = \sum_{i=1}^{n} \frac{1}{\psi'(\sigma_i)} (\psi(s_i) - \psi(s_{i-1})) \geq \sum_{i=1}^{n} (s_i - s_{i-1}) = \alpha - \delta,$$

because $\psi(s_i) - \psi(s_{i-1}) \geq \psi'(s_{i-1}) \cdot (s_i - s_{i-1})$ for all $i \in \overline{1,n}$. Taking now $\sigma_i' := s_i$, we have that $\psi'(\sigma_i') \in \partial\psi(s_i)$, and so $\Sigma(\Delta, (\sigma_i')) \leq \alpha - \delta$. It follows that $\int_\delta^\alpha \frac{1}{\psi'(s)} d\,\psi(s) = \alpha - \delta$. From (*) we get $\int_\gamma^\beta \frac{dt}{\psi_+'(\psi^{-1}(t))} = \alpha - \psi^{-1}(\gamma)$. Taking the limit for $\gamma \to 0$ we obtain the desired conclusion.

Exercise 2.17 Let

$$F : X \times Y \to \overline{\mathbb{R}}, \quad F(x,y) := \begin{cases} \|x\| & \text{if} \quad Tx = y, \\ +\infty & \text{if} \quad Tx \neq y. \end{cases}$$

It is obvious that F is convex and f is the marginal function associated to F. Therefore, f is convex. Furthermore, for $\lambda > 0$ and $y \in Y$ we have that

$$\begin{aligned} f(\lambda y) &= \inf\{\|x\| \mid Tx = \lambda y\} = \inf\left\{\lambda \cdot \left\|\lambda^{-1}x\right\| \mid T\left(\lambda^{-1}x\right) = y\right\} \\ &= \lambda \inf\{\|u\| \mid Tu = y\} = \lambda f(y). \end{aligned}$$

If T is open, there exists $\rho > 0$ such that $\rho U_Y \subset T(U_X)$. Therefore

$$\forall\, y \in \rho U_F,\ \exists x \in U_X \ : \ Tx = y,$$

and so $f(y) \leq 1$ for every $y \in \rho U_Y$; this implies that $\operatorname{dom} f = Y$ and f is continuous.

Exercise 2.18 By Exercise 1.15 there exists $\overline{x} \in \mathbb{R}^k$ such that $f(\overline{x}) \leq f(x)$ for every $x \in \mathbb{R}^k$. We assume, without loss of generality, that $\overline{x} = 0$ and $f(\overline{x}) = 0$. Let $(x,t) \in \overline{\text{co}}(\operatorname{epi} f)$. Then there exists a sequence $((x_n, t_n)) \subset \text{co}(\operatorname{epi} f)$ converging to (x,t). By the Carathéodory theorem (Exercise 1.1), for every $n \in \mathbb{N}$ there exist $\lambda_n^i \in [0,1]$, $(x_n^i, t_n^i) \in \operatorname{epi} f$ for $i \in \overline{1,k+2}$ such that $\sum_{i=1}^{k+2} \lambda_n^i = 1$ and $(x_n, t_n) = \sum_{i=1}^{k+2} \lambda_n^i (x_n^i, t_n^i)$. We may assume that for every $i \in \overline{1,k+2}$: $(\lambda_n^i) \to \lambda^i \in [0,1]$, $(t_n^i) \to t^i \in [0,\infty]$, $(\lambda_n^i t_n^i) \to \gamma^i \in [0,\infty]$, $\left(\lambda_n^i \left\|x_n^i\right\|\right) \to \eta^i \in [0,\infty]$ and $\left(\left\|x_n^i\right\|\right) \to \mu^i \in [0,\infty]$; moreover, if $\mu^i < \infty$, we may assume that $\left(x_n^i\right) \to x^i \in \mathbb{R}^k$. Of course, $\sum_{i=1}^{k+2} \lambda^i = 1$. First of all, because $\left(\sum_{i=1}^{k+2} \lambda_n^i t_n^i\right) \to t = \sum_{i=1}^{k+2} \gamma^i$, we have that $\gamma^i \in [0,\infty[$ for every $i \in \overline{1,k+2}$.

Assume that $\mu^i = \infty$. By our hypothesis we have that $\left(\left\|x_n^i\right\|^{-1} t_n^i\right) \to \infty$, which implies that $t^i = \infty$, $\lambda^i = 0$ and $\eta^i = 0$. Consider the sets $I := \{i \in \overline{1,k+2} \mid \eta^i > 0\}$ and $J := \overline{1,k+2} \setminus I$. Hence $\left(x_n^i\right) \to x^i \in \mathbb{R}^k \setminus \{0\}$, $\lambda^i > 0$, and so $t^i < \infty$, for $i \in I$; moreover, $(x^i, t^i) \in \operatorname{epi} f$ for $i \in I$, because f

is lsc. Let $\gamma := \sum_{i\in J}\gamma^i$ when J is nonempty and $\lambda := \sum_{i\in J}\lambda^i$ when I is nonempty. It follows that $(x,t) = (0,\gamma)$ if $I = \emptyset$, $(x,t) = \sum_{i\in I}\lambda^i(x^i,t^i)$ if $J = \emptyset$ and $(x,t) = \sum_{i\in I}\lambda^i(x^i,t^i) + (0,\gamma)$ if I, J are nonempty. It is obvious that $(x,t) \in \operatorname{co}(\operatorname{epi} f)$ if I or J is empty. If both of them are nonempty then $(x,t) = \sum_{i\in I}\lambda^i(x^i,t^i) + (1-\lambda)\left(0,(1-\lambda)^{-1}\gamma\right) \in \operatorname{co}(\operatorname{epi} f)$ if $\lambda \neq 1$ and $(x,t) = \sum_{i\in I}\lambda^i(x^i,t^i+\gamma) \in \operatorname{co}(\operatorname{epi} f)$ if $\lambda = 1$.

Exercise 2.19 If $\operatorname{dom} f \cap \operatorname{dom} g = \emptyset$ then $(f+\gamma g)(x) = \infty$ for all $\gamma > 0$ and $x \in X$, and the equalities are obvious. So assume that $\operatorname{dom} f \cap \operatorname{dom} g \neq \emptyset$. Let us begin with the first equality. Since $\overline{\operatorname{co}}(f+\alpha g) + \beta g \le f + \alpha g + \beta g = f + (\alpha+\beta)g$ we have that $\overline{\operatorname{co}}\left(\overline{\operatorname{co}}(f+\alpha g)+\beta g\right) \le \overline{\operatorname{co}}\left(f+(\alpha+\beta)g\right)$. In order to show the converse inequality it is sufficient to show that $\overline{\operatorname{co}}\left(f+(\alpha+\beta)g\right) \le \overline{\operatorname{co}}(f+\alpha g)+\beta g$, or equivalently, $\operatorname{epi}\left(\overline{\operatorname{co}}(f+\alpha g)+\beta g\right) \subset \overline{\operatorname{co}}\left(\operatorname{epi}\left(f+(\alpha+\beta)g\right)\right)$. So let (x,t) be in the first set; there exist $t_1, t_2 \in \mathbb{R}$ such that $t = t_1 + \beta t_2$ and $(x,t_1) \in \overline{\operatorname{co}}\left(\operatorname{epi}(f+\alpha g)\right)$ and $(x,t_2) \in \operatorname{epi} g$. Therefore $(x,t_1) = \lim_{i\in I}(x^i,t_1^i)$ with $(x^i,t_1^i) \in \operatorname{co}\left(\operatorname{epi}(f+\alpha g)\right)$ for all $i \in I$. Hence, for every $i \in I$, $(x^i,t_1^i) = \sum_{j=1}^{n_i}\lambda_j^i(x_j^i, s_j^i + \alpha r_j^i)$ for some $n_i \in \mathbb{N}$ and $\lambda_j^i \in \mathbb{P}$, $\sum_{j=1}^{n_i}\lambda_j^i = 1$, $(x_j^i,s_j^i) \in \operatorname{epi} f$, $(x_j^i,r_j^i) \in \operatorname{epi} g$ for $1 \le j \le n_i$. Because $(x^i)_{i\in I} \to x$ and $f \in \Gamma(X)$, we have that

$$f(x) \le \liminf_{i\in I} f\left(\sum\nolimits_{j=1}^{n_i}\lambda_j^i x_j^i\right) \le \liminf_{i\in I}\sum\nolimits_{j=1}^{n_i}\lambda_j^i f(x_j^i) \le \liminf_{i\in I}\sum\nolimits_{j=1}^{n_i}\lambda_j^i s_j^i,$$

and so $\lim_{i\in I}\left(f(x) + \gamma^i - \sum_{j=1}^{n_i}\lambda_j^i s_j^i\right) = 0$ and $(x, f(x)+\gamma^i) \in \operatorname{epi} f$ for $i \in I$, where $\gamma^i := \max\left(0, \sum_{j=1}^{n_i}\lambda_j^i s_j^i - f(x)\right)$. Because $\left(x_j^i, s_j^i + (\alpha+\beta)r_j^i\right)$ belongs to $\operatorname{epi}(f+(\alpha+\beta)g)$ for all $i \in I$, $1 \le j \le n_i$, and $\left(x, f(x)+\gamma^i+(\alpha+\beta)t_2\right) \in \operatorname{epi}\left(f+(\alpha+\beta)g\right)$, we have that

$$\begin{aligned}&\sum\nolimits_{j=1}^{n_i}\frac{\alpha}{\alpha+\beta}\lambda_j^i\left(x_j^i, s_j^i+(\alpha+\beta)r_j^i\right) + \frac{\beta}{\alpha+\beta}\left(x, f(x)+\gamma^i+(\alpha+\beta)t_2\right)\\ &\quad = \left(\frac{\alpha}{\alpha+\beta}x^i, t_i^i\right) + \frac{\beta}{\alpha+\beta}\left(x, f(x)+\gamma^i - \sum\nolimits_{j=1}^{n_i}\lambda_j^i s_j^i + (\alpha+\beta)t_2\right)\\ &\quad \in \operatorname{co}\left(\operatorname{epi}\left(f+(\alpha+\beta)g\right)\right).\end{aligned}$$

Taking the limit we obtain that $(x, t_1+\beta t_2) = (x,t) \in \overline{\operatorname{co}}\left(\operatorname{epi}\left(f+(\alpha+\beta)g\right)\right)$.

Assume that $h := \overline{\operatorname{co}}\left(f+(\alpha+\beta)g\right)$ is proper; then, by Theorem 2.3.4, we have that $\left(f+(\alpha+\beta)g\right)^{**} = h$. Because h is proper, from the first part we have $\overline{\operatorname{co}}\left(\overline{\operatorname{co}}(f+\alpha g)+\beta g\right)$ proper, too, and so $\overline{\operatorname{co}}(f+\alpha g)$ is proper; in the contrary case, by Proposition 2.2.5, $\overline{\operatorname{co}}(f+\alpha g)(x) = -\infty$ for all $x \in \operatorname{dom}\overline{\operatorname{co}}(f+\alpha g) \supset \operatorname{dom} f \cap \operatorname{dom} g$, and so $\overline{\operatorname{co}}(f+\alpha g)(x) + \beta g(x) = -\infty$ for $x \in \operatorname{dom} f \cap \operatorname{dom} g$. Therefore $(f+\alpha g)^{**} = \overline{\operatorname{co}}(f+\alpha g)$ and so $\left((f+\alpha g)^{**}+\beta g\right)^{**} = \left(f+(\alpha+\beta)g\right)^{**}$. Hence $\left((f+\alpha g)^{**}+\beta g\right)^{*} = \left(f+(\alpha+\beta)g\right)^{*}$. This relation is obvious when h is not proper.

Exercise 2.20 Assume that C is bounded and $A + C \subset B + C$. Then $s_{A+C} \le$

s_{B+C}, and so $s_A + s_C \le s_B + s_C$. Because C is bounded, as noted on page 79, $s_C(x^*) \in \mathbb{R}$ for every $x^* \in X^*$. From the preceding inequality we get $s_A \le s_B$. Using Theorem 2.4.14(vi) we obtain that $A \subset \overline{\text{co}}B$.

Exercise 2.21 (i) $\Rightarrow$ (ii) is obvious: $f(x) \le f(0) + L\|x\|$.

(ii) $\Rightarrow$ (iii) Let $u \in X$. Then

$$f_\infty(u) = \lim_{t\to\infty} t^{-1}(f(tu) - f(0)) \le \lim_{t\to\infty} t^{-1}(L\|tu\| + \alpha - f(0)) = L\|u\|.$$

(iii) $\Rightarrow$ (i) Let first $x \in \operatorname{dom} f$ and take $y \in X$. Then

$$f(y) - f(x) \le \sup_{t>0} t^{-1}(f(x + t(y-x)) - f(x)) = f_\infty(y-x) \le L\|x-y\|.$$

This shows that $\operatorname{dom} f = X$, and so the above inequality holds for all $x, y \in X$. Changing x and y we obtain that f is L-Lipschitz.

Exercise 2.22 Let $u \in X$ be fixed and consider $\gamma := \sup\{f(x+u) - f(x) \mid x \in \operatorname{dom} f\}$. The inequality $f_\infty(u) \ge \gamma$ follows immediately from Eq. (2.28) (see page 74). Hence the conclusion holds if $\gamma = \infty$. Assume that $\gamma < \infty$. Fix $x_0 \in \operatorname{dom} f$; then $x_0 + nu \in \operatorname{dom} f$ for every $n \in \mathbb{N}$. By Eq. (2.28) we have that $f(x_0 + ku) - f(x_0 + (k-1)u) \le \gamma$ for every $k \in \mathbb{N}$. Summing for k from 1 to n we get $f(x_0 + nu) - f(x_0) \le n\gamma$ Dividing by n and letting $n \to \infty$, we obtain that $f_\infty(u) \le \gamma$.

Exercise 2.23 Let $x_0 \in [f \le \lambda]$ be fixed. Take first $u \in [f \le \lambda]_\infty$. Then $f(x_0+tu) \le \lambda$ for every $t \ge 0$, and so $f_\infty(u) = \lim_{t\to\infty} t^{-1}(f(x_0+tu)-f(x_0)) \le 0$. Hence $[f \le \lambda]_\infty \subset [f_\infty \le 0]$. Let now $u \in [f_\infty \le 0]$. Then $f(x_0 + tu) - f(x_0) \le tf_\infty(u) \le 0$ for every $t \ge 0$, and so $x_0+tu \in [f \le \lambda]$ for $t \ge 0$. Hence $u \in [f \le \lambda]_\infty$.

Let $x^* \in \partial_\varepsilon f(x)$ be fixed and consider $u^* \in (\partial_\varepsilon f(x))_\infty$. Then $x^* + tu^* \in \partial_\varepsilon f(x)$ for every $t \ge 0$. It follows that $\langle x' - x, x^* + tu^*\rangle \le f(x') - f(x) + \varepsilon$ for all $x' \in \operatorname{dom} f$ and $t \ge 0$. Passing to the limit for $t \to \infty$ we get $\langle x' - x, u^*\rangle \le 0$ for every $x' \in \operatorname{dom} f$, and so $u^* \in N(\operatorname{dom} f; x)$. Conversely, if $u^* \in N(\operatorname{dom} f; x)$ then $\langle x' - x, u^*\rangle \le 0$ for every $x' \in \operatorname{dom} f$; this together with $\langle x' - x, x^*\rangle \le f(x') - f(x) + \varepsilon$ for $x' \in \operatorname{dom} f$ gives $\langle x' - x, x^* + tu^*\rangle \le f(x') - f(x) + \varepsilon$ for all $x' \in \operatorname{dom} f$ and $t \ge 0$. Therefore $x^* + tu^* \in \partial_\varepsilon f(x)$ for $t \ge 0$, and so $u^* \in (\partial_\varepsilon f(x))_\infty$.

Fix $(x^*, \lambda) \in \operatorname{epi} f^*$ and let $(u^*, \gamma) \in (\operatorname{epi} f^*)_\infty$. Then $(x^* + tu^*, \lambda + t\gamma) \in \operatorname{epi} f^*$, whence $\langle x, x^* + tu^*\rangle - f(x) \le \lambda + t\gamma$ for all $x \in \operatorname{dom} f$ and $t \ge 0$. Taking the limit for $t \to \infty$ we obtain that $\langle x, u^*\rangle \le \gamma$ for every $x \in \operatorname{dom} f$; hence $s_{\operatorname{dom} f}(u^*) \le \gamma$, i.e. $(u^*, \gamma) \in \operatorname{epi} s_{\operatorname{dom} f}$. Conversely, if $s_{\operatorname{dom} f}(u^*) \le \gamma$ then $\langle x, u^*\rangle \le \gamma$ for every $x \in \operatorname{dom} f$; this together with $\langle x, x^*\rangle - f(x) \le \lambda$ for $x \in \operatorname{dom} f$ gives $\langle x, x^* + tu^*\rangle - f(x) \le \lambda + t\gamma$ for all $x \in \operatorname{dom} f$ and $t \ge 0$. Therefore $f^*(x^* + tu^*) \le \lambda + t\gamma$ for $t \ge 0$, and so $(u^*, \gamma) \in (\operatorname{epi} f^*)_\infty$. Therefore $(\operatorname{epi} f^*)_\infty = \operatorname{epi} s_{\operatorname{dom} f}$, which shows that $(f^*)_\infty = s_{\operatorname{dom} f}$.

The formula $f_\infty = s_{\operatorname{dom} f^*}$ follows applying the preceding one for f^*.

Let $D := \Pr_Y(\operatorname{dom} F)$ and take $v^* \in Y^*$ such that $(F^*)_\infty(0, v^*) \le 0$; then, by what precedes, $s_{\operatorname{dom} F}(0, v^*) \le 0$, and so $\langle y, v^* \rangle \le 0$ for every $(x, y) \in \operatorname{dom} F$. Hence $\langle y, v^* \rangle \le 0$ for every $y \in D$, *i.e.* $v^* \in D^-$. Conversely, if $v^* \in D^-$ then $\langle y, v^* \rangle \le 0$ for every $y \in D$, and so $s_{\operatorname{dom} F}(0, v^*) \le 0$.

Exercise 2.24 (i) Since f_∞ is a lsc function, it is clear that K is a closed convex cone and so X_0 is a closed linear subspace of X. Let $u \in X_0$ and $x \in \operatorname{dom} f$. Taking into account Eq. (2.27), we have that

$$f(x+u) - f(x) \le f_\infty(u) \le 0, \quad f(x-u) - f(x) \le f_\infty(-u) \le 0.$$

Adding these two relations side by side, we get

$$0 \le 2\left(\tfrac{1}{2} f(x+u) + \tfrac{1}{2} f(x-u) - f(x)\right) \le f_\infty(u) + f_\infty(-u) \le 0,$$

and so $f(x+u) - f(x) = f_\infty(u) = 0$. Let now $x \notin \operatorname{dom} f$ and $u \in X_0$. If $x + u \in \operatorname{dom} f$, by what precedes we obtain that $x = (x+u) + (-u) \in \operatorname{dom} f$; hence $x + u \in \operatorname{dom} f$ and so $f(x+u) = f(x)$.

(ii) Note first that $\operatorname{dom} f^* \subset X_0^\perp$, and so $\overline{\operatorname{lin}(\operatorname{dom} f^*)} \subset X_0^\perp$. Indeed, let $x^* \in \operatorname{dom} f^*$; taking $x_0 \in \operatorname{dom} f$ and $u \in X_0$, we have

$$\langle x_0 + tu, x^* \rangle \le f(x_0 + tu) + f^*(x^*) = f(x_0) + f^*(x^*) \ \forall t \in \mathbb{R},$$

whence $\langle u, x^* \rangle = 0$. Hence $x^* \in X_0^\perp$.

To complete the proof it is sufficient to show that $(\operatorname{dom} f^*)^\perp \subset X_0$ ($\Leftrightarrow X_0^\perp \subset (\operatorname{dom} f^*)^{\perp\perp} = \overline{\operatorname{lin}(\operatorname{dom} f^*)}$). So, let $u \in (\operatorname{dom} f^*)^\perp$; then $s_{\operatorname{dom} f^*}(u) = 0$, and so, by the preceding exercise, $f_\infty(u) = 0$. Therefore $(\operatorname{dom} f^*)^\perp \subset K$, whence $(\operatorname{dom} f^*)^\perp \subset X_0$.

(iii) As X_0 is a closed linear subspace of X, the quotient space X/X_0 is a separated locally convex space, too (X/X_0 being endowed with the quotient topology). Moreover, its topological dual is $X_0^\perp$, and its weak* topology is the trace of w^* on $X_0^\perp$. When $x \in X$, its class $\widehat{x}$ is $x + X_0$. From (i) we obtain that $\widehat{f}$ is well defined. Moreover, $\operatorname{epi} \widehat{f} = \{(\widehat{x}, t) \mid (x, t) \in \operatorname{epi} f\} = \pi(\operatorname{epi} f)$, where π is the canonical projection of $X \times \mathbb{R}$ on $(X \times \mathbb{R})/(X_0 \times \{0\})$. Hence $\operatorname{epi} \widehat{f}$ is a convex set. Furthermore, because $\operatorname{epi} f + X_0 \times \{0\} = \operatorname{epi} f$, $\operatorname{epi} \widehat{f}$ is also closed. Hence $\widehat{f}$ is a proper lsc convex function.

The relation $\widehat{f}_\infty(\widehat{u}) = f_\infty(u)$ follows immediately from (i) and Eq. (2.27). Using (i) one obtains easily the other two relations.

Exercise 2.25 Replacing eventually f by $\varepsilon^{-1} f$, we may take $\varepsilon = 1$. Because $f_\infty(u) \le 0$, the mapping $\mathbb{R} \ni t \mapsto f(x + tu) \in \overline{\mathbb{R}}$ is nonincreasing for every $x \in X$ and $\operatorname{dom} f + \mathbb{R}_+ u = \operatorname{dom} f$, as seen in Remark 2.2.2. On the other hand, since $\lim_{t\to\infty} t^{-1}(f(x - tu) - f(x)) = f_\infty(-u) > 0$ for $x \in \operatorname{dom} f$, for every such x

there exists $t_x > 0$ such that

$$f(x - tu) \geq f(x) + t\gamma \quad \forall\, t \geq t_x, \tag{$^\circ$}$$

where $\gamma := \frac{1}{2} f_\infty(-u) > 0$. Consider $g : X \to \overline{\mathbb{R}}$ defined by $g(x) := \exp(-t)$ for $x = tu$ with $t \geq 0$ and $g(x) := \infty$ elsewhere. Consider $f_1 := f \Box g$. Hence $\operatorname{dom} f_1 = \operatorname{dom} f + \operatorname{dom} g = \operatorname{dom} f + \mathbb{R}_+ u = \operatorname{dom} f$. Let $x \in \operatorname{dom} f$; then there exists $\bar{t}_x \geq 0$ such that

$$f_1(x) = f(x - \bar{t}_x u) + \exp(-\bar{t}_x) \leq f(x) + g(0) = f(x) + 1; \tag{$^{\circ\circ}$}$$

this fact is obvious because, by ($^\circ$), $\lim_{t\to\infty} (f(x - tu) + \exp(-t)) = \infty$. Moreover, because $f(x) \leq f(x - tu)$ for every $t \geq 0$, we have that

$$f_1(x) = \inf\{f(x - tu) + \exp(-t) \mid t \geq 0\} \geq \inf\{f(x - tu) \mid t \geq 0\} = f(x).$$

Therefore $f \leq f_1 \leq f + 1$. Let us show that f_1 is lsc. Consider $(x_i)_{i\in I} \to x \in X$ with $(f_1(x_i))_{i\in I} \to \mu < \infty$. We may assume that $x_i \in \operatorname{dom} f_1 = \operatorname{dom} f$ for every $i \in I$. We must show that $f(x) \leq \mu$. By ($^{\circ\circ}$) we have that $f_1(x_i) = f(x_i - t_i u) + \exp(-t_i)$, where $t_i := \bar{t}_{x_i} \geq 0$. Suppose that $(t_i) \to \infty$. Then for every $t \geq 0$ there exists $i_t \in I$ such that $t_i \geq t$ for $i \succeq i_t$. Hence $f(x_i - t_i u) + \exp(-t_i) \geq f(x_i - tu)$, for $i \succeq i_t$, and so $\mu \geq f(x - tu)$ for every $t \geq 0$. This contradicts ($^\circ$). Therefore $(t_i)_{i\in I}$ has a subnet $(t'_j)_{j\in J}$ converging to $t \in \mathbb{R}_+$; hence $\mu \geq f(x - tu) + \varepsilon \exp(-t) \geq f_1(x)$ because f is lsc at $x - tu$. It follows that f_1 is lsc. Since $\inf f_1 = \inf f + \inf g = \inf f$ and $f_1(x) = f(x - \bar{t}_x u) + \exp(-\bar{t}_x) > f(x - \bar{t}_x u) \geq \inf f = \inf f_1$ for every $x \in \operatorname{dom} f_1$, we have that $\operatorname{argmin} f_1 = \emptyset$.

Exercise 2.26 As proved in Proposition 2.7.2, condition (ii) holds if one of conditions (iii)–(viii) is verified. As noted before this proposition, condition (ii) also holds for $\widetilde{F}$, where $\widetilde{F}(x, y) := F(x, y) - \langle x, x^* \rangle$ with $x^* \in X^*$. Taking $\varphi := F(\cdot, 0)$, using Theorem 2.7 1, we have

$$\begin{aligned}\varphi^*(x^*) &= \sup_{x\in X} (\langle x, x^* \rangle - F(x, 0)) = -\inf_{x\in X} \widetilde{F}(x, 0) = \min_{y^*\in Y^*} \widetilde{F}^*(0, y^*) \\ &= \min_{y^*\in Y^*} F^*(x^*, y^*).\end{aligned}$$

Therefore g is w^*-lsc and the values of g are attained.

It is obvious that $\iota_{\operatorname{dom} F}$ satisfies condition (ii) of Theorem 2.7.1, too. So, using once again this theorem, we have, as above, that

$$\begin{aligned}s_{\operatorname{dom}\varphi}(u^*) &= \sup_{x\in X} (\langle x, x^* \rangle - \iota_{\operatorname{dom}\varphi}(x)) = \sup_{x\in X} (\langle x, x^* \rangle - \iota_{\operatorname{dom} F}(x, 0)) \\ &= \min_{v^*\in Y^*} (\iota_{\operatorname{dom} F})^*(u^*, v^*) = \min_{v^*\in Y^*} s_{\operatorname{dom} F}(u^*, v^*)\end{aligned}$$

for every $u^* \in X^*$. From Exercise 2.23 we have that $s_{\operatorname{dom}\varphi} = (\varphi^*)_\infty = g_\infty$ and $s_{\operatorname{dom} F} = (F^*)_\infty$. So $g_\infty(u^*) = \min_{v^* \in Y^*} (F^*)_\infty(u^*, v^*)$ for every $u^* \in X^*$.

Let $D := \Pr_Y(\operatorname{dom} F)$ and $Y_0 = \operatorname{lin} D$. It is clear that condition (ii) of Theorem 2.7.1 implies that $0 \in {}^iD$, and so $Y_0^\perp = D^\perp = D^-$. The conclusion follows from the last part of Exercise 2.23.

Exercise 2.27 We consider that $a - b := a + (-b)$, the sum being taken in the sense of convex analysis: $(+\infty)+(-\infty) = +\infty$. We have that $\inf_{x\in X} (f(x) - g(x))$ equals

$$\inf_{x\in X} \left[f(x) - \sup_{x^*\in X^*} \left(\langle x, x^*\rangle - g^*(x^*)\right)\right] = \inf_{x\in X} \left[f(x) + \inf_{x^*\in X^*} \left(g^*(x^*) - \langle x, x^*\rangle\right)\right]$$
$$= \inf_{x\in X, x^*\in X^*} \left[f(x) + g^*(x^*) - \langle x, x^*\rangle\right] = \inf_{x^*\in X^*} \left[g^*(x^*) + \inf_{x\in X} \left(f(x) - \langle x, x^*\rangle\right)\right]$$
$$= \inf_{x^*\in X^*} \left[g^*(x^*) - \sup_{x\in X} \left(\langle x, x^*\rangle - f(x)\right)\right] = \inf_{x^*\in X^*} \left(g^*(x^*) - f^*(x^*)\right).$$

Exercise 2.28 Let $x \in \operatorname{dom} f$ be fixed and consider $\varphi : \mathbb{P} \to \overline{\mathbb{R}}$, $\varphi(t) := t^{-p} f(tx)$. Then for every $t \in I := \{t \in \mathbb{P} \mid \varphi(t) < \infty\}$ we have that $\varphi'_+(t) = t^{-p} f'(tx, x) - pt^{-p-1} f(tx)$ and $\varphi'_-(t) = -t^{-p} f'(tx, -x) - pt^{-p-1} f(tx)$.

(i) $\Rightarrow$ (ii) Since φ is nondecreasing we obtain that φ'_+ and φ'_- are nonnegative on I. In particular we get $\varphi'_-(1) = -f'(x, -x) - pf(x) \geq 0$.

(ii) $\Rightarrow$ (iii) This implication is obvious because $0 \leq f'(x, -x) + f'(x, x)$.

(iii) $\Rightarrow$ (i) For $x \in \operatorname{dom} f$ and $t \in I$ we have that $tx \in \operatorname{dom} f$, and so, by hypothesis, $pf(tx) \leq f'(tx, tx) = tf'(tx, x)$; hence $\varphi'_+(t) \geq 0$ for every $t \in I$. But φ is locally Lipschitz on $\operatorname{int} I$ as the product of two locally Lipschitz functions, and so $\varphi(t) - \varphi(s) = \int_s^t \varphi'(r)\, dr = \int_s^t \varphi'_+(r)\, dr \geq 0$ for all $s, t \in \operatorname{int} I$ with $s < t$. Hence φ is nondecreasing on $\operatorname{int} I$. Assume that $\overline{t} = \sup I \in I$. Since the mapping $I \ni t \mapsto f(tx)$ is continuous, we obtain that $\varphi(s) \leq \varphi(\overline{t})$ for $0 < s < \overline{t}$.

(ii) $\Rightarrow$ (iv) Let $(x, x^*) \in \operatorname{gr} \partial f$. Then $\langle -x, x^*\rangle \leq f'(x, -x) \leq -pf(x)$, whence the conclusion.

Assume now that f is continuous at 0; of course, $0 \in \operatorname{int}(\operatorname{dom} f)$ and f is continuous on $\operatorname{int}(\operatorname{dom} f)$ in this case. Moreover $\partial f(x) \neq \emptyset$ for every $x \in \operatorname{int}(\operatorname{dom} f)$, and so the implication (iv) $\Rightarrow$ (v) is obvious.

(v) $\Rightarrow$ (i) Consider $x \in \operatorname{int}(\operatorname{dom} f)$. Taking $x^* \in \partial f(x)$ furnished by our hypothesis, we have that $f'(x, x) \geq \langle x, x^*\rangle \geq pf(x)$. Hence $pf(x) \leq f'(x, x)$ for every $x \in \operatorname{int}(\operatorname{dom} f)$. Let now $x \in \operatorname{dom} f$. Then $tx \in \operatorname{int}(\operatorname{dom} f)$ for every $t \in \operatorname{int} I$. As in the proof of the implication (iii) $\Rightarrow$ (i) we obtain that the function φ is nondecreasing.

Exercise 2.29 Take first $x^* \in \bigcap_{i=1}^k \partial f(\overline{x}_i)$. Then $\langle x - \overline{x}_i, x^*\rangle \leq f(x) - f(\overline{x}_i)$ for $x \in \operatorname{dom} f$. Multiplying with λ_i, then adding side by side the corresponding inequalities, we get $\langle x - \overline{x}, x^*\rangle \leq f(x) - \overline{\operatorname{co}} f(\overline{x})$ for $x \in \operatorname{dom} f$. Take now $(x, t) = \sum_{j=1}^p \lambda_j (x_j, t_j) \in \operatorname{co}(\operatorname{epi} f)$ with $(x_j, t_j)_{j\in\overline{1,p}} \subset \operatorname{epi} f$ and $\lambda \in \Delta_p$. From the preceding inequality we get, as above, $\langle x-\overline{x},x^*\rangle \le t-\overline{\mathrm{co}}f(\overline{x})$. It follows that $\langle x-\overline{x},x^*\rangle \le t-\overline{\mathrm{co}}f(\overline{x})$ for all $(x,t)\in\overline{\mathrm{co}}(\operatorname{epi} f)$, and so $x^*\in\partial\overline{\mathrm{co}}f(\overline{x})$.

Conversely, assume that $x^*\in\partial\overline{\mathrm{co}}f(\overline{x})$. Fix $i_0\in\overline{1,k}$ and take $x\in\operatorname{dom} f$. Then $\sum_{i\neq i_0}\overline{\lambda}_i\,(\overline{x}_i,f(\overline{x}_i))+\overline{\lambda}_{i_0}\,(x,f(x))\in\operatorname{co}(\operatorname{epi} f)\subset\operatorname{epi}(\overline{\mathrm{co}}f)$, and so

$$\left\langle\sum\nolimits_{i\neq i_0}\overline{\lambda}_i\overline{x}_i+\overline{\lambda}_{i_0}x-\overline{x},x^*\right\rangle\le\sum\nolimits_{i\neq i_0}\overline{\lambda}_if(\overline{x}_i)+\overline{\lambda}_{i_0}f(x)-\overline{\mathrm{co}}f(\overline{x}),$$

whence $\langle x-\overline{x}_{i_0},x^*\rangle\le f(x)-f(\overline{x}_{i_0})$ because $\overline{\lambda}_{i_0}>0$. Therefore $x^*\in\bigcap_{i=1}^k\partial f(\overline{x}_i)$.

Exercise 2.30 Let $f_0:=f|_{X_0}$. Then $\partial f_0(0)=\{x^*|_{X_0}\mid x^*\in\partial f(0)\}$ and $f'(0,x)=f_0'(0,x)$ for $x\in X_0$, $f'(0,x)=\infty$ $x\in X\setminus X_0$. Since f_0 is continuous at 0 ($\in X_0$), we have that $f_0'(0,x)=\max\{\langle x,x^*\rangle\mid x^*\in\partial f_0(0)\}$ for all $x\in X_0$. Hence, for $\overline{x}\in X_0$,

$$\begin{aligned}f'(0,\overline{x})&=\max\{\langle\overline{x},x^*\rangle\mid x^*\in\partial f_0(0)\}=\sup\{\langle\overline{x},x^*\rangle\mid x^*\in\partial f_0(0)\}\\&=\sup\{\langle\overline{x},x^*|_{X_0}\rangle\mid x^*\in\partial f(0)\}=\sup\{\langle\overline{x},x^*\rangle\mid x^*\in\partial f(0)\}.\end{aligned}$$

If $\overline{x}\notin\overline{X_0}=X_0^{\perp\perp}$, there exists $\overline{x}^*\in X^*$ such that $\overline{x}^*|_{X_0}=0$ and $\langle\overline{x},\overline{x}^*\rangle\neq 0$. Taking $\overline{x}_0^*\in\partial f(0)$ ($\neq\emptyset$ by Theorem 2.4.12), we obtain that

$$\sup\{\langle\overline{x},x^*\rangle\mid x^*\in\partial f(0)\}\ge\sup\{\langle\overline{x},\overline{x}_0^*+t\overline{x}^*\rangle\mid t\in\mathbb{R}\}=\infty=f'(0,\overline{x}).$$

Let now $\overline{x}\in\overline{X_0}\setminus X_0$. Since f_0 is continuous at 0, by Theorem 2.2.11, f_0 is Lipschitz at 0, i.e. there exists $V_0\in\mathcal{N}^c_{X_0}$ such that $V_0\subset\operatorname{dom} f_0$ and $|f_0(x)-f_0(y)|\le p_{V_0}(x-y)$ for all $x,y\in V_0$, where p_{V_0} is the Minkowski gauge associated to V_0. It follows that $f_0'(0,x)\le p_{V_0}(x)$ for all $x\in X_0$. Replacing eventually V_0 by a subset, we may suppose that $V_0=V\cap X_0$, where $V\in\mathcal{N}^c_X$. So we have that $f_0'(0,x)\le p_V(x)$ for all $x\in X_0$. It follows that

$$\forall x\in X_0,\ \forall x^*\in\partial f(0)\ :\ \langle x,x^*\rangle=\langle x,x^*|_{X_0}\rangle\le f_0'(0,x)\le p_V(x).$$

As $\overline{x}\in\overline{X_0}$, there exists a net $(x_i)\subset X_0$ converging to $\overline{x}$. From the relation above, taking into account the continuity of $x^*\in\partial f(0)$ and p_V, we obtain that $\langle\overline{x},x^*\rangle\le p_V(x)$ for all $x^*\in\partial f(0)$, and so $\sup\{\langle\overline{x},x^*\rangle\mid x^*\in\partial f(0)\}\le p_V(\overline{x})<\infty$.

Exercise 2.31 We take $x\neq y$ (the other case being obvious). Consider the function $\varphi:\mathbb{R}\to\overline{\mathbb{R}}$, $\varphi(t)=f((1-t)x+ty)-t(f(y)-f(x))$. Because φ is convex, $\varphi(t)\le\varphi(0)=\varphi(1)$ for every $t\in[0,1]$. Since f is continuous on $\operatorname{int}(\operatorname{dom} f)$, φ is continuous on $[0,1]$. Therefore there exists $t_0\in[0,1]$ such that $\varphi(t_0)\le\varphi(t)$ for all $t\in[0,1]$. If $\varphi(t_0)<\varphi(0)$ then $t_0\in]0,1[$; otherwise $\varphi(t)=\varphi(0)$ for every $t\in[0,1]$ and take $t_0=1/2$ (for example). Take $z:=(1-t_0)x+t_0y$. From the choice of t_0 we obtain that

$$\forall t\in[0,1]\ :\ (t-t_0)(f(y)-f(x))\le f(z+(t-t_0)(y-x))-f(z).$$

Dividing by $t-t_0$ for $t \in]t_0, 1]$ and taking the limit for $t \to t_0$ we get $f(y)-f(x) \leq f'(z, y-x)$, while by dividing by $t-t_0$ for $t \in [0, t_0[$ and taking the limit for $t \to t_0$ we get $f(y) - f(x) \geq -f'(z, x-y)$. As

$$f'(z, y-x) = \max\{\langle y-x, z^*\rangle \mid z^* \in \partial f(z)\},$$
$$-f'(z, x-y) = \min\{\langle y-x, z^*\rangle \mid z^* \in \partial f(z)\},$$

we obtain the conclusion.

Exercise 2.32 (i) The equivalence of a) and b) is true for any multifunction $T : X \rightrightarrows X^*$, and follows immediately.

a) $\Rightarrow$ c) Suppose that ∂f is single valued on $\operatorname{dom}\partial f$. Assume that there exist distinct $x_0^*, x_1^* \in X^*$ and $\lambda \in]0,1[$ such that $[x_0^*, x_1^*] \subset \operatorname{Im}\partial f$ and $f^*(x_\lambda^*) = (1-\lambda)f^*(x_0^*) + \lambda f^*(x_1^*)$, where $x_\lambda^* := (1-\lambda)x_0^* + \lambda x_1^* \in \operatorname{Im}\partial f$. Let $x \in X$ such that $x_\lambda^* \in \partial f(x)$. Then, by Young–Fenchel inequality,

$$\begin{aligned} 0 &= f(x) + f^*(x_\lambda^*) - \langle x, x_\lambda^*\rangle \\ &= (1-\lambda)\left(f(x) + f^*(x_0^*) - \langle x, x_0^*\rangle\right) + \lambda\left(f(x) + f^*(x_1^*) - \langle x, x_1^*\rangle\right) \geq 0. \end{aligned}$$

It follows that $f(x)+f^*(x_i^*)-\langle x, x_i^*\rangle = 0$ for $i \in \{0,1\}$, whence the contradiction $\{x_0^*, x_1^*\} \subset \partial f(x)$.

c) $\Rightarrow$ a) Assume that ∂f is not single valued on $\operatorname{dom}\partial f$. Then there exist $x \in X$ and distinct $x_0^*, x_1^* \in \partial f(x)$. Of course, $[x_0^*, x_1^*] \subset \partial f(x) \subset \operatorname{Im}\partial f$. Hence $f(x) + f^*(x_\lambda^*) - \langle x, x_\lambda^*\rangle = 0$ for $\lambda \in [0,1]$, where $x_\lambda^* := (1-\lambda)x_0^* + \lambda x_1^*$. It follows that $f^*(x_\lambda^*) = (1-\lambda)f^*(x_0^*) + \lambda f^*(x_1^*)$, which shows that f^* is not strictly convex on $[x_0^*, x_1^*]$.

Assume now that f is continuous on $D := \operatorname{int}(\operatorname{dom} f)$ and $D \neq \emptyset$. Suppose there exists $\overline{x} \in \operatorname{dom}\partial f \setminus D$. By Theorem 1.1.3 there exists $x^* \neq 0$ such that $\langle x - \overline{x}, x^*\rangle \leq 0$ for every $x \in \operatorname{dom} f$. Taking $\overline{x}^* \in \partial f(\overline{x})$, we obtain that $\overline{x}^* + tx^* \in \partial f(\overline{x})$ for every $t > 0$, contradicting a).

(ii) Consider the topology w^* on X^*. Then $(f^*)^* = f$ and $\partial f^* = (\partial f)^{-1}$. By (i) we obtain that f is strictly convex on every segment $[x_0, x_1] \subset \operatorname{dom}\partial f$. But, by Theorem 2.4.9, $\operatorname{int}(\operatorname{dom} f) \subset \operatorname{dom}\partial f$. The conclusion follows.

Exercise 2.33 Let $(x^*, y^*) \in \partial f(x_0, y_0)$; it is obvious that $x^* \in \partial f(\cdot, y_0)(x_0)$ and $y^* \in \partial f(x_0, \cdot)(y_0)$, even without assuming that f is continuous at (x_0, y_0).

Let $x^* \in \partial f(\cdot, y_0)(x_0)$. Since f is continuous at $(x_0, y_0) \in \operatorname{dom} f$, the function $f'((x_0, y_0), \cdot)$ is finite, sublinear and continuous. Furthermore, since $x^* \in \partial f(\cdot, y_0)(x_0)$, we have that

$$\forall x \in X \ : \ \langle x, x^*\rangle \leq f(x_0 + x, y_0) - f(x_0, y_0),$$

and so

$$\forall x \in X \ : \ \langle x, x^*\rangle \leq f'((x_0, y_0), (x, 0)).$$

Consider the linear subspace $X \times \{0\}$ of $X \times Y$ and the linear functional $\varphi_0 : X \times \{0\} \to \mathbb{R}$, $\varphi_0(x,0) := \langle x, x^* \rangle$; taking into account the above inequality and the Hahn–Banach theorem, there exists a linear functional $\varphi : X \times Y \to \mathbb{R}$ such that

$$\varphi|_{X\times\{0\}} = \varphi_0 \text{ and } \forall\, (x,y) \in X \times Y \ : \ \varphi(x,y) \le f'((x_0,y_0),(x,y)).$$

Since $f'((x_0,y_0),\cdot)$ is continuous, from this inequality we get that φ is continuous, and so $\varphi \in \partial f(x_0,y_0)$. Taking $y^* := \varphi(0,\cdot) \in Y^*$ we obtain that $(x^*,y^*) \in \partial f(x_0,y_0)$. Therefore $x^* \in \mathrm{Pr}_{X^*}(\partial f(x_0,y_0))$. The conclusion follows.

Exercise 2.34 Let us consider the function

$$F : X \times \mathbb{R} \to \overline{\mathbb{R}}, \quad F(x,t) := \begin{cases} f_0(x) & \text{if} \quad f_1(x) \le t \\ \infty & \text{if} \quad f_1(x) > t, \end{cases}$$

and its associated marginal function $h : \mathbb{R} \to \overline{\mathbb{R}}$; h is a decreasing convex function such that $h(0) = v \in \mathbb{R}$ and $h^{**}(0) = v^*$. By Theorem 2.6.1(v) we have that $v^* = v$ if and only if h is lsc at 0. Since h is decreasing, we have:

$$\begin{aligned} v^* = v &\Leftrightarrow \forall\varepsilon > 0,\ \exists\delta > 0,\ \forall t \in]-\delta,\delta[\ :\ h(t) > y - \varepsilon \\ &\Leftrightarrow \forall\varepsilon > 0,\ \exists\delta > 0\ :\ h(\delta) > v - \varepsilon \\ &\Leftrightarrow \forall\varepsilon > 0,\ \exists\delta > 0\ :\ [f_1(x) \le \delta \Rightarrow f_0(x) > y - \varepsilon] \\ &\Leftrightarrow \forall\varepsilon > 0,\ \exists\delta > 0\ :\ [f_0(x) \le v - \varepsilon \Rightarrow f_1(x) > \delta] \\ &\Leftrightarrow \forall\varepsilon > 0\ :\ \inf\{f_1(x) \mid f_0(x) \le v - \varepsilon\} > 0, \end{aligned}$$

i.e. the conclusion holds.

Exercise 2.35 "$\Rightarrow$" If $x^* \le f$ then $f(x) \ge \langle x, x^* \rangle \ge -\|x\| \cdot \|x^*\|$ for every $x \in X$, and so the conclusion holds with $M = \|x^*\|$.

"$\Leftarrow$" Let $g : X \to \mathbb{R}$ be defined by $g(x) := M\|x\|$. The function g being continuous, from the Fenchel–Rockafellar duality formula we have

$$0 \le \inf_{x\in X} (f(x) + M\|x\|) = \max_{x^*\in X^*} (-f^*(x^*) - g^*(-x^*)) = \max_{\|x^*\|\le M} (-f^*(x^*)).$$

Since $-f^*(x^*) = \inf_{x\in X} (f(x) - \langle x, x^* \rangle)$, the conclusion follows.

Exercise 2.36 "$\Rightarrow$" If $-f_1 \le x^* + \alpha \le f_2$ then $f_1(x_1) \ge -\langle x_1, x^* \rangle - \alpha$ and $f_2(x_2) \ge \langle x_2, x^* \rangle + \alpha$, and so $f_1(x_1) + f_2(x_2) \ge \langle x_2 - x_1, x^* \rangle \ge -\|x_2 - x_1\| \cdot \|x^*\|$ for all $x_1, x_2 \in X$, and so the conclusion holds with $M = \|x^*\|$.

"$\Leftarrow$" Let $f, g : X \times X \to \overline{\mathbb{R}}$ be defined by $f(x_1,x_2) := f_1(x_1) + f_2(x_2)$ and $g(x_1,x_2) := M\|x_2 - x_1\|$. The function g being continuous, from the Fenchel–

Rockafellar duality formula we have

$$0 \le \inf_{x_1, x_2 \in X} (f(x_1, x_2) + g(x_1, x_2)) = \max_{x_1^*, x_2^* \in X^*} (-f^*(x_1^*, x_2^*) - g^*(-x_1^*, -x_2^*)) .$$

But $g^*(-x_1^*, -x_2^*) = 0$ if $x_1^* = -x_2^* \in U_{X^*}$, $g^*(-x_1^*, -x_2^*) = \infty$ otherwise. Therefore there exists $x^* \in X^*$ such that

$$\begin{aligned} 0 \le -f^*(-x^*, x^*) &= \inf_{x_1, x_2 \in X} (f_1(x_1) + f_2(x_2) + \langle x_1, x^* \rangle - \langle x_2, x^* \rangle) \\ &= \inf_{x_1 \in X} (f_1(x_1) + \langle x_1, x^* \rangle) + \inf_{x_2 \in X} (f_2(x_2) - \langle x_2, x^* \rangle) . \end{aligned}$$

Note that both infima are finite. Taking $\alpha := \inf_{x \in X} (f_1(x) + \langle x, x^* \rangle) \in \mathbb{R}$, we get the conclusion.

Exercise 2.37 The implication "$\Leftarrow$" follows immediately from the obvious inequalities $-2 \langle y, y^* \rangle \le 2 \|y\| \cdot \|y^*\| \le \|y\|^2 + \|y^*\|^2$.

"$\Rightarrow$" Consider the topology $\sigma(X, X')$ on X; so X is a separated locally convex space and $T : X \to Y$ is a continuous linear operator. Take $g : Y \to \mathbb{R}$ defined by $g(y) := \|y + y_0\|^2$. The function g being continuous, from the Fenchel–Rockafellar duality formula we have

$$0 \le \inf_{x \in X} (f(x) + \|Tx + y_0\|) = \max_{y^* \in Y^*} (-f^*(T^* y^*) - g^*(-y^*)) .$$

But

$$g^*(-y^*) = \langle y_0, y^* \rangle + \frac{1}{4} \|y^*\|^2 , \quad -f^*(T^* x^*) = \inf_{x \in X} (f(x) - \langle Tx, x^* \rangle)$$

(see eventually Theorems 2.3.1 and 3.7.2); the conclusion follows replacing y^* by $2y^*$.

Exercise 2.38 Taking $f : X \to \mathbb{R}$ defined by $f(x) := \|x - \overline{x}\|$ and $g := \iota_C$, we have that $d(\overline{x}, C) = \inf_{x \in X} (f(x) + g(x))$. We notice that $f^*(x^*) = \langle \overline{x}, x^* \rangle + \iota_{U_{X^*}}(x^*)$ and $g^* = \iota_{-C^+}$. Because f is continuous on X, applying the Fenchel–Rockafellar duality formula we get the first formula.

In order to prove the second formula take $f := \overline{x}^* + \iota_{U_X}$ and the same g as above. Notice that $f^*(x^*) = \|x^* - \overline{x}^*\|$ for $x^* \in X^*$. Because f is continuous at $0 \in \operatorname{dom} f \cap \operatorname{dom} g$, we can again use the Fenchel–Rockafellar duality formula. Hence $\inf_{x \in U_X \cap C} \langle x, \overline{x}^* \rangle = \max_{x^* \in C^+} (- \|x^* - \overline{x}^*\|)$. The conclusion follows.

Using the previous formulas we get

$$\sup_{x \in U_X \cap C} d(x, D) = \sup_{x \in U_X \cap C} \sup_{x^* \in U_{X^*} \cap D^+} (- \langle x, x^* \rangle) = \sup_{x^* \in U_{X^*} \cap D^+} d(x^*, C^+).$$

Exercise 2.39 The inclusions $[1,\infty[\cdot\partial f(\overline{x}) \subset \partial^{\leq} f(\overline{x}) \subset \partial^{<} f(\overline{x})$ are obvious (and true without convexity). Let $x^* \in \partial^{<} f(\overline{x})$ and $x \in [f \leq f(\overline{x})]$; there exists $x_0 \in X$ such that $f(x_0) < f(\overline{x})$. It follows that $(1-\lambda)x + \lambda x_0 \in [f < f(\overline{x})]$ for every $\lambda \in]0,1[$, and so

$$\langle (1-\lambda)x + \lambda x_0 - \overline{x}, x^* \rangle \leq f((1-\lambda)x + \lambda x_0) - f(\overline{x}) \leq (1-\lambda)f(x) + \lambda f(x_0) - f(\overline{x}).$$

Taking the limit for $\lambda \to 0$ we obtain that $\langle x - \overline{x}, x^* \rangle \leq f(x) - f(\overline{x})$, and so $x^* \in \partial^{\leq} f(\overline{x})$.

Let now $x^* \in \partial^{\leq} f(\overline{x})$; this means that $\overline{x}$ is an optimal solution of the convex problem

$(P)\qquad \min\ f(x) - \langle x, x^* \rangle,\ g(x) := f(x) - f(\overline{x}) \leq 0.$

By Theorem 2.9.3, there exists $\lambda \geq 0$ such that $0 \in \partial(f - x^* + \lambda g)(\overline{x}) = \partial((1+\lambda)f - x^*)$, *i.e.* $x^* \in (1+\lambda)\partial f(\overline{x})$. Therefore $x^* \in \partial^{<} f(\overline{x})$.

Exercise 2.40 1) It is obvious that (c) $\Rightarrow$ (a) and that (c) $\Rightarrow$ (b).

(b) $\Rightarrow$ (c) Let $x_0 \in \operatorname{dom} f$; therefore $\sum_{n=1}^{\infty} \lambda_n \|x_0 - a_n\|^2 < \infty$. Let $x \in X$ be fixed. Then

$$\|x - a_n\|^2 \leq (\|x - x_0\| + \|x_0 - a_n\|)^2 \leq 2(\|x - x_0\|^2 + \|x_0 - a_n\|^2).$$

So we obtain that

$$\begin{aligned} f(x) = \sum_{n=1}^{\infty} \lambda_n \|x - a_n\|^2 &\leq 2\|x - x_0\|^2 + 2\sum_{n=1}^{\infty} \lambda_n \|x_0 - a_n\|^2 \\ &\leq 2\|x - x_0\|^2 + 2f(x_0), \end{aligned} \qquad (^*)$$

which shows that $x \in \operatorname{dom} f$.

Taking $x_0 = 0$ in the above proof, we have that (a) $\Rightarrow$ (c).

2) Suppose that $\sum_{n=1}^{\infty} \lambda_n \|a_n\|^2 < \infty$. By 1) we have that $\operatorname{dom} f = X$. Consider the function

$$f_n : X \to \mathbb{R}, \quad f_n := \lambda_1 d_{a_1}^2 + \cdots + \lambda_n d_{a_n}^2,$$

where $d_a := d_{\{a\}} : X \to \mathbb{R}$, $d_a(x) := \|x - a\|$. It is obvious that $\lim_{n\to\infty} f_n(x) = f(x)$ for every $x \in X$. Since f_n is a convex function for every $n \in \mathbb{N}$, using Theorem 2.1.3(iii), we obtain that f is convex. Applying relation $(^*)$ for $x_0 = 0$, we obtain that $f(x) \leq 2 + 2f(0)$ for all $x \in U_X$. Since $\infty > f(x) \geq 0$ for every $x \in X$, it follows that f is finite and continuous on X.

3) Assume that $\sum_{n=1}^{\infty} \lambda_n \|a_n\|^2 < \infty$. First note that the series $\sum_{n=1}^{\infty} \lambda_n \|a_n\|$ is convergent; this follows immediately from the inequality $2\lambda_n \|a_n\| \leq \lambda_n + \lambda_n \|a_n\|^2$.

Secondly, recall that $\partial d_a^2(x) = 2d_a(x)\partial d_a(x) = 2\|x - a\| \cdot \partial \|\cdot\| (x - a)$ for all $x, a \in X$.

Let $x \in X$ be fixed and consider $x_n^* \in \partial\|\cdot\|(x - a_n)$ for every $n \geq 1$. Since $x_n^* \in U_{X^*}$ and $\lambda_n \|x - a_n\| \leq \lambda_n \|x\| + \lambda_n \|a_n\|$, we obtain that the series $\sum_{n\geq 1} 2\lambda_n \|x - a_n\| \, x_n^*$ is convergent; let x^* be its sum. Because $f(u) = \lim f_n(u)$ for every $u \in X$ and $\sum_{k=1}^n 2\lambda_k \|x - a_k\| \, x_k^* \in \partial f_n(x)$, we obtain immediately that $x^* \in \partial f(x)$.

Take now $x^* \in \partial f(x)$. Consider the function $h_n := f - f_n$; the function h_n is also convex and continuous. It is clear that $h_n = \lambda_{n+1} d_{a_{n+1}}^2 + h_{n+1}$, and so $\partial h_n(x) = \lambda_{n+1} \partial d_{a_{n+1}}^2(x) + \partial h_{n+1}(x)$ for all $n \geq 1$. As $x^* \in \partial f(x)$ and $f = \lambda_1 d_{a_1}^2 + h_1$ $\big($and so $\partial f(x) = \lambda_1 \partial d_{a_1}^2(x) + \partial h_1(x)\big)$, there exists $x_1^* \in \partial \|\cdot\| (x - a_1)$ such that $x^* - 2\|x - a_1\| x_1^* \in \partial h_1(x)$. As $h_1 = \lambda_2 d_{a_2}^2 + h_2$, there exists $x_2^* \in \partial\|\cdot\|(x - a_2)$ such that $x^* - 2\|x - a_1\| x_1^* - 2\|x - a_2\| x_2^* \in \partial h_2(x)$. Continuing in this way we obtain a sequence $(x_n^*)_{n\geq 1} \subset X^*$ such that

$$x_n^* \in \partial\|\cdot\|(x - a_n), \quad x^* - \sum\nolimits_{k=1}^{n} 2\|x - a_k\| x_k^* \in \partial h_n(x) \quad \forall n \in \mathbb{N}.$$

But for $n \in \mathbb{N}$ and $u \in U_X$ we have that

$$\begin{aligned}(h_n)'(x,u) &\leq h_n(x+u) - h_n(u) = \sum\nolimits_{k=n+1}^{\infty} \lambda_k \left(\|x + u - a_k\|^2 - \|x - a_k\|^2\right) \\ &\leq r_n \|u\|,\end{aligned}$$

where

$$r_n := \sum\nolimits_{k=n+1}^{\infty} \lambda_k \left(2\|x\| + 1 + 2\|a_k\|\right).$$

Of course, $(r_n) \to 0$. It follows that $(h_n)'(x,u) \leq r_n \|u\|$ for every $u \in X$, and so $\partial h_n(x) \subset r_n U_{X^*}$. Therefore $\left\|x^* - \sum_{k=1}^n 2\|x - a_k\| x_k^*\right\| \leq r_n$ for every $n \in \mathbb{N}$, whence we get $x^* = 2\sum_{n\geq 1} \|x - a_n\| x_n^*$. Hence the formula for $\partial f(x)$ holds.

Exercise 2.41 It is obvious that (d) $\Rightarrow$ (e) $\Rightarrow$ (a) $\Rightarrow$ (b) $\Rightarrow$ (c); moreover these implications hold for an arbitrary function f.

(c) $\Rightarrow$ (b) Let $x_0 \in X$ be such that $f(x_0) < \overline{\lambda}$ and $\lambda > \overline{\lambda}$. Consider $x \in [f \leq \lambda]$ and $t := (\lambda - \overline{\lambda})/(\lambda - f(x_0)) \in]0,1[$. Then

$$f(tx_0 + (1-t)x) \leq tf(x_0) + (1-t)f(x) \leq tf(x_0) + (1-t)\lambda = \overline{\lambda}.$$

Therefore

$$\frac{\lambda - \overline{\lambda}}{\lambda - f(x_0)} \cdot x_0 + \frac{\overline{\lambda} - f(x_0)}{\lambda - f(x_0)} \cdot x \in [f \leq \overline{\lambda}],$$

whence we obtain that

$$x \in \frac{\lambda - f(x_0)}{\overline{\lambda} - f(x_0)} \cdot [f \leq \overline{\lambda}] - \frac{\lambda - \overline{\lambda}}{\overline{\lambda} - f(x_0)} \cdot x_0.$$

Therefore $[f \leq \lambda]$ is bounded. The conclusion evidently holds for $\lambda \leq \overline{\lambda}$. (Note that we didn't use the convexity of f.)

(b) ⇒ (a) The proof is immediate by way of contradiction (without any condition on f).

(e) ⇒ (d) Since (e) is true, there exist $\alpha, \rho > 0$ such that $f(x) \geq \alpha\|x\|$ for every $x \in X$, $\|x\| \geq \rho$. Since $f \in \Gamma(X)$, by Theorem 2.2.6, there exist $x^* \in X^*$ and $\gamma \in \mathbb{R}$ such that $f(x) \geq \langle x, x^* \rangle + \gamma$ for every $x \in X$. Therefore

$$\forall x \in X,\ \|x\| \leq \rho \ :\ f(x) \geq -\|x\| \cdot \|x^*\| + \gamma \geq -\rho\|x^*\| + \gamma.$$

Taking $\beta := \min\{0, \gamma - \rho\|x^*\|\}$, condition (d) is satisfied.

(a) ⇒ (e) We show it by way of contradiction. Therefore there exists a sequence $(x_n) \subset X$ such that $(\|x_n\|) \to \infty$ and $\lim_{n\to\infty} f(x_n)/\|x_n\| = \mu \leq 0$. Therefore, for every $k \in \mathbb{N}$ there exists $n'_k \in \mathbb{N}$ such that

$$\forall n \in \mathbb{N},\ n \geq n'_k \ :\ \|x_n\| \geq k^2 \ \text{ and } \ f(x_n)/\|x_n\| \leq 1/k;$$

hence there exists an increasing sequence $(n_k)_{k\in\mathbb{N}} \subset \mathbb{N}$ such that $f(x_{n_k})/\|x_{n_k}\| \leq 1/k$ and $\|x_{n_k}\| \geq k^2$ for every k. We fix $\overline{x} \in \operatorname{dom} f$ and consider $t_k := \|x_{n_k}\|/k \geq k$ and $y_k := (1 - \frac{1}{t_k})\overline{x} + \frac{1}{t_k}x_{n_k}$. Then $(\|y_k\|) \to \infty$ and

$$f(y_k) \leq \left(1 - t_k^{-1}\right) f(\overline{x}) + t_k^{-1} f(x_{n_k}) \leq 1 + \max\{0, f(\overline{x})\},$$

a contradiction. Therefore the implication (a) ⇒ (e) is true.

(d) ⇒ (f) Let $g : X \to \mathbb{R}$, $g(x) := \alpha \cdot \|x\| + \beta$. The function g is convex; using Theorem 2.3.1 and Corollary 2.4.16 we have that $g^*(x^*) = \iota_{\alpha U_{X^*}}(x^*) - \beta$. Since $f \geq g$, we have that $f^* \leq g^*$, whence $\alpha U_{X^*} \subset \operatorname{dom} f^*$; the conclusion follows.

(f) ⇒ (d) Since f^* is w^*-lsc, f^* is $\|\cdot\|$-lsc, too; since $0 \in \operatorname{int}(\operatorname{dom} f^*)$, using Theorem 2.2.20 it follows that f^* is continuous at 0. Therefore there exist $\alpha, \beta \in \mathbb{R}$, $\alpha > 0$, such that $f^*(x^*) \leq \beta$ for every $x^* \in \alpha U_{X^*}$. Therefore $f^* \leq \iota_{\alpha U_{X^*}} + \beta$, whence, taking the conjugates, we obtain that $f(x) \geq f^{**}(x) \geq \alpha \cdot \|x\| - \beta$ for every $x \in X$.

Let $\dim X < \infty$. The implication (b) ⇒ (c′) is obvious. Assume that (c′) holds but (b) does not. Then there exist $\lambda \in \mathbb{R}$ and $(x_n) \subset [f \leq \lambda]$ such that $(\|x_n\|) \to \infty$. We may assume that $\left(\|x_n\|^{-1} x_n\right) \to u \neq 0$. It follows that $u \in [f \leq \lambda]_\infty$, and so, by Exercise 2.23, $u \in [f \leq \inf f]_\infty$, contradicting the boundedness of $[f \leq \inf f]$.

Suppose now that $p, q \in]1, \infty[$, $1/p + 1/q = 1$.

The implications (i) ⇒ (g) and (j) ⇒ (h) are obvious, while the proof of the implication (g) ⇒ (i) is similar to that of (e) ⇒ (d) above. The equivalence (g) ⇔ (h) follows immediately from the anti-monotonicity of the conjugation and using the relation $(\frac{1}{p}\|\cdot\|^p)^* = \frac{1}{q}\|\cdot\|^q$. Suppose that (h) holds. Then there exist $\alpha, \rho > 0$ such that $f^*(x^*) \leq \alpha\|x^*\|^q$ for every $x^* \in X^*$, $\|x^*\| \geq \rho$. Since $\operatorname{dom} f^*$ is a convex set, the preceding inequality shows that $\operatorname{dom} f^* = X^*$. Let

$x^* \in X^* \setminus \{0\}$, $\|x^*\| < \rho$; consider $x_1^* := \rho x^*/\|x^*\|$. Then

$$f^*(x^*) = f^*\left(\frac{\|x^*\|}{\rho}x_1^* + \left(1 - \frac{\|x^*\|}{\rho}\right)0\right) \le \frac{\|x^*\|}{\rho}f^*(x_1^*) + \left(1 - \frac{\|x^*\|}{\rho}\right)f^*(0)$$
$$\le \alpha\rho^q\frac{\|x^*\|}{\rho} + \left(1 - \frac{\|x^*\|}{\rho}\right)f^*(0) \le \alpha\rho^q + \max\{0, f^*(0)\}.$$

Taking $\beta := \alpha\rho^q + \max\{0, f^*(0)\}$, (j) is verified.

Exercise 2.42 Let $\gamma > 0$ be such that $|f_n(0)| \le \gamma$ for every $n \in \mathbb{N}$. By the preceding exercise there exists $r > 0$ such that $f(x) \ge 3\gamma$ for $x \in rS_X$. Since rU_X is a compact set, by Corollary 2.2.23 we have that (f_n) converges uniformly to f on rU_X, and so there exists $n_0 \in \mathbb{N}$ such that $|f_n(x) - f(x)| \le \gamma$ for $x \in rU_X$ and $n \ge n_0$; hence $f_n(x) \ge 2\gamma$ for $x \in rS_X$ and $n \ge n_0$. Let $x \in \mathbb{R}^m$ be such that $\|x\| > r$ and $n \ge n_0$. Then

$$2\gamma \le f_n\left(\frac{r}{\|x\|}x + \left(1 - \frac{r}{\|x\|}\right)0\right) \le \frac{r}{\|x\|}f_n(x) + \left(1 - \frac{r}{\|x\|}\right)f_n(0) \le \frac{r}{\|x\|}f_n(x) + \gamma,$$

whence $f_n(x) \ge r^{-1}\gamma\,\|x\|$. Taking $m := \inf f(\mathbb{R}^m)$, which is in $\mathbb{R}$ by the preceding exercise, we obtain that $f_n(x) \ge m - \gamma$ for $x \in rU_X$ and $n \ge n_0$. It follows that $f_n(x) \ge \alpha\,\|x\| + \beta$ for $x \in \mathbb{R}^m$ and $n \ge n_0$, where $\alpha := r^{-1}\gamma$ and $\beta := \min\{m - \gamma, 0\}$.

Exercise 2.43 The implication (i) $\Rightarrow$ (ii) is obvious. Assume that (ii) holds. Hence there exists $x_0 \in X$ and $x_0^* \in X^*$ such that

$$\langle x, x_0^*\rangle \ge \langle x_0, x_0^*\rangle + \|x - x_0\| \ge \|x\| - \beta \quad \forall x \in C, \tag{$\circ$}$$

where $\beta := \|x_0\| - \langle x_0, x_0^*\rangle$. If $\beta \le 0$, (i) obviously holds. Assume that $\beta > 0$. Since $0 \notin C$, there exist $x_1^* \in X^*$ and $\delta > 0$ such that

$$\langle x, x_1^*\rangle \ge \delta \quad \forall x \in C. \tag{$\circ\circ$}$$

Multiplying ($\circ\circ$) with $\gamma := \delta^{-1}\beta > 0$ and adding the result to ($\circ$) we get $\langle x, x_0^* + \gamma x_1^*\rangle \ge \|x\|$ for every $x \in C$.

Note that conditions (ii)–(iv) are invariant under translations; so we may assume, and we will, that $0 \notin C$.

(i) $\Rightarrow$ (iii) From (i) we have that $f := \iota_C + x_0^*$ satisfies condition (d) in Exercise 2.41, and so, by the same exercise, $0 \in \operatorname{int}(\operatorname{dom} f^*)$. Since $f^*(x^*) = \iota_C^*(x^* - x_0^*) = s_C(x^* - x_0^*)$, we have that $\operatorname{dom} f^* = x_0^* + \operatorname{dom} s_C$. Hence (iii) holds.

(iii) $\Rightarrow$ (i) Assume that (iii) holds, and take $-x_0^* \in \operatorname{int}(\operatorname{dom} s_C)$. Then $0 \in \operatorname{int}(\operatorname{dom} f^*)$ with f defined above. By the implication (f) $\Rightarrow$ (d) of Exercise 2.41 we have that $\langle x, x_0^*\rangle \ge \alpha\,\|x\| - \beta$ for every $x \in C$, for some $\alpha, \beta \in \mathbb{R}$, $\alpha > 0$.

Multiplying eventually with α^{-1}, we may suppose that $\alpha = 1$. As in the proof of the implication (ii) $\Rightarrow$ (i) above, we obtain that (i) holds.

(i) $\Rightarrow$ (iv) Let $c \in C$ be fixed and take $u \in C_\infty$, and $t \geq 0$; of course, $c+tu \in C$. From ($^\circ$) we obtain that

$$\langle c + tu, x_0^* \rangle \geq \|c + tu\| \geq t\,\|u\| - \|c\| \quad \forall t \geq 0;$$

letting $t \to \infty$, we get $\langle u, x_0^* \rangle \geq \|u\|$. Hence (a) holds. Moreover, if $(x_n) \subset C$ with $(\|x_n\|) \to \infty$ and $\left(\|x_n\|^{-1} x_n \right) \overset{w}{\to} u$, we have that $\left\langle \|x_n\|^{-1} x_n, x_0^* \right\rangle \geq 1$, whence $\langle u, x_0^* \rangle \geq 1$. Thus $u \neq 0$.

Assume now that X is a reflexive Banach space.

(iv) $\Rightarrow$ (i) Let $x_0^* \in X^*$ be such that $\langle u, x_0^* \rangle > 0$ for every $u \in C_\infty \setminus \{0\}$ and $x_1^* \in X^*$, $\delta > 0$ satisfy ($^{\circ\circ}$). It follows that $\langle u, x_1^* \rangle \geq 0$ for every $u \in C_\infty$. There exists $\eta > 0$ such that

$$\langle x, x_0^* \rangle \geq -\eta \quad \forall x \in C. \qquad (^{\circ\circ\circ})$$

Otherwise there exists $(x_n) \subset C$ such that $\langle x_n, x_0^* \rangle < -n$ for every $n \in \mathbb{N}$. Since $\|x_n\| \cdot \|x_0^*\| \geq n$, we have that $\|x_n\| \to \infty$. Passing to a subsequence if necessary (X being a reflexive space), we may suppose that $\|x_n\|^{-1} x_n \overset{w}{\to} u$. It follows that $u \in C_\infty$ (C being closed). By (b) we have that $u \neq 0$. Since $\left\langle \|x_n\|^{-1} x_n, x_0^* \right\rangle < 0$, we obtain that $\langle u, x_0^* \rangle \leq 0$, contradicting the choice of x_0^*. Hence ($^{\circ\circ\circ}$) holds. Replacing x_0^* by $\delta^{-1}(\eta + \delta)x_1^* + x_0^*$, we have that ($^{\circ\circ}$) still holds and $\langle u, x_0^* \rangle > 0$ for $u \in C_\infty \setminus \{0\}$ (*i.e.*, we may take $x_1^* = x_0^*$). Assume that C does not satisfy (i). Then for every $n \in \mathbb{N}^*$ there exists $x_n \in C$ such that $(n\delta \leq)\ \langle x_n, nx_0^* \rangle < \|x_n\|$. Hence $\|x_n\| \to \infty$. As above, we may assume that $\|x_n\|^{-1} x_n \overset{w}{\to} u$; hence $u \in C_\infty$, and so, by (b), $u \neq 0$. But, since $\left\langle \|x_n\|^{-1} x_n, x_0^* \right\rangle < n^{-1}$, we have that $\langle u, x_0^* \rangle \leq 0$, a contradiction.

Exercise 2.44 (a) $\Rightarrow$ (b) Suppose that (a) holds. Consider $x_0 \in X$ such that $f(x_0) < \lambda$. Let $x \in X$; it is clear that (b) is true for $\|x\| \leq \rho$. Consider that $\|x\| > \rho$; then $f(x) > \lambda$. Since for $f(x) = \infty$ the inequality (b) is obviously true, let $f(x) \in \mathbb{R}$. If $f(tx + (1-t)x_0) < \lambda$ for every $t \in]0,1[$ we have that $\|tx + (1-t)x_0\| \leq \rho$ for every $t \in]0,1[$, and so the contradiction $\|x\| \leq \rho$. Therefore there exists $t_1 \in]0,1[$ such that $f(t_1 x + (1-t_1)x_0) \geq \lambda$. By Theorem 2.1.12(vii), there exists $t_2 \in]0,1[$ such that $f(t_2 x + (1-t_2)x_0) < \lambda$. Since the function $]0,1[\ni t \mapsto f(tx + (1-t)x_0)$ is convex, it is continuous. So, there exists $\bar{t} \in]0,1[$ such that

$$\lambda = f(\bar{t}x + (1-\bar{t})x_0) \leq \bar{t}f(x) + (1-\bar{t})f(x_0).$$

But

$$\bar{t}\|x\| - (1-\bar{t})\|x_0\| \leq \|\bar{t}x + (1-\bar{t})x_0\| \leq \rho,$$

and so $\bar{t} \leq (\rho + \|x_0\|)/(\|x\| + \|x_0\|)$. Therefore

$$\lambda \leq \bar{t}(f(x) - f(x_0)) + f(x_0) \leq \frac{\rho + \|x_0\|}{\|x\| + \|x_0\|} \cdot (f(x) - f(x_0)) + f(x_0).$$

Thus

$$f(x) \geq f(x_0) + \frac{\|x\| + \|x_0\|}{\rho + \|x_0\|} \cdot (\lambda - f(x_0)) = \frac{\|x\| - \rho}{\rho + \|x_0\|} \cdot (\lambda - f(x_0)) + \lambda$$
$$\geq \frac{\|x\| - \rho}{2\rho} \cdot (\lambda - f(x_0)) + \lambda.$$

Since $f(x_0)$ can be taken as close to $\inf f$ as wanted, we have that

$$f(x) \geq \lambda + \frac{\|x\| - \rho}{2\rho} \cdot (\lambda - \inf f) \geq \inf f + \frac{\|x\| - \rho}{2\rho} \cdot (\lambda - \inf f),$$

which shows that the conclusion holds.

(b) $\Rightarrow$ (c) Set $\mu := (\lambda - \inf f)/(2\rho)$ and consider $g : X \to \mathbb{R}$, $g(x) := \mu \max\{0, \|x\| - \rho\}$; g is a continuous convex function. For computing g^* we could apply the conjugation formulas established in Section 2.8, but we proceed directly.

$$g^*(x^*) = \sup_{x \in X} (\langle x, x^* \rangle - \mu \max\{0, \|x\| - \rho\})$$
$$= \max\left\{\sup_{\|x\| \leq \rho} \langle x, x^* \rangle, \sup_{\|x\| \geq \rho} (\langle x, x^* \rangle - \mu(\|x\| - \rho))\right\}$$
$$= \max\left\{\rho\|x^*\|, \sup_{\|x\| \geq \rho} (\langle x, x^* \rangle - \mu\|x\|) + \rho\mu\right\}.$$

But $\sup_{\|x\| \geq \rho}(\langle x, x^* \rangle - \mu\|x\|) = \infty$ if $\|x^*\| > \mu$, while if $\|x^*\| \leq \mu$ then

$$\langle x, x^* \rangle - \mu\|x\| \leq \|x\| \cdot (\|x^*\| - \mu) \leq \rho(\|x^*\| - \mu),$$

and so

$$\sup_{\|x\| \geq \rho} (\langle x, x^* \rangle - \mu\|x\|) + \mu\rho \leq \rho \cdot \|x^*\|.$$

Therefore

$$g^*(x^*) = \max\{\rho \cdot \|x^*\|, \iota_{\mu U_{X^*}}(x^*)\}.$$

Since $f \geq \inf f + g$ and $\inf f = -f^*(0)$, by conjugation we obtain that $f^* \leq f^*(0) + g^*$, and so (c) holds.

(c) $\Rightarrow$ (b) Using (c) we obtain that $f^* \leq f^*(0) + g^*$. Since $g^{**} = g$, by conjugation we obtain that (b) holds.

Exercise 2.45 (a) The mentioned equivalence was established in Exercise 2.41.

Let $\mu > 0$ be such that $\liminf_{\|x\| \to \infty} f(x)/\|x\| > \mu$. Then there exists $\rho > 0$ such that $f(x) > \mu \cdot \|x\|$ for $\|x\| > \rho$. Since $f \in \Gamma(X)$, f is lower bounded on

μU_X, and so there exists $\gamma \in \mathbb{R}$ such that $f(x) \geq \mu\|x\| + \gamma$ for every $x \in X$. As in the proof of Exercise 2.41 we obtain that $f^* \leq \iota_{\mu U_{X^*}} - \gamma$. Therefore f^* is upper bounded on μU_{X^*}. Since $\mu \in]0, \liminf_{\|x\|\to\infty} f(x)/\|x\|[$ is arbitrary, we obtain that the inequality "$\leq$" of the relation to be proved holds.

Suppose now that $\mu > 0$ and f^* is upper bounded on μU_{X^*}. Then there exists $\gamma \in \mathbb{R}$ such that $f^* \leq \iota_{\mu U_{X^*}} - \gamma$. Taking the conjugates, we obtain that $f(x) \geq \mu\|x\| + \gamma$ for every $x \in X$, whence $\liminf_{\|x\|\to\infty} f(x)/\|x\| \geq \mu$. So we have obtained the inequality "$\geq$" of the desired relation.

Before beginning effectively the proof of the other assertions let us do some remarks. If $\mu > 0$ and

$$g : X \to \overline{\mathbb{R}}, \quad g(x) := f(x) + \mu\|x\|,$$

then, by using Theorem 2.8.7, we have that

$$g^*(0) = \min\{f^*(x^*) + \iota_{\mu U_{X^*}}(-x^*) \mid x^* \in X^*\} = \min\{f^*(x^*) \mid x^* \in \mu U_{X^*}\}.$$

Let $\mu > 0$ be such that $\liminf_{\|x\|\to\infty} f(x)/\|x\| > -\mu$. By what precedes, we get the existence of $\gamma \in \mathbb{R}$ such that $f(x) \geq -\mu\|x\| + \gamma$, *i.e.* $g(x) := f(x) + \mu\|x\| \geq \gamma$, for every $x \in X$. Therefore $0 \in \operatorname{dom} g^*$. By the preceding displayed relation we have that $\mu U_{X^*} \cap \operatorname{dom} f^* \neq \emptyset$, and so $d_{\operatorname{dom} f^*}(0) \leq \mu$; hence

$$-d_{\operatorname{dom} f^*}(0) \geq \min\left\{0, \liminf_{\|x\|\to\infty} f(x)/\|x\|\right\}.$$

Let now $(x_n^*) \subset \operatorname{dom} f^*$ be such that $(\|x_n^*\|) \to d_{\operatorname{dom} f^*}(0)$. Using the Young–Fenchel inequality, for every $n \in \mathbb{N}$ we have that

$$\forall x \in X: \ f(x) \geq \langle x, x_n^* \rangle - f^*(x_n^*) \geq -\|x\| \cdot \|x^*\| - f^*(x_n^*),$$

and so, dividing by $\|x\|$ and then passing to the limit inferior for $\|x\| \to \infty$, we get

$$\forall n \in \mathbb{N} \ : \ \liminf_{\|x\|\to\infty} f(x)/\|x\| \geq -\|x_n^*\|.$$

Taking now the limit, we have that

$$\liminf_{\|x\|\to\infty} f(x)/\|x\| \geq -d_{\operatorname{dom} f^*}(0).$$

So we obtained that

$$\min\left\{0, \liminf_{\|x\|\to\infty} f(x)/\|x\|\right\} = -d_{\operatorname{dom} f^*}(0). \qquad (*)$$

Let us prove now the other two assertions.

(b) If $\liminf_{\|x\|\to\infty} f(x)/\|x\| = 0$, by (*) we have that $0 \in \operatorname{cl}(\operatorname{dom} f^*)$; using (a) we have that $0 \notin \operatorname{int}(\operatorname{dom} f^*)$. Therefore $0 \in \operatorname{Bd}(\operatorname{dom} f^*)$. Conversely, if $0 \in \operatorname{Bd}(\operatorname{dom} f^*)$ we have $0 \notin \operatorname{int}(\operatorname{dom} f^*)$, and so, by (a), $\liminf_{\|x\|\to\infty} f(x)/\|x\| \leq 0$; using the fact that $d_{\operatorname{dom} f^*}(0) = 0$ and (*) we obtain $\liminf_{\|x\|\to\infty} f(x)/\|x\| = 0$.

(c) If $\liminf_{\|x\|\to\infty} f(x)/\|x\| < 0$, using (*) we obtain the equality from (c) and $0 < d_{\operatorname{dom} f^*}(0)$, *i.e.* $0 \notin \operatorname{cl}(\operatorname{dom} f^*)$. Conversely, if $0 \notin \operatorname{cl}(\operatorname{dom} f^*)$ then $d_{\operatorname{dom} f^*}(0) > 0$; using again relation (*), it follows that $\liminf_{\|x\|\to\infty} f(x)/\|x\| < 0$.

Exercise 2.46 Let $x = (x_n)_{n\geq 1} \in \ell^p$; then $(x_n) \to 0$, whence there exists $n_0 \geq 1$ such that $|x_n| \leq \frac{1}{2}$ for $n \geq n_0$. Since the series $\sum_{n\geq 1} n/2^n$ is convergent and $n|x_n|^n \leq n/2^n$ for $n \geq n_0$, it follows that the series $\sum_{n\geq 1} n|x_n|^n$ converges; therefore $x \in \operatorname{dom} f$.

The function $\ell^p \ni x \mapsto n|x_n|^n \in \mathbb{R}$ is obviously convex. It follows that the function

$$f_n : \ell^p \to \mathbb{R}, \quad f_n(x) := \sum\nolimits_{k=1}^{n} k|x_k|^k,$$

is convex, too. Since $f(x) = \lim f_n(x)$ for every $x \in \ell^p$, using Theorem 2.1.3(ii), we obtain that f convex.

Let $\rho \in]0,1[$ and $x = (x_n)_{n\geq 1} \in \rho U_{\ell^p}$. Then $|x_n| \leq \rho$ for every $n \geq 1$. It follows that

$$\forall\, x \in \rho U_{\ell^p} \;:\; f(x) \leq \sum\nolimits_{n\geq 1} n\rho^n \in \mathbb{R}.$$

The function f being convex, using Theorem 2.2.13, we obtain the continuity of f on $\operatorname{dom} f = \ell^p$.

Since $f(e_n) = n$, where $e_n = (0,\ldots,0,1,0,\ldots)$, 1 being on position n, f is not bounded on ρU_{ℓ^p} for every $\rho \geq 1$. Applying Exercise 2.45 (a) for $f^* : \ell^q = (\ell^p)^* \to \overline{\mathbb{R}}$, where $q = p/(p-1)$, since $(f^*)^* = f$, we have that $\liminf_{\|y\|\to\infty} f^*(y)/\|y\| = 1$.

Exercise 2.47 (a) Let $K := A(C)$; it is obvious that K is a convex set. Using Corollary 1.3.15 we have that K is also a closed set. Since the function $Y \ni y \mapsto \frac{1}{2}\|y\|^2 + \iota_K(y)$ is convex, lsc and coercive, and the space Y is reflexive, using Theorem 2.5.1(ii), there exists $\overline{y} \in K$ such that $\frac{1}{2}\|\overline{y}\|^2 \leq \frac{1}{2}\|y\|^2$ for every $y \in K$. Taking $\overline{x} \in C$ such that $\overline{y} = A\overline{x}$ we obtain that $\overline{x}$ is a solution of problem (P).

(b) Let $\overline{x} \in C$ and $f : X \to \mathbb{R}$, $f(x) := \frac{1}{2}\|Ax\|^2$. By the Pshenichnyi–Rockafellar theorem, $\overline{x}$ is a solution of (P) if and only if $\partial f(\overline{x}) \cap (-N(C;\overline{x})) \neq \emptyset$. Since $\partial(\frac{1}{2}\|\cdot\|^2)(y) = \{y\}$ for $y \in Y$, from Theorem 2.8.6 we have that $\partial f(x) = \{A^*(Ax)\}$ for $x \in X$. Therefore $\overline{x}$ is a solution of (P) if and only if $A^*(A\overline{x}) \in -N(C;\overline{x})$, *i.e.* $(\overline{x} \mid \overline{v}) \leq (x \mid \overline{v})$ for every $x \in C$, where $\overline{v} = A^*(A\overline{x})$.

Exercise 2.48 Using Exercise 1.5 (c) we have that $C + \ker A$ is a closed set, while using Exercise 2.47 (a) we have that (P) has optimal solutions. Suppose that there exists $\widetilde{x}$ with the mentioned property; then Slater's condition is satisfied for the problem

$$\min\ \tfrac{1}{2}\|Ax\|^2, \quad \langle x, \varphi_i\rangle - \alpha_i \leq 0,\ 1 \leq i \leq k.$$

The conclusion follows applying Corollary 2.9.4.

Exercise 2.49 Consider the function $f : X \to \mathbb{R}$, $f(x) := \|x - a_1\| + \|x - a_2\| + \|x - a_3\|$. It is obvious that the function f is convex, continuous and $\lim_{\|x\|\to\infty} f(x) = \infty$, *i.e.* f is coercive. The function f is even strictly convex. Indeed, for $x, y \in X$, we have that $\|x + y\| = \|x\| + \|y\|$ if and only if there exists $\lambda, \mu \geq 0$, $\lambda + \mu > 0$, such that $\lambda x = \mu y$ (*i.e.* x, y are on a half-line passing through the origin). Using this fact, if there exist $x, y \in X$, $x \neq y$ and $\lambda \in]0, 1[$ such that $f(\lambda x + (1 - \lambda)y) = \lambda f(x) + (1 - \lambda)f(y)$, then a_1, a_2, a_3 are on the straight-line determined by x, y, which is a contradiction. Therefore there exists a unique element $\overline{x} \in X$ such that $f(\overline{x}) \leq f(x)$ for every $x \in X$. Moreover $\overline{x}$ is characterized by $0 \in \partial f(\overline{x})$. But

$$\partial f(x) = \begin{cases} \sum_{1\leq i\leq 3} \|x - a_i\|^{-1}(x - a_i) & \text{if} \quad u \notin \{a_1, a_2, a_3\}, \\ \sum_{1\leq i\leq 3, i\neq j} \|x - a_i\|^{-1}(x - a_i) + U_X & \text{if} \quad x = a_j. \end{cases}$$

Denote $e_i := (\overline{x} - a_i)/\|\overline{x} - a_i\|$ when $\overline{x} \neq a_i$. If $\overline{x} \notin \{a_1, a_2, a_3\}$, using the relation $e_1 + e_2 + e_3 = 0$ we obtain that

$$\overline{x} = \left(\sum_{i=1}^{3} \frac{1}{\|\overline{x} - a_i\|}\right)^{-1} \sum_{i=1}^{3} \frac{a_i}{\|\overline{x} - a_i\|},$$

and so $\overline{x} \in \operatorname{co}\{a_1, a_2, a_3\}$. If $\overline{x} \in \{a_1, a_2, a_3\}$, of course $\overline{x} \in \operatorname{co}\{a_1, a_2, a_3\}$. Therefore the conclusion holds.

From $0 \in \partial f(\overline{x})$ we obtain that $\overline{x} = a_3$ if and only if $e_1 + e_2 \in U_X$, which is equivalent to $(e_1 \mid e_2) \leq -\frac{1}{2}$, *i.e.* $\angle(a_1, a_2, a_3) \geq 120°$. Suppose that $\overline{x} \notin \{a_1, a_2, a_3\}$. Then $e_1 + e_2 + e_3 = 0$, whence $(e_i \mid e_j) = -\frac{1}{2}$ for $i \neq j$. Therefore $\angle(a_i, \overline{x}, a_j) = 120°$ for $i \neq j$. Thus we obtained the known properties of Toricelli's point.

Exercise 2.50 For every $\mu \geq 0$ and $i \in \{1, 2, 3, 4\}$ we consider the function

$$f_\mu : X_i \to \mathbb{R}, \quad f_\mu(x) := \int_0^1 \left(tx(t) + \mu\sqrt{1 + (x(t))^2}\right) dt.$$

From Exercises 2.9 and 2.10 we have that f_μ is convex; moreover f is a $\mathcal{C}^2$ function on X_1 and X_3 and only Gâteaux differentiable on X_2 and X_4. In every case we have that

$$\forall x, u \in X_i \ : \ \nabla f_\mu(x)(u) = \int_0^1 \left(tu(t) + \mu\frac{x(t)u(t)}{\sqrt{1 + (x(t))^2}}\right) dt.$$

Moreover

$$v(P_1^\mu) \geq v(P_2^\mu), \qquad v(P_3^\mu) \geq v(P_4^\mu).$$

Let us take $i = 2$ and $\mu \geq 0$. The function to be minimized being convex and Gâteaux differentiable, $x_\mu \in X_2$ is a solution of (P_2^μ) if and only if $\nabla f_\mu(x_\mu) = 0$.

It is obvious that $\nabla f_0(x) \neq 0$ for every $x \in X_2$. So, let $\mu > 0$. The condition $\nabla f_\mu(x_\mu) = 0$ is equivalent to

$$\forall\, u \in X_2 \;:\; \int_0^1 \left(t + \mu \frac{x_\mu(t)}{\sqrt{1+(x_\mu(t))^2}} \right) u(t)\, dt = 0,$$

which is, obviously, equivalent to each of the following relations:

$$\mu \frac{x_\mu(t)}{\sqrt{1+(x_\mu(t))^2}} + t = 0 \text{ a.e. } t \in [0,1], \qquad x_\mu(t) = \frac{-t}{\sqrt{\mu^2 - t^2}} \text{ a.e. } t \in [0,1].$$

The function x_μ above is in X_2 if and only if $\mu \geq 1$. Therefore for $\mu \geq 1$ the problem (P_2^μ) has a unique optimal solution, x_μ, while for $\mu < 1$ the problem (P_2^μ) has no optimal solutions.

Let us determine $v(P_2^\mu)$ when $\mu \in [0,1[$. For every $\alpha \in]0,1[$ and $\beta \in]1,\infty[$ consider the function

$$x_{\alpha,\beta} : [0,1] \to \mathbb{R}, \quad x_{\alpha,\beta}(t) := \begin{cases} 0 & \text{if} \quad t \in [0, 1-\alpha], \\ -\frac{\beta}{\alpha}(t - 1 + \alpha) & \text{if} \quad t \in]1-\alpha, 1]. \end{cases}$$

An elementary computation shows that

$$f_\mu(x_{\alpha,\beta}) = \mu(1-\alpha) + \alpha\beta \left(\frac{\alpha}{6} - \frac{1}{2} + \frac{\mu}{2}\sqrt{1 + \frac{1}{\beta^2}} + \frac{\mu}{2\beta^2} \ln\left(\beta + \sqrt{1+\beta^2}\right) \right).$$

It follows that $f_\mu(x_{n^{-1},n^2}) \to -\infty$ for $\mu \in [0,1[$, and so $v(P_2^\mu) = -\infty$.

Let now $i = 1$. Since $x_{n^{-1},n^2} \in X_1$ we obtain that $v(P_1^\mu) = -\infty$ for $\mu \in [0,1[$. Doing again the calculations of the case $i = 2$ we obtain that (P_1^μ) has optimal solutions if and only if $\mu > 1$. The solution is unique and is the function x_μ determined above. Therefore $v(P_1^\mu) = v(P_2^\mu)$ for $\mu > 1$. For $\mu = 1$ we have already seen that $v(P_1^1) \geq v(P_2^1)$. Considering for every $n \in \mathbb{N}$ the function

$$x_1^n : [0,1] \to \mathbb{R}, \quad x_1^n(t) := \begin{cases} x_1(t) & \text{if} \quad t \in [0, 1 - \frac{1}{n}], \\ x_1(1 - \frac{1}{n}) & \text{if} \quad t \in]1 - \frac{1}{n}, 1], \end{cases}$$

we observe that $x_1^n \in X_1$ and $\lim_{n\to\infty} f_1(x_1^n) = f_1(x_1)$ (one can use Lebesgue's theorem). Therefore $v(P_1^1) = v(P_1^2)$.

Let $i = 3$. Similarly to the case $i = 2$, y_μ is an optimal solution for (P_3^μ) if and only if $\nabla f_\mu(y_\mu) = 0$. For $\mu = 0$ this is not possible, and so (P_3^0) has no optimal solutions. Let $\mu > 0$ and y_μ an optimal solution of (P_3^μ). The condition $\nabla f_\mu(y_\mu) = 0$ is equivalent to

$$\forall\, u \in X_3 \;:\; \int_0^1 \left(t + \mu \frac{y_\mu(t)}{\sqrt{1+(y_\mu(t))^2}} \right) u(t)\, dt = 0.$$

Recall the following result: if $x \in C[a,b]$ $(a,b \in \mathbb{R},\ a<b)$ and $\int_a^b x(t)u(t)\,dt = 0$ for every $u \in C[a,b]$ with $\int_a^b u(t)dt = 0$, then x is a constant function. Indeed, let $c := (b-a)^{-1}\int_a^b u(t)\,dt$ and $u := x - c$; by hypothesis we have that $\int_a^b (u(t) - c)^2 dt = \int_a^b (u(t)-c)u(t)\,dt = 0$ and so $x(t) = c$ for every $t \in [a,b]$.

Therefore there exists $\alpha \in \mathbb{R}$ such that

$$\forall t \in [0,1] \ : \ \mu \frac{y_\mu(t)}{\sqrt{1+(y_\mu(t))^2}} + t = \alpha;$$

it follows that

$$\forall t \in [0,1] \ : \ y_\mu(t) = \frac{\alpha - t}{\sqrt{\mu^2 - (a-t)^2}}.$$

Since $y_\mu \in C[0,1]$, we necessarily have that $\mu > |\alpha - t|$ for every $t \in [0,1]$, *i.e.* $1-\mu < \alpha < \mu$. It follows that $\mu > \frac{1}{2}$. In this case, $y_\mu \in X_3$ if and only if $\alpha = \frac{1}{2}$. So, for $\mu \le \frac{1}{2}$ the problem (P_3^μ) has no optimal solutions, while for $\mu > \frac{1}{2}$ the problem (P_3^μ) has a unique optimal solution, namely

$$y_\mu : [0,1] \to \mathbb{R}, \quad y_\mu(t) = \frac{\frac{1}{2} - t}{\sqrt{\mu^2 - (\frac{1}{2} - t)^2}}.$$

In a similar manner, but using this time the result: if $x \in L^\infty(a,b)$ $(a,b \in \overline{\mathbb{R}},\ a<b)$ and $\int_a^b x(t)u(t)\,dt = 0$ for every function $u \in L^1(a,b)$ with the property $\int_a^b u(t)\,dt = 0$, then x is constant a.e. (the proof being similar to that presented above), we obtain that (P_4^μ) has optimal solutions if and only if $\mu \ge \frac{1}{2}$; in this case the unique solution is y_μ defined above.

Let now $\mu < \frac{1}{2}$; for $\alpha \in]0,\frac{1}{2}[$, $\beta \in \mathbb{P}$ consider the function

$$y_{\alpha,\beta} : [0,1] \to \mathbb{R}, \quad y_{\alpha,\beta}(t) := \begin{cases} \beta & \text{if} \quad t \in [0,\alpha], \\ 0 & \text{if} \quad t \in]\alpha, 1-\alpha[, \\ -\beta & \text{if} \quad t \in [1-\alpha, 1]. \end{cases}$$

$y_{\alpha,\beta} \in X_4$ and

$$f(y_{\alpha,\beta}) = \alpha\beta\left(2\mu\sqrt{1+\beta^{-2}} - 1 + \alpha\right) + \mu(1-2\alpha).$$

Since $f_\mu(y_{n^{-1},n^2}) \to -\infty$, we have that $v(P_4^\mu) = -\infty$ for $\mu < \frac{1}{2}$. Taking into account that for every $(\alpha,\beta) \in]0,\frac{1}{2}[\times]0,\infty[$ the function $y_{\alpha,\beta}$ can be approximated pointwisely by a sequence $(y_{\alpha,\beta,k})_{k\in N} \subset X_3$ such that $|y_{\alpha,\beta,k}| \le |y_{\alpha,\beta}|$ and using eventually Lebesgue's theorem, we obtain that $v(P_3^\mu) = -\infty$ for $\mu < \frac{1}{2}$. Approximating similarly $y_{1/2}$, we obtain that $v(P_3^{1/2}) = v(P_4^{1/2})$. Elementary calculations show that

$$v(P_1^\mu) = v(P_2^\mu) = \begin{cases} -\infty & \text{if} \quad \mu \in [0,1[, \\ \frac{\mu^2}{2} \arcsin \frac{1}{\mu} + \frac{1}{2}\sqrt{\mu^2 - 1} & \text{if} \quad \mu \in [1, \infty[, \end{cases}$$

and

$$v(P_3^\mu) = v(P_4^\mu) = \begin{cases} -\infty & \text{if} \quad \mu \in [0, \frac{1}{2}[, \\ \mu^2 \arcsin \frac{1}{2\mu} + \frac{1}{2}\sqrt{\mu^2 - \frac{1}{4}} & \text{if} \quad \mu \in [\frac{1}{2}, \infty[. \end{cases}$$

The solution is complete.

Exercise 2.51 Recall that $x \in AC[0,1]$ if and only if x is derivable at almost every $t \in [0,1]$ and $x' \in L^1(0,1)$. Note that for $x \in X$, $\int_0^1 x(t)\,dt = -\int_0^1 tx'(t)\,dt$. Taking into account this fact, to problem (P) we associate the problems

(P_i) $\quad \min \int_0^1 tx(t)\,dt, \quad x \in X_i, \ \int_0^1 \sqrt{1 + (x(t))^2}\,dt \le L,$

$i = 3, 4$, the spaces X_3 and X_4 are those introduced in Exercise 2.50. Note that $v(P) = -v(P_3)$ if $X = C^1[0,1]$, and x is a solution of (P) if and only if x' is a solution of (P_3); if $X = AC[0,1]$ then $v(P) = -v(P_4)$ and x is a solution of (P) if and only if x' is a solution of (P_4).

Considering for $i \in \{3,4\}$ the functions

$$f, g : X_i \to \mathbb{R}, \quad f(x) := -\int_0^1 tu(t)\,dt, \quad g(x) := \int_0^1 \sqrt{1 + (u(t))^2}\,dt,$$

the problem (P_i) becomes

(P_i) $\quad \min f(x), \quad x \in X_i, \ g(x) \le L.$

The function f is linear and g is convex and Gâteaux differentiable; moreover the differential of g is given by $\nabla g(x)(u) = \int_0^1 xu/\sqrt{1+x^2}\,dt$ for $x, u \in X_i$ (see Exercises 2.9 and 2.10). Therefore the problems (P_i), $i \in \{3,4\}$, are convex problems.

If $L < 1$, then the problem (P_i) has no admissible solutions for every $i \in \{3,4\}$, while if $L = 1$, (P_i) has only an admissible solution, $x = 0$, which is the optimal solution, too.

Let $L > 1$ and $i = 4$. Since $g(0) = 1 < L$, Slater's condition is verified for the convex problem (P_4). Therefore an admissible solution $x \in X_4$ $(g(x) \le L)$ is optimal for (P_4) if and only if there exists $\lambda \ge 0$ such that

$$\lambda(g(x) - L) = 0, \quad \nabla f(x) + \lambda \nabla g(x) = 0.$$

Let $x \in X_4$ be an optimal solution of (P_4) and $\lambda \ge 0$ satisfying the above relations.

Since $\nabla f(x) \neq 0$ we obtain that $\lambda > 0$, and so $g(x) = L$. Therefore

$$\forall\, u \in X_4 \;:\; \int_0^1 \left(\frac{\lambda x(t)}{\sqrt{1+(x(t))^2}} + t \right) u(t)\, dt = 0.$$

As seen in the solution of Exercise 2.50, the above relation is possible only when $\lambda \geq 1$ and the sole function which satisfies it is

$$y_\lambda : [0,1] \to \mathbb{R}, \quad y_\lambda(t) = \frac{\frac{1}{2} - t}{\sqrt{\lambda^2 - (\frac{1}{2} - t)^2}}.$$

Since $g(y_\lambda) = L$ we have that $2\lambda \arcsin \frac{1}{2\lambda} = L$. But the function $]0,1] \ni t \mapsto \frac{1}{t} \arcsin t \in \mathbb{R}$ is increasing with range $]1, \pi/2]$. Since in convex programming the necessary condition is also sufficient, we obtain that for $L \in]1, \pi/2]$ the problem (P_4) has the unique solution y_λ, where $2\lambda \arcsin \frac{1}{2\lambda} = L$. The (unique) solution of problem (P) is, in this case, the function

$$x : [0,1] \to \mathbb{R}, \quad x(t) := \sqrt{\lambda^2 - (t - \tfrac{1}{2})^2} - \sqrt{\lambda^2 - \tfrac{1}{4}}, \qquad (^*)$$

with λ defined above.

For $L > \pi/2$ the problem (P_4) (and (P), too) has no optimal solutions. To compute the value of problem (P_4), we associate its dual problem. By Theorem 2.9.3 we have that $v(P_4) = v(D_4)$ (and (D_4) has optimal solutions), where

$$(D_4) \qquad \max\ \inf_{x \in X_4} (f(x) + \lambda(g(x) - L)), \quad \lambda \geq 0.$$

Taking into account the result obtained in solving Exercise 2.50, we have that

$$v(D_4) = \max \left\{ \lambda^2 \arcsin \tfrac{1}{2\lambda} + \tfrac{1}{2}\sqrt{\lambda^2 - \tfrac{1}{4}} - \lambda L \;\middle|\; \lambda \geq \tfrac{1}{2} \right\}.$$

If $i = 3$, the same argument as for the case $i = 4$ proves that (P_3) has a unique solution for $L \in]1, \pi/2[$, having the same expression, while for $L \geq \pi/2$ the problem has no optimal solutions. Moreover $v(P_3) = v(P_4)$ for every $L > 1$.

In conclusion, problem (P) has a unique solution for $L \in]1, \pi/2[$ and has no optimal solutions for $L \geq \pi/2$ when $X = C^1[0,1]$, and has a unique solution for $L \in]1, \pi/2]$ and has no optimal solutions for $L > \pi/2$ when $X = AC[0,1]$; in both cases the solution is given by formula $(^*)$. Moreover

$$v(P) = \begin{cases} \frac{1}{2}\sqrt{\lambda_L^2 - \frac{1}{4}} - \frac{1}{2} L \lambda_L & \text{if} \quad L \in]1, \pi/2], \\ \frac{\pi - 4L}{8} & \text{if} \quad L \in]\pi/2, \infty[, \end{cases}$$

where λ_L is the unique positive solution of the equation $2\lambda \arcsin \frac{1}{2\lambda} = L$ when $L \in]1, \pi/2]$.

Exercise 3.1 First recall that $\ell^1 := \{(x_k)_{k\geq 1} \subset \mathbb{R} \mid \sum_{k\geq 1} |x_k| < \infty\}$ is a Banach space with respect to the norm $\|x\|_1 := \sum_{k\geq 1} |x_k|$. Its dual is $\ell^\infty := \{(x_k)_{k\geq 1} \subset \mathbb{R} \mid (x_k) \text{ is bounded}\}$, the dual norm being $\|(x_k)\|_\infty := \sup_{k\geq 1} |x_k|$. Let $x = (x_k)_{k\geq 1} \in \ell^1$. Since $(x_k) \to 0$ we have that $\|x_k\| \leq 1$ for $k \geq k_0$, and so $|x_k|^{1+1/k} \leq |x_k|$ for $k \geq k_0$. It follows that the series defining $f(x)$ is convergent, and so $f(x) \in \mathbb{R}$. Moreover, because $|x_k| \leq 1$ $(k \geq 1)$ for $\|x\| \leq 1$, the above inequality shows that $f(x) \leq 1$ for $x \in U_{\ell^1}$; hence f is continuous if we succeed to show that f is (strictly) convex. Let us show that f is strictly convex. So, let $x, y \in \ell^1$ with $x \neq y$ and $\lambda \in]0,1[$. Of course, there exists $k_0 \geq 1$ such that $x_{k_0} \neq y_{k_0}$. Since the mapping $\varphi_\alpha : \mathbb{R} \to \mathbb{R}$, $\varphi_\alpha(t) := |t|^\alpha$, is strictly convex for every $\alpha > 1$ (see the examples after Theorem 2.1.10), we obtain that $|\lambda x_k + (1-\lambda)y_k|^{1+1/k} \leq \lambda |x_k|^{1+1/k} + (1-\lambda)|y_k|^{1+1/k}$, with strict inequality for $k = k_0$. Summing up these inequalities side by side we get $f(\lambda x + (1-\lambda)y) < \lambda f(x) + (1-\lambda)f(y)$. Let $x^* = (x'_k)_{k\geq 1} \subset (\ell^1)^* = \ell^\infty$ be in $\partial f(x)$ with $x = (x_k)_{k\geq 1}$. Let us fix $k_0 \in \mathbb{N}$. Taking $y_k := x_k$ for $k \neq k_0$ and $y_{k_0} = x_{k_0} + u$ in the definition of $\partial f(x)$ we get $|x_{k_0} + u|^{1+1/k_0} - |x_{k_0}|^{1+1/k_0} \geq u x'_{k_0}$ for every $u \in \mathbb{R}$, *i.e.* $x'_{k_0} \in \partial\varphi_{1+1/k_0}(x_{k_0})$. Using Theorem 3.7.2(iii) we obtain that $x'_{k_0} = (1 + \frac{1}{k_0})|x_{k_0}|^{1/k_0} \operatorname{sgn}(x_{k_0})$. Therefore $\partial f(x)$ (which is nonempty because f is continuous on ℓ^1) has a unique element x^* whose components are $x'_k := (1+\frac{1}{k})|x_k|^{1/k} \operatorname{sgn}(x_k)$. Taking into account Corollary 2.4.10 we obtain that f is Gâteaux differentiable.

Consider now $x = (x_k) \in \ell^1$ and show that f is not Fréchet differentiable at x. We distinguish two cases: 1) $\limsup_{k\to\infty} |x_k|^{1/k} > 0$ and 2) $\limsup_{k\to\infty} |x_k|^{1/k} = 0$.

1) In this case there exist $\eta > 0$ and an infinite set $P \subset \mathbb{N}$ such that $|x_k|^{1/k} \geq \eta$ for every $k \in P$. For $n \in \mathbb{N}$ consider $u^n \in \ell^1$ defined by $u^n_k := 0$ if $k < n$ or $k \in \mathbb{N} \setminus P$ and $u^n_k := -2x_k$ otherwise. It is obvious that $(\|u^n\|_1) \to 0$ for $n \to \infty$. But

$$\|\nabla f(x+u^n) - \nabla f(x)\| = 2 \sup_{k\in P, k\geq n} \left((1+k^{-1}) \cdot |x_k|^{1/k} \right) \geq 2\eta \quad \forall n \in \mathbb{N}.$$

Taking into account Corollary 3.3.6 we obtain that f is not Fréchet differentiable at x.

2) In this case $\left(|x_k|^{1/k}\right) \to 0$. For $n \in \mathbb{N}$ consider $u^n \in \ell^1$ defined by $u^n_k := k^{-2} \operatorname{sgn}(x_k)$ for $k \geq n$ and $u^n_k := 0$ otherwise; it is obvious that $(\|u^n\|_1) \to 0$. In this case

$$\begin{aligned}\|\nabla f(x+u^n) - \nabla f(x)\| &= \sup_{k\geq n} \left((1+k^{-1}) \cdot \left| (|x_k| + k^{-2})^{1/k} - |x_k|^{1/k} \right| \right) \\ &\geq \lim_{k\to\infty} \left((1+k^{-1}) \cdot \left| (|x_k| + k^{-2})^{1/k} - |x_k|^{1/k} \right| \right) = 1\end{aligned}$$

for every $n \in \mathbb{N}$. As above we obtain again that f is not Fréchet differentiable at

Exercise 3.2 Because $|x_n| \le \|x\|$ for every n, we have that $0 \le f(x) \le \|x\|$ for every $x \in \ell^\infty$. Because $\ell^\infty \ni x \mapsto |x_n| \in \mathbb{R}$ is convex, by Theorem 2.1.3(iii) f is convex (even sublinear). It follows that f is continuous, being bounded above on U_{ℓ^∞}.

Let $\overline{x} \in \ell^\infty$ be fixed. If $f(\overline{x}) = 0$ consider $u := (1, \ldots, 1, \ldots) \in \ell^\infty$. Then for any $t > 0$ we have that $f(\overline{x} + tu) = t = f(\overline{x} - tu)$. and so $\lim_{t\downarrow 0} t^{-1}\big(f(\overline{x} + tu) + f(\overline{x} - tu) - 2f(\overline{x})\big) = 2 > 0$.

Assume that $f(\overline{x}) > 0$ and consider an infinite subset P of $\mathbb{N}$ such that $\lim_{n\in P} |x_n| = f(\overline{x})$; without loss of generality we assume that $x_n > \frac{1}{2}f(\overline{x})$ for $n \in P$ (replacing eventually $\overline{x}$ by $-\overline{x}$). Let P_1, P_2 be infinite disjoint subsets of P such that $P = P_1 \cup P_2$. Let $u \in \ell^\infty$ be such that $u_n := 1$ for $n \in P_1$ and $u_n := 0$ for $n \in \mathbb{N} \setminus P_1$. Let $0 < t < \frac{1}{2}f(\overline{x})$. Then $|x_n + tu| = x_n + t$ for $n \in P_1$ and $|x_n + tu| = |x_n|$ for $n \in \mathbb{N} \setminus P_1$. It follows that $f(\overline{x} + tu) = \limsup_{n\in\mathbb{N}} |x_n + tu| \ge \lim_{n\in P_1}(x_n + t) = f(\overline{x}) + t$. Since $|x_n - tu| = |x_n|$ for $n \in \mathbb{N}\setminus P_1 \supset P_2$, we have that $\limsup_{n\in\mathbb{N}\setminus P_1} |x_n - tu| = f(\overline{x})$. Because $|x_n - tu| = x_n - t$ for $n \in P_1$, it follows that $\limsup_{n\in\mathbb{N}} |x_n - tu| = f(\overline{x})$. Therefore $\lim_{t\downarrow 0} t^{-1}\big(f(\overline{x} + tu) + f(\overline{x} - tu) - 2f(\overline{x})\big) \ge \lim_{t\downarrow 0} t^{-1}\big(f(\overline{x}) + t + f(\overline{x}) - 2f(\overline{x})\big) = 1 > 0$.

Hence, in both cases there exists $u \in \ell^\infty$ such that $\lim_{t\downarrow 0} t^{-1}\big(f(\overline{x} + tu) + f(\overline{x} - tu) - 2f(\overline{x})\big) > 0$. By Theorem 3.3.2 f is not Gâteaux differentiable at $\overline{x}$.

Exercise 3.3 Because ρ_f is nondecreasing on $[0, \infty[$, it is obvious that the three situations in (ii) are equivalent. Also for the equivalence of the three situations in (iii) it is sufficient to show that $\widetilde{\rho} : [0, \infty[\to [0, \infty]$ defined by

$$\widetilde{\rho}(s) := \inf\{\tfrac{1}{2}f(x) + \tfrac{1}{2}f(y) - f(\tfrac{1}{2}x + \tfrac{1}{2}y) \mid x, y \in \operatorname{dom} f,\ \|x - y\| = s\},$$

is nondecreasing (of course, $\inf\emptyset = +\infty$). Indeed, taking $0 < s' < s$ and $x, y \in \operatorname{dom} f$ with $\|x - y\| = s$, then applying Theorem 2.1.5(vii) for $\lambda = 1/2$, $\varphi(t) := f(x + t(y - x))$ and $0 < \frac{s'}{s} < 1$ we obtain that

$$\tfrac{1}{2}f(x) + \tfrac{1}{2}f(x + \tfrac{s'}{s}(y - x)) - f(x + \tfrac{s'}{2s}(y - x)) \le \tfrac{1}{2}f(x) + \tfrac{1}{2}f(y) - f(\tfrac{1}{2}x + \tfrac{1}{2}y),$$

which shows that $\widetilde{\rho}(s') \le \widetilde{\rho}(s)$ because $\|\frac{s'}{s}(y - x)\| = s'$.

Let now $x, y \in \operatorname{dom} f$ with $\|x - y\| = t > 0$ and $\lambda \in]0, 1[$; we may assume that $\lambda \in]0, 1/2]$. Set $z := (1 - \lambda)x + \lambda y$; then

$$\begin{aligned} f(z) &= f\left((1 - 2\lambda)x + 2\lambda(\tfrac{1}{2}(x + y))\right) \le (1 - 2\lambda)f(x) + 2\lambda f(\tfrac{1}{2}(x + y)) \\ &\le (1 - 2\lambda)f(x) + 2\lambda\left(\tfrac{1}{2}f(x) + \tfrac{1}{2}f(y) - \widetilde{\rho}(\|x - y\|)\right) \\ &= (1 - \lambda)f(x) + \lambda f(y) - 2\lambda\widetilde{\rho}(t) \\ &\le (1 - \lambda)f(x) + \lambda f(y) - 2\lambda(1 - \lambda)\widetilde{\rho}(t). \end{aligned}$$

Therefore $2\widetilde{\rho}(t) \le \rho_f(t) \le 4\widetilde{\rho}(t)$ for $t \ge 0$. These inequalities show that (iii) $\Leftrightarrow$ (i). Since the implications (i) $\Rightarrow$ (ii) $\Rightarrow$ (iii) are obvious, the equivalence of these three conditions is proved.

(iv) Fixing $x \in \operatorname{dom} f$ in the preceding proof we get the conclusion.

Exercise 3.4 Let $x \in \ell^1$ and $t > 0$. Consider $e^n \in \ell^1$ with $e^n_k := 1$ if $k = n$ and $e^n_k := 0$ otherwise. Then $\rho_{f,x}(t) \le 4\left(\frac{1}{2}f(x+te^n) + \frac{1}{2}f(x) - f(x+\frac{1}{2}te^n)\right)$ for every $n \in \mathbb{N}$, and so

$$0 \le \rho_{f,x}(t) \le 4\lim_{n\to\infty}\left(\tfrac{1}{2}\left|x_n+t\right|^{1+1/n} + \tfrac{1}{2}\left|x_n\right|^{1+1/n} - \left|x_n+\tfrac{1}{2}t\right|^{1+1/n}\right) = 0.$$

Because f is Gâteaux differentiable, we have (see page 201) that

$$\vartheta_{f,x}(t) = \inf\{D_f(y,x) \mid \|y-x\| = t\} \quad \forall x \in \ell^1,\ t \ge 0,$$

where $D_f(y,x) := f(y) - f(x) - \langle y-x, \nabla f(x)\rangle$.

Note first that $\|\overline{y}-x\| = \|y-x\|$ and $D_f(y,x) \ge D_f(\overline{y},x)$ for any $y \in \ell^1$, where $\overline{y} \in \ell^1$ is defined by $\overline{y}_k := y_k$ if $\operatorname{sgn}(y_k - x_k) = \operatorname{sgn}(x_k)$ and $\overline{y}_k := 2x_k - y_k$ otherwise. It is obvious that $|\overline{y}_k - x_k| = |y_k - x_k|$ for every $k \ge 1$, and so $\|\overline{y}-x\| = \|y-x\|$. Then

$$D_f(y,x) - D_f(\overline{y},x) = f(y) - f(\overline{y}) - \langle y-\overline{y}, \nabla f(x)\rangle = \sum\nolimits_{k\ge 1}\theta_k,$$

where $\theta_k := |y_k|^{1+1/k} - |\overline{y}_k|^{1+1/k} - (1+\frac{1}{k})\operatorname{sgn}(x_k)\cdot|x_k|^{1/k}(y_k - \overline{y}_k)$. It is sufficient to show that $\theta_k \ge 0$ for every $k \ge 1$. Of course, $\theta_k = 0$ if $\operatorname{sgn}(y_k - x_k) = \operatorname{sgn}(x_k)$ or $x_k = 0$. So, assume that this is not the case. Then $u_k := 1 - y_k/x_k \ge 0$, $p_k := 1 + \frac{1}{k} \in]1,2]$, $\overline{y}_k = 2x_k - y_k$ and

$$\theta_k = |x_k|^{p_k}\left(|1-u_k|^{p_k} - |1+u_k|^{p_k} + 2p_k u_k\right) = |x_k|^{p_k}\varphi_{p_k}(u_k),$$

because $y_k - \overline{y}_k = 2(y_k - x_k)$ and $\operatorname{sgn}(y_k - x_k) = -\operatorname{sgn}(x_k)$, where $\varphi_p(t) := |1-t|^p - |1+t|^p + 2pt$ for $t \ge 0$. Using Exercise 2.7 we have that φ_p is nondecreasing for $p \in]1,2]$, and so $\varphi_{p_k}(u_k) \ge 0$. Hence $\theta_k \ge 0$ for every $k \ge 1$, and so $D_f(y,x) \ge D_f(\overline{y},x)$.

Assume now that $x \in \ell^1 \setminus A$ and $t > 0$. Of course, $\limsup_{k\to\infty}|x_k|^{1/k} = 1$. Consider $u^n := \operatorname{sgn}(x_n)\cdot e^n$. Then $\vartheta_{f,x}(t) \le f(x+tu^n) - f(x) - t\langle u^n, \nabla f(x)\rangle = (|x_n|+t)^{1+1/n} - |x_n|^{1+1/n} - t(1+\frac{1}{n})|x_n|^{1/n}$. It follows that

$$0 \le \vartheta_{f,x}(t) \le \liminf_{n\to\infty}\left((|x_n|+t)^{1+1/n} - |x_n|^{1+1/n} - t(1+\tfrac{1}{n})|x_n|^{1/n}\right) = 0.$$

Fix $x \in A$ and take $\gamma := 1 - \limsup_{k\to\infty}|x_k|^{1/k} > 0$. Consider the function

$$\phi : [0,\gamma[\to \mathbb{R}, \quad \phi(s) := \sum\nolimits_{k\ge 1}\left(\frac{ks}{k+1} + |x_k|^{1/k}\right)^k.$$

The function ϕ is well-defined, continuous, increasing and $\phi(0) = \|x\|$. Indeed, as

$$\limsup_{k\to\infty}\left(\frac{ks}{k+1} + |x_k|^{1/k}\right) = s + \limsup_{k\to\infty}|x_k|^{1/k} < 1 \quad \forall\, s \in [0, \gamma[,$$

by the Cauchy–Hadamard test of convergence we have that the series defining $\phi(s)$ is convergent. Moreover, for any $k \geq 1$ the function ϕ_k defined by $\phi_k(s) := \left(\frac{ks}{k+1} + |x_k|^{1/k}\right)^k$ is continuous and increasing on $\mathbb{R}_+$, and so ϕ is increasing on $[0, \gamma[$. Also, for $s_0 \in]0, \gamma[$, we have that $0 \leq \phi_k(s) \leq \phi_k(s_0)$ for all $s \in [0, s_0]$ and $k \geq 1$, and so the series defining ϕ is uniformly convergent on $[0, s_0]$. It follows that ϕ is continuous on $[0, \gamma[$. It is obvious that $\phi(0) = \|x\|$. Let $t_0 := \phi(\gamma/2) - \|x\| > 0$.

Take now $t \in]0, t_0]$; we have that $\phi(0) < t + \|x\| \leq \phi(\gamma/2)$, and so there exists (a unique) $s := s_t \in]0, \gamma/2]$ such that $\phi(s) = t + \|x\|$. We intend to show that $\vartheta_{f,x}(t) = D_f(w, x)$, where $w := w_t \in \ell^1$ is defined by

$$w_k := \operatorname{sgn}(x_k) \cdot \left(\frac{ks_t}{k+1} + |x_k|^{1/k}\right)^k \quad \forall\, k \geq 1.$$

First note that $\operatorname{sgn}(w_k - x_k) = \operatorname{sgn}(x_k)$ and $|w_k - x_k| = \left(\frac{ks_t}{k+1} + |x_k|^{1/k}\right)^k - |x_k|$, whence $\|w - x\| = t$. This also implies that $w = (w - x) + x \in \ell^1$.

Take $y \in \ell^1$ with $\|y - x\| = t$. Replacing, if necessary, y by $\overline{y}$ as above, we may assume that $\operatorname{sgn}(y_k - x_k) = \operatorname{sgn}(x_k)$ for $k \geq 1$. Then, by the convexity of f,

$$D_f(y, x) - D_f(w, x) = f(y) - f(w) - \langle y - w, \nabla f(x)\rangle \geq \langle y - w, \nabla f(w) - \nabla f(x)\rangle.$$

But

$$(\nabla f(w) - \nabla f(x))_k = \left(1 + \frac{1}{k}\right)\operatorname{sgn}(x_k)\left(\frac{ks_t}{k+1} + |x_k|^{1/k} - |x_k|^{1/k}\right) = s\cdot\operatorname{sgn}(x_k).$$

Hence

$$\begin{aligned}
D_f(y, x) - D_f(w, x) &\geq s \sum\nolimits_{k\geq 1} \operatorname{sgn}(x_k)\,(y_k - w_k) \\
&= s\left(\sum\nolimits_{k\geq 1} \operatorname{sgn}(x_k)\,(y_k - x_k) - \sum\nolimits_{k\geq 1} \operatorname{sgn}(x_k)\,(w_k - x_k)\right) \\
&= s\left(\sum\nolimits_{k\geq 1} |y_k - x_k| - \sum\nolimits_{k\geq 1} |w_k - x_k|\right) = 0.
\end{aligned}$$

It follows that $\vartheta_{f,x}(t) = D_f(w_t, x) > 0$, the last inequality because f is strictly convex and $w_t \neq x$. As $\vartheta_{f,x}$ is nondecreasing we obtain that $\vartheta_{f,x}(t) > 0$ for every $t > 0$.

Exercise 3.5 (i) Taking $f_n : X \to \mathbb{R}$, $f_n(x) := \sum_{k=1}^n 2^{-k}\langle x, x_k^*\rangle^2$, we have that f_n is convex, $0 \leq f_n(x) \leq \|x\|^2$ and $(f_n(x)) \to f(x)$ for every $x \in X$. Therefore f

is convex, nonnegative and continuous (being bounded above on U_X). Let $x \in X$, $u \in S_X$ and $t \in \mathbb{R} \setminus \{0\}$. Then

$$\left| \frac{f(x+tu) - f(x)}{t} - \sum_{n\geq 1} 2^{-n+1} \langle x, x_n^* \rangle \cdot \langle u, x_n^* \rangle \right| = |t| \sum_{n\geq 1} 2^{-n} \langle u, x_n^* \rangle^2 \leq |t| .$$

Hence f is uniformly Fréchet differentiable and $\nabla f(x) = \sum_{n\geq 1} 2^{-n+1} \langle x, x_n^* \rangle \cdot x_n^*$ for every $x \in X$.

It is obvious that $g := \sqrt{f}$ is quasiconvex and $g(\lambda x) = |\lambda| \, g(x)$ for all $x \in X$ and $\lambda \in \mathbb{R}$. Using Corollary 2.2.3 we obtain that g is sublinear, and so g is a semi-norm.

For $x \in X$ we have that

$$f(x) = 0 \Leftrightarrow g(x) = 0 \Leftrightarrow [\forall n \in \mathbb{N} \,:\, \langle x, x_n^* \rangle = 0] \Leftrightarrow x \in A^{\perp}.$$

Therefore g is a norm if and only if $A^{\perp} = \{0\}$ if and only if $X^* = A^{\perp\perp} = w^*$-cl(lin A).

For $x, y \in X$ and $\lambda \in]0,1[$ we have (by a simple calculation) that

$$\lambda f(x) + (1-\lambda) f(y) - f(\lambda x + (1-\lambda) y) = \lambda(1-\lambda) \cdot f(x-y).$$

So, f is strictly convex if and only if $f(z) = 0 \Rightarrow z = 0$, which is equivalent to $\sqrt{f}$ be a norm.

Of course, if A is $\|\cdot\|$-dense in S_{X^*} then lin A is $\|\cdot\|$-dense in X^*, which implies that lin A is w^*-dense in X^*.

(ii) Assume that X^* is separable; then S_{X^*} is separable, too; hence there exists $A := \{x_n^* \mid n \in \mathbb{N}\} \subset S_{X^*}$ dense in S_{X^*}. Consider f defined in (i). From (i) we have that f is strictly convex, and so is $\|\cdot\|^2 + f$. As in (i) we obtain that $\|\cdot\|_1 := \sqrt{\|\cdot\|^2 + f}$ is a norm on X. Because $\|\cdot\| \leq \|\cdot\|_1 \leq \sqrt{2}\,\|\cdot\|$, $\|\cdot\|_1$ is equivalent to $\|\cdot\|$. Since $\|\cdot\|_1^2$ is strictly convex, from Theorem 3.7.2(v) we have that $(X, \|\cdot\|_1)$ is strictly convex.

Recall that X is separable when X^* is separable. Indeed, if X^* is separable there exists $A := \{x_n^* \mid n \in \mathbb{N}\} \subset S_{X^*}$ dense in S_{X^*}. For every $n \in \mathbb{N}$ consider $x_n \in S_X$ such that $\langle x_n, x_n^* \rangle > 1/2$ and take $B := \{x_n \mid n \in \mathbb{N}\}$. Let $x^* \in S_{X^*}$; there exists $n \in \mathbb{N}$ such that $\|x^* - x_n^*\| < 1/2$. Then $\langle x_n, x^* \rangle = \langle x_n, x_n^* \rangle + \langle x_n, x^* - x_n^* \rangle \geq 1/2 - \|x^* - x_n^*\| > 0$, which shows that $x^* \notin B^{\perp}$. Therefore $B^{\perp} = \{0\}$, and so X is separable.

Assume now that X is reflexive and separable. Because X^* is separable, there exists an equivalent norm $\|\cdot\|_1$ on X such that $(X, \|\cdot\|_1)$ is strictly convex. Using Theorems 3.7.3 and 3.7.2(iv) we obtain that $\frac{1}{2}(\|\cdot\|_1^*)^2$ is Gâteaux differentiable. Taking $h(x^*) := \sum_{n\geq 1} 2^{-n} \langle x_n, x^* \rangle^2$, where $\{x_n \mid n \in \mathbb{N}\} \subset S_X$ is $\|\cdot\|$-dense in S_X, by (i) we have that $k := h + \frac{1}{2}(\|\cdot\|_1^*)^2$ is strictly convex and Gâteaux differentiable. Using again Theorem 3.7.2, it follows that $\sqrt{k}$ is a norm on X^* equivalent to $\|\cdot\|_1^*$ such that $(X^*, \sqrt{k})$ is smooth and strictly convex. Then the dual norm of $\sqrt{k}$ on $X^{**} = X$, denoted by $\|\cdot\|_2$, is equivalent to the initial norm and $(X, \|\cdot\|_2)$ is smooth and strictly convex.

Exercise 3.6 The fact that $\mathbb{P} \ni t \mapsto t^{-p}\varphi(t)$ is nondecreasing means that $\varphi(ct) \le c^p\varphi(t)$ for all $c \in]0,1[$ and $t > 0$. Let $0 < s_1 < s_2$ and $\varphi^e(s_1) < \mu$. Then there exists $t < \mu$ such that $\varphi(t) \ge s_1$. Let $c := (s_1/s_2)^{1/p} \in]0,1[$. Then

$$s_1 \le \varphi(t) = \varphi\left(c\left(c^{-1}t\right)\right) \le c^p\varphi\left(c^{-1}t\right) = \frac{s_1}{s_2}\varphi\left(c^{-1}t\right),$$

and so $s_2 \le \varphi(c^{-1}t)$. It follows that $\varphi^e(s_2) \le c^{-1}t < (s_2/s_1)^{1/p}\mu$. As $\mu > \varphi^e(s_1)$ was arbitrary, we obtain that $\varphi^e(s_2) \le (s_2/s_1)^{1/p}\varphi^e(s_1)$. Hence $\mathbb{P} \ni t \mapsto t^{-1/p}\varphi^e(t)$ is nonincreasing. The proof is similar for the other cases.

Exercise 3.7 Note that $g = \iota_C \square \|\cdot\|$, and so, by Theorem 2.3.1(ix) and Corollary 2.4.16 we have that $g^* = \iota_{U_{X^*}} + \iota_C^*$. Moreover, if $x_0 \in X \setminus C$ and $\overline{x} \in P_C(x_0)$, then $g(x_0) = \iota_C(\overline{x}) + \|x_0 - \overline{x}\|$. Applying Corollaries 2.4.7 and 2.4.16 we get the formula for $\partial g(x_0)$.

Let us consider the extension ψ_0 of ψ to $\mathbb{R}$ such that $\psi_0(t) := 0$ for $t < 0$. Then $\psi_0 \in \Gamma(\mathbb{R})$ is nondecreasing. If $\operatorname{dom}\psi = \{0\}$ the formula $(\psi \circ d_S)^* = \iota_S^* + \psi^{\#} \circ \|\cdot\|$ is obvious. In the contrary case condition (iii) of Theorem 2.8.10 holds. Let $x^* \ne 0$; applying this theorem we get:

$$\begin{aligned}
(\psi \circ d_C)^*(x^*) &= \sup_{x\in X}(\langle x, x^*\rangle - \psi_0(d_C(x))) = \min_{\lambda\ge 0}\left(\psi_0^*(\lambda) + (\lambda d_C)^*(x^*)\right) \\
&= \min_{\lambda>0}\left(\psi^{\#}(\lambda) + \lambda d_C^*(\lambda^{-1}x^*)\right) \\
&= \min_{\lambda>0}\left(\psi^{\#}(\lambda) + \lambda\left(\iota_C^*(\lambda^{-1}x^*) + \iota_{U_{X^*}}(\lambda^{-1}x^*)\right)\right) \\
&= \min_{\lambda>0}\left(\psi^{\#}(\lambda) + \iota_C^*(x^*) + \iota_{\lambda U_{X^*}}(x^*)\right) = \iota_C^*(x^*) + \min_{\lambda\ge\|x^*\|}\psi^{\#}(\lambda) \\
&= \iota_C^*(x^*) + \psi^{\#}(\|x^*\|),
\end{aligned}$$

because $\psi^{\#}$ is nondecreasing; for $x^* = 0$ the equality is obvious. Since $\psi \circ d_C$ is continuous and convex, there other formula follows from the bi-conjugate theorem.

Exercise 3.8 By Theorem 3.10.1, there exists $c > 0$ such that $f^*(x^*) = f^*(0) + \iota_S^*(x^*)$ for all $x^* \in cU_{X^*}$. Because S is bounded, ι_S^* is continuous, and so subdifferentiable at any $x^* \in X^*$. From the preceding relation we obtain that $\partial f^*(x^*) = \partial\iota_S^*(x^*) \subset S$ for every $x^* \in X^*$ with $\|x^*\| < c$. Hence, for $x^* \in cB_{X^*}$ there exists $x \in S$ such that $x \in \partial f^*(x^*)$, or, equivalently, $x^* \in \partial f(x)$. Therefore $cB_{X^*} \subset \partial f(S)$.

Exercise 3.9 The implications (i) $\Rightarrow$ (ii) $\Leftrightarrow$ (iii) $\Rightarrow$ (iv) are obvious, while the implication (iii) $\Rightarrow$ (i) follows from Corollary 3.4.2 applied to h and $\psi(t) := \alpha^{-1}t$ for $t \in [0, \delta]$.

Assume that S is compact and (iv) holds. Then for every $z \in S$ there exist $\delta_z, \alpha_z > 0$ such that $d_S(x) \leq \alpha_z h(x)$ for every $x \in D(z, \delta_z)$. Since S is compact and $S \subset \cup_{z \in S} B(z, \delta_z/2)$, there exists a finite set $S_0 \subset S$ such that $S \subset \cup_{z \in S_0} B(z, \delta_z/2)$. Let $\alpha := \max\{\alpha_z \mid z \in S_0\} > 0$ and $\delta := \min\{\delta_z/2 \mid z \in S_0\} > 0$. Consider $x \in X$ with $d_S(x) \leq \delta$. Since S is compact, there exists $z \in S$ such that $\|x - z\| \leq \delta$. For this z, there exists $z_0 \in S_0$ such that $z \in B(z_0, \delta_{z_0}/2)$. Hence $\|x - z_0\| \leq \|x - z\| + \|z - z_0\| < \delta + \delta_{z_0}/2 \leq \delta_{z_0}$. By the choice of δ_{z_0} and α_{z_0} we have that $d_S(x) \leq \alpha_{z_0} h(x) \leq \alpha h(x)$. The proof is complete.

Exercise 3.10 1) Note that $f_B(\cdot, \mu) = f \Box \mu \|\cdot\|$. Therefore $f_B(\cdot, \mu)$ is a convex function and $\operatorname{dom} f_B(\cdot, \mu) = X$. So, if $f_B(\cdot, \mu)$ takes the value $-\infty$ then it is identical $-\infty$. Suppose that $f_B(\cdot, \mu)$ does not take the value $-\infty$; hence the function is finite. Let us prove that $f_B(\cdot, \mu)$ is Lipschitz with Lipschitz constant μ. Let $x_1, x_2 \in X$. For $\varepsilon > 0$ there exists $y_1 \in X$ such that

$$f_B(x_1, \mu) > f(y_1) + \mu\|x_1 - y_1\| - \varepsilon,$$

and so

$$f_B(x_2, \mu) - f_B(x_1, \mu) < f(y_1) + \mu\|x_2 - y_1\| - f(y_1) - \mu\|x_1 - y_1\| + \varepsilon \leq \mu\|x_2 - x_1\| + \varepsilon.$$

Since $\varepsilon > 0$ is arbitrary, we obtain that

$$f_B(x_2, \mu) - f_B(x_1, \mu) \leq \mu\|x_2 - x_1\|.$$

Interchanging x_1 and x_2 we obtain that $f_B(\cdot, \mu)$ is Lipschitz, with Lipschitz constant μ.

It is obvious that for $0 < \mu_1 < \mu_2$ one has $f_B(\cdot, \mu_1) \leq f_B(\cdot, \mu_2) \leq f$. Let us prove now the last statement of 1). Note that, since $f \in \Gamma(X)$, there exists $x^* \in X^*$ and $\gamma \in \mathbb{R}$ such that

$$\forall x \in X \ : \ f(x) \geq \langle x, x^* \rangle + \gamma.$$

Let $\overline{x} \in X$ be fixed, and $\mathbb{R} \ni \lambda < f(\overline{x})$. Since f is lsc at $\overline{x}$, there exists $\rho > 0$ such that $f(x) > \lambda$ for $x \in D(\overline{x}, \rho)$. Let $\overline{\mu} := \max\{\|x^*\|, (\lambda + \|x^*\| - \gamma - \langle \overline{x}, x^* \rangle)/\rho\}$.

For $\mu \geq \overline{\mu}$ we have

$$\begin{aligned}
f_B(\overline{x},\mu) &= \min\left\{\inf_{y\in D(\overline{x},\rho)}(f(y)+\mu\|\overline{x}-y\|),\ \inf_{y\notin D(\overline{x},\rho)}(f(y)+\mu\|\overline{x}-y\|)\right\}\\
&\geq \min\left\{\inf_{y\in D(\overline{x},\rho)}(\lambda+\mu\|\overline{x}-y\|),\ \inf_{y\notin D(\overline{x},\rho)}(\langle y,x^*\rangle+\gamma+\mu\|\overline{x}-y\|)\right\}\\
&\geq \min\left\{\lambda, \inf\{\langle \overline{x},x^*\rangle+t\langle u,x^*\rangle+\gamma+\mu t \mid u\in S,\ t\geq\rho\}\right\}\\
&\geq \min\left\{\lambda, \inf\{\langle \overline{x},x^*\rangle+\gamma+t(\overline{\mu}-\|x^*\|) \mid t\geq\rho\}\right\}\\
&\geq \lambda.
\end{aligned}$$

Therefore $\lim_{\mu\to\infty} f_B(\overline{x},\mu) = f(\overline{x})$. Since $\overline{x}$ is arbitrary, we have the desired conclusion.

2) Let $x_0 \in X$ be fixed and $g : X \to \overline{\mathbb{R}}$, $g(x) := f(x)+\mu\|x-x_0\|$. It is clear that

$$\partial g(x_0) = \partial f(x_0) + \mu U_{X^*}$$

when $x_0 \in \operatorname{dom} f$.

(a) $\Rightarrow$ (b) From the relation $\partial f(x_0)\cap \mu U_{X^*} \neq \emptyset$ we have that $x_0 \in \operatorname{dom} f$ and $0 \in \partial g(x_0)$. Therefore

$$\forall x \in X \ :\ f(x_0) = g(x_0) \leq g(x),$$

and so $f(x_0) = f_B(x_0,\mu)$.

(b) $\Rightarrow$ (c) By what was proved above and using the hypothesis we have that $-\infty < f(x_0) = f_B(x_0,\mu) < \infty$. Therefore $f_B(x_0,\mu) \in \mathbb{R}$. From this, taking into account 1), it follows that $f_B(\cdot,\mu)$ is convex and continuous. Therefore $\partial f(x_0,\mu) \neq \emptyset$. Since $f_B(\cdot,\mu)$ is an exact convolution at $x_0 = x_0 + 0$, using Corollary 2.4.7, we have that $\partial f_B(x_0,\mu) = \partial f(x_0)\cap \mu U_{X^*}$.

(c) $\Rightarrow$ (a) Since $f_B(x_0,\mu) \in \mathbb{R}$, by 1) we have that $f_B(\cdot,\mu)$ is convex and continuous. Therefore $\partial f(x_0,\mu) \neq \emptyset$, and so (a) is true.

Exercise 3.11 1) Since $f \in \Gamma(X)$, there exist $x^* \in X^*$ and $\gamma \in \mathbb{R}$ such that

$$\forall x \in X \ :\ f(x) \geq \langle x,x^*\rangle + \gamma. \qquad (°)$$

Therefore for every $x \in X$,

$$\forall y \in X \ :\ f_{\mu,x}(y) := f(y)+\tfrac{\mu}{2}\|x-y\|^2 \geq \tfrac{\mu}{2}\|x-y\|^2 - \|x^*\|\cdot\|x-y\| + \langle x,x^*\rangle+\gamma; \quad (^*)$$

this implies that $f_M(x,\mu) > -\infty$ for every $x \in X$. Since $f_M(\cdot,\mu) = f\Box\frac{\mu}{2}\|\cdot\|^2$, we have that $\operatorname{dom} f_M(\cdot,\mu) = \operatorname{dom} f + X = X$; hence $f_M(\cdot,\mu)$ is finite. Since f is convex and $\frac{\mu}{2}\|\cdot\|^2$ is convex and continuous, it follows that $f_M(\cdot,\mu)$ is convex and continuous. The inequalities $f_M(\cdot,\mu_1) \leq f_M(\cdot,\mu_2) \leq f$ are obvious for $0 < \mu_1 < \mu_2$. The proof of $\lim_{\mu\to\infty} f_M(x,\mu) = f(x)$ for every $x \in X$ is completely analogue to that of Exercise 3.10.

Let $x^* \in \partial f(x)$. Then

$$\forall\, y \in X \;:\; \langle y - x, x^* \rangle + \tfrac{\mu}{2}\|y - x\|^2 \le f(y) + \tfrac{\mu}{2}\|y - x\|^2 - f(x).$$

Taking the infimum with respect to y in both members of the inequality we obtain $-\frac{1}{2\mu}\|x^*\|^2 \le f_M(x,\mu) - f(x)$, *i.e.* our conclusion. (In this case relation $\lim_{\mu\to\infty} f_M(x,\mu) = f(x)$ is immediate.)

2) From (*), the function $f_{\mu,x}$ is coercive; being also lsc, since X is a reflexive Banach space, there exists an element $x_\mu \in X$ such that

$$f_M(x,\mu) = f(x_\mu) + \tfrac{\mu}{2}\|x_\mu - x\|^2. \qquad (^{**})$$

(If X is strictly convex, using Theorem 3.7.2(v), the function $\frac{\mu}{2}\|\cdot\|^2$ is strictly convex, whence $f_{\mu,x}$ is strictly convex, too; in this case x_μ is unique.) Furthermore, using Corollary 2.4.7, we have

$$\partial f_M(x,\mu) = \partial f(x_\mu) \cap \mu\Phi_X(x - x_\mu). \qquad (^{***})$$

Let us consider the operator $J_\mu : X \to X$, $J_\mu(x) := x_\mu$. Relation (**) proves the formula of the statement. Let $x \in \operatorname{dom} f$ be fixed, and $\mu > 0$. By (°) and (**) we have that

$$\langle J_\mu(x), x^* \rangle + \gamma + \tfrac{\mu}{2}\|J_\mu(x) - x\|^2 \le f_M(x,\mu) \le f(x),$$

and so, for $\eta := f(x) - \gamma - \langle x, x^* \rangle$, we have that

$$\forall\, \mu > 0 \;:\; \tfrac{\mu}{2}\|J_\mu(x) - x\|^2 - \|x^*\| \cdot \|J_\mu(x) - x\| - \eta \le 0.$$

Thus we have

$$\forall\, \mu > 0 \;:\; \|J_\mu(x) - x\| \le \mu^{-1}\left(\|x^*\| + \sqrt{\|x^*\|^2 + 2\mu\eta}\right).$$

Taking the limit for $\mu \to \infty$ we obtain that $\lim_{\mu\to\infty} J_\mu(x) = x$. By (**) we have that $f(J_\mu(x)) \le f_M(x,\mu) \le f(x)$ for every $\mu > 0$. Since f is lsc and $\lim_{\mu\to\infty} J_\mu(x) = x$, we obtain that

$$f(x) \le \liminf_{\mu\to\infty} f(J_\mu(x)) \le \limsup_{\mu\to\infty} f(J_\mu(x)) \le f(x),$$

and so $\lim_{\mu\to\infty} f(J_\mu(x)) = f(x)$.

3) Let $x^* \in \partial f(x)$ and $x_\mu^* \in \partial f_M(x,\mu)$ for $\mu > 0$ (x_μ^* exists because $f_M(\cdot,\mu)$ is continuous). Using (***) we have that

$$\mu\langle x - J_\mu(x), x_\mu^* \rangle = \mu^2\|x - J_\mu(x)\|^2 = \|x_\mu^*\|^2$$

and, taking into account 1) and (**),

$$\langle x - J_\mu(x), x_\mu^* \rangle \le f(x) - f(J_\mu(x)) \le \tfrac{1}{2\mu}\|x^*\|^2 - \tfrac{\mu}{2}\|x - J_\mu(x)\|^2.$$

Using the above relations we obtain that $\|x^*_\mu\| \le \|x^*\|$. Since $x^* \in \partial f(x)$ is arbitrary, we have that $\|x^*_\mu\| \le d_{\partial f(x)}(0)$.

Let now $(\mu_i)_{i\in I} \to \infty$ and $x^* = w^* - \lim x^*_{\mu_i}$ (such a convergent net exists because U_{X^*} is w^*-compact). Because $x^*_\mu \in \partial f(J_\mu(x))$ and $(J_\mu(x)) \xrightarrow{\|\cdot\|} x$, by Theorem 2.4.2(ix) we have that $x^* \in \partial f(x)$. Moreover, because the norm of X^* is w^*-lsc, we have that $d_{\partial f(x)}(0) \le \|x^*\| \le \liminf_{i\in I} \|x^*_{\mu_i}\|$. From the first part we get $\lim \|x^*_{\mu_i}\| = d_{\partial f(x)}(0) = \|x^*\|$, and so $x^* \in P_{\partial f(x)}(0)$. Therefore $\lim_{\mu\to\infty} \|x^*_\mu\| = d_{\partial f(x)}(0)$.

4) Assume that X is smooth. By Theorem 3.7.2(iv), the duality mapping Φ_X is single-valued. From (***) we have that $\partial f_M(x,\mu) = \mu\Phi_X(x - J_\mu(x))$. Using Corollary 2.4.10 we obtain that $f_M(\cdot,\mu)$ is Gâteaux differentiable and $\nabla f_M(x,\mu) = \mu\Phi_X(x - J_\mu(x))$ for every $x \in X$.

Assume now that $\partial f(x) \ne \emptyset$. The space X being reflexive, by Theorem 3.7.3, X^* is strictly convex. It follows that $P_{\partial f(x)}(0)$ is a singleton $\{x^*_0\}$. From 3) we obtain that x^*_0 is the sole limit point of the net $(\nabla f_M(x,\mu))$ for $\mu \to \infty$, and so $x^*_0 = w^* - \lim_{\mu\to\infty} \nabla f_M(x,\mu)$. Because $(\|\nabla f_M(x,\mu)\|) \to \|x^*_0\|$, we obtain that $(\nabla f_M(x,\mu)) \xrightarrow{\|\cdot\|} x^*_0$ when X^* has the Kadec–Klee property.

5) Let now X be locally uniformly smooth and consider $x \in X$; from 4) we have that $f_M(\cdot,\mu)$ is Gâteaux differentiable and $\nabla f_M(x,\mu) = \mu\nabla f_2(x - J_\mu(x))$, where $f_2 := \frac{1}{2}\|\cdot\|^2$. Then for $y \in X$ we have

$$\begin{aligned}
0 &\le f_M(y,\mu) - f_M(x,\mu) - \langle y - x, \nabla f_M(x,\mu)\rangle \\
&\le f(J_\mu(x)) + \tfrac{\mu}{2}\|y - J_\mu(x)\|^2 - f(J_\mu(x)) - \tfrac{\mu}{2}\|x - J_\mu(x)\|^2 \\
&\qquad\qquad - \langle y - x, \mu\nabla f_2(x - J_\mu(x))\rangle \\
&= \mu\big(f_2(y - J_\mu(x)) - f_2(x - J_\mu(x)) - \langle y - x, \nabla f_2(x - J_\mu(x))\rangle\big).
\end{aligned}$$

From Theorem 3.7.4 we obtain that $f_M(\cdot,\mu)$ is Fréchet differentiable at x. Using Theorem 3.3.2 we obtain that $\nabla f_M(\cdot,\mu)$ is continuous.

Exercise 3.12 Let $\varphi \in \mathcal{A}_0 \cap \mathcal{N}_0$ and $\psi = \overline{\text{co}}\varphi$. It is obvious that $0 \le \psi \le \varphi$, and so $\psi(0) = 0$. The implication 2) $\Rightarrow$ 1) is clear.

3) $\Rightarrow$ 2) Let $2\alpha := \liminf_{x\to\infty} \varphi(x)/x > 0$; then there exists $\rho > 0$ such that $\varphi(x) \ge \alpha x$ for every $x \ge \rho$. Let $x \ge 0$ be such that $\psi(x) = 0$; hence $(x,0) \in \text{epi}\,\psi = \overline{\text{co}}(\text{epi } f) \subset \mathbb{R}^2$. Using Carathéodory's theorem (Exercise 1.1), for every $i \in \{1,2,3\}$ there exist the sequences (λ^i_n), (x^i_n), $(t^i_n) \subset \mathbb{R}_+$ such that: $\lambda^1_n + \lambda^2_n + \lambda^3_n = 1$, $\varphi(x^i_n) \le t^i_n$ for all $n \in \mathbb{N}$ and $1 \le i \le 3$, $\left(\sum_{i=1}^3 \lambda^i_n x^i_n\right) \to x$ and $\left(\sum_{i=1}^3 \lambda^i_n t^i_n\right) \to \psi(x) = 0$. Without loss of generality, we assume that

$$(x^i_n) \to x^i, \quad (t^i_n) \to t^i, \quad (\lambda^i_n) \to \lambda^i, \quad (\lambda^i_n x^i_n) \to y^i, \quad (\lambda^i_n t^i_n) \to \tau^i,$$

with $x^i, t^i, y^i, \tau^i \in \overline{\mathbb{R}}_+$ and $\lambda^i \in [0,1]$ for $i \in \{1,2,3\}$. Of course, $\lambda^1 + \lambda^2 + \lambda^3 = 1$, $y^1 + y^2 + y^3 = x$ and $\tau^1 + \tau^2 + \tau^3 = 0$. Therefore $\tau^1 = \tau^2 = \tau^3 = 0$. If $\lambda^i > 0$, since $(\lambda^i_n t^i_n) \to 0$, we have that $t^i = 0$, and so $(\varphi(x^i_n)) \to 0$. Since φ

is nondecreasing, from our hypothesis we obtain that $(x_n^i) \to x^i = 0$, which, at its turn, implies that $y^i = 0$. If $\lambda^i = 0$ and $y^i > 0$ then $(x_n^i) \to \infty$. Therefore there exists $n_i \in \mathbb{N}$ such that $x_n^i \geq \rho$ for every $n \geq n_i$. Thus, for $n \geq n_i$ we have that $t_n^i \geq \varphi(x_n^i) \geq \alpha x_n^i$, and so $\lambda_n^i t_n^i \geq \alpha \lambda_n^i x_n^i$. So we get the contradiction $0 = \tau^i \geq \alpha y^i > 0$. Therefore $y^1 = y^2 = y^3 = 0$, and so $x = 0$.

1) $\Rightarrow$ 3) Suppose that $\liminf_{x\to\infty} \varphi(x)/x = 0$; then there exists $(x_n) \subset \mathbb{R}_+$ such that $(x_n) \to \infty$ and $\varphi(x_n) = \alpha_n x_n$ with $(\alpha_n) \to 0$. Let $x \in \mathbb{P}$; there exists $n_x \in \mathbb{N}$ such that $x_n \geq x$ for $n \geq n_x$. Set $\lambda_n := x/x_n \in]0,1[$ for $n \geq n_x$; then

$$\lambda_n(x_n, \varphi(x_n)) = (x, \alpha_n x) \in \operatorname{co}(\operatorname{epi}\varphi) \subset \operatorname{epi}\psi.$$

Hence $(x, 0) \in \operatorname{epi}\psi$. Therefore $\psi(x) = 0$ for every $x > 0$. The solution is complete.

Exercise 3.13 Let $x \in X$ and $x_1 \in K$. By Theorem 3.8.4(vi) we have that

$$\begin{aligned} x_1 = P_K(x) &\Leftrightarrow \forall y \in K \ : \ (x - x_1 \mid x_1 - y) \geq 0 \\ &\Leftrightarrow \forall y \in K \ : \ (x - x_1 \mid x_1) \geq (x - x_1 \mid y) \\ &\Leftrightarrow \forall y \in K \ : \ (x - x_1 \mid x_1) \geq 0 \geq (x - x_1 \mid y) \\ &\Leftrightarrow x - x_1 \in K^-, (x - x_1 \mid x_1) = 0. \end{aligned}$$

Let $x_1 = P_K(x)$. By the above characterization we have that

$$x_2 := x - x_1 \in K^-, \ x - x_2 = x_1 \in (K^-)^- = K, \ (x - x_2 \mid x_2) = (x_1 \mid x - x_2) = 0.$$

By the same characterization we have that $x_2 = P_{K^-}(x)$. Therefore the conclusion is true.

Suppose now that $x = x_1 + x_2$ with $x_1 \in K$, $x_2 \in K^-$ and $(x_1 \mid x_2) = 0$. Then $x_1 \in K$, $x - x_1 \in K^-$ and $(x - x_1 \mid x_1) = 0$. Therefore, using again the above characterization, we have that $x_1 = P_K(x)$, and so $x_2 = P_{K^-}(x)$.

Exercise 3.14 (a) Using the equivalence (i) $\Leftrightarrow$ (ii) of Theorem 1.3.16 with $T = \mathrm{Id}_Y$ and $\mathcal{C} = A$, there exists $m > 0$ such that $\|A^*y\| \geq m\|y\|$ for every $y \in Y^* = Y$. It follows that

$$\|Ty\| \cdot \|y\| \geq \langle Ty, y\rangle = \langle A^*y, A^*y\rangle = \|A^*y\|^2 \geq m^2\|y\|^2.$$

So, $\|Ty\| \geq m^2\|y\|$ for every $y \in Y$. Of course, this relation implies that T is injective. Since $T^* = T$, the preceding inequality can be written as $\|T^*y\| \geq m^2\|y\|$ for every $y \in Y$. Using again the equivalence (i) $\Leftrightarrow$ (ii) of Theorem 1.3.16 with T replaced by Id_Y and $\mathcal{C}$ replaced by our T, we obtain that T is surjective, and so T is bijective.

(b) Consider the function $f : X \to \overline{\mathbb{R}}$, $f(x) = \frac{1}{2}\|x\|^2 + \iota_C(Ax + y_0)$. It is obvious that f is coercive and strictly convex. Using Theorem 2.5.1 and Proposition 2.5.6 we obtain that f has a unique minimum point $\overline{x} \in X$. By Corollary 2.8.5 there exists $v \in N(C; A\overline{x} + y_0)$ such that $\overline{x} = -A^*v$. Of course, $\overline{y} := A\overline{x} + y_0 \in C$, and so $A\overline{x} = -(A \circ A^*)v = -Tv$, whence $\overline{y} - y_0 = -Tv$. It follows that $T^{-1}(\overline{y} - y_0) = -v$, and so $(A^* \circ T)(\overline{y} - y_0) = -A^*v = \overline{x}$. Since $\langle y - \overline{y}, v\rangle \le 0$ for every $y \in C$, we have that

$$\forall y \in C \; : \; \langle y - \overline{y}, \overline{y} - (\overline{y} - \rho T^{-1}\overline{y} + \rho T^{-1}y_0)\rangle \ge 0,$$

where $\rho > 0$. By Theorem 3.8.4(vi) we have that $\overline{y} = P_C(\overline{y} - \rho T^{-1}\overline{y} + \rho T^{-1}y_0)$.

Conversely, if $\overline{y} = P_C(\overline{y} - \rho T^{-1}\overline{y} + \rho T^{-1}y_0)$ for $\rho > 0$ and $\overline{x} = (A^* \circ T^{-1})(\overline{y} - y_0)$, taking $v := -T^{-1}(\overline{y} - y_0)$ we have that $\overline{x} = -A^*v$, $A\overline{x} + y_0 = \overline{y} \in C$ and $v \in N(C; A\overline{x} + y_0)$. Using again Corollary 2.8.5 we obtain that $\overline{x}$ is the solution of (P).

Exercise 3.15 Applying Exercise 2.41 we find $\alpha, \beta \in \mathbb{R}$, $\alpha > 0$, such that $g(x) \ge \alpha\|x\| + \beta$ for every $x \in X$. Consider $((x_n, x_n^*)) \subset \operatorname{gr} \partial g$ with $(x_n^*) \overset{\|\cdot\|}{\to} 0$. Let $\overline{x} \in \operatorname{dom} g$ be fixed. Then

$$\alpha\|x_n\| + \beta - \|x_n - \overline{x}\| \cdot \|x_n^*\| \le g(x_n) + \langle \overline{x} - x_n, x_n^*\rangle \le g(\overline{x}).$$

There exists $n_0 \in \mathbb{N}$ such that $\|x_n^*\| \le \alpha/2$ for $n \ge n_0$, and so $\|x_n - \overline{x}\| \le 2\alpha^{-1}(g(\overline{x}) + \alpha\|\overline{x}\| - \beta)$ for $n \ge n_0$. Hence the sequence (x_n) is bounded. Assume that the conclusion is false. Then there exist an increasing sequence $(n_k) \subset \mathbb{N}$ and $t > \inf g$ such that $g(x_{n_k}) \ge t$ for every $k \in \mathbb{N}$. Because X is reflexive and (x_n) is bounded, we may assume that $(x_{n_k}) \overset{w}{\to} x \in X$. As $(x_n^*) \overset{\|\cdot\|}{\to} 0$, by Theorem 2.4.2(ix), we have that $0 \in \partial g(x)$, and so $g(x) = \inf g$. The inequality $t + \langle x - x_{n_k}, x_{n_k}^*\rangle \le g(x_{n_k}) + \langle x - x_{n_k}, x_{n_k}^*\rangle \le g(x)$ yields the contradiction $t \le \inf g$.

Exercise 3.16 (a) Because g^* is w^*-lsc, g^* is w-lsc, too. Since X^* is a Banach space it follows that g^* is continuous at 0, and so $0 \in \operatorname{int}(\operatorname{dom} g^*)$. By the preceding exercise we get $g \in \mathcal{F}$.

(b) Let K and X_0 be as in Exercise 2.24 (with f replaced by g). Taking into account Exercise 2.24, we obtain that $\widehat{g}$ has the properties in (a), and so $\widehat{g}$ has good asymptotic behavior. Let $\left((x_n, x_n^*)\right)_{\mathbb{N}} \subset \partial g$ be such that $(\|x_n^*\|) \to 0$. Then, by the same Exercise 2.24, we have that $\left((\widehat{x_n}, x_n^*|_{X_0^\perp})\right)_{\mathbb{N}} \subset \partial\hat{g}$. Since $(\|x_n^*|_{X_0^\perp}\|) \to 0$, we have that $\left(\hat{g}(\widehat{x_n})\right) = \left(g(x_n)\right) \to \inf \hat{g} = \inf g$. Hence $g \in \mathcal{F}$.

Exercise 3.17 Consider $f := g + \iota_A$. Then $f_\infty(\overline{u}) = g_\infty(\overline{u}) + \iota_{A_\infty}(\overline{u}) = -\alpha^{-1} < 0$. Using Theorem 3.10.13 we get the existence of an $\alpha > 0$ satisfying the conclusion (maybe not our α). In fact, for $x \in \operatorname{dom} f \setminus C = (A \cap \operatorname{dom} g) \setminus C$ and

$t := \alpha \cdot f(x)$ we have that $f(x + t\overline{u}) \leq f(x) + tf_\infty(\overline{u}) = 0$. So $x + t\overline{u} \in C$, whence $d_C(x) \leq \|x + t\overline{u} - x\| = t = \alpha \cdot f(x)$.

Exercise 3.18 It is obvious that $\inf g_1 = -1$, $\operatorname{argmin} g_1 = \mathbb{R}_+ \times \{0\}$, $\inf g_2 = -\infty$, $\nabla g_1(x,y) = \left(\frac{x}{\sqrt{x^2+y^2}} - 1, \frac{y}{\sqrt{x^2+y^2}}\right)$ for $(x,y) \neq (0,0)$ and $\nabla g_2(x,y) = (-2, 2y)$ for $(x,y) \in \mathbb{R}^2$. Using Corollary 3.10.9 we obtain that

$$\begin{aligned} r_{g_1}(t) = k_{g_1}(t) &= \inf\left\{\|\nabla g_1(x,y)\| \mid \sqrt{x^2+y^2} - x - 1 = t\right\} \\ &= \inf\left\{\sqrt{\frac{2t+2}{x+1+t}} \;\middle|\; x \geq \frac{t+1}{2}\right\} = 0 \quad \forall t > -1, \end{aligned}$$

and

$$r_{g_2}(t) = k_{g_2}(t) = \inf\left\{\|\nabla g_2(x,y)\| \mid y^2 - 2x - 1 = t\right\} = \left\{2\sqrt{1+y^2} \mid y \in \mathbb{R}\right\} = 2$$

for any $t \in \mathbb{R}$. Note that $g_2(x,y) = g_1(x,y) \cdot c(x,y)$, where $c(x,y) := \sqrt{x^2+y^2} + x + 1 \geq 1$. It follows that $g_3(x,y) = g_2(x,y) \geq 0$ if $y^2 - 2x \geq 1$ and $g_3(x,y) = g_1(x,y) \leq 0$ if $y^2 - 2x < 1$; moreover, $[g_1 = 0] = [g_2 = 0] = [g_3 = 0] = \{(x,y) \in \mathbb{R}^2 \mid y^2 = 2x + 1\}$. From the very definition of $r_f = r_f^1$ and k_f we obtain that $r_{g_3}(0) = r_{g_1}(0)$ and $k_{g_3}(0) = k_{g_2}(0)$.

Bibliography

Adly, S., Ernst, E. and Théra, M. (2001a) "A characterization of convex and semicoercive functionals", J. Convex Anal. **8**(1), 127–148.

Adly, S., Ernst, E. and Théra, M. (2001b) "Stabilité de l'ensemble des solutions d'une inéquation variationnelle non coercive", C. R. Acad. Sci. Paris Sér. I Math. **333**, 409–414.

Amara, C. (1998) Direction de majoration, ensembles sous-sériellement convexes. Applications à l'optimisation, PhD thesis, University of Avignon, Avignon, France.

Amara, C. and Ciligot-Travain, M. (1999) "Lower CS-closed sets and functions", J. Math. Anal. Appl. **239**(2), 371–389.

Anger, B. and Lembcke, J. (1974) "Hahn-Banach type theorems for hypolinear functionals", Math. Ann. **209**, 127–151.

Asplund, E. and Rockafellar, R. T. (1969) "Gradients of convex functions", Trans. Amer. Math. Soc. **139**, 443–467.

Attouch, H. and Beer, G. (1993) "On the convergence of subdifferentials of convex functions", Arch. Math. (Basel) **60**(4), 389–400.

Attouch, H. and Brézis, H. (1986) Duality for the sum of convex functions in general Banach spaces, *in* "Aspects of mathematics and its applications", North-Holland, Amsterdam, pp. 125–133.

Attouch, H. and Théra, M. (1996) "A general duality principle for the sum of two operators", J. Convex Anal. **3**(1), 1–24.

Aubin, J.-P. and Frankowska, H. (1990) Set-valued analysis, Birkhäuser Boston Inc., Boston, MA.

Auslender, A., Cominetti, R. and Crouzeix, J.-P. (1993) "Convex functions with unbounded level sets and applications to duality theory", SIAM J. Optim. **3**(4), 669–687.

Auslender, A. and Crouzeix, J.-P. (1988) "Global regularity theorems", Math. Oper. Res. **13**(2), 243–253.

Auslender, A. and Crouzeix, J.-P. (1989) "Well-behaved asymptotical convex functions", Ann. Inst. H. Poincaré Anal. Non Linéaire **6**(suppl.), 101–121. Analyse non linéaire (Perpignan, 1987).

Aussel, D., Corvellec, J.-N. and Lassonde, M. (1995) "Mean value property and subdifferential criteria for lower semicontinuous functions", Trans. Amer. Math. Soc. **347**(10), 4147–4161.

Azé, D. (1994) "Duality for the sum of convex functions in general normed spaces", Arch. Math. (Basel) **62**(6), 554–561.

Azé, D. (1997) Eléments d'analyse convexe et variationnelle, Ellipses, Paris.

Azé, D. and Penot, J.-P. (1995) "Uniformly convex and uniformly smooth convex functions", Ann. Fac. Sci. Toulouse Math. (6) **4**(4), 705–730.

Azé, D. and Rahmouni, A. (1996) "On primal-dual stability in convex optimization", J. Convex Anal. **3**(2), 309–327.

Barbu, V. and Da Prato, G. (1985) "Hamilton–Jacobi equations in Hilbert spaces: variational and semigroup approach", Ann. Mat. Pura Appl. (4) **142**, 303–349 (1986).

Barbu, V. and Precupanu, T. (1986) Convexity and optimization in Banach spaces, second edn, D. Reidel Publishing Co., Dordrecht.

Bauschke, H. H., Borwein, J. M. and Combettes, P. L. (2001) "Essential smoothness, essential strict convexity, and Legendre functions in Banach spaces", Commun. Contemp. Math. **3**(4), 615–647.

Bauschke, H. H., Borwein, J. M. and Tseng, P. (2000) "Bounded linear regularity, strong CHIP, and CHIP are distinct properties", J. Convex Anal. **7**(2), 395–412.

Borwein, J. and Kortezov, I. (2001) "Some generic results on nonattaining functionals", Set-Valued Anal. **9**(1-2), 35–47. Wellposedness in optimization and related topics (Gargnano, 1999).

Borwein, J. M. (1983) "Adjoint process duality", Math. Oper. Res. **8**(3), 403–434.

Borwein, J. M. and Lewis, A. S. (1992) "Partially finite convex programming. I. Quasi relative interiors and duality theory", Math. Programming **57**(1, Ser. B), 15–48.

Borwein, J. M. and Lewis, A. S. (2000) Convex Analysis and Nonlinear Optimization. Theory and Examples, Springer-Verlag, New York.

Borwein, J. M. and Preiss, D. (1987) "A smooth variational principle with applications to subdifferentiability and to differentiability of convex functions", Trans. Amer. Math. Soc. **303**(2), 517–527.

Borwein, J. M. and Vanderwerff, J. D. (1995) "Convergence of Lipschitz regularizations of convex functions", J. Funct. Anal. **128**(1), 139–162.

Borwein, J. M. and Vanderwerff, J. D. (2000) Convex functions of Legendre type in general Banach spaces, Technical report, CECM, Simon Fraser University, Barnaby.

Bourbaki, N. (1964) Eléments de mathématique. Topologie générale, Hermann, Paris.

Burke, J. V. and Ferris, M. C. (1993) "Weak sharp minima in mathematical programming", SIAM J. Control Optim. **31**(5), 1340–1359.

Butnariu, D. and Iusem, A. N. (2000) Totally convex functions for fixed points computation and infinite dimensional optimization, Kluwer, Dordrecht.

Caristi, J. (1976) "Fixed point theorems for mappings satisfying inwardness conditions", Tran. Am. Math. Soc. **215**, 241–251.

Cârjă, O. (1989) "Range inclusion for convex processes on Banach spaces; applications in controllability", Proc. Amer. Math. Soc. **105**(1), 185–191.

Cârjă, O. (1998) Elements of nonlinear functional analysis, Editura Universităţii "Al.I.Cuza", Iaşi. Romanian.

Castaing, C. and Valadier, M. (1977) Convex analysis and measurable multifunctions, Springer-Verlag, Berlin. Lecture Notes in Mathematics, Vol. 580.

Cioranescu, I. (1990) Geometry of Banach spaces, duality mappings and nonlinear problems, Kluwer Academic Publishers Group, Dordrecht.

Clarke, F. H. (1983) Optimization and nonsmooth analysis, John Wiley & Sons Inc., New York. A Wiley-Interscience Publication.

Combari, C., Laghdir, M. and Thibault, L. (1994) 'Sous-différentiels de fonctions convexes composées", Ann. Sci. Math. Québec **18**, 554–561.

Combari, C., Laghdir, M. and Thibault, L. (1999) "On subdifferential calculus for convex functions defined on locally convex spaces", Ann. Sci. Math. Québec **23**(1), 23–36.

Cominetti, R. (1994) "Some remarks on convex duality in normed spaces with and without compactness", Control Cybernet. **23**(1-2), 123–138. Parametric optimization.

Coodey, M. and Simons, S. (1996) "The convex function determined by a multifunction", Bull. Austral. Math. Soc. **54**(1), 87–97.

Cornejo, O., Jourani, A. and Zălinescu, C. (1997) "Conditioning and upper-Lipschitz inverse subdifferentials in nonsmooth optimization problems", J. Optim. Theory Appl. **95**(1), 127–148.

Correa, R., Jofré, A. and Thibault, L. (1994) "Subdifferential monotonicity as characterization of convex functions", Numer. Funct. Anal. Optim. **15**(5& 6), 531–535.

Crouzeix, J.-P. (1980) "Conditions for convexity of quasiconvex functions", Math. Oper. Res. **5**(1), 120–125.

Crouzeix, J.-P. (1981) Continuity and differentiability properties of quasi-convex functions on $\mathbb{R}^n$, *in* S. Schaible and W. Ziemba, eds, "Generalized Concavity in Optimization and Economics", Academic Press, pp. 109–130.

Crouzeix, J.-P., Ferland, J. A. and Schaible, S. (1992) "Generalized convexity on affine subspaces with an application to potential functions", Math. Programming **56**(2, Ser. A), 223–232.

Deng, S. (1997) "Computable error bounds for convex inequality systems in reflexive Banach spaces", SIAM J. Optim. **7**, 274–279.

Deng, S. (1998) "Global error bounds for convex inequality systems in Banach spaces", SIAM J. Control Optim. **36**(4), 1240–1249.

Deville, R., Godefroy, G. and Zizler, V. (1993) Smoothness and renormings in Banach spaces, Longman Sci. Tech., Harlow. Lecture Notes in Mathematics.

Vol. 580.

Diestel, J. (1975) Geometry of Banach spaces, Vol. 485 of Lecture Notes in Mathematics, Springer-Verlag, Berlin.

Dieudonné, J. (1966) "Sur la séparation des ensembles convexes", Math. Ann. **163**, 1–3.

Dolecki, S. and Angleraud, P. (1996) "When a well behaving convex function is well-conditioned", Southeast Asian Bull. Math. **20**(2), 59–63.

Dontchev, A. L. and Zolezzi, T. (1993) Well-posed optimization problems, Springer-Verlag, Berlin.

Efimov, N. V. and Stechkin, S. B. (1959) "Support properties of sets in Banach spaces and Chebyshev sets", Dokl. Akad. Nauk SSSR **127**(2), 254–257.

Ekeland, I. (1974) "On the variational principle", J. Math. Anal. Appl. **47**, 324–353.

Ekeland, I. and Temam, R. (1974) Analyse convexe et problèmes variationnels, Dunod. Collection Études Mathématiques.

Eremin, I. I. and Astafiev, N. N. (1976) Introduction to the theory of linear and convex programming, Nauka, Moskow. (Russian).

Giles, J. R. (1982) Convex analysis with application in the differentiation of convex functions, Pitman (Advanced Publishing Program), Boston, Mass.

Hamel, A. (1994) "Remarks to equivalent formulation of Ekeland's variational principle", Optimization **31**, 233–238.

Hiriart-Urruty, J.-B. (1982) ε-subdifferential calculus, *in* "Convex analysis and optimization (London, 1980)", Pitman, Boston, Mass., pp. 43–92.

Hiriart-Urruty, J.-B. (1998) Optimisation et analyse convexe, Presses Universitaires de France, Paris.

Hiriart-Urruty, J.-B. and Lemaréchal, C. (1993) Convex analysis and minimization algorithms. I, Springer-Verlag, Berlin. Fundamentals.

Hiriart-Urruty, J.-B., Moussaoui, M., Seeger, A. and Volle, M. (1995) "Subdifferential calculus without qualification conditions, using approximate subdifferentials: a survey", Nonlinear Anal. **24**(12), 1727–1754.

Hiriart-Urruty, J.-B. and Phelps, R. R. (1993) "Subdifferential calculus using ϵ-subdifferentials", J. Funct. Anal. **118**(1), 154–166.

Holmes, R. B. (1972) A course on optimization and best approximation, Springer-Verlag, Berlin. Lecture Notes in Mathematics, Vol. 257.

Holmes, R. B. (1975) Geometric functional analysis and its applications, Springer-Verlag, New York. Graduate Texts in Mathematics, No. 24.

Hörmander, L. (1955) "Sur la fonction d'appui des ensembles convexes dans un espace localement convexe", Ark. Mat. **3**, 181–186.

Ioffe, A. and Tikhomirov, V. (1974) Theory of extremal problems, Nauka, Moskow.

Jameson, J. (1972) "Convex series", Proc. Cambridge Philos. Soc. **72**, 37–47.

Jeyakumar, V. and Wolkowicz, H. (1992) "Generalizations of Slater's constraint qualification for infinite convex programs", Math. Program. **57**, 85–101.

Joly, J.-L. and Laurent, P.-J. (1971) "Stability and duality in convex minimization problems", R.I.R.O. **2**, 3–42.

Jourani, A. (2000) "Hoffman's error bound, local controllability, and sensitivity analysis", SIAM J. Control Optim. **38**(3), 947–970.

Kassara, K. (2000) "Feedback spreading control laws for semilinear distributed parameter systems", Systems Control Lett. **40**(4), 289–295.

Kelley, J. (1955) General topology, Van Nostrand, Princeton.

Klatte, D. and Li, W. (1999) "Asymptotic constraint qualifications and global error bounds for convex inequalities", Math. Programming **84**(5), 137–160.

Kosmol, P. and Müller-Wichards, D. (2001) Homotopy methods for optimization in orlicz spaces, *in* V. Bulatov and V. Baturin, eds, "Proceedings of the 12th Baikal International Conference on Optimization Methods and their Applications", Institute of System Dynamics and Control Theory of SB RAS, Irkutsk, pp. 224–230.

Kusraev, A. G. and Kutateladze, S. S. (1995) Subdifferentials: theory and applications, Kluwer Academic Publishers Group, Dordrecht. Translated from the Russian.

Kutateladze, S. S. (1977) "Formulas for the computation of subdifferentials", Dokl. Akad. Nauk SSSR **232**(4), 770–772.

Kutateladze, S. S. (1979a) "Convex operators", Uspekhi Mat. Nauk **34**(1(205)), 167–196. Russian.

Kutateladze, S. S. (1979b) "Convex ε-programming", Dokl. Akad. Nauk SSSR **245**(5), 1048–1050.

Laurent, P. (1972) Approximation et optimisation, Hermann, Paris.

Lemaire, B. (1985) Subdifferential of a convex composite functional. Application to optimal control in variational inequalities, *in* "Nondifferentiable Optimization", Springer-Verlag, Sopron, 1984.

Lemaire, B. (1989) "Quelques résultats récents sur l'algorithme proximal", Séminaire d'Analyse Numérique, Toulouse.

Lemaire, B. (1992) "Bonne position, conditionnement et bon comportement asymptotique", Semin. Anal. Convexe, Univ. Sci. Tech. Languedoc, Expo. 5.

Levitin, E. S. and Polyak, B. T. (1966) "Convergence of minimizing of sequences in the conditional-extremum problem", Dokl. Akad. Nauk SSSR **168**, 997–1000. Russian.

Lewis, A. S. and Pang, J.-S. (1998) Error bounds for convex inequality systems, *in* J.-P. Crouzeix, J.-E. Martinez-Legaz and M. Volle, eds, "Proceedings of the 5th International Symposium on Generalized Convexity held in Luminy, June 17–21, 1996", Kluwer Academic Publishers, Dordrecht, pp. 75–110.

Li, W. and Singer, I. (1998) "Global error bounds for convex multifunctions and applications", Math. Oper Res. **23**(2), 443–462.

Lifšic, E. A. (1970) "Ideally convex sets", Funkcional. Anal. i Priložen. **4**(4), 76–77.

Lions, P.-L. and Rochet, J.-C. (1986) "Hopf formula and multitime Hamilton-Jacobi equations", Proc. Amer. Math. Soc. **96**(1), 79–84.

Luc, D. T. (1993) "On the maximal monotonicity of subdifferentials", Acta Math. Vietnam. **18**(1), 99–106.

Luc, D. T. and Swaminathan, S. (1993) "A characterization of convex functions", Nonlinear Anal. **20**(6), 697–701.

Marti, J. T. (1977) Konvexe Analysis, Birkhäuser Verlag, Basel.

Moreau, J. (1965) "Proximité et dualité dans un espace de Hilbert", Bull. Soc. Math. France **93**, 273–299.

Moussaoui, M. and Seeger, A. (1994) "Sensitivity analysis of optimal value functions of convex parametric programs with possibly empty solution sets", SIAM J. Optim. **4**(3), 659–675.

Moussaoui, M. and Volle, M. (1997) "Quasicontinuity and united functions in convex duality theory", Comm. Appl. Nonlinear Anal. **4**(4), 73–89.

Penot, J.-P. (1996a) "Conditioning convex and nonconvex problems", J. Optim. Theory Appl. **90**(3), 535–554.

Penot, J.-P. (1996b) "Subdifferential calculus without qualification assumptions", J. Convex Anal. **3**(2), 207–219.

Penot, J.-P. (1998a) Are generalized derivatives useful for generalized convex functions?, *in* "Generalized convexity, generalized monotonicity: recent results (Luminy, 1996)", Kluwer Acad. Publ., Dordrecht, pp. 3–59.

Penot, J.-P. (1998b) Well-behavior, well-posedness and nonsmooth analysis, *in* "Proceedings of the 4th International Conference on Mathematical Methods in Operations Research and 6th Workshop on Well-posedness and Stability of Optimization Problems (Sozopol, 1997)", Vol. 12, pp. 141–190.

Penot, J.-P. and Bougeard, M. L. (1988) "Approximation and decomposition properties of some classes of locally D.C. functions", Math. Programming **41**(2 (Ser. A)), 195–227.

Penot, J.-P. and Volle, M. (1990) "Inversion of real-valued functions and applications", Z. Oper. Res. **34**(2), 117–141.

Penot, J.-P. and Zălinescu, C. (2000) "Harmonic sum and duality", J. Convex Anal. **7**(1), 95–113.

Phelps, R. R. (1989) Convex functions, monotone operators and differentiability, Vol. 1364 of Lecture Notes Math., Springer-Verlag, Berlin.

Phelps, R. R. (1993) Convex functions, monotone operators and differentiability, Vol. 1364 of Lecture Notes Math., second edn, Springer-Verlag, Berlin.

Poliquin, R. A. (1990) "Subgradient monotonicity and convex functions", Nonlinear Anal. **14**, 305–317.

Polyak, B. T. (1966) "Existence theorems and convergence of minimizing of sequences in extremum problems with constraints", Dokl. Akad. Nauk SSSR **166**, 287–290. Russian.

Polyak, B. T. (1980) Sharp minimum, Technical report, Institute of Control Sciences, Moskow, USSR.

Precupanu, A. (1976) Mathematical Analysis. Real functions, Editura Didactică şi Pedagogică, Bucharest. Romanian.

Precupanu, T. (1992) Linear topological spaces and elements of convex analysis,

Editura Academiei Române, Bucharest. Romanian.

Robinson, S. M. (1976) "Regularity and stability for convex multivalued functions", Math. Oper. Res. **1**(2), 130–143.

Rockafellar, R. T. (1966) "Characterization of the subdifferentials of convex functions", Pacific J. Math. **17**, 497–510.

Rockafellar, R. T. (1969) "Local boundedness of nonlinear, monotone operators", Michigan Math. J. **16**, 397–407.

Rockafellar, R. T. (1970) Convex analysis, Princeton University Press, Princeton, N.J. Princeton Mathematical Series, No. 28.

Rockafellar, R. T. (1974) Conjugate duality and optimization, Society for Industrial and Applied Mathematics, Philadelphia, Pa. Lectures given at the Johns Hopkins University, Baltimore, Md., June, 1973, Conference Board of the Mathematical Sciences Regional Conference Series in Applied Mathematics, No. 16.

Rockafellar, R. T. (1976) "Monotone operators and the proximal point algorithm", SIAM J. Control Optimization **14**(5), 877–898.

Rockafellar, R. T. (1981) The theory of subgradients and its applications to problems of optimization, Heldermann Verlag, Berlin. Convex and nonconvex functions.

Seeger, A. and Volle, M. (1995) "On a convolution operation obtained by adding level sets: classical and new results", RAIRO Rech. Opér. **29**(2), 131–154.

Shioji, N. (1995) "On uniformly convex functions and uniformly smooth functions", Math. Japonica **41**, 641–655.

Simons, S. (1990) "The occasional distributivity of $\circ$ over $\overset{e}{+}$ and the change of variable formula for conjugate functions", Nonlinear Anal. **14**(12), 1111–1120.

Simons, S. (1994a) "Subtangents with controlled slope", Nonlinear Anal. **22**(11), 1373–1389.

Simons, S. (1994b) "Swimming below icebergs", Set-Valued Anal. **2**(1-2), 327–337. Set convergence in nonlinear analysis and optimization.

Simons, S. (1998a) Minimax and monotonicity, Vol. 1693 of Lecture Notes Math., Springer-Verlag, Berlin.

Simons, S. (1998b) "Sum theorems for monotone operators and convex functions", Trans. Amer. Math. Soc. **350**(7), 2953–2972.

Singer, I. (1979) "A Fenchel-Rockafellar type duality theorem for maximization", Bull. Austral. Math. Soc. **20**(2), 193–198.

Soloviov, V. (1993) "Duality for nonconvex optimization and its applications", Anal. Math. **19**(4), 297–315.

Soloviov, V. (1997) "Dual extremal problems and their applications to minimax estimation problems", Russian Math. Surveys **52**(4), 685–720.

Stoer, J. and Witzgall, C. (1970) Convexity and Optimization in Finite Dimensions I, Vol. 163 of Die Grundlehren der Mathematischen Wissenschaften, Springer Verlag, Berlin.

Sullivan, F. (1981) "A characterization of complete metric spaces", Proc. Am.

Math. Soc. **83**, 345–346.

Thibault, L. (1995a) A generalized sequential formula for subdifferentials of sums of convex functions defined on Banach spaces, *in* "Recent developments in optimization (Dijon, 1994)", Springer, Berlin, pp. 340–345.

Thibault, L. (1995b) "A note on the Zagrodny mean value theorem", Optimization **35**, 127–130.

Thibault, L. and Zagrodny, D. (1995) "Integration of subdifferentials of lower semicontinuous functions on Banach spaces", J. Math. Anal. Appl. **189**(1), 33–58.

Toland, J. F. (1978) "Duality in nonconvex optimization", J. Math. Anal. Appl. **56**, 399–415.

Ursescu, C. (1975) "Multifunctions with convex closed graph", Czechoslovak Math. J. **25(100)**(3), 438–441.

Vainberg, M. M. (1964) Variational methods for the study of nonlinear operators, Holden-Day Inc., San Francisco, Calif. With a chapter on Newton's method by L. V. Kantorovich and G. P. Akilov. Translated and supplemented by Amiel Feinstein.

Vladimirov, A. A., Nesterov, Y. E. and Chekanov, Y. N. (1978) "On uniformly convex functionals", Vestnik Moskov. Univ. Ser. XV Vyčisl. Mat. Kibernet. **3**, 12–23. Russian.

Vlasov, L. P. (1972) "Several theorems on Chebyshev sets", Mat. Zametki **11**, 135–144.

Volle, M. (1992) "Some applications of the Attouch-Brézis condition to closedness criterions, optimization, and duality", Sém. Anal. Convexe **22**, Exp. No. 16, 15.

Volle, M. (1993) "Sous-différentiel d'une enveloppe supérieure de fonctions convexes", C. R. Acad. Sci. Paris Sér. I Math. **317**(9), 845–849.

Volle, M. (1994) "Sur quelques formules de dualité convexe et non convexe", Set-Valued Anal. **2**(1-2), 369–379. Set convergence in nonlinear analysis and optimization.

Walkup, D. W. and Wets, R. J.-B. (1967) "Continuity of some convex-cone-valued mappings", Proc. Amer. Math. Soc. **18**, 229–235.

Willard, S. (1971) General topology, Addison-Wesley, Reading.

Zagrodny, D. (1988) "Approximate mean value theorem for upper subderivatives", Nonlinear Anal. **12**, 1413–1428.

Zălinescu, C. (1978) "A generalization of the Farkas lemma and applications to convex programming", J. Math. Anal. Appl. **66**(3), 651–678.

Zălinescu, C. (1980) "On an abstract control problem", Numer. Funct. Anal. Optim. **2**(6), 531–542 (1981).

Zălinescu, C. (1983a) "Duality for vectorial nonconvex optimization by convexification and applications", An. Ştiinţ. Univ. "Al. I. Cuza" Iaşi Secţ. I a Mat. (N.S.) **29**(3, suppl.), 15–34.

Zălinescu, C. (1983b) "On uniformly convex functions", J. Math. Anal. Appl. **95**(2), 344–374.

Zălinescu, C. (1984) Duality for vectorial convex optimization, conjugate operators and subdifferentials. The continuous case. Unpublished. Presented at the Conference "Mathematical Programming — Theory and Applications", Eisenach.

Zălinescu, C. (1986) "Letter to the editor: on J. M. Borwein's paper: "Adjoint process duality"", Math. Oper. Res. **11**(4), 692–698.

Zălinescu, C. (1987) "Solvability results for sublinear functions and operators", Z. Oper. Res. Ser. A-B **31**(3), A79–A101.

Zălinescu, C. (1989) Stability for a class of nonlinear optimization problems and applications, *in* "Nonsmooth optimization and related topics (Erice, 1988)", Plenum, New York, pp. 437–458.

Zălinescu, C. (1992a) "A note on d-stability of convex programs and limiting Lagrangians", Math. Programming **53**(3, Ser. A), 267–277.

Zălinescu, C. (1992b) "On some open problems in convex analysis", Arch. Math. (Basel) **59**(6), 566–571.

Zălinescu, C. (1998) Mathematical programming in infinite dimensional normed spaces, Editura Academiei, Bucharest, Romania. (Romanian).

Zălinescu, C. (1999) "A comparison of constraint qualifications in infinite-dimensional convex programming revisited", J. Austral. Math. Soc. Ser. B **40**(3), 353–378.

Zălinescu, C. (2001) Weak sharp minima, well-behaving functions and global error bounds for convex inequalities in Banach spaces, *in* V. Bulatov and V. Baturin, eds, "Proceedings of the 12th Baikal International Conference on Optimization Methods and their Applications", Institute of System Dynamics and Control Theory of SB RAS, Irkutsk, pp. 272–284.

Index

Symbols and Notations

$\overline{A}$	denotes	the closure of the set $A \subset (X, \tau)$
$A+B$, $A-B$	—	the sets $\{a+b \mid a \in A,\ b \in B\}$, $\{a-b \mid a \in A,\ b \in B\}$
Af	—	the function defined by $(Af)(y) = \inf\{f(x) \mid Ax = y\}$
A^i	—	the algebraic interior of A (p. 3)
A_∞	—	the cone $\bigcap_{t>0} t(A-a)$ with $a \in A$ when $A = \overline{\mathrm{co}}\, A$
A^+, A^{++}	—	the sets $\{x^* \in X^* \mid \forall x \in A\ :\ \langle x, x^* \rangle \geq 0\}$ and $(A^+)^+$
A°, $A^{\circ\circ}$	—	the sets $\{x^* \in X^* \mid \forall x \in A\ :\ \langle x, x^* \rangle \geq -1\}$ and $(A^\circ)^\circ$
$A^\perp$, $A^{\perp\perp}$	—	the sets $\{x^* \in X^* \mid \forall x \in A\ :\ \langle x, x^* \rangle = 0\}$ and $(A^\perp)^\perp$
${}^iA$	—	the relative algebraic interior of A (p. 3)
${}^{ib}A$	—	${}^iA$ if the parallel subspace to $\mathrm{aff}\, A$ is barreled, $\emptyset$ otherwise
${}^{ic}A$	—	${}^iA$ if $\mathrm{aff}\, A$ is closed, $\emptyset$ otherwise
$\mathcal{A}$	—	the set $\{\varphi : \mathbb{R}_+ \to \overline{\mathbb{R}}_+ \mid \varphi(0) = 0\}$
$\mathcal{A}_0$	—	the set $\{\varphi \in \mathcal{A} \mid \varphi(t) = 0 \Leftrightarrow t = 0\}$
$\mathrm{aff}\, A$	—	the affine hull of the set A (p. 2)
$\mathrm{argmin}\, f$	—	the set $\{x \in X \mid f(x) = \inf f\}$ for $f : X \to \overline{\mathbb{R}}$
$B(x, \varepsilon)$	—	the set $\{y \in (X, d) \mid d(y, x) < \varepsilon\}$
B_X	—	the set $B(0, 1)$ in a normed space $(X, \|\cdot\|)$
$\mathcal{B}_X$	—	the class of all bounded subsets of the tvs X
$\mathrm{Bd}\, A$	—	the set $\mathrm{cl}\, A \setminus \mathrm{int}\, A$
$\mathfrak{C}^*$	—	the adjoint of the convex process $\mathfrak{C}$ (p. 26)
$\mathcal{C}(A; a)$	—	the cone $\mathrm{cl}\,\big(\mathrm{cone}(A-a)\big)$
$C[a, b]$	—	the nvs of real continuous functions on the interval $[a, b]$
$\mathcal{C}(X)$	—	the set of convex and Lipschitz functions on $(X, \|\cdot\|)$
$\mathrm{cl}\, A$	—	the closure of the set $A \subset (X, \tau)$
$\mathrm{co}\, A$, $\overline{\mathrm{co}} A$	—	the convex hull of the set A (p. 2) and $\overline{\mathrm{co}\, A}$, respectively
$\overline{\mathrm{co}} f$	—	the lsc convex hull of the function f (p. 63)
$\mathrm{cone}\, A$, $\overline{\mathrm{cone}} A$	—	the conic hull of the set A (p. 2) and $\overline{\mathrm{cone}\, A}$, respectively
$\overline{D}f(x, \cdot)$	—	the upper Dini directional derivative (p. 58)
$D(x, \varepsilon)$	—	the set $\{y \in (X, d) \mid d(y, x) \leq \varepsilon\}$
d	—	metric or distance

$\operatorname{diam} A$ — the diameter of $A \subset (X,d)$, *i.e.* $\sup\{d(x,y) \mid x,\, y \in A\}$
$\dim A$ — the dimension of the parallel space to $\operatorname{aff} A$
$\operatorname{dom} f$ — the set $\{x \in X \mid f(x) < \infty\}$ for the function $f : X \to \overline{\mathbb{R}}$
$\operatorname{dom} H$ — the set $\{x \in X \mid H(x) < \infty\}$ for the operator $H : X \to Y^\bullet$
$\operatorname{dom} \mathcal{R}$ — the set $\{x \in X \mid \mathcal{R}(x) \neq \emptyset\}$ for the multifunction $\mathcal{R} : X \rightrightarrows Y$
$d_A(x)$, $d(x,A)$ — the number $\inf\{d(x,a) \mid a \in A\}$
$\operatorname{epi} f$ — the set $\{(x,t) \in X \times \mathbb{R} \mid f(x) \le t\}$ for the function $f : X \to \overline{\mathbb{R}}$
$\operatorname{epi}_s f$ — the set $\{(x,t) \in X \times \mathbb{R} \mid f(x) < t\}$ for the function $f : X \to \overline{\mathbb{R}}$
$\operatorname{epi} H$ — the set $\{(x,y) \in X \times Y \mid H(x) \le y\}$ for $H : X \to Y^\bullet$
$f|_A$ — the restriction of the function $f : X \to Y$ to the set $A \subset X$
$\overline{f}$ — the lsc hull of the function f: $\operatorname{epi} \overline{f} = \operatorname{cl}(\operatorname{epi} f)$
f_∞ — the recession function $f \in \Gamma(X)$: $\operatorname{epi} f_\infty = (\operatorname{epi} f)_\infty$
f_+ — the function $f \vee 0 = \max\{f, 0\}$
$f'(t)$, $f''(t)$ — the derivative and the second derivative of f at t, respectively
$f'_+(t)$, $f'_-(t)$ — the right and left derivative of f at t, respectively
$f'(a,\cdot)$ — the directional derivative of f at a
$f^\dagger(a,\cdot)$ — the Clarke–Rockafellar directional derivative of f at a (p. 171)
$f'_\varepsilon(a,\cdot)$ — the ε-directional derivative of f at a
f^*, f^{**} — the function $f^*(x^*) := \sup\{\langle x, x^*\rangle - f(x) \mid x \in X\}$; $(f^*)^*$
$f_1 \Box f_2$ — $f_1 \Box f_2(x) := \inf\{f_1(x_1) + f_2(x_2) \mid x_1, x_2 \in X,\ \ x_1 + x_2 = x\}$
$f_1 \Diamond f_2$ — $f_1 \Diamond f_2(x) := \inf\{f_1(x_1) \vee f_2(x_2) \mid x_1, x_2 \in X,\ \ x_1 + x_2 = x\}$
$[f \le \lambda]$, $[f < \lambda]$ — the sets $\{x \in X \mid f(x) \le \lambda\}$, $\{x \in X \mid f(x) < \lambda\}$
f_φ — the function defined by $f_\varphi(x) := \int_0^{\|x\|} \varphi(t)\, dt$ with $\varphi \in N_0$
$\operatorname{gr} \mathcal{R}$ — the set $\{(x,y) \in X \times Y \mid y \in \mathcal{R}(x)\}$ for $\mathcal{R} : X \rightrightarrows Y$
$\operatorname{gr} T$ — the set $\{(x, T(x)) \mid x \in X\}$ for the operator $T : X \to Y$
(Hx), (Hwx) — see p. 14
$H_{x^*,\alpha}$ — the set $\{x \in X \mid \langle x, x^*\rangle = \alpha\}$
$H^<_{x^*,\alpha}$, $H^>_{x^*,\alpha}$ — the sets $\{x \in X \mid \langle x, x^*\rangle < \alpha\}$, $\{x \in X \mid \langle x, x^*\rangle > \alpha\}$
$H^\ge_{x^*,\alpha}$, $H^\le_{x^*,\alpha}$ — the sets $\{x \in X \mid \langle x, x^*\rangle \ge \alpha\}$, $\{x \in X \mid \langle x, x^*\rangle \le \alpha\}$
Id_X — the function from X into X defined by $\mathrm{Id}_X(x) = x$
$\operatorname{Im} T$ — the set $\{T(x) \mid x \in X\}$ for the operator $T : X \to Y$
$\operatorname{Im} \mathcal{R}$ — the set $\bigcup_{x \in X} \mathcal{R}(x)$ for the multifunction $\mathcal{R} : X \rightrightarrows Y$
$\inf g$, $\inf_X g$ — the number $\inf_{x \in X} g(x)$, where $g : X \to \overline{\mathbb{R}}$
$\operatorname{int} A$ — the interior of the set $A \subset (X,\tau)$
$\operatorname{int}_Y A$ — the interior of $A \subset Y$ w.r.t. the topology on $Y \subset (X,\tau)$
$\ker T$ — the set $\{x \in X \mid T(x) = 0\}$ for $T \in L(X,Y)$
lcs — separated locally convex space
$L^1(0,1)$ — the space of (classes of) Lebesgue integrable functions on $]0,1[$
$L(X,Y)$ — the space of linear operators $T : X \to Y$
$\mathcal{L}(X,Y)$ — the space of continuous linear operators from X into Y
$\liminf_{i \in I} \lambda_i$ — $\sup_{j \in I} \inf_{i \succeq j} \lambda_i \in \overline{\mathbb{R}}$ for the net $(\lambda_i)_{i \in I} \subset \overline{\mathbb{R}}$
$\liminf_{x \to a} f(x)$ — $\sup_{V \in \mathcal{N}(a)} \inf_{x \in V} f(x) \in \overline{\mathbb{R}}$ for $f : (X,\tau) \to \overline{\mathbb{R}}$
$\limsup_{i \in I} \lambda_i$ — $\inf_{j \in I} \sup_{i \succeq j} \lambda_i \in \overline{\mathbb{R}}$ for the net $(\lambda_i)_{i \in I} \subset \overline{\mathbb{R}}$

$\limsup_{x\to a} f(x)$	—	$\inf_{V\in\mathcal{N}(a)}\sup_{x\in V} f(x)\in\overline{\mathbb{R}}$ for $f:(X,\tau)\to\overline{\mathbb{R}}$
$\operatorname{lin} A$	—	the linear hull of the set A (p. 2)
lsc	—	lower semicontinuous
ℓ^p $(p\geq 1)$	—	the space $\{(x_n)\subset\mathbb{R}\mid \|(x_n)\|_p := \left(\sum_{n\geq 1}\lvert x_n\rvert^p\right)^{1/p}<\infty\}$
ℓ^∞	—	the space $\{(x_n)\subset\mathbb{R}\mid \|(x_n)\|_\infty := \sup_{n\geq 1}\lvert x_n\rvert<\infty\}$
$\mathbb{N}$	—	the set of positive integers $\{1,2,\ldots\}$
$N(A;a)$	—	the cone $-(\mathcal{C}(A;a))^+$ for $a\in A$
N_0, N_1, N_2	—	$N_k := \{\varphi\in\mathcal{A}\mid \mathbb{P}\ni t\mapsto t^{-k}\varphi(t)\in\overline{\mathbb{R}}_+$ is nondecreasing$\}$
$\mathcal{N}(x)$, $\mathcal{N}_X(x)$	—	the class of all neighborhoods of the element x $(x\in X)$
$\mathcal{N}_X$	—	the set $\{U\in\mathcal{N}(0)\mid U$ balanced and closed$\}$, X being a tvs
$\mathcal{N}_X^c$	—	the set $\{V\in\mathcal{N}_X\mid V$ convex$\}$
$\mathbb{P}$	—	the set $]0,\infty[$ of positive real numbers
$\mathcal{P}$	—	a family of semi-norms
$P_C(x)$	—	the set $\{c\in C\mid \forall c'\in C\ :\ \|x-c\|\leq\|x-c'\|\}$
Pr_X	—	the function from $X\times Y$ onto X defined by $\mathrm{Pr}_X(x,y)=x$
$p_A(x)$	—	the number $\inf\{t\geq 0\mid x\in tA\}$
p, q, r	—	functions from $\varsigma_{X\times X^*}$ into X, X^* and $\mathbb{R}$, respectively (p. 270)
$\operatorname{qri} A$	—	the quasi relative interior of A (p. 15)
$\mathbb{R}$, $\overline{\mathbb{R}}$	—	the set of real numbers and $\mathbb{R}\cup\{-\infty,+\infty\}$, respectively
$\mathbb{R}_+$, $\overline{\mathbb{R}}_+$	—	the sets $[0,\infty[$ and $\mathbb{R}_+\cup\{+\infty\}$, respectively
$\mathbb{R}^E$, $\overline{\mathbb{R}}^E$	—	the classes of functions from E into $\mathbb{R}$ and $\overline{\mathbb{R}}$, respectively
$\mathcal{R}: X\rightrightarrows Y$	—	a multifunction from X into Y
$\mathcal{R}^{-1}$	—	$\mathcal{R}^{-1}: Y\rightrightarrows X$, $\mathcal{R}^{-1}(y) := \{x\mid y\in\mathcal{R}(x)\}$ for $\mathcal{R}: X\rightrightarrows Y$
$\mathcal{R}\circ\mathcal{S}$	—	the composition of the multifunctions $\mathcal{R}$ and $\mathcal{S}$ (p. 12)
$\mathcal{R}+\mathcal{S}$	—	the sum of the multifunctions $\mathcal{R}$ and $\mathcal{S}$ (p. 12)
$\mathcal{R}(A)$	—	the set $\{y\in Y\mid \exists x\in A\ :\ (x,y)\in\mathcal{R}\}$
$\mathcal{R}^{-1}(B)$	—	the set $\{x\in X\mid \exists y\in B\ :\ (x,y)\in\mathcal{R}\}$
$\operatorname{rec} A$	—	the set $\{u\in X\mid \forall a\in A\ :\ a+u\in A\}$
$\operatorname{ri} A$	—	the set $\operatorname{rint} A$ if $\operatorname{aff} A$ is closed, $\emptyset$ otherwise
$\operatorname{rint} A$	—	the relative interior of A, *i.e.* $\operatorname{int}_{\operatorname{aff} A} A$
S_X	—	the set $U_X\setminus B_X$
$S(f,C)$	—	the set $\{\overline{x}\in C\mid \forall x\in C\ :\ f(\overline{x})\leq f(x)\}$
$S(P)$	—	the set $S(f,C)$ for the problem: $(P)\quad \min f(x),\quad x\in C$
$S_\varepsilon(f,C)$	—	the set $\{\overline{x}\in C\mid \forall x\in C\ :\ f(\overline{x})\leq f(x)+\varepsilon\}$
$S_\varepsilon(P)$	—	the set $S_\varepsilon(f,C)$ for the problem: $(P)\quad \min f(x),\quad x\in C$
s_A	—	the support function of $A\subset X$ (see p. 79)
$\operatorname{supp}\mu$	—	the set $\{x\in E\mid \mu(x)\neq 0\}$ for a function $\mu\in\varsigma_E$
T^*	—	the operator $T^*: Y^*\to X^*$ defined by $T^*y^* := y^*\circ T$
$T_C(M,x)$	—	the Clarke tangent cone of M at x (p. 169)
tvs	—	topological vector space
$(t_n)\downarrow 0$	—	$(t_n)\subset]0,\infty[$, $(t_n)\to 0$
U_X	—	the set $D(0,1)$ in the normed space $(X,\|\cdot\|)$
usc	—	upper semicontinuous
$v(f,C)$	—	the number $\inf\{f(x)\mid x\in C\}$

$v(P)$ — $v(f, C)$ for the problem: $(P) \quad \min f(x), \quad x \in C$
w — the topology $\sigma(X, X^*)$ on the lcs X
w^* — the topology $\sigma(X^*, X)$ on the topological dual X^* of X
w.r.t. — with respect to
X', X^* — the spaces $L(X, \mathbb{R})$ and $\mathcal{L}(X, \mathbb{R})$, respectively
X/X_0 — the quotient space of X w.r.t. the linear subspace $X_0 \subset X$
(X, d) — a metric space
(X, τ) — a topological space
$x_i \to x$, $(x_i) \to x$, $x = \lim x_i$ — the net (x_i) converges to x
$(x_i) \xrightarrow{w} x$ — the net (x_i) converges to x for the topology w of X
$(x_i^*) \xrightarrow{w^*} x^*$ — the net (x_i^*) converge to x^* for the topology w^* of X^*
$(x_{n_k})_{k \in \mathbb{N}}$, (x_{n_k}) — a subsequence of (x_n)
$[x, y]$, $[x, y[$ — $\{(1-\lambda)x + \lambda y \mid \lambda \in [0, 1]\}$ and $\{(1-\lambda)x + \lambda y \mid \lambda \in [0, 1[\}$
$]x, y[$ — the set $\{(1-\lambda)x + \lambda y \mid \lambda \in]0, 1[\}$
$\langle x, x^* \rangle$ — the image of $x \in X$ by $x^* \in X^*$
$Y^\bullet$ — the set $Y \cup \{\infty\}$
$y_1 \leq_Q y_2$, $y_1 \leq y_2$ — the fact that $y_2 - y_1 \in Q$, for $Q \subset Y$ convex cone
$0 \cdot f$ — the function $\iota_{\operatorname{dom} f}$
$\overline{1, n}$ — the set $\{1, \ldots, n\}$ for $n \in \mathbb{N}$
$\|\cdot\|$, $\|x\|$ — a norm, the norm of the element x
∂f — the Fenchel subdifferential of the function f (p. 80)
$\partial_\varepsilon f$ — the ε-subdifferential of the function f (p. 82)
$\partial_C f$ — the Clarke subdifferential of the function f (p. 175)
$\overline{\partial}$ — abstract subdifferential (p. 178)
$\partial f / \partial x_i$ — the partial derivative of the function f with respect to x_i
$\partial^2 f / \partial x_i \partial x_j$ — $\partial(\partial f / \partial x_i) / \partial x_j$
Γ — the class of lsc convex functions $f \in \mathcal{A}$
Γ_0, Γ_0^2 — the sets $\Gamma \cap \mathcal{A}_0$, $\{\varphi \in \Gamma_0 \mid \liminf_{t \to \infty} t^{-2} \varphi(t) > 0\}$
$\Gamma(X)$ — the class of lsc proper convex functions $f : X \to \overline{\mathbb{R}}$
$\Gamma^*(X^*)$ — the class of w^*-lsc proper convex functions $h : X^* \to \overline{\mathbb{R}}$
Δ_n — the set $\{(\lambda_1, \ldots, \lambda_n) \in \mathbb{R}_+^n \mid \lambda_1 + \cdots + \lambda_n = 1\}$ for $n \in \mathbb{N}$
δ_u $(u \in E)$ — the element of ς_E for which $\delta_u(x) := 1$ if $x = u$, 0 otherwise
Φ_X — the duality mapping of the normed space $(X, \|\cdot\|)$ (p. 230)
φ_A $(A \subset X \times \mathbb{R})$ — the function defined by $\varphi_A(x) := \inf\{t \mid (x, t) \in A\}$
$\varphi^\#$ — the conjugate of $\varphi \in \mathcal{A}$: $\varphi^\#(t) := \sup\{ts - \varphi(s) \mid s \geq 0\}$
φ^e, φ^h — quasi-inverses of $\varphi \in \mathcal{A}$ (p. 188)
χ_M — the convex function associated to $M \subset X \times X^*$ (p. 270)
ι_A — the indicator function of $A \subset X$ (see p. 41)
$\Lambda \cdot A$, λA — the sets $\{\lambda a \mid \lambda \in \Lambda,\ a \in A\}$, $\{\lambda\} \cdot A$
$\Lambda(X)$ — the class of proper convex functions $f : X \to \overline{\mathbb{R}}$
Ω_0, Ω_1 — $\Omega_k := \{\varphi \in \mathcal{A} \mid \lim_{t \downarrow 0} t^{-k} \varphi(t) = 0\}$ for $k \in \{0, 1\}$
Π_f — a class of functions associated to $f \in \Gamma(X)$ (p. 101)
ψ_f — the conditioning gage of f (p. 196)
ρ_f — the gage of uniform convexity of f (p. 203)

$\rho_{f,r}$	—	the gage of uniform convexity of f on rU_X (p. 221)
$\rho_{f,\overline{x}}$	—	the gage of uniform convexity of f at $\overline{x}$ (p. 201)
Σ_1, Σ_1^2	—	the sets $\Gamma \cap \Omega_1$, $\{\sigma \in \Sigma_1 \mid \limsup_{t\to\infty} t^{-2}\varphi(t) < \infty\}$
σ_f, σ_f^0	—	gages of uniform smoothness of f (p. 204)
$\widetilde{\sigma}_f$	—	see p. 210
$\sigma_{f,A}$, $\sigma_{f,A}^0$	—	gages of uniform smoothness of f on A (p. 204)
$\sigma_{f,r}$, $\sigma_{f,r}^0$	—	gages of uniform smoothness of f on rU_X (p. 221)
$\sigma_{f,\overline{x}}^0$	—	the gage of smoothness of f at $\overline{x}$ (p. 191)
σ_x	—	the gage of smoothness of the norm at x (p. 231)
σ_X	—	the gage of uniform smoothness of the norm of X (p. 231)
ς_E	—	the set $\left\{\mu : E \to [0,\infty[\;\middle|\; \operatorname{supp}\mu \text{ is finite}, \sum_{x\in E}\mu(x) = 1\right\}$
$\prod_{i\in I} X_i$	—	the set $\left\{u : I \to \bigcup_{i\in I} X_i \;\middle|\; \forall i \in I \;:\; u(i) \in X_i\right\}$
τ_d	—	the topology associated with a metric d
$\tau_{\mathcal{P}}$	—	the topology generated by the family of semi-norms $\mathcal{P}$
Θ_p	—	function of the form $\Theta_p(x) := \sum_{n\geq 0}\mu_n \lVert x - u_n\rVert^p$ (p. 31)
$\vartheta_{f,\overline{x}}$, $\vartheta_{f,B}$	—	see pages 201, 225, respectively
$\nabla f(a)$	—	the gradient (the differential) of the function f at a
$\nabla^2 f(a)$	—	the second order differential of the function f at a
$\forall$	—	for all, for every
$\exists$	—	there exists (at least one)
$\vee$	—	maximum of extended real numbers or logical "or"
$\wedge$	—	minimum of extended real numbers or logical "and"
$:=$, $=:$	—	$a := b$, $b =: a$ means a is equal to b by definition
$\square$	—	the end of a proof or of a theorem without proof

www.ingramcontent.com/pod-product-compliance
Ingram Content Group UK Ltd.
Pitfield, Milton Keynes, MK11 3LW, UK
UKHW020401150926
8624IPUK00004B/273

9 789812 380678